Probabilities and Materials

NATO ASI Series

Advanced Science Institutes Series

A Series presenting the results of activities sponsored by the NATO Science Committee, which aims at the dissemination of advanced scientific and technological knowledge, with a view to strengthening links between scientific communities.

The Series is published by an international board of publishers in conjunction with the NATO Scientific Affairs Division

A Life Sciences	Plenum Publishing Corporation
B Physics	London and New York
C Mathematical	Kluwer Academic Publishers
and Physical Sciences	Dordrecht, Boston and London
D Behavioural and Social Sciences	
E Applied Sciences	
F Computer and Systems Sciences	Springer-Verlag
G Ecological Sciences	Berlin, Heidelberg, New York, London,
H Cell Biology	Paris and Tokyo
I Global Environmental Change	

NATO-PCO-DATA BASE

The electronic index to the NATO ASI Series provides full bibliographical references (with keywords and/or abstracts) to more than 30000 contributions from international scientists published in all sections of the NATO ASI Series.
Access to the NATO-PCO-DATA BASE is possible in two ways:

– via online FILE 128 (NATO-PCO-DATA BASE) hosted by ESRIN,
Via Galileo Galilei, I-00044 Frascati, Italy.

– via CD-ROM "NATO-PCO-DATA BASE" with user-friendly retrieval software in English, French and German (© WTV GmbH and DATAWARE Technologies Inc. 1989).

The CD-ROM can be ordered through any member of the Board of Publishers or through NATO-PCO, Overijse, Belgium.

Series E: Applied Sciences - Vol. 269

Probabilities and Materials
Tests, Models and Applications

edited by

D. Breysse

L.M.T. Cachan
and
University of Marne-la-Vallée,
France

Springer-Science+Business Media, B.V.

Proceedings of the NATO Advanced Research Workshop on
PROBAMAT: Probabilities and Materials – Test, Models and Applications
Cachan, France
November 23–25, 1993

A C.I.P. Catalogue record for this book is available from the Library of Congress.

ISBN 978-94-010-4500-1   ISBN 978-94-011-1142-3 (eBook)
DOI 10.1007/978-94-011-1142-3

TABLE OF CONTENTS

Acknowledgements

We would like to thank the members of the Scientific Committee and of the Local Organizing Committee for their help before, during and after the Workshop.

We also thank the scientific associations which have supported the Workshop, namely : NATO, CNRS, RILEM and AFREM, Mécamat, the GRECO Géomatériaux, AUGC and the GdR Physique des Milieux Hétérogènes. All the documents published in French are issued thanks to the Délégation Générale à la Langue Française and to AFREM.

Our consideration goes also towards the Direction des Etudes et Recherches of EDF, and more specifically to Mr. Marc Lasne, and to Usinor-Sacilor, for their scientific interest and their financial support.

Finally, we cannot forget D. Baptiste, M. Brusin, F. Darve, M. Fickelson, Y. Malier, G. Pijaudier-Cabot and all the participants whose active contribution was the condition for a fruitful scientific meeting.

PROBAMAT - ENS CACHAN - 23.24.25 nov 93

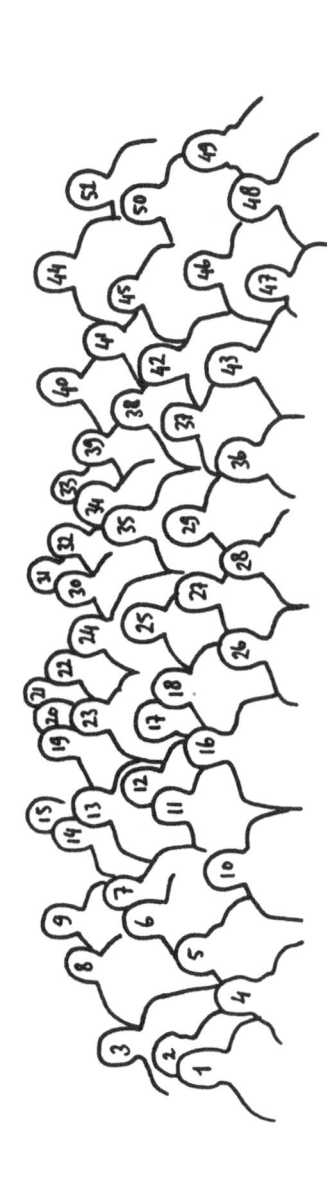

1 L. Davenne
2 H. Carre
3 D. Quenard
4 P. Castera
5 P. Morlier
6 J.P. Vilotte
7 A. Mebarki
8 E. Ringst
9 D. Rosowsky
10 P.L. Bourdeau

11 P. Kittl
12 Riesch Opperman
13 A. Carpinteri
14 B. Muhunthan
15 F. Hild
16 G.I. Schueller
17 M. Yasumura
18 Taheri
19 B. Chiaia
20 P. Renaudin

21 K. Sab
22 M. Gorelik
23 J. Lamon
24 C. Cherubini
25 C.S. Chang
26 D. Breysse
27 L. Faravelli
28 D. Jeulin
29 M. Lemaire
30 A. Chudnovsky

31 H. Mihashi
32 P. Rossi
33 N. Kachanov
34 S. Ghosh
35 A. Baratta
36 F. Casciati
37 M. Soulie
38 B. Kunin
39 B. Kunin
40 K. Touati

41 ?
42 G. Regnier
43 M. Ostoja
44 A. Bolle
45 J. Carmeliet
46 G. Frantziskonis
47 ?
48 ?
49 J. van Mier
50 G. Auvinet
51 H. Einstein

Introduction to the Workshop ProbaMat

LMT/ENS - Cachan, november 23 - 25, 1993

Probabilities and Materials

Tests, Models and Applications

by

Maurice LEMAIRE

Institut Français de Mécanique Avancée

Campus des Cézeaux, BP 165 F-63175 Aubière

Opening this workshop is for me an honour and a responsibility and I shall try to assume it as best I can. I don't wish to get now to the heart of the debate which will certainly be relevant as you may have noticed when reading the subjects of the abstracts, but rather to propose an idea or two on the role of probabilities and also statistics in engineering. For that, I shall show what are, to my mind, the stakes and the open questions of scientific research in our field.

1 To work for tomorrow and for uncertainty

Is it necessary to come back to a philosophical discussion as to whether hazard exists, or, if what we call hazard is nothing but the result of our ignorance and our incapability to take into account all the initial conditions of a process ? This discussion is certainly vain today and it is probably more pragmatic to believe that between the present level of knowledge and physical reality, there is always a gap which cannot be represented by any available model. Moreover, we know that very small perturbations of an initial state can lead to very large potential differences in the consequences, as is proved by the instability phenomena, for example in meteorology or in structural mechanics.

On the other hand, methods based on random modelization give information about certain definite phenomena. That is the case in geostatistics, for example, where the ore content is considered as a unique realization of a random function.

1

D. Breysse (ed.), Probabilities and Materials, 1–8.

© *1994 Kluwer Academic Publishers.*

A realization of a random function is a definite event, but we have to take into account sometimes, not a limited likelihood of realization, but rather an uncertainty in a model or an objective function. This is where the field of the fuzzy sets comes in.

Faced with this incapacity to control the set of data which we nevertheless theoretically have at our disposal, since they are before us and all we have to do is read it, we shall try to make a bet and should frequently win.

To introduce this idea, Blaise PASCAL says the following words :

"Or, quand on travaille pour demain et pour l'incertain, on agit avec raison : car on doit travailler pour l'incertain, par la règle des partis qui est démontrée".

The first part of this sentence is very simple. We work for tomorrow through the studies that we make which contribute to scientific progress, to increase our knowledge of phenomena ; we work with uncertainty, because we will never have complete data. B. Pascal tells us that we are right to do so and he justifies it by the expected winnings.

We must work with uncertainty because it is the reality of our environment, but it implies a chance of success and a risk of failure. It is therefore important to estimate what success brings and what we fear when there is failure. Answering the knight de Méré, Pascal demonstrates how to divide, how to give everybody the share which is rightly his, in a deal subject to risk and expected profit. The rule of sharing allows everybody to justify his commitment in relation with his expectations.

B. Pascal's *pensée*, here applied to commercial stakes, may also be applied to technical stakes. In our conquest for knowledge, in our wish to design increasingly daring buildings and machines, we take risks in the hope of expected progress to benefit all humanity.

What remains to be known is whether the control of uncertainty is sufficient so that the risks can be calculated accurately and remain acceptable. Otherwise we should be in dangerous waters.

It is in this context that I wish to set this workshop. Through your scientific contribution, you are able to give *further precision* to our knowledge so that increasingly daring *bets* will be won.

2 To control uncertainty

The control of uncertainty seems to me to be the essential object of our action. Faced with impossibility of establishing the exact parameters of physical models, we must accept to take into account their variability and to make a guess to their consequences.

The scientific community has been led for many years to give answers to the various questions raised. Not all of the latter will be subject to discussion in this workshop, but all contribute to the intended goal and we shall try to show how they are ordered in a consistent way. They can be classified as follows :

– spatial variability,

- time variability,
- sensitivity analysis,
- reliability analysis.

2.1 Spatial variability

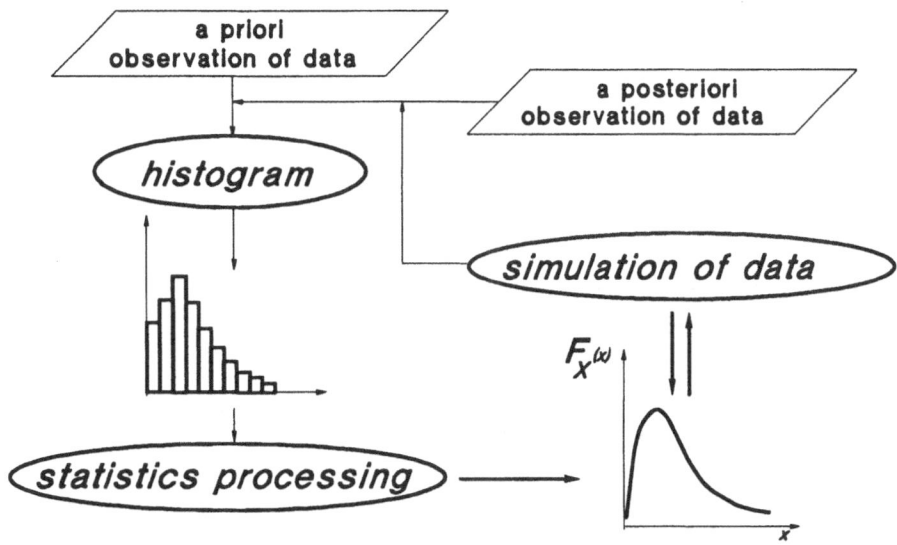

Figure 1 : random variable

First, let us recapitulate on the approach by random variable (figure -1-). Even if it is well known, we cannot however cast aside the context of its definition. The knowledge of a random variable goes through the databases *a priori* concerning the accumulated experience in similar situations and the *a posteriori* experience, for example, during manufacturing control, with which it is possible to carry out a Bayesian approach. A statistics processing allows us to identify a probability density function. It is the basic process indispensable if we are to make use of the probability methods.

In many situations, and particularly for the problems which concern us today, it isn't the knowledge of a variable that we try to find, but that of a stochastic field.

In this situation (figure -2-), it is first of all necessary to make a careful observation of the space through a randomized block design, which allows us to obtain a maximum amount of information with a minimum amount of trials. The spatial variability is then identified by a process and its spatial correlation functions. This field is certainly a major topic in this workshop where several studies are presented, concerning various

4

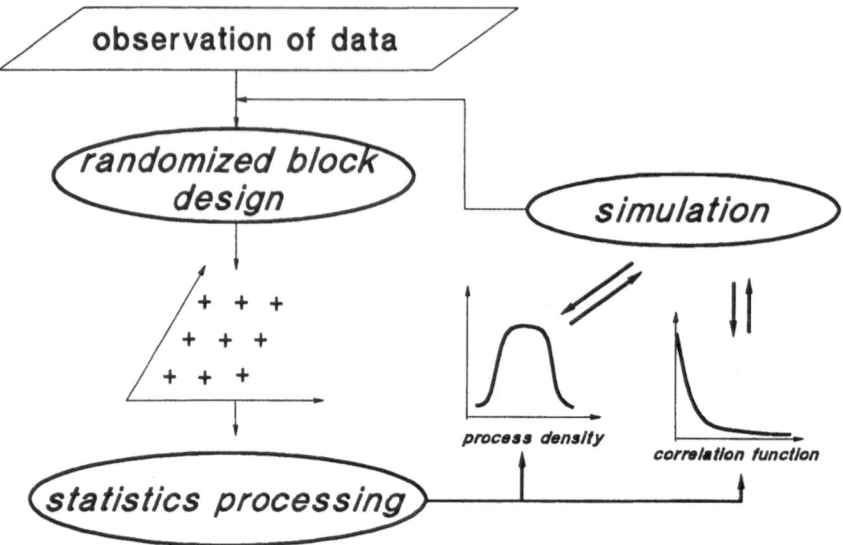

Figure 2 : stochastic field

materials : soils, rocks, wood, concrete... some authors put strong stress on the necessity of having available databases. A systematic effort to obtain information and feedback is to be implemented.

We must finally stress the question of the scale effect. The principles of continuum mechanics express relations with a geometrical scale implying an average effect taking into account the heterogeneous nature of materials. According to the observed phenomena and the considered materials, the parameters of the models are defined for a macroscopic volume that is difficult to define. This question is also raised during the numerical solution of a continuum mechanics problem by discretization.

2.2 Time variability

Some variables, particularly in modelization of natural phenomena, are time functions and are represented by stochastic processes (figure -3-). From observations of realizations generally in very limited number, an identification of the process is carried out. It is then that we use the stochastic representation of the turbulent speed of the wind and the acceleration due to an earthquake.

In the case of repeated events, we use the statistics of extremes.

This field concerns the modelization of actions more than that of materials, which is topic of this workshop.

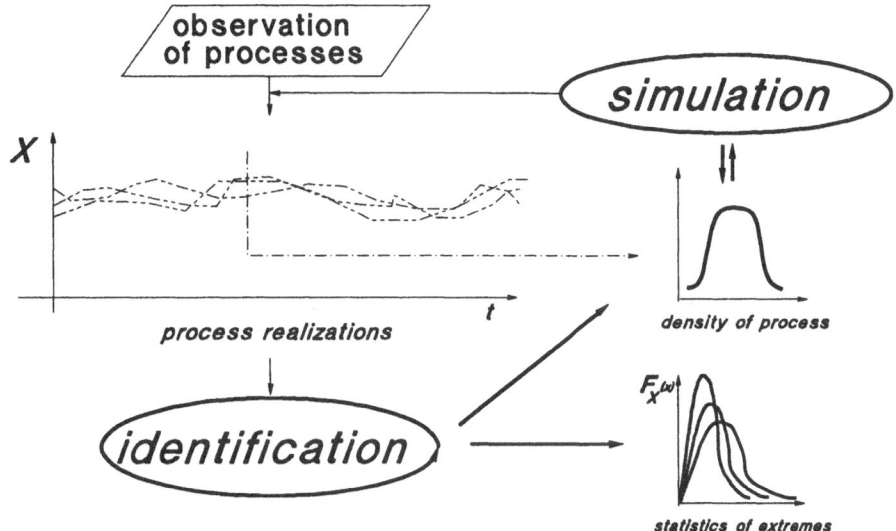

Figure 3 : process

2.3 Sensitivity analysis

Concerning sensitivity analysis (figure -4-), we couple uncertainty and a mechanical model. Uncertainties result from random data as input which propagates itself through the mechanical model. The goal is to determine the statistical properties of the mechanical model output. It is then possible to know if the variabilities of the input data are amplified or absorbed by the mechanical filter.

Such approaches are proposed when there is a strong sensitivity of the mechanical output to random defects. Two of the most significant cases are these of fracture mechanics and structural buckling. It is also the field of application of "stochastic finite element" which can be coupled with reliability method.

2.4 Reliability analysis

A further dimension is added when a suitable answer is required between needs (or internal forces) and resources (or resistances). The needs are random variables dependent on random data through a mechanical model calculating an internal force that the material must satisfy. This internal force depends on action data (applied forces) and also on state data (mechanical and geometrical parameters).

The resources are the expected availability of the material, *i.e.* strength. The elementary principle for design consists in defining resources that exceed needs, it presents a probability of failure and it is the role of a reliability model to evaluate this.

6

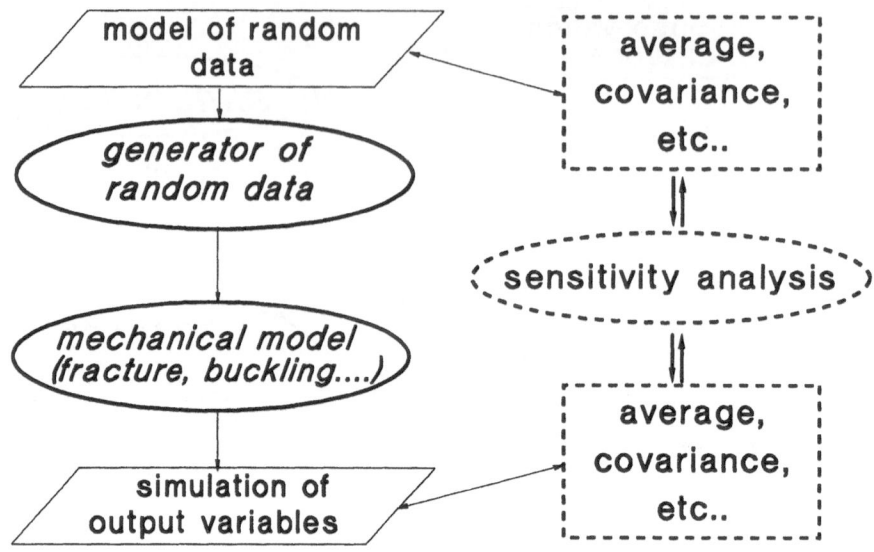

Figure 4 : sensitivity analysis

If we dispose of "random-number generators", – *i.e.* numerical tools able to create particular samplings of random variables or stochastic processes – it is then the role of a reliability model to stimulate random-number generators in order to calculate the probability of failure. The most famous method is the Monte Carlo simulation which, in principle, is easy to use in principle, but is unsuited as soon as internal force or resistance models result from the application of complex programs.

3 Tools

If, up to now, the accent has been put on the context of the control of uncertainty, the latter requires the development of methodological and numerical tools.

At present, it is the theory of probability, which must never be separated from the theory of statistics, that remains the essential tool of uncertainty modelization (but is it really uncertainty or imprecision?). The fractal theory now brings a new approach of uncertainty, as some contributions demonstrate in this workshop, and in the future probably, the theory of possibilities and the fuzzy logic will bring a new insight into uncertainty and imprecision.

Today, the tools used are probability, to identify and to simulate the stochastic fields, to make statistics of extremes. The randomized block designs must now allow us to select the physical or numerical trials. Response surfaces also produce a way of reducing the call to numerical simulations which are too expensive.

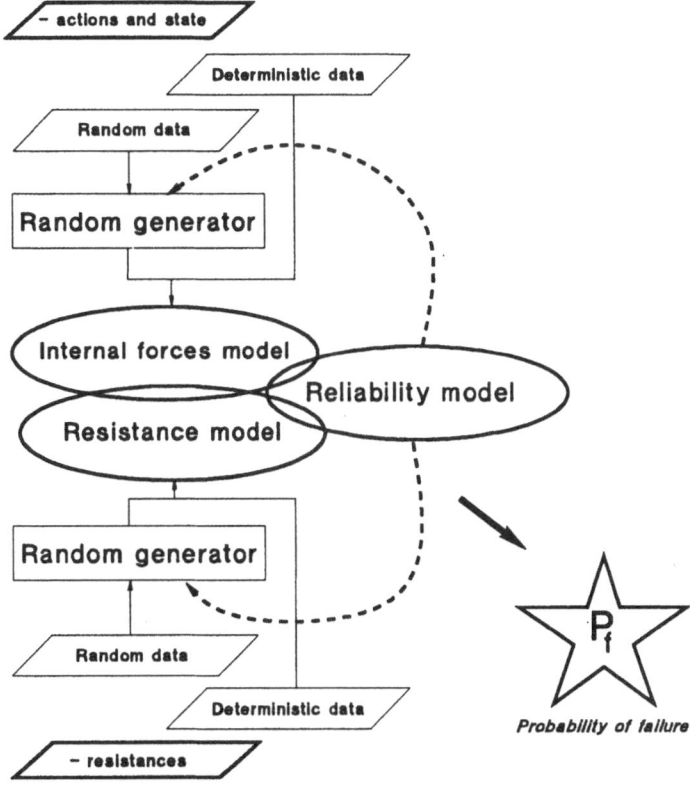

Figure 5 : "coupling" mechanics and reliability models

The way from the knowledge of hazard to the sensitivity analysis and to the reliability analysis requires research into methods which limit the realization of the mechanical calculation. To this end, we stress yet again the necessity of building response surfaces and of carrying out methods of conditional simulations which today are being highly developed.

The main tools for mechanical calculations are today founded on the finite element method. Coupling a finite element code and software for probability methods are not yet included in tested solutions. The stochastic finite elements remain limited to sensitivity analysis. An important effort in the area of implementation will be necessary to transfer the result of research on randoms in materials and structures to design.

4 Applications

The research which takes into account the randoms in mechanics of materials and structures is not only academic but also involves technical and economical stakes.

As far as technical stakes are concerned, we are confronted with the design and the realization of increasingly daring buildings or machines. We must evaluate uncertainty, and thereby control the risks. Important projects in offshore engineering and nuclear construction nowadays call for random modelization.

As far as economical stakes are concerned, we search to build at the lowest cost and with an acceptable risk. The search for a better design implies to keep risks at their current level. Even if it is not realistic to believe that the probability methods will be soon be widely available, they will, on the other hand, be used to calibrate the rules and thus give information to the designers. It will be an indirect repercussion of the experience of this research for them.

5 Conclusion

After having set up a general framework introducing the probability and statistics in engineering, I have to congratulate the scientific committee of this workshop on its initiative. It seems to me important to underline the fact that a limited topic was fixed, whilst accepting methodological contributions. No scientific discipline can be submitted to too strict a plan and the wealth arises from pluridisciplinarity.

We have here specialists in materials, soils, woods or concrete, some are attentive to the modelization of stochastic fields or some believe that the probability approach is an additional contribution to the understanding of fracture mechanics, others demonstrate how the approaches using fractals, entropy... are able to lead to a new scientific point of view.

I am sure that this workshop is a significant contribution to materials science, in the reality of uncertainty.

Acknowledgements : *thanks go to Sarah Davey for rereading this article.*

PROBABILITY AND MATERIALS: DATA SUBJECTIVE AND OBJECTIVE MODELING
A Synthesis Paper

P.L. BOURDEAU
Purdue University, West Lafayette, Indiana, U.S.A., and
Swiss Federal Institute of Technology, Lausanne,
Switzerland

Abstract
A brief synthesis is presented for the contributed papers
of Session A on "Data: Subjective and Objective Modeling"
of the International Workshop PROBAMAT. Then, some of the
critical issues of geotechnical site data modeling are
discussed.
Keywords: Data analysis, Statistics, Probability,
Correlation, Autocorrelation, Information, Soil, Rock,
Wood, Concrete, Material properties, Scale effect,
Heterogeneity, Randomness.

1 Introduction

Session A of the International Workshop PROBAMAT was
devoted to the characterization and assessment of material
property data, with emphasis on subjective and objective
modeling techniques. Six papers including a keynote lecture
were contributed to address this topic in the session. In
the first section of this report a brief synthesis of these
contributions is presented. Then, in the second section,
the critical problem of geotechnical and geoenvironmental
site characterization is discussed in the light of
probabilistic data analysis.

2 Synthesis of contributed papers

2.1. Objective modeling of data

Because material data are input quantities of the
analytical or numerical models used to predict the
performances of constructed facilities, their definitions
are dictated by the models. Therefore it is natural, in an
attempt to classify these data and their major attributes,
to refer to the model classification proposed by Breysse in
this workshop (Breysse, 1993). This classification based on
the material scales in which the model is formulated and

9

D. Breysse (ed.), Probabilities and Materials, 9–16.

the level of formulation at which randomness is introduced, is recalled in Table 1.

In Table 2 the main features of each contributed paper are presented including the class of related models, types of data discussed, sources of uncertainty affecting these data and their major probabilistic attributes.

scale of phenomeno-logy	micro-description	micro-macro transition	macro-description	Type
MACRO	no meaning	no meaning	determinist	D1
MACRO	no meaning	no meaning	probabilist	P1
MICRO	determinist	determinist	determinist	D2
MICRO	determinist	probabilist	probabilist	P2
MICRO	probabilist	determinist	determinist	D3
MICRO	probabilist	probabilist	probabilist	P3

Table 1. Classification of models (after Breysse, 1993)

author	Eins-tein	Favre	Cheru-bini	Soulie	Rossi	Caste-ra
material	rock	soil	soil	soil (clay)	conc-rete	wood
class of model	P1 or P3	P2	P1	P1	P1	P1

Table 2. Principal attributes of contributed papers

author	Eins-tein	Favre	Cheru-bini	Soulie	Rossi	Caste-ra
material property	indi-vid-ual joint prop-ert-ies; joint geom-etry and surf-ace para-mete-rs on the scale of the rock mass	grain and fabr-ic inde-xes; macr-osco-pic cons-titu-tive para-mete-rs	water conte-nt; state of compa-ction; compr-essib-ility; stren-gth	water conte-nt; shear stren-gth	tens-ile stre-ngth	dens-ity; bend-ing resi-stan-ce; modu-lus of elas-tici-ty
source of uncertain-ty	spati-al hete-roge-neity and disc-onti-nuity measur-em-ent error model unce-rtain ty; samp-ling	stru-ctur-al rand-omne-ss; cons-titu-tive model measur-em-ent error	spati-al heter-ogene-ity; measu-rement error; sampl-ing	spati-al heter-ogene-ity; measu-rement error; sampl-ing	stru-ctur-al rand-omne-ss; spat-ial hete-roge-neity measur-em-ent error	biol-ogic-al patt-erns; tree age; stru-ctur-al hete-roge-neity measur-em-ent error samp-ling

Table 2. Principal attributes of contributed papers (cont'd)

author	Eins-tein	Favre	Cheru-bini	Soulie	Rossi	Caste-ra
expected value	stat-isti-cs	regr-essi-on (mac-ro para-mete-rs)		statis-tics	stat-isti-cs	stat-isti-cs
scatter	stat-isti-cs	regr-essi-on (mac-ro para-mete-rs)	datab-ase of coeff-icien-ts of varia-tion	statis-tics	stat-isti-cs	stat-isti-cs
distrbuti-on	stat-isti-cs			statis-tics		stat-isti-cs and model fitt-ing
cross-correlati-ons	betw-een indi-vidu-al join-ts	tran-siti-on micro - macro				betw-een resi-stan-ce and defo-rmab-ility
scale effect	frac-tals		datab-ase of fluct-uation dista-nce	obser-ved; fract-al model	obse-rved and sign-ific-ant	
spatial variabili-ty	auto-corr-elat-ion		autoc-orrel-ation	observ-ed trend; variog-ram		

Table 2. Principal attributes of contributed papers (cont'd)

It is noted that three of the four contributions on geotechnical data include spatial variability as a major attribute of soil and rock masses properties. Some implications of this important feature are discussed in the next section.

2.2. Subjective Assessment

Although in civil engineering practice material properties are often assessed subjectively on the basis of personal or collective experience, little has been done to date, in comparison to the amount of research work published on objective data modeling, to formalize this line of approach. Among the contributions discussed herein, this subject is addressed by Einstein (1993) in relationship with the description of rock masses and tunneling works.

3 Challenges of Probabilistic Site Characterization in Geotechnical and Geoenvironmental Engineering

Site characterization and identification of in-situ soil properties are required for any geotechnical or geoenvironmental analysis, and the quality of this analysis is directly related to the quality and reliability of the data. In spite of the emergence of probabilistic and reliability concepts for the modeling of geotechnical systems and of computer-aided visualization techniques for geotechnical site characterization, practitioners take little advantage of modern data analysis methods that can provide the necessary statistical foundation to the models. Advanced geological site visualization software currently used in the practice lack such statistical basis and most often perform interpolation of spatially variable data from a deterministic perspective. This situation originates in the qualitative and inductive nature of the geologic methodology that is the primary step of every site investigation. A second cause, perhaps more determinant, is the limited number of geotechnical and geoenvironmental test data usually available to characterize the engineering soil properties for a given project.

Within such a context, the reluctance of geotechnical engineers to perform detailed statistical analyses of their data and to use probabilistic design methods is not surprising. Furthermore, when probabilistic methods are applied in spite of the difficulties mentioned above, insufficient analysis of the available data often leads to meaningless results. The following examples illustrate this particular situation.

3.1. Reliability of empirical correlations

Empirical correlations are often used in geotechnical practice to relate properties that are simple and inexpensive to measure to more complex soil characteristics. A number of such correlation-based relationships is available in the geotechnical literature (e.g. Carter and Bentley, 1991). However, a measure of the reliability of these relationships is seldom presented. Consequently they are used as deterministic recipes in spite of their purely statistical origin and meaning. In order to remedy to such misuse of correlation data, it would be necessary to revisit the databases that were used to develop the empirical relationships and to reformulate the results in probabilistic terms. Another promising approach proposed by Favre at al. (1993) consists of relating statistically the necessary macroscopic soil properties such as the constitutive parameters to the index properties of the grains and the granular fabric. It is expected that more development will be undertaken along this line.

3.2. Autocorrelation of soil properties varying in space

The autocorrelation function attached to an in-situ soil property quantifies statistically the degree of dependency between values taken by this spatially variable quantity at different location within the soil mass (e.g. Lumb, 1974; Alonso and Krizek, 1975; Tang, 1984). In geostatistics, the mathematical counterpart of an autocorrelation function is a variogram (e.g. Journel and Huijbregts, 1978). In general, the autocorrelation of a parameter decreases with the distance separating two points (e.g. Matsuo and Asaoka, 1977; Tang, 1979; Vanmarke, 1983). Identifying the characteristics of this decay is important to the mapping of soil properties and should be an essential ingredient of the modeling of geotechnical and geoenvironmental systems (e.g. Beacher, 1984; Wu et al., 1989). However, in most analyses, engineers do not have the means to evaluate this function and soil parameters are assumed by default perfectly autocorrelated. It has been suggested that this kind of simplification leads to overestimating the probability of failure of geotechnical systems when the autocorrelated parameters involved represent the soil resistance (Favre, 1988; Bolle, 1988; Mostyn and Soo, 1992). This opinion has been supported by the contributions of Cherubini (1993) and Bolle (1993) in this workshop.

In consequence, it is critical for researchers to contemplate the following tasks:
> (a) help expanding the available database of autocorrelation data;

(b) investigate for a broad range of geotechnical and geoenvironmental typical problems the impact of autocorrelations on the computed reliabilities, and validate these results;
(c) establish guidelines to help practitioners of probabilistic analysis methods to take into account the autocorrelated structure of soil properties whenever it is necessary.

4 References

Alonso, E.E. and Krizek, R.J. (1975), Stochastic Formulation of Soil Properties, **Proceedings, 2nd International Conference on Application of Statistics and Probability in Soil and Structural Engineering**, Aachen, Germany, Vol. 1, pp. 9-32.

Baecher, G.B. (1984), Need for Probabilistic Characterization (Just a few more tests and we'll be sure!), **Proceedings, Symposium on Probabilistic Characterization of Soil Properties**, Atlanta, ASCE, pp. 1-18.

Bolle, A. (1988), **Approche probabiliste en mécanique des sols avec prise en compte de la variabilité spatiale**, Thèse de doctorat ès sciences techniques No 743, Swiss Federal Institute of Technology, Lausanne, Switzerland.

Bolle, A. (1993), How to Manage the Spatial Variability of Natural Soils, **in this book**.

Breysse, D. (1993), Failure and Probabilities at Various Scales. A Synthesis Paper, **in this book**.

Carter, M. and Bentley, S.P. (1991), **Correlations of Soil Properties**, Pentech Press, London.

Castera, P.and Morlier, P. (1993), Variability of the Mechanical Properties of Wood: Randomness and Determinism, **in this book**.

Cherubini, C. (1993), The Variability of Geotechnical Parameters, **in this book**.

Einstein, H.H. (1993), Applications of Probabilistic Approaches in Geotechnical Engineering and Engineering Geology, Keynote Lecture, **in this book**.

Favre, J.-L. (1988), Properties of Soil: Large Uncertainties - A Minor Aspect of Foundation Reliability, Special Lecture, **Proceedings, Symposium on Reliability-Based Design in Civil Engineering**, Lausanne, Vol. 2, pp. 51-66.

Favre, J.-L. and Rahma, A. (1993), Classification Logic and Correlations Between Soil Parameters: Application to an Elasto-plastic Law, **in this book**.

Journel, A.G. and Huijbregts, C.J. (1978), **Mining Geostatistics**, Academic Press, Inc., London.

Lumb, P. (1974), Applications of Statistics in Soil

Mechanics, in **Soil Mechanics – New Horizons**, I.K. Lee, Ed., Newnes-Butterworths, London.

Matsuo, M. and Asaoka, A. (1977), Statistical Model Identification of Undrained Strength of Saturated Clay, **Proceedings, 9th International Conference on Soil Mechanics and Foundation Engineering**, Tokyo, Special Session No 6.

Mostyn, G.R. and Soo, S. (1992), The Effect of Auto-correlation on the Probability of Failure of Slopes, **Proceedings, 6th Australian-New Zealand Conference on Geomechanics "Risk – Identification, Evaluation and Solutions"**, Christchurch, New Zealand, pp. 542-546.

Rossi, P., Wu, X., Le Maou, F. and Belloc, A. (1993), Experimental Study on the Scale Effect on Concrete in Tension, **in this book**.

Soulie, M. and Masengo, E. (1993), Small Scale Variability in Clay Samples, **in this book**.

Tang, W.H. (1979), Probabilistic Evaluation of Penetration Resistance, **Journal of the Geotechnical Engineering Division**, ASCE, Vol. 105, No GT10, pp. 1173-1191.

Tang, W.H. (1984), Principles of Probabilistic Characterization of Soil Properties, **Proceedings, Symposium on Probabilistic Characterization of Soil Properties**, Atlanta, ASCE, pp. 74-89.

Vanmarke, E.H. (1973), **Random Fields**, MIT Press, Cambridge, Massachusetts.

Wu, T.H., Tang, W.H., Sangrey, D.A. and Baecher, G.B. (1989), Reliability of Offshore Foundations – State of the Art, **Journal of Geotechnical Engineering**, ASCE, Vol. 115, No 2, pp. 157-178.

INTRODUCTORY REMARKS TO SESSION 2: TOOLS

F. CASCIATI
Department of Structural Mechanics, University of Pavia, Pavia, Italy

Abstract
What the session could be guessed to be and what the session actually is. This report first provides a brief review of the problems which still need research, development or, more simply, a theoretical systematization. The conference contributions, then, emphasize which is the real problem of current general interest: how to make a good use of numerical simulation techniques.
Keywords : Experiment Design, Fractal Dimension, Scaling Law, Simulation, Uncertainty Modelling.

1 Introduction

When the organizing committee members met in October 1992 and considered the idea of a conference session on Tools, the main goal was to create an open forum for the scientists working on material properties modelling, independently of the specific approach they were adopting.

Such a session would have been predicted to become either a container for contributions covering a broad spectrum of open problems or a check point for that currently is the main unsolved problem for people working in material modelling. The tremendous job of the conference chairman was successful in directing the session toward the second, and more fascinating, character. It is therefore task of the chairman, in introducing the section, to provide a bird's-eye-view of the aspects which presently are the embers under the cinder: may be tomorrow they will be completely forgotten; may be they will attract the competitive activity of several research groups.

D. Breysse (ed.), Probabilities and Materials, 17–22.
© 1994 *Kluwer Academic Publishers.*

2 Modelling: what, why and how?

This conference is mainly devoted to standard classical materials for use in Civil and Mechanical Engineering. However, one cannot say nothing on the large research effort which is presently ongoing within the topic of **smart materials**. In extreme synthesis, one can distinguish three main classes of such materials, Rogers (1990):

1. shape memory alloys, which offer a *plastic* branch in their constitutive law, i.e. the possibility of achieving large values of local deformation; however one can annihilate that will be the final distortion just by heating the material, Boller and Brand (1993). Shape memory polymers are also available and investigated;

2. electro-rheological materials, which are able to change their effective stiffness and damping properties with changes in an applied electric field; these changes occur in the interval of few milliseconds, Ehrgott and Masri (1992);

3. piezoelectric rubber: particles of piezoelectric ceramic material (of a brittle nature) are embedded in a synthetic rubber: when subjected to a mechanical strain, it produce a voltage.

Some new materials, not actually smart but for use in smart systems, have also been proposed for manifacturing passive dampers: they offer a special hysteretic loop enabling to locally dissipate energy under cycling loading.

The tools to be employed in the area of intelligent adaptive systems cover an impressive broad range. They go from applied chemistry processes to identification algorithms; from numerical simulation to uncertainty modelling. These materials are presently the object of specialist investigations but their applications cover a so large field that perhaps tomorrow they will become part of any engineer background.

Back to standard materials, this conference focus attention on disordered media, i.e. media with a non-isotropic and a non-homogeneous character. In this area, **homogeneization techniques**, Sab (1992), are the ones preferred by the analyst who makes intensive use of general-purpose structural analysis deterministic codes. The fascinating feature is their character of filter of the spatial data-scatter toward a reliable deterministic solution. The round-table organized by Dr. Vilotte also discusses, among other topics, the pros and cons of such an approach (see its report within this volume).

Homogeneization techniques, however, remove those unsymmetries and/or imperfections which cause the selection of a bifurcation branch rather than another. To prevent this information to vanishing, one must pursue a modelling of such imperfections. Assume first that one would like to bypass the formulation of any mathematical model: this goal can be achieved by relying on **artificial intelligence tools** as expert systems and neural networks, Casciati and Faravelli (1991).

Uncertainty treatment, then, plays an important role owing to the fact the information in the knowledge base is subjective, incomplete and imprecise. Moreover, this induce uncertainty in the validity of the conclusion achieved by any logical network. There are a number of alternative techniques from different disciplines for handling uncertainty:

- the logical models use structured symbolic information about uncertainty and avoid numerical assessments;

- the linguistic models quantify the extent to which the imprecise language statements match the propositon to be expressed (*fuzzy reasoning*);

- the legal models are based on the *theory of evidence*: belief and plausability measures are selected to represent ignorance and uncertainty;

- the statistical engineering models make use of *probability calculus* and the *Bayesian approach* provides a method for updating the probability of a hypothesis given an observation of evidence.

The best way of **uncertainty treatment** in the different branches of material mechanics is still matter of discussion and, hence, its selection represents an open problem.

Due to the large development of the applications of probabilistic methods in structural engineering, Augusti et al. (1984), there are several research groups working into the probabilistic framework. The **spatial distribution** which characterizes the material properties, then, suggests different levels of sophistication in this modelling:

- each material property is specified by a single random variable in a sort of global homogeneization;

- each material property is specified by a vector of independent random variables: each entry of this random vector corresponds to a local homogeneization over a volume of size larger than the actual correlation length;

- each material property is specified by a random field; the discretization of the continuum, then, leads one to the introduction of random vectors with correlated entries.

The last case gives rise to the so called **stochastic finite elements**, Shinozuka (1987), Faravelli and Bigi (1990), Ghanem and Spanos (1991), and **stochastic boundary elements**, Burczynski (1993), Casciati et al. (1993). They can be regarded either as a perturbed version of the deterministic scheme or as repeated analyses of the basic problem with data which are realizations of random fields. Here, another problem arises when the step adopted in the discretization of the random field is compared with the discretization mesh required by the tool used for the numerical structural analysis. In particular one can assign property values element by element or, for instance, Gauss point by Gauss point, Chamis (1987).

Tools are also the ways of dealing with **statistical data** in both to provide them an interpretation and to extend their validity from the laboratory to the applications. For this purposes, regression analyses, experiment design and response surface are today in competition with probabilistic casual networks and neural networks, all providing a mapping between input and output quantities.
The five papers which will be presented in the session all cover a special aspect of this last topic: provided that numerical simulation is a powerful tool of investigation, the main interest lies in the plan of (numerical) experiments to be carried out and in the generalization of their results.

3 Numerical simulation aspects

The five papers which form the program of this session are:

1. A. CARPINTERI and B. CHIAIA, Fractal, Renormalization Group Theory and Scaling Laws for Strength and Toughness of Disordered Materials;

2. L. FARAVELLI, Blocking Problems in the Analysis of Random Fields;

3. I. LAALAI and K. SAB, Size Effect and Stochastic Nonlocal Damage in Quasi-brittle Materials;

4. G. REGNIER and B. SOULIER, Experimental Designs for an Experimental Modelling of Material or Structural Tests;

5. S. ROUX, Role of Disorder in Brittle Fracture.

The five papers are easily grouped into two classes: one dealing with size scale effect and the other with experimental design. Nevertheless, the five papers have a common root: leaving one-dimensional problems toward two- and three-dimensional situations, laboratory experiments are difficult to conceive and to perform. One substitutes laboratory with computer and experiment with simulation. The main questions to be answered are then:

- how to plan these numerical simulations?

- how to make the results of these numerical experiments robust?

An answer to the second question is related to the so-called scale effects, to which three papers are devoted. **Carpinteri and Chiaia** provide a general presentation of the governing parameters (tensile strength and fracture energy) and their fractal nature. Fractal schemes, in fact, are ruled by dimensions which act as exponents in scaling laws. Two sets of laboratory experiments are presented to support the theoretical treatment, the goal being the identification of the physical quantities that are invariant with a size scaling. The abstract ends with a significant statement very well explained during the presentation of the paper: "Variations in the fractal dimension of fracture surfaces produce variations in the physical dimension of thoughness and not ... in the measure of thoughness".

The other two papers have a manifest goal: to validate a model of simulation which introduces a probabilistic definition of the dominant parameter. The paper by **Roux** considers the brittle failure of a disordered medium: the idea is to regard it as a lattice the bonds of which have a strength with a probabilistic definition. Numerical simulations produce results which are in agreement with the expected scaling law and this makes the simulations reliable, even if, as Roux underlines, one should be careful in extrapolating their limit of validity. A resulting multifractal spectrum provides information on the shape of the probability distribution of the tensile strength to be selected as input.

Laalai and Sab approach the problem of fracture failure of notched and unnotched concrete specimens. Their objective is to find both the volume size effect and the structure size effect (when damage is not concentrated in a relatively small zone around the crack tip). They select the elastic energy of the damaged system as the reference parameter and introduce a Weibull distribution for its limit value. As in Roux paper, the results are convincing, but in both papers the intrinsic reasons for selecting one parameter rather than another, one distribution rather than another are not explained in an objective manner.

The paper by **Faravelli** starts with an overview of tools for the analysis of samples of the data arising from laboratory and/or computer tests. Response surface techniques turn out to be ready to incorporate uncertainties. In particular the introduction of random fields and large random vectors leads one to consider the problem of blocking. It represents the way by which experiments affected with spatially distributed uncertainty should be planned. Different blocking techniques are illustrated and their results are compared within a numerical example.

Regnier and Soulier consider the efficiency of the choice of an experiment plan which is sub-optimal: D-optimal designs are presented as an alternative to the classical orthogonal test matrix for which $k\mathbf{X}^T\mathbf{X} = \mathbf{I}$, $\mathbf{X}$ being the plan matrix. Given the initial experiments, one adds or deletes one or more tests to achieve an increase of the determinant of the matrix $\mathbf{X}^T\mathbf{X}$.

4 Conclusions

The single value of all the contributions to this session has been determined by the possibility given to the single scientists of providing the best of their research experience. Nevertheless, this report emphasizes the broad spectrum of open problems concerning the adequacy of some tools for handling uncertainty in the field of material mechanics.

Acknowledgement - This report has been partially supported by a grant from the Engineering Committee of the Italian Research Council (CNR - 93.02201.07).

References

Far from being complete or extensive, this list of references only provides the key for entering the topics sketched in the present report.

Augusti G., Baratta A. and Casciati F. (1984), **Probabilistic Methods in Structural Engineering**, Chapman & Hall, London

Boller C. and Brand W. (1993), Some Basic Ideas on the Design of Adaptive Aircraft Structures Using Shape Memory Alloys, in Breitbach E. et al. (eds.), **Proceedings of 4th Int. Conference on Adaptive Structures**, in print for Technomic

Burczyński T. (1993), Stochastic Boundary Elements Methods: Computational Methodology and Applications, **Proceedings of the IUTAM Symposium on Probabilistic Structural Mechanics**, San Antonio, Texas

Casciati F. and Faravelli L. (1991), **Fragility Analysis of Complex Structural Systems**, Research Studies Press, Taunton

Casciati F., Faravelli L. and Callerio A. (1993), Alcune Considerazioni sugli Elementi di Contorno Stocastici (in Italian), **Proceedings of the Meeting of the Italian Group of Stochastic Mechanics**, Taormina

Chamis C.C. (1987), Probabilistic Structural Analysis Methods for Space Propulsion System Components, **Probabilistic Engineering Mechanics**, 2, 100-110

Ehrgott R.C. and Masri S.F. (1992), Use of Electro-Rheological Materials in Intelligent Systems, in Housner G.W. et al. (eds.), **Proceedings of the US-Italy-Japan Workshop/Symposium on Structural Control and Intelligent Systems**, USC CE-9210, 87-100

Faravelli L. and Bigi D. (1990), Stochastic Finite Elements for Crash Problems, **Structural Safety**, 7, 113-130

Ghanem R.G. and Spanos P.D. (1991), **Stochastic Finite Elements: A Spectral Approach**, Springer Verlag

Rogers C.A. (1990), An Introduction to Intelligent Material Systems and Structures, in Chong K.P. et al. (eds.), **Intelligent Structures**, Elsevier, 3-41

Sab K. (1992), On the Homogenization and the Simulation of Random Materials, **Eur. J. Mech., A/Solids**, 11, n. 5, 585-607

Shinozuka M. (ed.) (1987), **Stochastic Mechanics**, Vol. I, Columbia University

MODELS: MICRO-MACRO
The Synthetic Paper

M. OSTOJA-STARZEWSKI
Department of Materials Science and Mechanics
Michigan State University
East Lansing, Michigan, USA

Abstract

The problem of relating mechanical response of materials to their microstructure is common to many areas. It is relevant in mechanics of polycrystalline, granular (and other type) media, in problems of effective moduli and problems of crack propagation and damage formation/evolution. We thus have seven papers in this session. In the following, rather than providing here a review of each of these papers, we give a discussion of salient features and techniques common to various disciplines that these papers represent. This is accompanied by an account of state of the art and future challenges in micromechanics. It is hoped that the reader will see potential applications in study of other types of materials, e.g. cellular, fibrous, biological.

Keywords: microstructure, micromechanics, granular media, effective moduli, crack propagation, experimental methods, analytical methods, numerical methods, scale effects.

1. Need for Probabilistic Micromechanics

While it is impossible to say when did micromechanics begin, it is clear that the pace of research in this field has especially been increasing over the last two decades. As a result, considerable progress has recently been made in the basic understanding of processing, microstructure, and mechanical response of materials, see e.g. (Mura, 1982; Nemat-Nasser and Horii, 1993). Most of this work, however, treats the materials in a deterministic manner without accounting for the random nature of their geometry, physical properties and interactions. Thus, many micromechanics models are still, essentially, deterministic approximations of stochastic phenomena. Examples of phenomena which need to be treated in a probabilistic manner are: effective constitutive laws of random media, with fracture and damage being of primary importance, evolution of microstructures, and scale-dependence in all such problems. It should be noted here that although statistical studies in micromechanics of solids have not yet become as popular as the deterministic ones, they date back at least some three decades, see e.g. (Beran, 1968; Chudnovsky, 1977, Axelrad, 1978). In this paper we give a brief account of the present accomplishments and outstanding challenges in micromechanics of random media. We try to give an objective, synthetic viewpoint with some key references but without any attempt at a complete list of all the works.

23

D. Breysse (ed.), Probabilities and Materials, 23–28.

2. State of the Art and Future Challenges of Probabilistic Micromechanics

The material of this section is a reflection of some results of a recent workshop on the subject (NSF Workshop, 1993). First we give here a list of accomplishments in micromechanics of random media.

i) Development and calculation of the following statistical descriptors which arise in bounds and estimates on effective physical properties of two-phase composites and mono-phase polycrystals with static microstructure:
- n-point probability functions;
- statistics on granular media;
- orientation distributions;
- interfacial surface statistics;
- clustering, connectedness, percolation, scale-dependence;

Advances in the above items have been made using theoretical, numerical and experimental methods, with the first three being well established.

ii) The following property/structure relationships have been obtained for two-phase composites and mono-phase polycrystals, with the object of going beyond mean-field estimates:
- three-point estimates for linear materials, which invlove conductivity, elasticity, flow and diffusion/reaction processes;
- estimates for periodic microstructure;
- property/structure relationships computed by numerical simulations;
- estimates for coupled properties;
- optimal bounds on constitutive response.

iii) Some progress has been made in:
- statistical fracture mechanics;
- physically nonlinear properties;
- random heterogeneous materials with dissipation.

Many of the future challenges of probabilistic micromechanics lie in:

i) Identification of the experimentally accessible morphological information which is essential to the determination of the macroscopic response.

ii) Multiscale statistical properties and response fields.

iii) Modelling of stochastic evolution of microstructures.

iv) Damage-related properties such as strength, creep, toughness, and material instabilities.

v) Mechanics of granular materials, especially of multi-phase particulates such as concrete, asphalt, coated suspensions, and sintering.

3. Current Advances - Seven Session Papers

In an attempt to give a synthesis of advances presented by seven contrtibuted works, we discuss them according to several criteria; these are:
- geometric aspects
- experimental methods
- numerical methods
- analytical methods
- effective response

- scale effects

The stage for synthesis is set in the form of Table 1 below, which points out (with an X) the novel features presented by each paper. These features are discussed later, in some detail, in six subsections bearing the titles of six criteria. In the discussion references are also given to other relevant works so as to give a more complete picture of the state of the art techniques; space limitations here preclude any attempt at the completeness.

Table 1:

	Auvinet	Chameau & Muhunthan	Chang	Ghosh	Huet	Jeulin	Kachanov Tsukrov Shapiro
Geometric Aspects		X				X	X
Experiment. Methods		X					
Analytical Methods	X		X		X	X	X
Numerical Methods				X			
Effective Response	X		X	X		X	X
Scale Effects					X	X	

3.1 Geometric aspects

The choice of a good geometric model is fundamental in any micromechanics study. Classically, a lot of the past, and present, work has been based on three rather simple models of: a polycrystal, a granular medium, and inclusion-matrix composites. Thus, in the first case, a spatial or planar system of grains - such as the Poisson-Voronoi mosaic - has been used (Stoyan et al, 1982; Kumar et al, 1992). In the second case, as can be seen in a number of works (Shen et al, 1993), a spatial or planar system of disks of spherical (or ellipsoidal) shapes forms the basis of the model. Finally, in case of matrix-inclusion systems, the idea is to embed the inclusions in a deterministic (i.e. periodic) or random fashion in a matrix.

There is, on the other hand, another more general formulation based on mathematical morphology theory, which has been developed by the French school (Mathéron, 1967, 1975; Serra, 1982; Jeulin, 1986, 1989, 1993, 1994). It is based on the concepts of Boolean sets and functions, and provides a wealth of other, more complex models. It is important to note here that the Poisson point process is fundamental to all of these models.

A special application of random geometry to soil mechanics is discussed in the paper by Chameau and Muhunthan, whose objective is to measure, and then relate the soil fabric to the effective response of the material.

3.2 Experimental methods

Besides their obvious role of verifying any theory, experiments are fundamental in setting up a theoretical model of the microstructure. Thus, the reference is made here first to the preceding section, whereby two papers - Jeulin and Muhunthan - need to be mentioned.

3.3 Analytical methods

The classical references on analytical methods in micromechanics of composites are (Willis, 1981), Hashin (1983), Torquato (1991). Jeulin presents a review of methods for statistical fracture mechanics, that are closely related to the Boolean set and function methodology.
Kachanov et al give an account of the results obtained by the conformal mapping and potential methods; see also (Kachanov et al, 1994) in (Ostoja-Starzewski and Jasiuk, 1994).
Finally, both Auvinet and Chang give an account of their methods developed for granular media.

3.4 Numerical methods

With the rapid increase in capability of computers, there has been associated a development and increasing sophistication of numerical methods. First and foremost are the simulation methods of microstructures based on either finite elements (FE) or finite differences (FD). In our session, a novel example of FE is presented in the paper by Ghosh and Moorthy. An example of an FD application can be found in (Jasiuk et al, 1992). Both FE and FD are employed to model crack formation and breakdown in microstructures (Herrmann and Roux, 1990). It should be noted here that, recently, boundary element methods (BE) have also been used (Thorpe and Hetherington, 1992). A simulation method, related to FD, had been developed in mechanics of granular media, it is called a 'discrete element method', see (Shen et al, 1992).
Finally, we note that the FE and FD methods can be generalized to a stochastic setting, whereby one gets SFE and SFD. While work on the SFE problems, where 'S' is due to random material heterogeneities, has been going on since the early eighties (see e.g. Kleiber and Hien, 1992), a lack of a link to material microstructure was a major obstacle; theses issues are discussed in (Ostoja-Starzewski, 1993).

3.5 Effective response

Huet develops a theory of linear elastic and viscoelastic response of concrete-type materials, which is intrinsically scale-dependent, see section below.
Kachanov et al discuss a vast subject of effective moduli of solids with cracks, see also (Kachanov et al, 1994) and (Kachanov, 1993).
Chang as well as Auvinet advance new methods for setting up of effective constitutive laws of granular media; (Shen et al, 1992) is the main general reference in that area.

3.6 Scale Effects, Meso- and Macro-Response

Huet gives a philosophy and theoretical aspects invloved in setting up of a hierarchy of scale-dependent bounds on effective moduli. Applications in solid and geomechanics are discussed. Scale-dependence signifies that a whole range of length-scales, from micro through meso to macro,

is covered. This type of approach has recently been furthered by Ostoja-Starzewski (1993, 1994) in deriving random continuum approximations, that are crucial in formulating the SFE methods. The importance of scale effects in relation to computational micromechanics is discussed at length in (Breysse, Fokwa and Drahy, 1994).
Scale effects in the context of crack propagation in random microstructures are discussed by Jeulin.

Acknowledgement

This work was supported in part by the NSF under Grant No. MSS-9202772.

References

Auvinet, G (1993) Random stress fields within granular media, this book.

Axelrad, D.R. (1978) *Micromechanics of Solids*, PWN/Elsevier, Warsaw/Amsterdam.

Beran, M.J. (1968) *Statistical Continuum Theories*, J. Wiley, New York.

Breysse, D., Fokwa, D. and Drahy, F. (1994) Spatial variability in concrete: nature, structure and consequences, in (Ostoja-Starzewski and Jasiuk, 1994).

Chameau, J.L. and Muhunthan, B. (1993) Modelling granular fabric by tensors and their statistical test, this book.

Chang, C.S. and Liao, C.L. (1994) Estimates of elastic modulus for media of randomly packed granules, in (Ostoja-Starzewski and Jasiuk, 1994).

Chang, C.S. (1993) Micromechanical model for randomly packed granules, this book.

Chudnovsky, A. (1977) *The Principles of Statistical Theory of Long Term Strength*, Novosibirsk (in Russian).

Ghosh, S and Moorthy, S. (1993) A Voronoi cell finite element model for random heterogeneous media, this book.

Hashin, Z. (1983) Analysis of composite materials - a survey, J. Appl. Mech., Vol. 50, pp. 481-505.

Herrmann, H.J. and Roux, S. (1990) *Statistical models for the fracture of disordered media*, Elsevier (North-Holland).

Huet, C. (1990) Application of variational concepts to size effects in elastic heterogeneous bodies, *J. Mech. Phys. Solids* 38(6), 813-841.

Huet, C. (1993) Experimental characterization, micromechanical simulation and spatio-stochastic approach of concrete behaviors below the representative volume, this book.

Jasiuk, I., Chen, J. and Thorpe, M.F. (1992) Elastic moduli of composites with rigid sliding inclusions, *J. Mech. Phys. Solids*, Vol. 40 (2), pp. 373-391.

Jeulin, D. (1994) Fracture statistics models and crack propagation in random media, in (Ostoja-Starzewski and Jasiuk, 1994).

Jeulin, D. (1986) Study of spatial distributions in multicomponent structures by image analysis, *Acta Stereol.* Vol. 5 (2), pp. 232-239.

Jeulin, D. (1989) Some aspects of mathematical morphology for physical applications, *Physica A*, Vol. 157, pp. 13-20.

Jeulin, D. (1993) Morphological models for random media, this book.

Jeulin, D. (1994) *Morphological Models of Random Structures*, CRC Series on Continuum Modelling and Discrete Systems, in preparation.

28

Kachanov, M. (1993) Elastic solids with many cracks and related problems, *Adv. Appl. Mech.* Vol. 30, in press.

Kachanov, M., Tsukrov, I. and Shafiro, B. (1994) Effective moduli of solids with cavities of various shapes, in (Ostoja-Starzewski and Jasiuk, 1994).

Kachanov, M., Tsukrov, I. and Shafiro, B. (1993) Effective properties of solids with randomly located defects, this book.

Kleiber, M. and Hien, T.D. (1992) The Stochastic Finite Element Method, J. WIley, New York.

Kumar, S., Kurtz, S.K., Banavar, J.R. and Sharma, M.G. (1992) Properties of a three-dimensional Poisson-Voronoi tessellation: a Monte Carlo study, *J. Stat. Phys.*, Vol. 67 (3/4), pp. 523-550.

Mathéron, G. (1967) *Elements pour une Theorie des Milieux Poreux*, Masson, Paris.

Mathéron, G. (1975) *Random Sets and Integral Geometry*, J. Wiley, New York.

Moorthy, S., Ghosh, S. and Liu, Y. (1994) Voronoi cell finite element model for thermoelastoplastic deformation in random heterogenous media, in (Ostoja-Starzewski and Jasiuk, 1994).

Mura, T. (1982) *Micromechanics of Defects in Solids*, Martinus Nijhoff Publishers, Hague.

Nemat-Nasser, S and Horii, M. (1993) *Micromechanics: Overall Properties of Heterogeneous Solids*, Amsterdam, North-Holland.

NSF Workshop (1993) on "The statistical characterization of material microstructure and its relation to material performance," The Catholic University of America, Washington, D.C.; final report by M.J. Beran and J.J. McCoy.

Ostoja-Starzewski, M. (1993) Micromechanics as a basis of random elastic continuum approximations, *Probab. Engng. Mech.*, Vol. 8(2), pp. 107-114, 1993.

Ostoja-Starzewski, M. (1993) Micromechanics as a basis of stochastic finite elements and differences - an overview, *Appl. Mech. Rev.* (Special Issue: *Mechanics Pan-America 1993*, M.R.M. Crespo da Silva and C.E.N. Mazzilli, Eds.), Vol. 46 (11, Part 2), pp. S134-S147.

Ostoja-Starzewski, M. (1994) Micromechanics as a basis of continuum random fields, *Appl. Mech. Rev.* (Special Issue: *Micromechanics of Random Media*), in press.

Ostoja-Starzewski, M., and Jasiuk, I., Eds. (1994) *Appl. Mech. Rev.* (Special Issue: *Micromechanics of Random Media*), Vol. 47 (1, Part 2).

Serra, J. (1982) *Image Analysis and Mathematical Morphology*, Academic Press, London.

Serra, J., Ed. (1988) *Image Analysis and Mathematical Morphology*, Vol. 2, Academic Press, London.

Shen, H.H., Satake, M., Mehrabadi, M., Chang, C.S. and Campbell, C.S. (1992) *Advances in Micromechanics of Granular Materials*, Proceedings of the 2nd US-Japan Seminar on Micromechanics of Granular Materials, *Stud. Appl. Mech.*, Vol. 31, Elsevier; also *Mech. Mater.* (Special Issue: *Mechanics of Granular Materials*), Vol. 16 (1-2), 1993.

Stoyan, D. Kendall, W.S. and Mecke, J. (1987) *Stochastic Geometry and Applications*, J. Wiley, New York.

Torquato, S. (1991) Random heterogeneous media: microstructure and improved bounds on effective properties, *Appl. Mech. Rev.* Vol. 44 (2), pp. 37-75.

Willis, J.R. (1981) Variational and Related Methods for the Overall Properties of Composites, *Adv. Appl. Mech.*, Vol. 21, pp. 2-78.

FAILURE AND PROBABILITIES AT VARIOUS SCALES
A Synthetic Paper

D. BREYSSE
Laboratoire de Mécanique et de Technologie, ENS-CNRS-Univ. Paris 6, France.

Abstract

Heterogeneities are the source of the apparent erraticism of fracture surfaces. They also cause the scatter of macroscopic properties like strength or fracture energy. Lastly, they play a great role in size-effect or in localization. The purpose of this paper is to investigate at what scale can various models account for the presence of heterogeneities and the different possibilities (implicit or explicit) of their description. The different existing models are classified regarding these criteria and their merits are compared. The paper focusses on the identification of parameters, since the real predictive power of the models is very dependent on the efficiency and objectivity of identification. In the last part, alternative approaches, seeking universality and/or fractal properties are reviewed.

Keywords, brittle fracture, descriptive power, heterogeneity, identification, micro-scale, macro-scale, models, parameters, predictive power, size-effect, scatter

1. Introduction.

Many materials exhibit a significant degree of variability of their intrinsic properties regarding failure : failure strength, time to failure, energy of failure. The existence of this scatter has given force to the concept of statistical strength, which became the basis for the analysis of structural reliability (Weibull, 1951, Freudenthal, 1968). It has soon been noticed that the main source for variability lies in the existence of material inhomogeneities, which can be found at various scales in any material (microcracks, pores, impurities, local defects, ...). It was shown that the Weibull-type distribution of properties, originally presented as a phenomenological description of nature (Weibull, 1952) could be naturally deduced from mechanical assumptions and some description of the microstructural randomness. In certain cases, for instance in perfectly brittle fracture, it has even been possible to draw a close relationship between the probability density function (pdf) of the defect size and the macroscopic properties (Phoenix, 1978, Jayatilaka, 1979, Hild, 1992).

Taking into account the microdefects makes also possible a good description of size effects, the basic idea being that the structural failure results from a combination of one/several/many dependent (or independent) local failures. The predictions can have a qualitative nature (Jayatilaka and Trustrum, 1977, explaining the difference between "ductile" compressive and "brittle" tensile strengths of concrete by different combinations of local events) or, under certain conditions, a quantitative nature. In any case, the role of defects during the failure process remains widely unexplored. For instance, it has been shown (Jeulin, 1990) that using a Weibull pdf was not compatible with experimentally measured size effects and that the size effect depends on details of the microstructure like the particle shape. Another important question is the way the defects act or interact during the failure process. In uniaxial tensile tests on concrete or mortars, it has been noticed that the fracture energy is first consumed by diffuse microcracking and in several non-critical fracture process zones before the maximum load is reached. It is only in a second stage that one critical fracture process zone develops (Berthaud et al, 1991). This fact can

D. Breysse (ed.), Probabilities and Materials. 29–38.
© 1994 *Kluwer Academic Publishers.*

be numerically reproduced by using a probabilistic damage model for the continuum (Breysse and Schmitt, 1991). One has here two different stages : in the beginning, a "crowd" of defects interplay when, in the following, the evolution is driven by a unique defect. Chudnovsky and Gorelik *[in this book]* remark that a parallel exists with the crack propagation problem where one can distinguish two extreme cases, in which the defects play quite different roles : *"The first one is called cooperative fracture, where the intenisty of damage formed as a response to the stress-concentration at the tip is much greater than the intensity of preexisting heterogeneity. The exact defect locations, sizes and orientations may vary from one specimen to another but the behavior is highly reproducible. In this case, a deterministic macro-response is obtained. The second extreme mode of failure is when the crack propagation is controlled by the preexisting field of defects and when it does not change - by itself - this field. In this case, it results in a highly irregular crack pattern and a wide scatter of the main fracture parameters."*

Frantziskonis *[in this book]* suggests that the loading rate and the crack velocity play a great role on the development of fracture, since the way the stress redistributes (or not) around the crack tip will change with them. It is to bridge the gap between Linear Fracture Mechanics (LFM) and Statistical Strength Theory (STT) that Chudnovsky and Kunin (1987) have developed the Statistical Fracture Mechanics (SFM), of which they detail two applications in this book.

The last question that will be adressed is the following : when one knows that the existence of defects makes that a "material property" changes with the specimen size or with the boundary conditions, as it is shown by Van Mier et al *[in this book]*, what can be told about the experimental measurements and how can they be used for modelling ? If the influence of inhomogeneities on various measurements (like elastic parameters) can be accounted for using homogenization techniques, this is not the case for fracture properties, when the existence of inhomogeneities provokes major changes in the nature of phenomena themselves (Jeulin, 1992).

2. Scale and type of description

2.1. Example : Crack propagation : where is the randomness ?

When studying fracture, the randomness can appear at both micro and macro scales :

- at a macro-scale, where one can consider a pdf for any parameter (strength, fracture energy, ...) (Rossi and Piau, 1988),
- at a micro-scale, where Stroeven, 1988, has noticed that the crack propagation in concrete reveals a stochastic character.

As a consequence, for a given problem, let us say, propagation of a fatigue crack, one can always choose the scale at which the randomness is introduced. The choice of this scale corresponds to the fact that one does not want to look at lower scales and that the resulting response of all phenomena playing at lower scales (whatever they are) is phenomenologically accounted for by a set of random events and a probabilistic description of these events (Breysse, 1991a). For instance, when studying the remaining life time of a specimen in which a single fatigue crack propagates, different possibilities can be chosen (Table 1).

		conditions for propagation	
		deterministic	random
crack tip	deterministic	deterministic	random
position	random	random	random

Table 1. Different possibilities for considering randomness.

The randomness of life time can result :
- from the probabilistic description of the initial crack tip position (or of the defect size),
- from the probabilistic description of the propagation (corresponding to the fact that the material in front of the tip is not perfectly known),
- or from a combination of these descriptions.
The descriptive possibilities of these different approaches are comparable. The criteria for one choice instead of another must (normally) be related to the amount of available information on experimental data which can feed the models. Of course, the very first question, answer to which is beyond the scope of this paper, is : "Do we need any non-deterministic model ?", knowing that, in any case, a probabilistic model needs a larger amount of data than a deterministic one.

2.2. Randomness at various scales in the fracture process : models and parameters
Following the above remarks, one can classify the models for failure into five categories according to whether they use a deterministic or a probabilistic description for three key-items : micro-description, transition from micro to macro-scale, macro-description. Table 2 separates the models between MACROmodels and MICROmodels depending on the scale the phenomenological behavior is assumed at (d holds for deterministic, r for random). Some examples are given below for each category.

scale of phenomenology	micro-description	micro-macro transition	macro-description	class
MACRO	no meaning	no meaning	d	D1
MACRO	no meaning	no meaning	r	P1
MICRO	d	d	d	D2
MICRO	d	r	r	P2
MICRO	r	r or d	r or d	P3 or D3

Table. 2. Different classes of micro and macro models.

- D1 is the category of all usual deterministic models for failure, like, for instance, Damage Mechanics models (Mazars, 1984, Lemaître and Chaboche, 1985, Bazant and Pijaudier-Cabot, 1987). These models need a varying number of parameters whose value is identified after mechanical tests on specimens even if, in certain cases (f.i. for nonlocal damage), some of these parameters are said to be related to microstructural information (Mazars, 1987);
- D2 groups deterministic micromechanical models (Horii and Nemat-Nasser, 1986), where one studies the way a given perfectly known defect propagates (or stops) in a perfectly known matrix. In this case, the requested data concern the defect itself (shape, size, orientation) and the matrix (energy release rate, elastic properties, ...);
- P1 is the class of macroscopical probabilistic models like, in the case of concrete, those developed by Rossi and Piau (1988), Breysse (1990) or Wu (1991). The parameters of the models are comparable to those of class D1, excepted the fact that tests have to be repeated to identify first and second moments of the distributions. Rossi has recently proposed a way to identify some parameters directly from material data (like the aggregate size) (Rossi et al 1992);
- P2 is the class regrouping many models like those of Bogdanoff and Kozin (1980) and Newby (1987) (for fatigue crack growth, Newby has shown that describing the failure process by a Markov chain leads to results identical to those obtained when the well-known Paris-Erdogan law is randomised (class P1)), Mihashi (1983) (who used a

Markov process describing the failure of concrete) or the Statistical Fracture Mechanics theory (see in this book papers by Chudnovsky and Gorelik and by Kunin). The major problems are the identification of the micro-parameters used in the models (i.e., in the case of SFM theory, a micro random field $\gamma(x)$ and a characteristic length r). One can note that to choose a value for r (with the assumption that properties are uncorrelated at a range higher than r) amounts to fix implicitly a material scale (which will depend on the material microstructure);

- In the last class, one can put all the models in which the microstructure is explictly randomised. Interestingly, the macro-response can be random (Van Mier [ibid], Haddad, 1986, Hild and Marquis, 1992) or, for some aspects, deterministic. For instance, studying a set of many interacting random cracks, Fond (1992) found that, depending on the way the friction/sliding between the crack lips is accounted for, the global failure can be either due to the propagation of a unique (or few) crack(s) or to a large number of cracks. This can be compared to the two extreme modes quoted in the introduction for crack propagation. In the case of what has been called cooperative fracture, the macroscopical fracture tends to be deterministic, even if the micro-behavior is random. Initial disorder is progressively erased during the evolution. The emerging response comes from an average of the micro-scale responses, when in the other extreme case, the macro-response is driven by micro-events which are very far from the average. The main difficulty with this class of models is the high number of requested parameters since these parameters have to describe both the initial microstructure (pdf of defect size and defect orientation for instance) and its evolution. Furthermore, micro-mechanical measurements are very uneasy to perform.

2.3. Descriptive and predictive models

Whatever the model he uses, an engineer is only interested in its predictive power, i.e. its ability , once the parameters of the model have been identified, to predict the response in new situations. Any model whose parameters cannot be related to a given measurement or whose parameters can be fitted only using an a posteriori reverse engineering technique is of little value for practical use. Theoretically, the descriptive power of a model (i.e. its ability to obtain any response, once the parameters have been estimated), increases with the complexity of the model, therefore, the number of parameters. Normally, one can expect the following order of increasing quality :

$$D1 < D2 \quad ; \quad D2 < P2 \quad ; \quad D1 < P1 < P2 < P3$$

In practice, the fact that some parameters are difficult and/or impossible to identify can modify this order. This would not be true, it would not exist any scientific debate on the deterministic/probabilistic dilemna but only conferences about "how the deterministic models disappeared from the face of earth ?". In consequence, one has, in any case, to find the model whose predictive power is better. Let us detail this question for the texts presented in this chapter.

3. About the identification of material parameters

3.1. Models of class D1 and P1

In their paper, Van Mier et al [ibid] show that a key properties like fracture energy deeply depend on the experimental conditions since it is estimated at 40.0, 35.3 and 29.8 N/m for three different boundary conditions (in a second example, it can even be divided by four !). This fact confirms results previously obtained by Breysse (1991b) who shown the influence of boundary conditions on the tensile strength (as being an obvious consequence of the material heterogeneity). However, this casts some doubts on any "classical" model which would use measured values of σ_t or of G_f without any care. For instance, it seems highly dubious that a finite element deterministic model using the "measured" value of σ_t for the local strength would give a correct result, as soon as the role of heterogeneities cannot be neglected. In the same line, it is unlikely that, if the heterogeneity is a key factor for fracture, using Fracture Mechanics with a local Griffith criterion would give a correct response, since Chudnovsky and Gorelik [*ibid*] show that G_{Ic} is by no means a material parameter. In their paper, they demonstrate that, even if, at a micro-scale, the Griffith criterion drives the material behavior, at a macro-scale, the conventional fracture toughness G_{Ic} departs from the mean value of the specific fracture energy $2 <\gamma>$. The existence of heterogeneities changes the mode of response and the average response is not the response of the average. Therefore, a great care must be taken when simulations need parameters whose value have been identified at a different scale.

3.2. Models of class P2 : an example

SFM, as it is developed in the two papers yet quoted (Chudnovsky and Gorelik, Kunin), has the purpose to develop a material/mechanical description rich enough to write explicitily any macro-scale quantity (probability of crack initiation/arrest, distribution of crack length, of critical load, of toughness or of time to failure, ...) as a function of a "crack propagator" which depends in a probabilistic sense of the "true material parameters" : the mean value of the local random field γ and its standard deviation. An additional assumption is that the correlation length r is very small compared to the specimen size D. The two papers focuss on :
- the controlled discontinuous crack growth in brittle materials (Kunin): the crack path is assumed to be rectilinear and the growth results from successive random jumps and random waiting times. The first moments of several macro-measurement (crack length, time to failure, ...) are explicited and related to microstructural measurements,
- the fracture energy accompanying a failure, the crack path being tortuous (and derived from a Wiener process).
The descriptive power of these approaches is impressive, all the difficulty being to identify the data requested by the models. Kunin says us that "*even for a macroscopically homogeneous solid, the first two moments and covariances for the random entries (local specific fracture energy* $\gamma(\underline{x}, \underline{n})$) *amount to 12 empirical constants in the 2D case and to 42 constants in the 3D case*". The difficulty is somewhat overcomed by using the crack trajectories themselves as a part of the characterization of $\gamma(\underline{x}, \underline{n})$, with the assumption of local minimal values. It however remains to identify the parameters of the distribution of γ (i.e. three constants for a Weibull distribution), one having always two possibilities for that :
- the first is to directly relate the parameter to a microstructural measurement,
- the second is an 'inverse fitting', comparing the macro-response to the prediction.
Regarding the correlation length r, Chudnovsky and Gorelik suggest to deduce it from direct measurements when Kunin shows that it may be -indirectly but simply- related to the time to failure scatter. Some others assumptions made in the model can be discussed, like the description of the random material parameters with a Poisson process when one should better see a mechanical description of the properties.

In any case, these models are very promising. A further stage of development would be to examine the assumptions regarding the correlation length, for instance ;
- what if the local correlation length r of the random field is not very small when compared to the specimen size, as it is in concrete ? In this case, not only fluctuations, but also correlations are to be accounted for and Frantziskonis [*ibid*] shows their great influence on the failure process,
- what if this correlation length could change with time ? For instance, the random waiting time, assumed by Kunin as phenomenological, corresponds to subcritical events, occuring at a lower scale and probably accompanied by a progressive microstructural change of the material. The study of the microstructure around the crack tip would make possible the introduction of more material data and push downward the scale of the phenomenological description.

3.3. Models of class P3 : an example

In this chapter, Van Mier et al [*ibid*] present an example of a model describing the progressive failure of a disordered material. The similarities with previously published models make us rank this model among those called "numerical concrete" (Zaitsev, and Wittmann, 1981, Roelfstra et al 1985, Stankovski et al, 1990). The principle consists in mapping the material topology/geometry and properties on a numerical support (usually finite elements) and to perform numerical simulations on this model.

These approaches have brought some interesting results like the influence of the relative properties of phases (aggregate-matrix-transition area) on the crack(s) propagation process and on macroscopical measurements. However, they suffer from remaining generally limited to 2D computations (unable to describe some mechanisms) and to small specimens. The use of a discrete lattice (bars or beams) on which the properties are mapped (instead of a 2D continuum) makes possible to generate a medium with a higher number of degrees of freedom and therefore to reproduce the response of a specimen containing a great number of heterogeneities (aggregates). However, the equivalence between the lattice and the continuum is not always clearly treated, particularly out of the elastic domain.

Let us detail on Table 3 the methodology of identification suggested by Van Mier et al in their paper.

	parameter	value	method for identification
beam geometry	length l	random, $> 1/3 \, D_{min}$	mapping
	height h	0.68 l	reverse engineering on global stiffness
	width b	constant	no importance
elastic properties	E_i (3 phases)	E_i	?
	ν	0.16	equivalence in the elastic domain
material strength	σ_{ti} (3 phases)	σ_{ti}	?
traction/bending coupling	α	0.05	reverse engineering on fracture energy

Table 3. Identification of parameters in numerical concrete.

This combination of tuning procedures (h, ν, α) and assumptions (what about the real local values for the cement-aggregate interface ? what about D_{min} when the aggregate size goes continuously from 0 to D_{max} ?) puts some limits on the predictive power of this kind of model. In practice, this kind of models is, for the moment, limited to the description of laboratory specimens or structures of limited size, for which the user owns a large amount of information. The tuning phase is, in this case, followed by a semi-predictive phase, during which the user is able to draw further information. As a simple example, in their paper, Van Mier et al clearly indicate why the boundary conditions have a great influence on the specimen response. They also point the fact that the crack bridging has important consequences and give tracks for material modelling.

4. Other tracks. On the universality of fracture

Trying to better understand fracture (for a better prediction) can be done through other ways. The increasing number of references to fractal aspects of failure is representative of such approaches. The basic concept is that the failure process are, in a certain way, independent of the material and that observing the failure of various materials (or post-mortem specimens) can bring useful information. These studies can be classified in three parts:
- measurement of fractal dimensions, for instance for crack patterns,
- models relating micro-material parameters to crack patterns,
- models relating fractal dimensions to macro-material parameters.
Since fracture and fractal have the same latin root, it is generally considered that advances in these three domains will be a progress towards a full understanding of the failure process, even if Mandelbrot himself (1989) feared that the fractal could be limited to the description of physics even if the mechanics of fractals is only in its first years (Panagiotopoulos et al, 1993).

If the erraticism of fracture surfaces is a well documented field (Ziegeldorf, 1983), problems arise as soon as one wants to quantify the fracture patterns. The apparent surface of failure masks the existence of many multiple cracks which close once the major crack appears. The measurements generally gives fractal dimensions for the crack profile around 1.20 (1.20+/- 0.1 for Schmittbuhl et al, 1993), this value seeming to be universal (rocks, plaster, bakelite, ...). Concerning the relation with fracture energy, the fractal dimension appears to be positively or negatively related to the fracture toughness (Xie Heping, 1989, and results quoted by Saouma and Barton, 1992). It is possible that the correlation changes with the ductile/brittle character of failure. Carpinteri and Ferro (1992) have recently correlated the fracture energy to the fractal dimension of the fracture surface, but their fractal dimension (2.38 for the crack surface) is well above usual results. They consider that the apparent size effect is only due to this fractal character. Few works have been devoted to the relation between micro-material parameters and crack patterns. Issa et al (1992) have shown that the fractal dimension of the surface seems to increase with the aggregate size (from 2.10 to 2.22 when the size varies from 4.7 mm to 37.5 mm).
As it has been noted by Schmittbuhl et al (1993), this "universality" would imply that no information can be deduced on the mode of fracture, being given only the roughness index. On the other hand, it would also mean that physical consequences derived from the geometry of cracks can be transposed to a wide class of different situations.

Another (semi-empirical) method consists in studying in which the failure properties are modified when the failure is "helped". Wittmann et al [1994] compare the failure of grooved and ungrooved cylinders, and their tests are a step towards a better understanding of the relations between crack tortuosity, material data (aggregate size) and mechanical parameters (fracture energy, material strength). The fact that the crack can -or cannot- meander around aggregates influences the failure. The authors highlight two competitive phenomena acting in a grooved specimen:
- the major crack tends to be more linear and it has less liberty to reach weak zones in the material,
- the major crack tends to be more linear and the effects of aggregate interlocking and shear retention are reduced.
These two phenomena seem to exactly complement each other and this would be the reason why the fracture energy is the same for both grooved and ungrooved specimens. Micro-measurements of the crack tortuosity are planned and their results will be a useful contribution to the knowledge of failure.

Frantziskonis [*ibid*] tries to identify the value of "universal" constants which would characterize, whatever the material and for a given set of loading conditions (f.i. volumic dilatation), the transition between a failure with a single crack to a multiple crack failure. Assuming that the constitutive response of the micro-medium for brittle materials is linear, the transition appears to be related to the ξ_0/ξ ratio, where the material correlation lengths (initial value and value after evolution) appear. The universal value is respectively equal to 1.02 (volumic dilatation), $\sqrt{2}$ (pure shear) and 0.88 (uniaxial tension). Upper bounds from the fractal dimension of the fracture surface are deduced from the results, for instance $D = 1.24$ appears in dilatation, which is not too far from experimental results. However, these theoretical results have to be confronted to many experimental data.

Conclusions

Fracture is one of the best illustrations for the role of chance in physical phenomena. Therefore the use of probabilistic models seems natural. The fundamental question that has to be answered is : "what is the use of probabilistic models of failure for engineering purposes ?". It has been shown how and why the practical use of these models remains generally limited. Their main weakness does not lie in their descriptive power, since they are far above their deterministic companions, but in their predictive power, i.e. in their ability, once their parameters have been identified following a non-subjective procedure, to predict the material/structural response in new situations. Efforts have to be done to relate micro-material parameters of the models to micro-measurements, and it is only under this condition that probabilistic models will convince the users of their full interest.

At present, the main field of application of these models remains the simulation/prediction of the response of laboratory specimens (or test structures) for which one possesses in general far much more information than in an engineering context. In this case, the richness of the description is a key factor and, since the quantitative prediction is often less important than the qualitative one, the probabilistic models are a well suited tool. Their use will give a better knowledge upon phenomena acting at various scales, and make one able to build MACROmodels whose phenomenology can result from these previous analysis. Another possibility is, for a given problem, to combine deterministic and probabilistic models, depending on the quantity of information one needs and on the quantity of available data (Breysse, 1994).

References

Bazant Z.P., Pijaudier-Cabot G. (1987) Measurement of characteristic length of nonlocal continuum, *Rep. n87-12/498m*, Center for concrete and geomaterials, Evanston, USA.

Berthaud Y., Ringot E., Schmitt N. (1991) Experimental measurements of localization for tensile tests on concrete, Int. Conf. on *Fracture Prosesses in Brittle Disordered Material*, Noordwijk, The Netherlands.

Bogdanoff J.L., Kozin F. (1980) A new cumulative damage model, *J. Appl. Mech.*, 47, pp.40-44.

Breysse D. (1990) Probabilistic formulation of damage-evolution law of cementitious composites, *J. Eng. Mech. ASCE*, 116, 7, pp.1389-1410.

Breysse D., Schmitt N. (1991) A test for delaying localization in tension. Numerical interpretation through a probabilistic approach, *Cem. Concr. Res.*, 21, pp. 963-974.

Breysse D. (1991a) Méthodes pour l'analyse multi-échelles des matériaux et des structures hétérogènes, *Habil. dir. rech.*, LMT Cachan, ENS/CNRS/Univ. Paris 6.

Breysse D. (1991b) Numerical study of the effect of inhomogeneities and boundary conditions on the tensile strength of concrete, *ACI Mat. J.*, 88, 5, pp. 489-498.

Breysse D. (1994) Spatial variability in concrete: nature, structure and consequences, *Appl. Mech. Rev.*, 47 (1, Part 2), in press.

Carpinteri A., Ferro G. (1992) Apparent tensile strength and fictitious fracture energy of concrete: a fractal geometry approach to related size-effects, *FDCR-2*, Vienna.

Chudnovsky A., Kunin B.I. (1987) A probabilistic model of brittle crack formation, J. Appl. Phys., 62 (10), pp. 4124-4129.

Chudnovsky A., Gorelik M. (1993) Statistical fracture mechanics - basic concepts and numerical realization, *in this book*.

Fond C. (1992) Interactions entre fissures et cavités circulaires dans des milieux élastiques plans, *Ph. D Thesis*, LMT Cachan/CNRS/Univ. Paris 6, France.

Frantziskonis G. (1993) Crack pattern related universal constants, *in this book*.

Freudenthal A.M. (1968) Statistical approach to brittle fracture, in *Fracture : an advanced treatise*, Ed. Liebowitz, Academic Press, New-York, Ch. 6, pp. 591-619.

Haddad Y.M. (1986) A stochastic approach to the internal damage in a structure solid, *Theor. Appl. Fract. Mech.*, 6, pp. 175-185.

Hild F. (1992) De la rupture des matériaux à comportement fragile, *Ph.D. Diss.*, LMT Cachan, ENS-CNRS-Univ. Paris 6.

Hild F., Marquis D. (1992) A statistical approach to the rupture of brittle materials, *Eur. J. Mech. A/solids*, 11, 6, pp. 753-765.

Horii H., Nemat-Nasser S. (1986) Brittle failure in compression: splitting, faulting and brittle-ductile transition, *Phil. Trans. Royal Soc. London*, 319, 1549, pp. 337-374.

Issa M.A., Hammad A.M., Chudnovsky A. (1992) Fracture surface characterization of concrete, *ASCE Conf.*, College Station, TX, pp. 127-130.

Jayatilaka A. (1979) Statistical approaches to brittle failure, in *Fracture of engineering brittle materials*, Applied Science Publ., London.

Jayatilaka A., Trustrum K. (1977) Statistical approach to brittle failure, *J. Mat. Sci.*, 12, pp. 1426-1430, pp. 2043-2048 and 13, pp. 455-457.

Jeulin D. (1990) Random fields models for fracture statistics, Actes du 32ème Colloque de Métallurgie, INSTN, *Ed. de la Revue de Métallurgie*, n°4, pp.9-13..

Jeulin D. (1992) Modèles probabilistes pour la Structure et le Comportement Mécanique des Matériaux", *Bull. Mécamat*, 6, pp. 15-24, feb. 1992.

Kunin B.I. (1993) A stochastic model for controlled discontinuous crack growth in brittle materials, *in this book*.

Lemaître J., Chaboche J.L. (1985) *Mécanique des matériaux solides*, Ed. Dunod, Paris.

Mandelbrot B. (1989) *Les objets fractals. Forme, hasard et dimension*, Ed. Flammarion, Paris

38

Mazars J. (1984) Application de la mécanique de l'endommagement au comportement non linéaire et à la rupture du béton de structure, *Doct. Etat.*, LMT Cachan, ENS/CNRS/Univ. Paris 6.

Mazars J. (1987) Endommagement des structures en béton armé ou précontraint. Stratégie de Modélisation, d'Identification et de Numérisation., *AFREM-LMT-DAEI final report*.

Mihashi H. (1983) A stochastic theory for fracture of concrete, in *Fracture Mechanics of concrete*, ed. F.H. Wittmann, Elsevier Publ., The Netherlands.

Newby M.J. (1987) Markov models for fatigue crack growth, *Eng. Fract. Mech.*, 27, 4, pp. 477-482.

Panagiotopoulos P.D., Panagouli O.K., Mistakidis E.S. (1993), Fractal geometry and fractal material behaviour in solids and structures, *Archiv. Appl. Mech.*, 63, pp. 1-24.

Phoenix S.L. (1978) Composite materials: testing and design, *Fifth Conf. ASTM*, Philadelphia.

Roelfstra P.E., Sadouki H., Wittmann F.H. (1985) Le béton numérique, *Mat. Str.*, 118, pp. 309-317.

Rossi P., Piau J.M. (1988) The usefulness of statistical models to describe damage and fracture in concrete, France-US Workshop on *Strain localization and size-effect due to cracking and damage*, LMT Cachan, France.

Rossi P., (1992) Effet d'échelle sur le comportement du béton en traction, *Bull. Liaison LPC*, 182, pp. 11-20.

Saouma V.E., Barton C.C. (1992) private communication.

Schmittbuhl J., Gentier S., Roux S. (1993), Field measurements of the roughness of fault surfaces, *Geoph. Res. Lett.*, 20, 8, pp. 639-641.

Stankovski T., Runesson K., Sture S., Willam K.J. (1990) Simulation of progressive failure in particle composites, in *Micromechanics of failure of quasi-brittle materials*, pp. 285-294, ed. SP Shah, SE Swartz, ML Wang, Elsevier Publ.

Stroeven P. (1988) Characterization of microcracking in concrete, Res. Civ. Eng. Mat., *Annual Report, T.U. Delft*, The Netherlands.

Van Mier J.G.M., Vervuurt A., Schlangen E. (1993) Crack growth simulations in concrete and rock, *in this book*.

Weibull W. (1951) A statistical distribution of wide applicability, *J. Appl. Mech.*, 18, pp. 293-297.

Weibull W. (1952) A survey of 'Statistical effects' in the field of material failure, *Appl. Mech. Rev.*, 5, 11, pp. 449-451.

Wittmann F.H., Slowik V., Alvaredo A.M. (1994), Probabilistic aspects of fracture energy of concrete, *Materials and Structures*, to be published.

Wu X. (1991), Modélisation numérique de la fissuration du béton à partir d'une approche stochastique, *Ph. D. Diss.*, ENPC, Paris.

Xie Heping (1989) The fractal effect of irregularity of crack branching on the fracture toughness of brittle materials, *Int. J. Fract.*, 41, pp. 267-274.

Zaitsev Y.B., Wittmann F.H. (1981) Simulation of crack propagation and failure of concrete, *Mat. Constr.*, 83, pp. 357-364.

Ziegeldorf S. (1983) Phenomenological aspects of the fracture of concrete, in *Fracture Mechanics of Concrete*, pp. 31-41, ed. F.H. Wittmann, Elsevier.

PROBABILISTIC AND STOCHASTIC MODELS
A Synthetic Paper

D. JEULIN and P. ROSSI*
Centre de Géostatistique, Ecole des Mines de Paris, Fontainebleau, France.
* Laboratoire Central des Ponts et Chaussées, Paris, France.

Abstract

Probabilistic models for the failure of materials are a first step towards reliability models that account for different scales of defects present in materials. This paper stresses the following points developped during this session: the description and simulation of the microstructure, the statistical analysis of mechanical data, models of damage evolution and of brittle fracture.

Keywords: brittle fracture, damage, stochastic process, random set, weakest link.

1 Introduction

The papers presented in this session introduce tools and models that can be organized in four separate parts: 1) Description and simulation of the microstructure. 2) Statistical analysis of mechanical data. 3) Model of damage evolution. 4) Models of brittle fracture.

2 Description and simulation of the microstructure

When modeling random media, it is interesting to use in a first step a probabilistic approach of the microstructure: Coster et al. apply models derived from the theory of random sets (Matheron 1967, 1975; Serra 1982) to the study of various sintered carbide materials: WC-Co samples, and various Titanium carbides (TiC-Co, TiC-Ni). Three types of models (the Boolean model with Poisson polyhedra or spherical grains, the Poisson mosaic model, and the Dead Leaves model with Poisson grains) are tested from measurements by image analysis of the probability for a segment of length l, a pair or a triple of points separated by the distance h, to belong to one of the phases of the materials. The method enables us to test from available information the validity of a given model, and to estimate its parameters (usually two or three in the present case). Moreover, the models are defined in the three-dimensional space, but can be evaluated from data obtained by image analysis on polished sections. This interesting stereological property is due to the underlying Poisson point process used in these models (and in all the other models used in this session, even if this stochastic process appears in an implicit form!). From obtained data, the most appropriate

39

D. Breysse (ed.), Probabilities and Materials, 39–43.

model to reproduce the microstructure of the studied materials seems to be the Boolean model, using random primary grains growing from Poisson germs.

In the paper the random models are used as descriptive tools, the parameters of the models giving a full summary of the complex microstructure of the studied materials. Their evolution with the sintering time can be useful to describe the coarsening kinetics of the materials.

A complementary use of these models, which is not developed in this paper, is in a predictive way: it is possible to derive an estimation of morphological data concerning the three-dimensional arrangement of the medium, such as its connectivity number (or Euler-Poincaré parameter) $N_v - G_v$

or as the statistical distribution of distances from a random point outside the primary grains to their boundary. On a different level, they provide mathematical models of correlation functions of the material properties (elastic moduli, or any physical property of interest), that can be introduced in a model of estimation of the effective properties of these two phase media, as theoretically proposed by many authors, such as Matheron (1967), Beran (1968), Kröner (1971), or Willis (1981).

More general models of this type can be used to describe multiphase materials, or continuously varying properties (scalar or tensors), as proposed in Jeulin (1994) and the included references.

3 Statistical analysis of mechanical data

Statistical analysis of data is of importance in preliminary investigations before modeling and simulating materials with heterogeneous physical properties. In his paper, Rosowsky provides interesting and useful results on the mechanical properties of an industrial material, lumber, with a statistical characterization of random variables, the strength and stiffness of lumber. The two variables are positively correlated, and this must be considered in simulations. The author uses a normal transformation, simulates pairs of correlated normal variables, and then performs a back transformation. Corrections must be given to recover the correct coefficient of correlation. It should be interesting to consider in details the correlation clouds of the initial data to see whether other models of bivariate distributions (such as for instance distributions proposed in Jeulin 1993) could be used. With a large amount of data, this bivariate distribution could be estimated and used in Monte Carlo simulations of the pairs of material properties.

The author models the creep behavior of lumber, considered as a viscoelastic material, using a conventional analog model (combination of Maxwell and Kelvin elements). This provides means of simulation for reliability analysis. The background of creep behavior is therefore connected to stochastic processes by a discrete model; these aspects, not developed in the present paper, are described in the enclosed references.

4 Model of damage evolution

For materials such as concrete, damage processes occur before fracture. Mihashi developed probabilistic models for the description of the effects of cycles of freezing and thawing, or of cycling loads. The author mentions three scales in his approach: micro, meso and macro. The evolution of damage is described by a Markov process with a rate depending on information at the different scales: microstructure, stress concentrators such as aggregates or cracks, loading conditions. The basic cell of the microstructure is an element made of n units, each one containing a single crack and being prone to damage. A survival probability is given for a single element as a function of a rate of occurence of damage. This last one is a function of the studied process. For example in the case of freezing and thawing cycles, it is a function of temperature, water content, heterogeneity, and number of cycles. The basic cells are combined in a way depending on the damage process (they are put in parallel for the freezing and thawing problem). This approach is interesting, but introduces a lot of arbitrary assumptions (and of parameters), that require many separate experiments to be validated. The separation between different scales is mainly a matter of. presentation of the model. This last one remains basically macroscopic.

5 Models of brittle fracture

Statistical models for brittle materials are based on the presence of defects in homogeneous media. The models proposed in this session (Lamon 1994, Kittl et al. 1994) are derived from the well-known Weibull model. They share the following assumptions (not always given in an explicit form): a microcrack is initiated at the weakest defect in the specimen B of a material, and propagates to cause the fracture of the specimen (weakest link assumption); The non fracture events of two separate elementary volume elements are independent (in probability). If this assumption is satisfied at any scale, including for infinitesimal volumes dV, the defects in the microstructure must be points of a Poisson point process; the resulting model is a particular case of Boolean random function (Jeulin 1993, 1994), which is a generalization of the Boolean random set used in section 1 of this presentation. For this type of model, we get:

$$P\{\text{no fracture of B}\} = \exp - V(B) \, \phi(\sigma) \qquad (1)$$

In (1) $\phi(\sigma)$ is an increasing function of the fracture criterion (usually the maximal principal stress). In the case of the Weibull distribution model, $\phi(\sigma)$ is the following power law:

$$\phi(\sigma) = \left(\frac{\sigma - \sigma_0}{\sigma_u} \right)^m \quad \text{for } \sigma \geq \sigma_0, \text{ else } 0 \qquad (2)$$

In his paper, J. Lamon applies this model to multiaxial loading. The fracture criterion is the maximum in a strain energy release rate in the direction of propagation (combined mode I and mode II). This criterion is converted into an equivalent tensile stress which follows the Weibull distribution. The author stresses the strong sensitivity of failure predictions (mainly concerning low probability of failures) to the statistical parameters of the model, and therefore the importance of their correct experimental determination. Intervals of confidence of the parameters of the model, depending on the method of estimation, should be helpful for practical applications in reliability analysis. Here, the estimation is fulfilled by means of the combination of fracture strength measurements and fractographic examinations, using various loading conditions: 3-point and 4-point bending, biaxial flexure. This is illustrated by an applications to a ceramic material. Difficulties in the case of shearing effects are mentioned but the limitations of the model in multiaxial stress states are not detailed: in particular, friction effects of the sliding parts during the fracture process should be expected, and are not accounted for in the model. Similarly, the evolution of existing flaws or of damage, mentioned in the conclusion, should require a different approach.

Kittl et al. propose derivations from the Weibull model in the case of a non uniform stress field. In one of the models, a function $\phi(\sigma)$ different from (2) is proposed:

$$\phi(\sigma) = \left(\frac{\sigma - \sigma_0}{\sigma_u - \sigma} \right)^m \quad \text{for } \sigma_0 \le \sigma \le \sigma_u$$

$$\begin{array}{ll} 0 & \text{for } \sigma \le \sigma_0 \\ \infty & \text{for } \sigma \ge \sigma_u \end{array}$$

(3)

This model is very similar to the Weibull distribution model, except that it is defined for a finite range of stresses ($\sigma_0 \le \sigma \le \sigma_u$).

To modify the size effect involved in equation (1), the authors propose other formulations of this equation, replacing the term V(B) by f(V) as follows:

$$f(V) = \frac{1}{1-\lambda} (V/V_0)^{1-\lambda} \quad \text{for } 0 \le \lambda \le 1$$

$$\ln (V/V_m) \quad \text{for } \lambda = 1$$

$$(V/V_0)^{1-\lambda} \quad \text{for } \lambda > 1$$

(4)

Various size effects are obtained from this model: the strength of the materials decreases with its volume for $\lambda < 1$. Low size effects are expected with $\lambda = 1$, while the strength is expected to increase with the volume of specimens when $\lambda > 1$. We should mention that for these models fails the traditional assumption on the statistical independence of separate elements of volumes. Justifications of these empirical derivations are provided from various models: the case $\lambda = 1$ can be obtained with a critical concentration of cracks, and the case $\lambda > 1$ with a set of parallel bars. The affirmation that there is no size effect in the case of the cement paste should be moderated, and limited to the case of scales much larger than the hydrates contained in the material. A different derivation of equation (4) can be proposed for the case $\lambda < 1$, as illustrated by Watson and Smith (1985) for fiber composites with random diameters: starting, with equation (1), the parameter $\theta = 1/\sigma_u^m$ can be considered as a volume intensity of defects. For some materials, the parameter θ may vary from one specimen to the other, and can be considered as a random variable, with Laplace transform Φ. After randomization of equation (1) and using the power law function of equation (2), it comes:

$$P\{\text{no fracture of B}\} = \Phi(V(B) \ (\sigma - \sigma_0)^m) \tag{5}$$

Using a stable distribution (Feller 1971) with index $\alpha < 1$, and with Laplace transform $\Phi(x) = \exp(-ax^\alpha)$, results into equation (4) when $\alpha = 1 - \lambda$ and $\lambda < 1$.

The other cases cannot be derived from a randomization of equation (1), and therefore other assumptions on the fracture criterion should be introduced for their derivation.

Acknowledgements: We are indebted to the authors of the presentation of this session, for their comments on an earlier version of this synthetic paper.

References

Beran M.J. (1968) **Statistical Continuum Theories,** J. Wiley, New York.

Coster M., Quenec'h J.L., Chermant J.L., and Jeulin D. (1994) Probabilistic models and image analysis: tools to describe liquid phase sintered materials, **this volume.**

Feller W. (1971) **An introduction to Probability Theory and its applications,** vol. 2, J. Wiley, New York.

Jeulin D. (1993) Morphological Models for Fracture Statistics, in **Materials Science Forum Vol. 123-125,** Proc. 7th Symp. on Continuum Models of Discrete Systems, Paderborn, Germany, June 1992, pp. 505-514.

Jeulin D. (1994) Morphological random media for micromechanics, **this volume.**

Kittl P., Diaz G., Martinez V. (1993) Applicability of a Weibullian model of fracture by application of a slowly gradual load, **this volume.**

Kröner E. (1971) **Statistical Continuum Mechanics,** Springer Verlag.

Matheron G. (1967) **Eléments pour une théorie des milieux poreux,** Masson, Paris.

Matheron G. (1975) **Random sets and integral Geometry,** J. Wiley, New York.

Lamon J. (1994) Probabilistic failure predictions in brittle materials under multiaxial loading, **this volume.**

Mihashi H. (1994) A stochastic model to describe probabilistic aspects of damage accumulation in concrete, **this volume.**

Rosowsky D. (1994) Stochastic modeling of lumber properties, **this volume.**

Serra J. (1982) **Image analysis and Mathematical Morphology,** Academic Press, London.

Watson A.S., Smith R.L. (1985) An examination of statistical theories for fibrous materials in the light of experimental data, **J. Mater. Sci.** vol. 20, pp. 3260-3270.

Willis J.R. (1981) Variational and related methods for the overall properties of composites, **J. of Applied Mechanics,** vol. 21, pp. 1-78.

NUMERICAL SIMULATION AND PRACTICAL ASPECTS OF MATERIAL MODELING - AN INTRODUCTION

G.I. SCHUËLLER
Institute of Engineering Mechanics, University of Innsbruck, Innsbruck, Austria

Abstract
Some basics of numerical simulation as well as more recent developments are discussed. In particular selective and Monte Carlo Simulation techniques, such as importance and adaptive sampling as well as the so-called "Double and Clump" procedure are reviewed with respect to their computational merits. Finally, a brief introduction to the papers as presented in this session is given.
Keywords: Monte Carlo simulation, variance reduction techniques, importance sampling, adaptive sampling, "double and clump"

1 Introduction

Modern developments in Structural Mechanics, focus on a realistic modeling of loading as well as structural and, particularly, material parameters. Naturally, this implies the consideration of the statistical uncertainties involved in most of the problems encountered in structural mechanics. It is a well known fact that the so-called deterministic analysis utilizes selectively only part of the available information on the parameters involved. In other words only representative (e.g. so-called minimum or maximum) values are used. Consequently, modern methods of structural mechanics may be considered as part of an effort of information processing, where, based on applicable mechanical models, the entire spectrum of values - known or estimated - of certain parameters are utilized. This, of course, requires on one hand the development of sophisticated concepts, mechanical and probabilistic models, and on the other hand of new, efficient computational procedures. In this context it is important to stress the fact that the increase in sophistication of material and load models must not be traded off by simplifying the mechanical models. This is even more important when solutions to practical, i.e. real world problems have to be developed. An additional advantage is the fact that the analysis is consistent in the sense that load, mechanical and safety analyses are carried out, at least approximately, at the same level of sophistication. This generally can certainly not be claimed for "deterministic" analyses. Most important, however, the increased effort which is required to carry out this more realistic analysis provides a *quantitative* information on the structural reliability or risk of failure.

Monte Carlo simulation proved to be a most powerful - and in many cases the only - tool of analysis. Since - despite the development of high speed computers, the computational efforts for direct simulation may be very high - and for some reliability problems even insurmountable - in recent years several variance reduction techniques

D. Breysse (ed.), Probabilities and Materials, 45–50.
© 1994 *Kluwer Academic Publishers.*

have been developed and advanced. In this context the application of the response surface technique proved to be also instrumental.

2 Simulation Techniques

2.1 General Remarks
In this section first the basics of simulation procedures are discussed. Subsequently, more recent developments, particularly in the area of variance reduction techniques or selective Monte Carlo simulation are reviewed.

2.2 Direct Monte Carlo Simulation
The direct Monte Carlo Simulation (MCS) is known as the most general approach in stochastic mechanics. The procedure has several particular advantages over all other existing approaches. The most prominent feature is certainly its general applicability. In fact MCS is utilized quite often to verify results obtained by other approaches. Furthermore, MCS is conceptionally simple, which helps to make the procedure applicable also for complex structural models. This feature makes the approach attractive, especially for practicing engineers, which might not be specialists in stochastic mechanics, since all well established tools used for a deterministic analysis are immediately applicable. Moreover, there are several additional advantages to mention. In contrast to all analytical procedures, the accuracy of MCS is independent of the type of material laws., i.e. whether it is linear of nonlinear. This makes the approach especially suitable for strongly nonlinear structural systems, when modeling collapse conditions, as so often required for reliability assessment. Another advantage is the fact that MCS is independent of the dimensionality of the problem since the achieved accuracy depends on the sample size only. It appears that the latter aspect is not well recognized in the literature. Direct MCS is sometimes characterized as a crude procedure, requiring a substantial amount of computer time and resources. For many reasons, this characterization of the procedure can be hardly justified.

For time variant problems, for example, the state vectors generated by MCS might be seen as discrete representation of the joint distribution of the response. The accuracy of the representation of this distribution depends only on the sample size and the quality of the random number generator. Assume for the time being that the randomness is confined to loading only, as represented by a stochastic process. Among various procedures to generate a stochastic process by MCS, the most frequently applied model is based on the representation of a stationary process

$$X(t) = \sum_k a_k \cos(\omega_k t) + b_k \sin(\omega_k t) \tag{1}$$

where a_k and b_k are normally distributed independent random variables related to the two sided spectral density function $S_{XX}(\omega)$

$$E\{a_k^2\} = E\{b_k^2\} = S_{XX}(\omega_k) \cdot \Delta\omega_k \tag{2}$$

where $\Delta\omega_k$ is reasonably small frequency interval with the center ω_k and the entire spectrum is divided by $\Delta\omega_k$. Ignoring the randomness of the amplitudes $A_k = \sqrt{a_k^2 + b_k^2}$, the stationary process may be represented by

$$X(t) = \sum_k \sqrt{2S_{XX}(\omega_k) \cdot \Delta\omega_k} \cdot \cos(\omega_k t + \varphi) \tag{3}$$

where only the phase angles φ are considered to be random, independent and uniformly distributed between 0 and 2π. In eq. (4) the amplitudes A_k are considered deterministic, i.e. $A_k = \sqrt{2S_{XX}(\omega_k) \cdot \Delta\omega_k}$ (see e.g. Shinozuka, 1972), since it is claimed that due to the central limit theorem, X(t) approaches a normal distribution in any case. (The technique described above can be extended for non stationary cases by multiplying the generated realizations by a deterministic modulating function. As a further extension, an evolutionary spectrum can be generated (see e.g. Kameda, 1975; Scherer et al., 1982).)

Another approach adopted in the present applications, is based on white noise w(t) represented by

$$E\{w(t)w(t+\tau)\} = I(t) \cdot \delta(\tau) \tag{4}$$

where $\delta(\tau)$ denotes Dirac's impulse function and I(t) the white noise intensity. This non-physical process with infinite amplitudes can be approximated by a continuous function with randomly distributed amplitudes $A_j = A(t_j) = A(j \cdot \Delta t)$ interpolated linearly within time steps Δt. The amplitudes are assumed to be independent and normally distributed with zero mean and standard deviation $\sigma_{A_j} = \sqrt{I(t_j)/\Delta t}$ (see e.g. Priestley, 1981).

3 Selective Monte Carlo Simulation Techniques

3.1 Introductory Remarks
In case direct MCS is utilized for the reliability assessment on nonlinear structures, the information in the low probability tails is generally insufficient for a reliable estimate of failure probabilities in the important range $10^{-4} < pf < 10^{-7}$. Using for example a sample size N=1000, most likely not even a single realization will fall in the failure domain. In other words, a considerable amount of computational effort is spent to obtain samples in a range which is not relevant. This drawback of direct MCS has been commonly recognized, and so-called variance reduction techniques have been developed (see e.g. Schuëller et al., 1989 for a review in context with structural reliability). All these procedures increase the density of Monte Carlo realizations in the region of interest, i.e. in the region which contributes most to the total failure probability. In other words, efficient variance reduction techniques are required to reduce the number of simulations. These variance reduction methods are utilized for nonlinear static and dynamic reliability problems involving a limited set of important random variables which may be determined by *prior* sensitivity analyses. It should be noted, however, that these procedures are not readily extended to problems involving a Markovian type of loading processes. The difficulty there is the determination of the sampling strategy, since a sequence of random variables needs to be analyzed for a Markov process.

Among the various variance reduction simulation techniques available such as stratified sampling, latin hypercube sampling, antithetic variates, directional sampling, importance sampling, adaptive sampling, etc. (see e.g. Schuëller et al., 1989 and the references cited therein for further details) only the latter two procedures are discussed here in some detail. A variance reduction technique can be most advantageously applied for such cases where prior information about the problem i.e. region of simulation is utilized. Instead of simulating the samples randomly over the entire range of each variable, it is concentrated at the important regions. Therefore, the highest efficiency in variance reduction can be accomplished if the prior information about the problem to be analyzed is included in the simulation process. This information can be obtained e.g. from direct simulation or other approximate methods.

3.2 Importance Sampling

The basic idea of this technique (see e.g. Rubinstein, 1981) is to concentrate the distribution of the sampling points in the region of most "importance" i.e. when applied to structural reliability problems, to the part which mainly contributes to the failure probability instead of spreading out the sampling points evenly over the whole range of definition of the involved parameters, i.e. the samples are simulated from importance sampling density function $h_Y(X)$ instead of the original joint probability density function $f_X(X)$ (Schuëller and Stix, 1987). Numerically this yields to:

$$p_f = \frac{1}{N} \sum_{j=1}^{N} I(g(x_j) \le 0) \frac{f_x(x_j)}{h_Y(x_j)} \tag{5}$$

where
  N = number of simulations and
  $g(x)$ = the limit state function of the system considered such that
    ≤ 0 denotes failure state
    $>$ denotes safe state
  $I(\cdot)$ = an indicator function defined such that
    = 1 if $g(x) \le 0$
    = otherwise

It can be shown that the optimal importance sampling density function - which gives the minimum variance i.e. the variance for $p_f = 0$ - can be obtained if its density ordinate is proportional to $1/p_f$ times the original joint probability density function in the failure domain. The ordinate of this importance sampling density function must be equal to zero in the safe domain. Although p_f is usually not known in advance, the variance of p_f can be essentially reduced if the importance sampling density function is chosen to be similar to the shape of the original probability density function (Rubinstein, 1981).

Applying statistical analysis the uncertainty of the estimator (eq. (5)) due to finite number of simulation can be estimated in terms of the variance S_1^2:

$$S_1^2 = \frac{1}{N} \sum_{j=1}^{N} \left[I(g(x_j) \le 0) \frac{f_x(x_j)}{h_Y(x_j)} \right]^2 - \left[\frac{1}{N} \sum_{j=1}^{N} I(g(x_j) \le 0) \frac{f_x(x_j)}{h_Y(x_j)} \right]^2 \tag{6}$$

The so-called standard error (S_{IE}) now yields $S_{IE} = \sqrt{S_1^2 / N}$. An iterative procedure to find the optimal importance sampling density function based on the concept of an Iterative Fast Monte Carlo (IFM) procedure has been developed (Bucher, 1988) and proved to be very efficient. This method will be discussed in more detail below.

3.3. Adaptive Sampling
This Iterative Fast Monte Carlo (IFM) procedure (Bucher, 1988) utilizes the results obtained from importance sampling to adapt the sampling density to the specific problem. As stated above, the standard error of the estimated failure probability reduces to zero, if the sampling density $h_Y(x)$ is chosen to be the original density conditional to the failure domain D_f

$$h_Y(x) = f_X(x \mid x \in D_f) \tag{7}$$

In practice, this is an impossible choice. However, using an iterative procedure, eq. (7) can be satisfied at least in terms of first and second moments, so that

$$E_h(x) = E_f(x \mid x \in D_f) \tag{8}$$

$$E_h(xx^T) = E_f(xx^T \mid x \in D_f) \tag{9}$$

In the above equations the indices f and h refer to the respective densities.

From a starting procedure based on Monte Carlo simulation initial estimates of $E_h(x)$, $E_h(xx^T)$ are obtained. Based on these values, importance sampling is carried out and from the results $E_h(x)$, $E_h(xx^T)$ are updated. Since mean and covariances uniquely define a multinormal density, this type is chosen as sampling density. This method is particularly suitable for problems with nondifferentiable limit state functions, such as multi-failure-mode problems or "noisy" limit states (due to numerical errors). Since only the characteristic function of the failure domain is required, possible problems encountered when applying optimization procedures to find the design point may easily be avoided. It should be noted, that the general concept of adaptive sampling is not restricted to any particular type of joint probability density function.

3.4 Other Methods
Most recently selective Monte Carlo simulation techniques were developed such that the samples in the "interesting" region are doubled, while they are "clumped" in the "uninteresting" region (see e.g. Pradlwarter, 1992; Pradlwarter et al., 1994). It appears that this procedure is somehow related to the "Russian Roulette and Splitting" method. The latter is described in some detail in Kahn, 1956.

4 Applications

In this session the five presentations refer in part to further theoretical developments as well as practical applications of simulation techniques and material modeling.
   The keynote lecture given by *M.Shinozuka* addresses various methods to simulate stochastic fields. In this context applications to laminated orthotropic composites are treated. This contribution closes by introducing a stochastic finite element approach based on the concept of weighted integrals.
   As in theoretical developments the need for realistic, i.e. physically sound probabilistic models of the various input, and in particular the parameters of the respective material laws, are needed, the remaining contributed papers to this session deal with the development of such models, partially substantiated by experimental results. The spatial variability and correlation of fiber reinforced ceramics, natural soils and masonry is in the focus of interest of the contribution by *F. Hild, A. Bolle*

and *A. Baratta* et al. respectively. Finally a contribution by *A. Poitou* et al. is concerned with the statistical modeling for the flow of short fibers composites in terms of numerical modeling as it is most instrumental in defining the mechanical behavior of the composite.

References

Bucher, C.G.: "Adaptive Sampling - An Iterative Fast Monte Carlo Procedure", *Structural Safety*, 1988, 5(2), pp. 119-126.

Kahn, H.: "Use of Different Monte Carlo Sampling Techniques", Symp. on MC Meth., H.A. Mayer (Ed.), John Wiley and Sons, New York, 1956.

Kameda H.: "Evolutionary spectra of seismogram by multifilter", *J. of the Engineering Mechanics Division*, ASCE, No. 191, No. EM6, 1975, pp. 787-801.

Pradlwarter, H.J.: "A Selective MC Simulation Technique for Non-linear Structural Reliability", Proc. ASCE Specialty Conference, Denver, Colorado, July 8-10, ASCE, New York, USA, 1992, pp. 69-72.

Pradlwarter, H.J., Schuëller, G.I., Melnik-Melnikov, P.G.: "Reliability of MDOF-Systems", to appear: *Probabilistic Engineering Mechanics*, 1993.

Priestley, M.B.: "Spectral Analysis and Time Series", Vol. I and II, Academic Press Inc. Ltd., London, 1981.

Rubinstein, R.Y.: "Simulation and the Monte Carlo Method", John Wiley and Sons, New York, 1981.

Scherer, R.J.., Riera, J.D., Schuëller, G.I.: "Estimation of Time Dependent Frequency content of Earthquakes Accelerations", *Nuclear Engr. Design*, Vol. 71, 1982, pp. 301-310.

Schuëller, G.I., Bucher, C.G., Bourgund, U., Ouypornprasert, W.: "On Efficient Computational Schemes to calculate Structural Failure Probabilities", *Probabilistic Engineering Mechanics*, Vol. 4, No. 1, 1989, pp. 10-18.

Schuëller, G.I., Stix, R.: "A Critical Appraisal of Methods to Determine Failure Probabilities", *Structural Safety*, 1987, 4(4), pp. 293-309.

Shinozuka M., Jan, C.-M.: "Digital Simulation of Random Processes and Its Applications", *J. of Sound and Vibration*, Vol. 25, No. 1, 1972, pp. 111-128.

APPLICATION OF PROBABILISTIC APPROACHES IN GEOTECHNICAL ENGINEERING AND ENGINEERING GEOLOGY

H. H. EINSTEIN
Massachusetts Institute of Technology, Cambridge, Massachusetts

Abstract
Typical geotechnical and engineering geology problems such as performance prediction of jointed rock slopes, tunnel cost and time prediction and landslide assessment and mitigation are used to illustrate the effect of uncertainties and how to handle them with probabilistic methods. This is done by discussing the typical steps in decision making under uncertainty, namely information collection, modelling, updating and decisions. Emphasis is put on the fact that both objective and subjective methods in collecting information and modelling are applicable.
Keywords: Uncertainty, Decision Making, Information Collection, Modelling, Updating, Objective and Subjective Probabilistic Methods.

1 Introduction

Uncertainty, and considering uncertainty in their decisions is what geotechnical engineers and engineering geologists are facing day in and day out. While most engineers in their design decisions and, as a matter of fact, most decision makers have to consider uncertainties, the complex uncertainties of the geologic environment are particularly demanding.

For instance, the stability of rock slopes is governed by joints and other discontinuities since it is along these discontinuities that rock blocks may slide and in which water pressure may build up. At best, the geotechnical engineer will have a two-dimensional exposure of discontinuity traces on the surface. However, the instability and water pressure buildup usually occur in the interior of the rock mass about which one has no direct information. Discontinuities in rock masses govern also the flow of groundwater and thus of contaminants in it. However, the geometry and the mechanical characteristics are only certain where they have been observed; otherwise they have to be inferred.

Tunneling through soil or rock is affected by the ground conditions about which again only limited information is available. For shallow tunnels, one may have exploratory borings at 100 m intervals while in deep lying tunnels, the borings are much more widely spaced. Geophysical profiling is possible but, at present, the resolution is limited. Based on this limited information, designers and contractors have to determine the structural characteristics and construction procedures which are the basis of cost and time estimates. Uncertainty in tunneling is not only caused by the lack of information between borings but also affects the relation

51

D. Breysse (ed.), Probabilities and Materials, 51–68.

between ground conditions on the one hand and the structural requirements and construction consequences on the other hand. Such uncertainties in tunneling and in many other geotechnical applications often prevent one from making a final design decision before starting construction. This has led to what has become a classic approach of dealing with uncertainty in geotechnical engineering, the so called Observational Method (Terzaghi,1961, Peck, 1969) which is shown schematically in Figure 1.

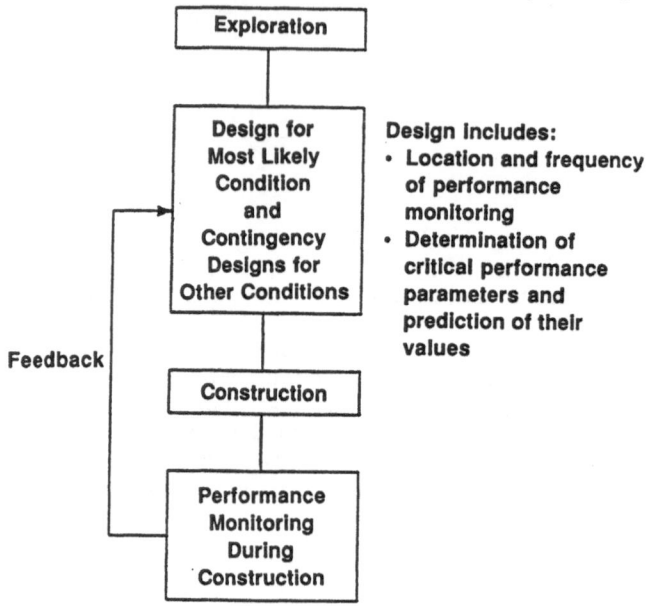

Knowledge of design parameters is updated
through monitoring and feedback into design.

Fig. 1. Observational Method.

Landslides are natural hazards which cause severe damage and often also the loss of life, but many of them are notoriously unpredictable. In some cases empirical evidence of creep movements exists, such as bent trees or hummocky terrain but it is difficult to say if such movements will continue or not. In other cases sudden slides occur in areas which have been stable for some time; there may be some historic landslides in similar terrain but relating them to the occurrence of a new slide is problematic. It is not only the inherent uncertainty in the landslide hazard which makes it difficult to handle; in most cases the hazard has to be translated into a risk (property damage, injuries, loss of life) and management or mitigation procedures have to be devised.

Many other examples of uncertainty in geotechnical engineering and engineering geology exist but these three examples are fairly characteristic and will be further discussed in this paper. What they show is the range and complexity of uncertainty related issues:

Sources of Uncertainty

• Uncertainty is not only related to limited information and measurement errors but also to the relations between geologic/geotechnical characteristics and performance.

Character of Collected Information and of Performance Predictions

• The collected information may be objective or subjective or a mixture. In other words, the characteristics may be measured properties or interpretations of indicators based on experience. Similarly, the performance models may be straightforward analytical relationships or empirical relations.

Updating

• Various levels of information exist and one usually starts with coarse information which is refined (updated) as one proceeds. Very often, if not in most cases, one will have to use information revealed during construction and adapt the design through the Observational Method.

Decisions

• Decisions have to reflect uncertainty and make use of methods ranging from basic risk analysis to approaches such as the Observational Method. The latter is an example of what one may call risk management and a number of other management or mitigation procedures exist. Most decisions affect humans and, therefore, cannot be made on merely a technical basis.

What interests us in the context of this symposium is the use of probabilistic approaches to formally deal with uncertainties. As a matter of fact, the sequence "information collection-modelling-updating-decisions" is known as the decision cycle in decision making under uncertainty (Raiffa and Schlaifer, 1964). It is this sequence which will be discussed in more detail (Sections 3, 4, 5) after some initial comments on sources of uncertainty (Section 2).

2. Sources of Uncertainty

The author and his co-workers have discussed sources of geotechnical and engineering geologic uncertainties at a number of occasions (Baecher, 1982; Einstein and Baecher, 1982; Einstein et al., 1988) and have come up with the following:

1. Geologic uncertainty or spatial variability of geologic characteristics.
2. Parameter uncertainty caused by systematic or random measurement errors and statistical fluctuation.
3. Load uncertainties caused by uncertain or partially known service conditions.
4. Model uncertainty reflecting the fact that neither analytical nor empirical models can be entirely accurate.
5. Omission uncertainty.

Given that uncertainties 2 through 5 are not unique to geotechnical engineering and engineering geology, it will be uncertainty type 1 which will be mostly discussed in this paper.

3. Information Collection and Modelling

3.1 General
One of the main issues both regarding the collection procedure and establishing relations (models) between ground characteristics (information) and performance is their objective or subjective character. The possibility and need to use subjective information and subjective relations is what distinguishes geotechnical engineering and engineering geology from most other engineering disciplines. As will be shown, it is possible to use exclusively subjective information and relations, exclusively objective information and relations or combinations.

It will also become clear in the following that information collection not only involves the act of sampling information but also the building of a model of the geologic conditions with this information. The geologic/geotechnical model will then have to be related to another model representing the performance. It is thus logical that information collection and modelling of any kind are discussed in the same context.

3.2 Objective information collection and representation
Joint geometry
Joints are characterized by their orientation (attitude), size, location (or spacing between joints) on the scale of the rock mass, Fig. 2a and by the characteristics of the joint surface, aperture and possible joint filling on the scale of the individual joint (Figure 2b). Usually, joints are exposed only on outcrop surfaces and the characterization is two-dimensional through their joint traces. It is relatively straightforward to sample joint trace geometry and characterize it by distributions. In particular, on the rock mass scale this has been done extensively and much has been written about different distributional forms (see e.g. Dershowitz, et al., 1988; Baecher et al, 1977). Also, fractal representation has been used for this purpose (Barton and Larson, 1981; La Pointe and Barton, 1992). On the scale of the individual joint, statistical and probabilistic descriptions have also been introduced including auto-correlation of surface geometry and cross-correlation of two surfaces (Roberds, 1979; Roberds et al, 1990). Similarly, fractal descriptions have been applied to joint surfaces, (e.g. Brown and Scholtz, 1987).

Relating the statistically sampled information to geometric spatial models of joint patterns is where the difficulties occur. Models such as the well known Baecher (1977) disk model (Figure 3) and polygonal models (Veneziano, 1979; Dershowitz, 1984) are based on homogeneous Poisson processes for joint location and are not capable of representing the usual clustering of joints. This was remedied by models proposed by Long et al, (1987) and particularly by the hierarchical model (Figure 4) as well as extensions of the Baecher model by Dershowitz (1993). While the hierarchical model of Lee et al. (1990), Xiaomeng (1993) accurately represent the geometry, they are still two-dimensional; the extended Baecher model by Dershowitz is three dimensional but does not represent clusters as well.

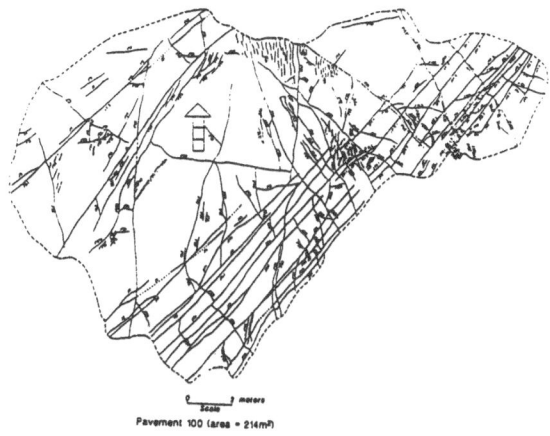

Fig. 2a. Joints on an outcrop.

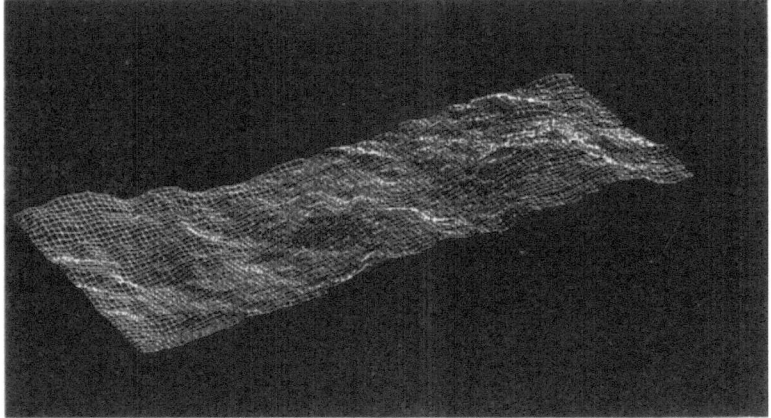

Fig. 2b. Natural joint surface in Granodiorite (Iwano, 1992).

Landslide Occurrence

In areas where landslides are very frequent, it is possible to measure or objectively estimate landslide occurrence. Observations on the slopes on the eastern shore of Lake Geneva have shown that small slides affecting roughly 100 m^2 occur yearly for each 10 km^2 area. Another example is in the area of Villars where sudden slope instabilities have been recorded at intervals of 60 to 80 years, i.e., approximately a 10^{-2}/year probability of failure. - The famous Flims landslide and similar slides in the Alps can be associated with the end of glaciation which means that they occurred roughly 10000 years ago and that one could talk about a 10^{-4}/year probability of failure. It is obvious, however, that conditions at the time of glacial retreat were quite different from today and that such a recurrence interval based approach would have to be modified by additional information.

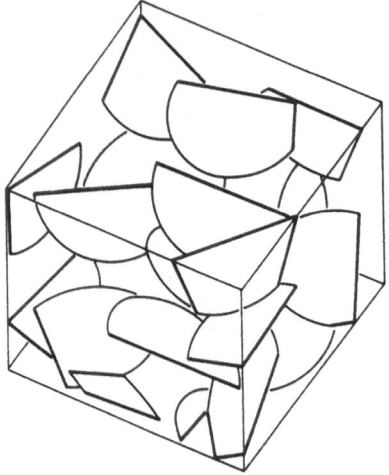

Fig. 3.   Baecher Disk Model.

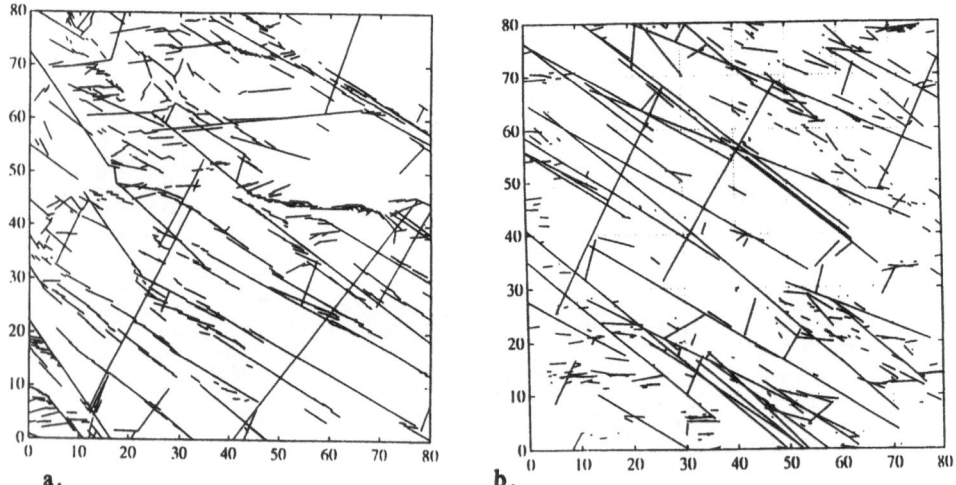

a.   b.

Fig. 4.   Hierarchical fracture trace model (Xiaomeng, 1993).
a. Actual trace map from Odling (1991) Total 672 traces
b. Simulated trace map using hierarchical fracture trace model
   Scale = "80" = 16 m

As a matter of fact, all these objectively obtained landslide frequencies are most useful if they are not used directly but if they serve as a basis for updating (see Section 4).

3.3 Subjective information collection and representation

As discussed in Section 1, the geologic conditions along a tunnel profile are known at few points only and the remainder has to be estimated.   This uncertainty about he geologic conditions starts with the extent of so called homogeneous zones which have to be interpreted from surface exposures and borings.   In the case of the new Gotthard Base tunnel (Figure 5) the

estimated lengths of these zones differed by as much as 500 to 2000 meters.
In each homogeneous zone one will then have to estimate the probabilities of
a number of geologic/geotechnical parameters such as lithology, joint
intensity, water inflow leading to so called probabilistic parameter profiles
(Figure 6).

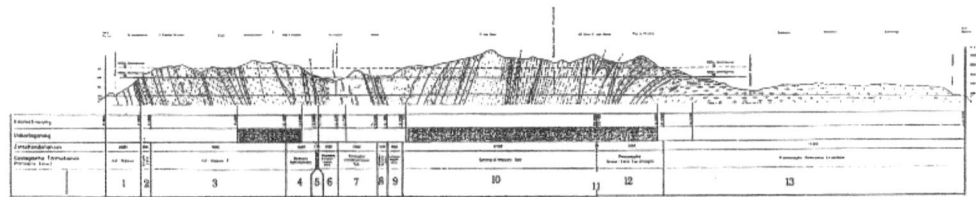

Fig. 5.   Homogeneous zones (numbered 1-13) along Gotthard Base Tunnel.

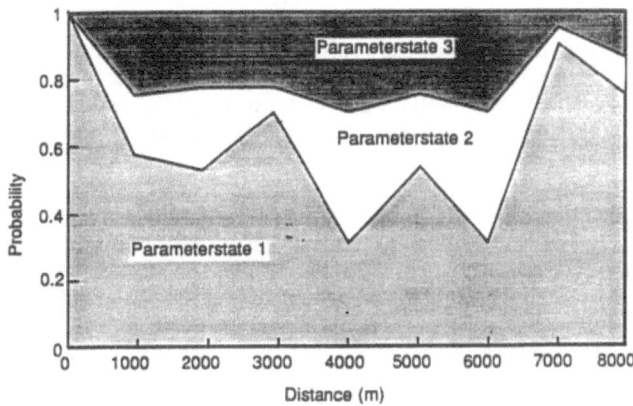

Fig. 6.   Probabilistic parameter profile.

3.4 Objective performance models (relations)

Objective performance models can be in form of explicit relations, in which
a geologic information model (obtained as in 3.1 or 3.2) is used to establish an
analytical model of ground behavior in which the parameters are entered in
probabilistic form (entire distributions, moments or other expressions of
uncertainty).   The resulting performance uncertainty is then calculated by
propagating the parameter uncertainties through the analytical model.   The
rock slope stability example below belongs to this category.   Another
possibility is to relate statistically sampled parameters directly to
performance statistics; this has been done in a number of landslide
predictions and a corresponding example will be shown, also.

<u>Rock Slope Stability (Reliability) Analysis</u>

Such analyses are usually conducted in two steps (see e.g. Glynn, 1979;
Einstein, et al., 1992).   One first assesses the kinematic probability of failure
i.e. the probability that a number of joints can be combined to form a block
or wedge that can kinematically move.   Once the kinematically unstable

blocks or wedges have been identified, one examines in a second step the kinetic probability of failure. In the simplest case, assumming frictional resistance only, such a kinetic probability would be simply:

$$P_{kinetic}=P[(\alpha-\Phi_j)>0] \qquad (1)$$

where α=the inclination of a joint and Φ_j, the joint friction angle.
The overall probability of failure of an individual wedge will then be:

$$P_f=P[kinetic|kinematic] \cdot P_{kinematic} \qquad (2)$$

This is straightforward and so is determining the probability of failure of a slope containing a number of independent wedges or blocks:

$$P_{slope}=(1-(1-P_f)^n) \qquad (3)$$

However, in most cases, the blocks are dependent and require the application of approaches using systems reliability. Also, what has been simplified in the preceding lines is the fact that parts of rock masses may become unstable (fail) even if they are bounded by so called impersistent joints. In other words, intact rock bridges between joints have to fail. Past (Einstein et al, 1983) and recent work (Reyes, 1991) has made it possible to deal with this problem and to include the associated uncertainty to some extent.

Landslide Probability
In areas where landslides are frequent, it is tempting and often successful to relate the frequency of landslides to parameters characterizing the geology, topography, vegetation and climate (rainfall). One of the best known approaches of this kind is the one by Carrara et al., (1977) and Carrara, (1984) using discriminant analysis. Jones (1961) did a regression analysis on the landslide frequency along Lake Roosevelt (the artificial lake on the Columbia river behind Grand Coulee dam). This was expanded in a Master's thesis by Carpenter (1984) under the supervision of Veneziano and the author applying logistic regression analysis. A total of 24 variables (6 related to topography and the drawdown characteristics of the lake, 5 related to ground properties, 9 "space" variables and 4 time variables) observed over a total period of 30 years were related to landslide frequency and allowed us to determine the landslide probabilities shown in Fig. 7.

3.5 Subjective performance models (relations)
As in the objective relations, it is possible to subjectively relate e.g. a (geologic) information model (which in turn may have been objectively or subjectively derived) to a performance model (tunnel example below) or to directly relate indicators to performance (landslide example below).

Tunnel Ground Classes and Construction Classes
Geologic/geotechnical parameters along a future tunnel alignment can be estimated (see Section 3.2). The ground conditions at a particular location of the tunnel will thus be represented by a combination of parameter states e.g. "lithology-granite, jointing-dense, water inflow-medium". Such combinations can be grouped in so called groundclasses, where a groundclass represents construction performance. The association of

parameter states with groundclasses is done subjectively and it is common to associate several parameter state combinations with a ground class. Given the uncertainty in geologic parameters as expressed by the parameter profile in Figure 6, one will analogously obtain a probabilistic groundclass profile (Figure 8). When relating this uncertainty to construction cost and time, one includes in addition, the uncertainties in construction, i.e. the fact that construction cost and time vary even if the geologic conditions are constant. The combination of these uncertainties can be presented in so called scatter-grams such as the one obtained with the 'DAT', the Decision Aids for Tunneling (Einstein, et al., 1992) and shown in Figure 9 for the Gotthard Base tunnel project.

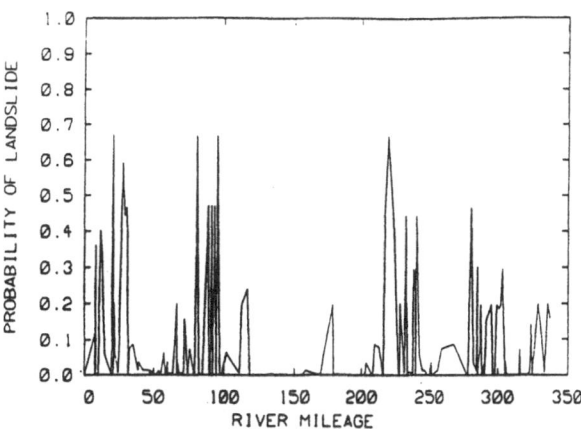

Fig. 7. Landslide probabilities along Lake Roosevelt.

Landslide Probability
An interesting subjective approach for creep type landslides was developed by Noverraz for the area of Villars (DUTI, 1985; Bonnard and Noverraz, 1984). Noverraz observed existing creep movements and a number of indicators such as distorted fences, hummocky surfaces, cracks in pavements. Based on this experience, he related this to predictions of further development of these creep slides (acceleration, continuing at same rate, deceleration) and the probability of such developments. This is indicated in Figure 10 by arrows of different direction and length. The author, through some formal questioning, was able to transfer these subjective graphical expressions into numerical expressions of probability (see DUTI, 1985).

3.6 Collection of information and modelling - final comments
The result of the procedures discussed above are performance probabilities, be they cost distributions, time distributions or probabilities of failure; be they derived directly or indirectly via a geologic model; and be they obtained subjectively or objectively. The resulting performance probabilities form the basis for making a decision. The possible decisions may involve collecting further information (through updating), selecting a design alternative or a mitigating procedure as will be discussed in the following sections.

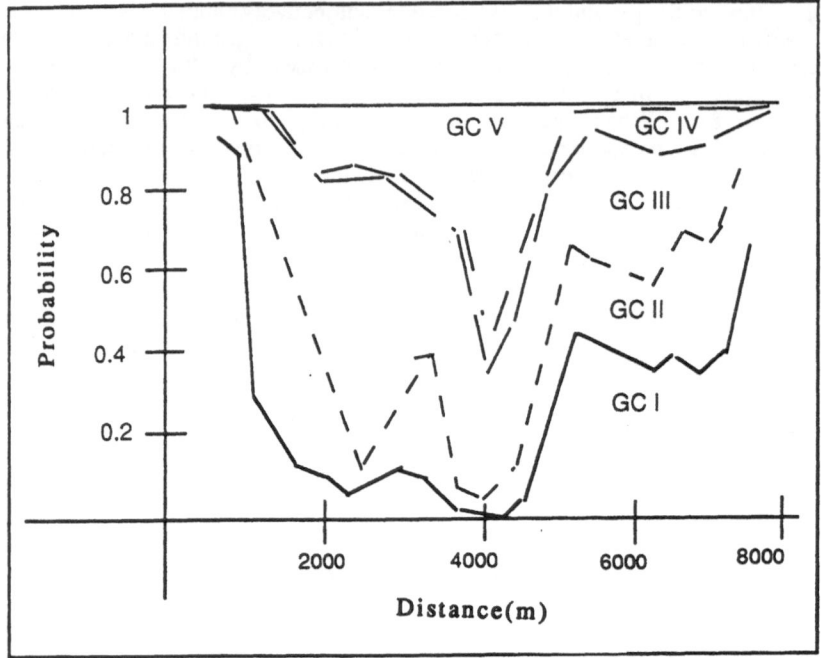

Fig. 8. Probabilistic groundclass (GC) profile.

4. Updating

4.1 General

The prevalence of uncertainty in geotechnical engineering and engineering geology has led to approaches by which uncertainties can be reduced long before they were used in other engineering disciplines. Such approaches existed without formal probabilistic representation of uncertainty. The observational method, discussed in Section 1 (Figure 1) is such an approach. It has been widely applied and it has found its way into the Eurocodes (Eurocode 7:Geotechnics). Nevertheless, it is also possible and increasingly common to formally update the probabilistically expressed uncertainties. This is usually done through Bayesian updating in a preposterior analysis:

$$P'[B_j|A] = \frac{P[A|B_j] 1 P^o[B_j]}{\sum_{j=1}^{n} P[A|B_j] P[B_j]} \tag{4}$$

where $P'[B_j|A]$ =   the posterior probability of outcome B_j given that information A has been obtained

P[A|B$_j$]   =   the probability that information A is observed if B$_j$ actually exists (the so called likelihood function)

P^o[B$_j$]   =   the prior probability of B$_j$

Expression in denominator =   normalizing function

The prior probabilities and likelihood functions can be subjectively or objectively determined. The preposterior analysis and the updating can be performed in different phases of the project:

- When considering if additional information collection e.g. exploration is worthwhile and before actually performing such an exploration
- After additional information collection has taken place by updating the information (parameter states) in a geologic model or in the performance model
- After performance has been observed in reality by updating not only the parameters in the model but the model itself.

The following examples will illustrate these updating processes.

4.2 Deciding on additional information collection - "virtual exploration"

The author and his co-workers have shown at several occasions (Einstein et al., 1978 and 1987) how one can decide if the cost of an additional boring is worthwhile. For instance the mean cost of building a tunnel is estimated considering existing uncertainties (prior probabilities of geologic conditions). One then imagines drilling an additional exploration boring; the likelihood functions are estimates of the boring result one would expect from particular actual conditions. For instance, the likelihood function would express the probability of encountering different joint intensities (high, medium, low) in the boring, given that the actual conditions were low intensity jointing. Such a likelihood function is largely objective since one can obtain it from observing a number of boring logs and comparing them to the actual conditions encountered when excavating a tunnel. Other likelihood functions such as relating water in a boring to water inflow in a tunnel would involve considerable judgement and are thus largely subjective. Once the likelihood functions for all the relevant parameters have been established one conducts the preposterior analysis and obtains posterior (updated) parameter probabilities. These are used to calculate an updated mean cost for construction of the tunnel. If the new (posterior) mean cost is smaller than the prior one, and if the difference is greater than the cost of the exploratory boring, one would perform this exploration.

    Instead of using cost as a criterion, one can use other criteria to decide on the value of additional information. In radioactive waste storage, rock mass conductivity is an important performance criterion.

62

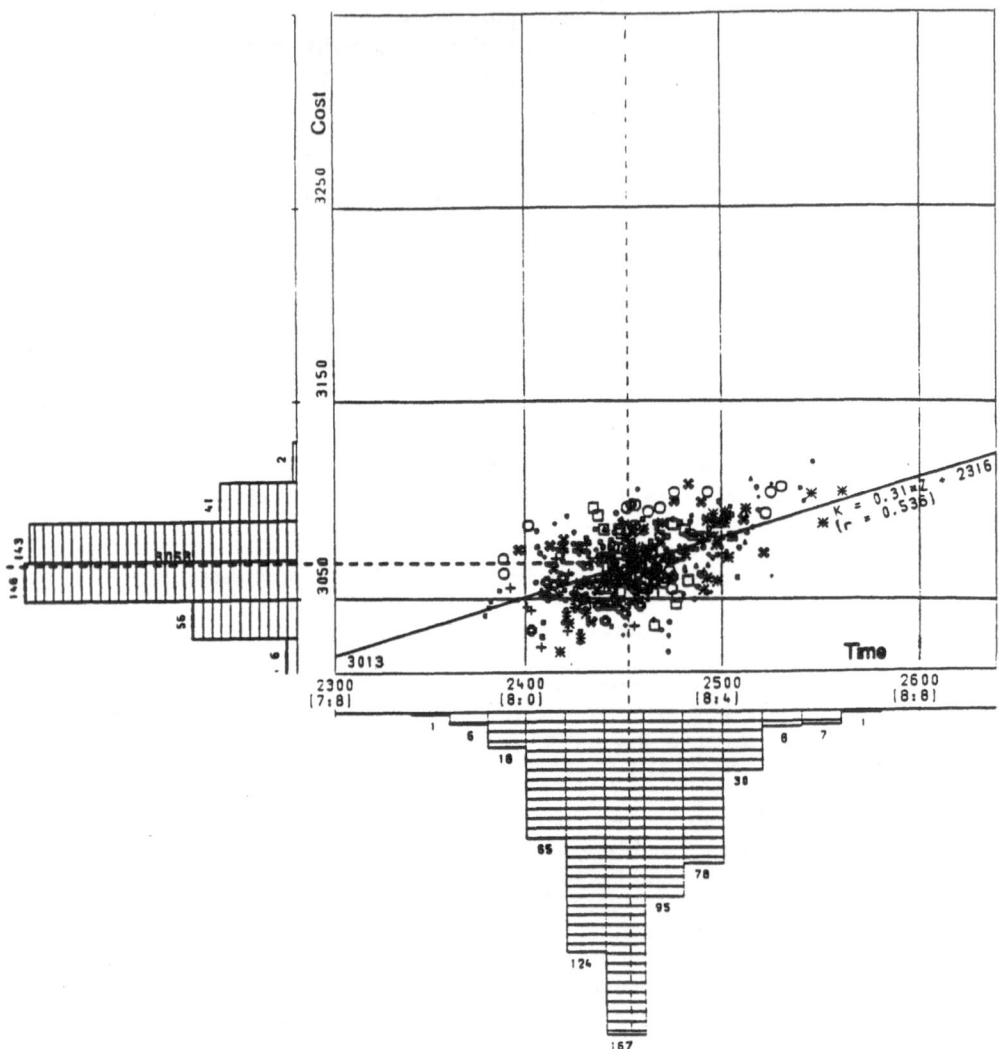

Fig. 9.   Time-Cost-Scattergram.
(Time in working days and years: months; 1 year-300 working days)
Points reflect geologic and construction uncertainty.

One can express one's uncertainty about conductivity of the rock mass
around a planned storage facility by Hasover's and Lind's (1974) reliability
index:

$$\beta = \frac{E(K) - K_c}{\sigma K} \qquad\qquad (3)$$

where K_c the critical conductivity that must not be exceeded and the $E(K)$,
$\sigma(K)$ the best estimate and standard deviation of the rock mass conductivities.
In our considerations regarding a nuclear waste storage facility we had

three possible geologic interpretations based on prior information and had to chose between two locations for an additional exploratory boring. We estimated the observations that one would obtain in each boring (i.e. we conducted two virtual explorations) and expressed this in terms of E(K), σ_K for each boring. The virtual boring showing the greater β, that is the one producing a more reliable prediction, was then chosen.

Fig. 10. Example of a DUTI Danger Map (from DUTI, 1985). Circled numbers designate individual slide, other numbers (e.g. 20-40) are estimates of depths; double arrows indicate relative acceleration/deceleration tendency; single arrows indicate acceleration tendency relative to maintaining a constant rate of movement.

4.3 Including additional information

This is the most obvious application of Bayesian updating. In our tunnel - cost and - time estimation with the DAT (Einstein, et al., 1992), for instance, we make cost and time estimates based on prior information (length of homogeneous zones, parameter state probabilities and ground class probabilities). The DAT also allow one to consider additional information from borings. What one does is not to directly use the information observed in the boring, e.g. joint intensity, but to again formulate a likelihood function. This likelihood function not only expresses the probability that the observation represents reality at the boring location but it allows one to express how well the boring information represents the geologic conditions over a certain length of tunnel. This aspect of the likelihood function is largely subjective.

Another possibility to update tunnel geology is to use observed geologic/geotechnical conditions in the excavated part of the tunnel to modify the predictions in the as yet unexcavated part. This possibility has been implemented in a construction management support tool for tunneling (see e.g. Kim, et al., 1985; Einstein et al., 1987).

When discussing landslide probabilities based on historical records or general observations (Section 3.2), it was mentioned that such probabilities are often used as prior information. The engineering geologist will use more specific information on smaller areas to update the prior probabilities. Again, the additional information is formulated as a likelihood function.

4.4 Model updating

With the procedures discussed above, one updates the variables in a model, e.g. the geologic parameters in the geologic model or the variables in the performance model which depend on the geologic parameters. The model itself, that is the structure or the "coefficients" of the model are not affected. It is likely that the model itself needs to be changed. In other words, if the performance such as the observed deformation in a tunnel deviates from the predictions, it may be caused by geologic/geotechnical parameters being different from those assumed in the prediction or it may be a model inadequacy (e.g. assumption of linear behavior rather than non linear behavior). Updating of the model itself is also possible as shown by Rollin (1979), who developed a Bayesian updating procedure for a General Multivariate Normal Model which he used to express the performance relations, i.e. the relation between geology and performance. - What one does in the observational method is actually a combination of information updating and model updating although usually not with formal probabilistic approaches.

5 Decisions

The performance probabilities (performance distributions) obtained and possibly updated as described in sections 3 and 4 are used to make the engineering decisions. Such a decision can, for instance, be the selection of a design alternative or of a particular hazard/risk management procedure. Decisions can be made using the performance probability directly, using them in a financial risk analysis or using them in a multi-attribute (utility) analysis.

Tunnel Cost and Time

Scattergrams as shown in Figure 9 can be used directly to decide on the preferred alternative. Further evaluations may include computing the probability of exceeding a certain cost and time and doing this in a utility analysis. It is also possible to transform time into cost. - Another interesting application of the scattergram is to use it as a contract basis: In order to provide an incentive for efficient construction by the contractor one does want to leave some of the risk (both positive and negative) with the contractor. On the other hand, it would be unfair to contractor to have him/her assume the burden of extensive unfavorable geologic conditions;

similarly the owner should benefit from very favorable conditions. One could, therefore, formulate a contract such that the extreme time/cost risks are shared by the owner and the contractor.

Slope Stability

The classic and most straightforward application of risk analysis is in slope reliability analysis for mining operations. The probability of failure 'Pf' is multiplied by the cost consequence 'C' of a failure (damage, cleanup) to obtain risk:

$$R=P_f C \qquad\qquad (6)$$

Flattening a slope will decrease the probability of failure 'Pf' (possibly also the cost consequences 'C') producing a reduced risk R_{red}. The difference in risk, $R-R_{red}$, has to be compared with the increased cost of flattening a slope (excavation cost, disposal cost of waste material). Such an analysis can again be done on the basis of utilities rather than cost directly. Multiattribute utility analysis is required if factors other than cost, such as environmental effects, need to be considered.

Landslides

One way to consider landslides is entirely analogous to the slope failure risk analysis above. After determining the original landslide hazard, i.e. the probability of sudden failure or of accelerating creep, this probability can be multiplied with the cost consequences. This was discussed in detail by the author (see Einstein, 1988) including the possibility to distinguish different hazards affecting the same area (sudden landslide, creep, and also floods, etc.) and different consequences occuring in the same area, e.g. destruction of buildings, temporary loss of pasture, temporary interruption of roads. In addition, it is possible to represent the fact that the consequence of a landslide may not materialize with a 100% probability even if the landslide takes place. This so called vulnerability (a term introduced by Varnes, 1984) can be considered in form of a conditional probability. This and similar formal assessments of landslide hazards and risks have been applied in a variety of ways and by many authorities (Humbert (1977), LPC (1978), PER (1985, a,b), Brabb (1984)).

   The logical extension of landslide hazard and risk assessment is to consider hazard (risk) management and mitigation measures in the same manner. One can distinguish 1) passive countermeasures such as zoning, 2) constructive or active countermeasures such as drainage or retaining walls and 3) warning devices.

   Zoning in essence reduces the cost consequences by eliminating or reducing the number/type of structures that can be damaged. - Constructive (active) countermeasures reduce the hazard/probability (e.g. drainage increases slope stability). The modification of the landslide probability is again introduced through Bayesian updating with the likelihood function expressing the probability that the countermeasure is successful. The cost consequences may also be reduced by constructive countermeasures and so may the vulnerability. - Warning devices will usually affect the cost consequences in that property may be removed from the danger zone.

   All these countermeasures lead to a reduced risk and the risk reduction can be compared to the cost of the countermeasures. Although cost consequences were mentioned, it has to be understood that in most cases, it is better to use utilities rather than cost directly. This allows one to better include risk aversion and risk proneness of different participants. Finally, if life is involved, one may not want to transform death or injuries into costs but deal with them directly either in multi-attribute form or through optimization (see Vanmarcke and Bohnenblust, 1982).

6 Conclusions

Geotechnical engineering and engineering geology are strongly affected by uncertainties. Not surprisingly, approaches to deal with uncertainties such as the Observational Method have been used long before formal probabilistic approaches. Nevertheless, the decision cycle, "information collection-modelling-updating-decision", with formal probabilistic approaches is now well established. Information collection procedures, as well as information - and performance models on the basis of objective and subjective probabilistic approaches exist in such typical geotechnical problem areas as slope reliability analysis, landslide hazard assessment and mitigation and tunnel construction and time estimation.

7 References

Baecher, G.B. (1982) Playing the odds in rock mechanics, **Proc. 23rd U.S. Symp. on Rock Mechanics.**

Baecher, G.B. Lanney, N.A. and Einstein, H.H. (1977) Statistical description of rock properties and sampling, **Proceedings of the 18th U.S. Symposium on Rock Mechanics** 5C1-8.

Barton, C.C. Larson, E. (1981) Fractal geometry of two-dimensional fracture networks at Yucca Mountain, Southwest Nevada. **Proceedings Int'l. Symp. Fundamental of Rock Joints** (ed O. Stephanson).

Bonnard, C. Noverraz, F. (1984) Instability risk maps. From detection to the administration of landslide prone areas. **Int'l. Symp. on Landslides** Toronto, Vol. 2, pp. 511-522.

Brabb, E.E. (1984) Innovative approaches to landslide hazard and risk mapping. **IVth Int'l. Symp. on Landslides.** Toronto, Vol. 1, pp. 307-323.

Brown, R.S. Scholtz, C.H. (1985) Broad bandwith study of the topography of natural rock surfaces. **J. of Geophys. Research,** 90 (B14), 575-582.

Carpenter, J.H. (1984) Landslide risk along Lake Roosevelt. **Master's Thesis,** Massachusetts Institute of Tech., Cambridge, MA 125 pp.

Carrara, A. (1984) Landslide hazard mapping: aims and methods. **Mouvements de Terrains.** Association Française de Géographie Physique. Colloque de CAEN, pp. 1142-151.

Carrara, A. Publiese-Carratelli, E. Merenda, L. (1977) Computer based data bank and statistical analysis of slope instability phenomena. **Z geomorph. N.F.** 21, No. 2, 187-222.

Dershowitz, W.S. (1984) Rock joint systems. **Ph.D. Dissertation,** Massachusetts Institute of Tech., Cambridge, MA.

Dershowitz, W.S. (1993) Geometric conceptual models for fractured rock masses:Implicating for groundwater flow and rock deformation, **Proc. EUROCK-93.**

DUTI (1985) Détection et Utilisation des Terains Instables. Rapport Final. Ecole Polytechnique Fédérale de Lausanne, Switzerland.

Einstein, H.H. Labreche, D.A. Markow, M.J. Baecher, G.B. (1978) Decision analysis applied to rock tunnel exploration, Eng. Geology. 12, 143-161.

Einstein, H.H. (1988) Landslide risk assessment procedure. Proceedings 5th Int'l. Symp. on Landslides.

Einstein, H.H. Baecher, G.B. (1982) Probabilistic and statistical methods in engineering geology, I. Problem statement and introduction to solution. Rock Mechanics. Suppl. 12, pp. 47-61.

Einstein, H.H. Baecher, G.B. Veneziano, D. O'Reilly, K.P. (1983) The effect of discontinuity persistence on rock slope stability. Int'l. J. of Rock Mechanics and Mining Sciences., 20, No. 5.

Einstein, H.H. Baecher, G.B. Veneziano, D.V. (1988). Quantitative exploration planning, NAGRA Report, IB 88-61.

Einstein, H.H. Salazar, G.F. Kim, Y.N. Ioannou, P.S. (1987) Computer based decision support systems in underground construction, Proceedings, RETC.

Einstein, H.H. Dudt, J.P. Halabe, V.B. Descoeudres, F. (1992) Decision aids for tunneling, Monograph by Swiss Federal Office of Transportation.

Einstein, H.H. Lee, J.-S. (1992) Topological slope stability analysis using a stochastic fracture geometry model, Proceedings Fractured and Jointed Rock Masses Conference.

Glynn, E.F. (1979) A probabilistic approach to the stability of rock slopes. Ph.D. Dissertation, Massachusetts Institute of Tech., Cambridge, MA.

Hasover, A.M. Lind, N.C. (1974) Exact and invariant second moment code format, J. of Eng. MEchanics, ASCE, 100, No. EMI.

Humbert, M. (1977) La cartographie ZERMOS. Modalités et établissement des cartes des zones exposées à des risques liés aux mouvements du sol et du sous-sol. Bulletin du B.R.G.M. (deuxième série) Section III, No. 1/2, pp. 5-8.

Iwano, M. (1992) Interim Ph.D. Thesis Presentation., Massachusetts Institute of Tech., Cambridge, MA.

Jones, F.O. (1961) Landslides along the Columbia River valley, Northeastern Washington. U.S. Geological Survey Professional Paper 367, U.S. Government Printing Office, Washington, D.C.

Kim, Y. Einstein, H.H. Logcher, R.D. (1985) Decision support system for tunneling, ASCE Fall Convention.

LaPointe, P.R. Barton, C.C. (1992) Shortcourse on fractals in connection with 33rd U.S. Symp. on Rock Mechanics. Sta Fe.

Lee, J.S. Veneziano, D. Einstein, H.H. (1990) Hierarchical fracture trace model. Proc. 31st. U.S. Symp. on Rock Mechanics.

Long, J.C.S. Bilaux, D. Hestir, K. Chiles, J.-P. (1987) Some geostatistical tools for incorporating spatial structure in fracture network modelling. Proc. 6th Int'l. Congress of the ISRM. 171-176.

LPC (1978) Eboulements et chutes de pierres sur les routes. Méthode de Cartographie. Groupe d'Etudes des Falaises (GEF) Laboratoire Central des Ponts et Chaussées. Rapport de Recherche LPC, No. 80.

Odling, N. (1993) Personal Communication.

Peck, R.B. (1969) Advantages and limitations of the observational method in applied soil mechanics, 9th Rankine lecture, Géotechnique, Vol. 19, pp. 171-187.

68

PER (1985a) Catalogue de mesures de prévention, Mouvements de Terrains. Plan d'exposition aux risques. **Premier Ministre-Délégation aux risques majeurs**. 443 p.

PER (1985b) Mise en oeuvre des plans d'exposition aux risques naturels prévisibles. Rapport administratif et technique provisoire. **Premier Ministre-Délégation aux risques majeurs**. 443 p.

Raiffa, H. (1964) **Applied Statistical Decision Theory**. Harvard Business School, Cambridge,MA.

Reyes, O. (1991) Experimental study and analytical modelling of fracture coalescence in brittle materials. **Ph. D. Thesis**, Massachusetts Institute of Tech., Cambridge, MA.

Roberds, W.D. (1979) Numerical modelling of jointed rock. **Sc.D. Thesis**, Massachusetts Institute of Tech., Cambridge, MA.

Roberds, W.D. Iwano, M. Einstein,H.H. (1990) Probabilistic mapping of rock joint surfaces. **Proc. of the Int'l. Symp. on Rock Joints**, Loen.

Rollin, F. (1979) The updating of information during tunnel construction. **Master's Thesis**, Massachusetts Institute of Tech., Cambridge, MA.

Terzaghi, K. (1961) Past and future of applied soil mechanics. **J. of the BSCE**, April.

Vanmarcke, E.H. Bohnenblust, H. (1982) Methodology for integrated risk assessment for dams. **MIT Research Report**, R-82-11.

Varnes, D.J. (1984) Landslide hazard zonation; a review of principles and practice. **Natural Hazards 3. UNESCO**, 63 pp.

Veneziano, D. (1979) Probabilistic model of joints in rock. **Internal Report,M.I.T.**

Xiaomeng, Y. (1993) Stochastic modelling of rock fracture geometry. **Master's Thesis**, Massachusetts Institute of Tech., Cambridge, MA.

THE VARIABILITY OF GEOTECHNICAL PARAMETERS

C. CHERUBINI
Institute of Engineering Geology and Geotechnics,
Technical University, Bari, Italy.

Abstract
Natural soils are often characterized by a remarkable
variability of their physico - mechanical properties. This
work reports on several studies and discusses a large
number of data concerning the variation coefficients of
certain geotechnical parameters of fundamental importance
and/or most commonly used in computation models. However,
for the purpose of probabilistic evaluations, the use of
this coefficients of variation results into evaluating
failure probabilities that are too high compared to the
corresponding frequency observed in actual pratice.
Obviously, such a state of affairs causes lack of
confidence in any kind of probabilistic approach.
    The discrepancies just described can be overcome by
utilizing probabilistic models that take into due account
the spatial variability of geotechnical data: the
fluctuation scale, in particular, must be well known.
While it can be evaluated by a number of procedure, we
still only have few nonhomogeneous data available.
    The communication reports, and comments on, some
values of fluctuation scale, for certain geotechnical
characteristics.
Keywords: Soil, Rock, Coefficient of Variation,
Fluctuation Scale.

1 Introduction

Since the use of probabilistic and semiprobabilistic
methods in geotechnics have been steadily expanding in
recent years, it has now become essential to possess a
good knowledge of the geotechnical parameters employed in
computation models. It is the purpose of this
communication to update the state of knowlwdge on the
variability of the most important geotechnical
characteristic, particularly those obtained from
laboratory tests.

D. Breysse (ed.), Probabilities and Materials, 69–80.
© 1994 *Kluwer Academic Publishers.*

2   In situ and laboratory geotechnical tests

Parameters for direct use in evaluating stability or calculating deformation can be obtained in geotechnics by two major classes of methodology:

- by laboratory tests, allowing specific control of boundary conditions and practically "direct" reading of the investigated parameters;
- by in situ tests which, instead, always require "transformation models" for passing from measured quantity to the parameters one is trying to define.

In either case, however, the investigated soil volumes represent only a minor part of the size of volume actually subjected to variations of stress.

The results of in situ tests are affected essentially by measurement - related uncertainties resulting from sampling, specimen preparation and from the test itself.

For sake of simplicity and brevity, only the variabilities of geotechnical properties observed in laboratory tests are discussed here.

Thus, the standard deviation (or the variance, or the coefficient of variation) we are going to examine will either be derived from the soil's intrinsic variability, from sampling disturbance, and from testing procedures.

Let σ be the standard deviation "read" from the tests, σrp that due to "repeatability", and σh the standard deviation related to the intrinsic variability of the investigated geotechnical parameter; we can then write (Rethati, 1989):

$$\sigma h = \sqrt{(\sigma^2 - \sigma rp^2)}$$

σrp is only roughly known and appears to be varying by some 20 - 30% of σ.

Of course, for σrp to be reduced to negligible levels with respect to σ, well defined and universally accepted Test Standards are essential. At present, this is practically true only for some geotechnical tests.

3   Variability of geotechnical data

Table 1 shows the most important statistical data relative to the coefficients of variation of some geotechnical soil properties and of the rock's uniaxial compression and tensile strength. Part of the data are quoted from the work of Kulhawy et al. (1991) while others have been collected and selected by the writer (Cherubini et.al., 1993). Figures 1a, b, c show the Box-plots (Tukey, 1977) developed for the C.V. of the some properties.

	COEFFICIENT OF VARIATION (%) OF									
	e	W_P	W_L	W	Υ	φ'	C_c	C_u	σ_c	σ_t
NUM.	26	64	64	48	28	31	13	55	34	27
MEAN	18.4	14.7	16.2	25.1	6.9	12.2	35.6	43.2	34.8	34.8
TRIMEAN	17.7	13.5	14.5	23.2	6.4	11.2	33.2	38.7	32.2	32.0
MEDIAN	18	13	13.5	23.5	6	11	30	37	38	28
MIN	7	5	3	6	2	3	17	12	5	7
MAX	32	45	46	55	16	33	55	145	102	102
Q_L	14	8	7.5	15.5	4	8	26	29	24	23
Q_U	21	20	23.5	30.5	9.5	15	47	52	42	48
STANDARD DEV.	6.4	8.1	9.9	11.9	3.6	6.1	12.7	23.3	19.5	23.9

Table 1 - A synthetic presentation of the main C.V. statistical indicators for certain physical and mechanical properties of soil and rocks.

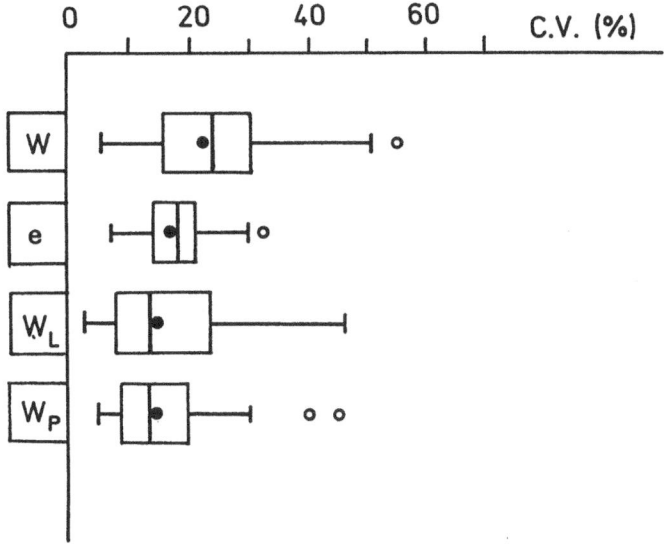

Fig. 1a - Box-plots of variation coefficients of certain physical properties of soils.

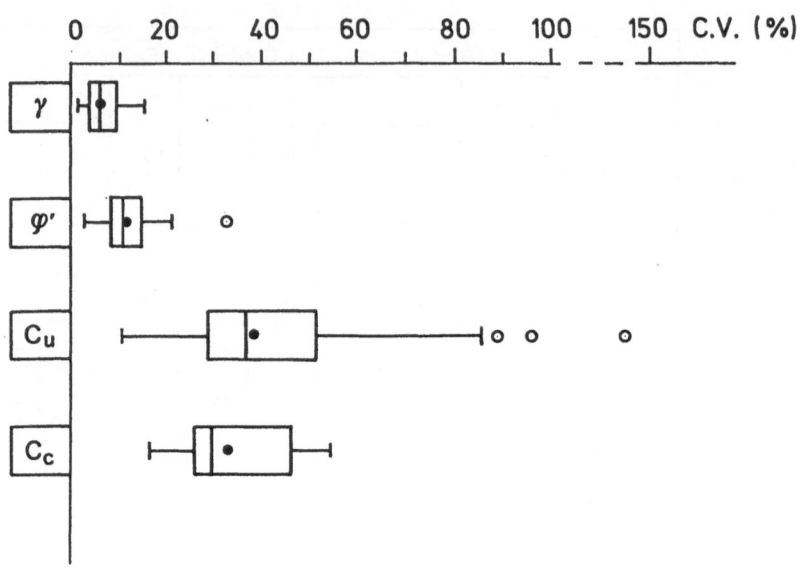

Fig. 1b - Box-plots of variation coefficients of certain mechanical properties of soils.

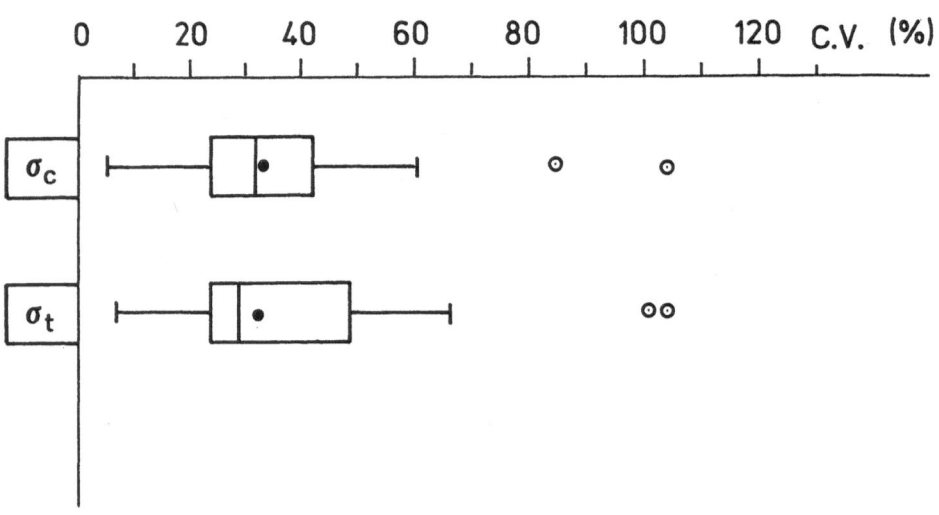

Fig. 1c - Box-plots of variation coefficients of certain mechanical properties of rocks.

Both table 1 and the Box-plots clearly point out that certain index properties such as void ratio, water content and liquid and plastic limits generally exhibit characteristic varabilities that are relatively moderate (apart from certain outliers) and belong to pratically the same order of magnitude (ranging from a few percent points to a peak value of 55%, the latter corresponding to the natural water content).

These geotechnical parameters are all easy to determine: however, although they are characterized by well standardized testing procedures, they are not used directly in computation models. Instead, they are utilized as basic properties for geotechnical parameters that are used directly in the computation models (friction angle, compression index, etc. etc.).

Matters are different for the variability of the second investigated group including physico-mechanical parameters and magnitudes used directly in computation models.

Unit weight variability is rather limited (between 2 and 16%); the friction angle $\emptyset$ is generally characterized by somewhat higher coefficients of variations (between 3 and 33%). However, by observing how the coefficients of variation of $\emptyset$ are distributed with respect to the corresponding mean values, one can see that the higher coefficients of variation relate to the lowest mean values of $\emptyset$, up to 30°. For $\emptyset>30°$, the C.V. are contained between 3 and 15% (fig. 2).

High C.V. values are found for the compression index Cc (between 17 and 55%) whereas the lowest values observed for undrained cohesion is 12% and the peak value is extremely high (145%).

However, the variability of the latter parameter should be considered from two different points of view:

a) like with the friction angle, here again C.V. values diminish as mean Cu increases (See fig. 3);
b) the very high variability observed with lower Cu values can be explained by considering that such levels usually belong to normally consolidated soils which are characterized by Cu increasing with effective vertical stress, hence with depth. It may well be then, that in some instances variability has been evaluated without first eliminating this trend from the computations. This is, however, a mere hypothesis and ought to be adequately verified.

In order to provide a full picture of the state of knowledge on the variability of the soils' geotechnical parameters, we would point out that data on effective cohesion have been collected by the writer: unfortunately, these data are too few for statistical processing. An

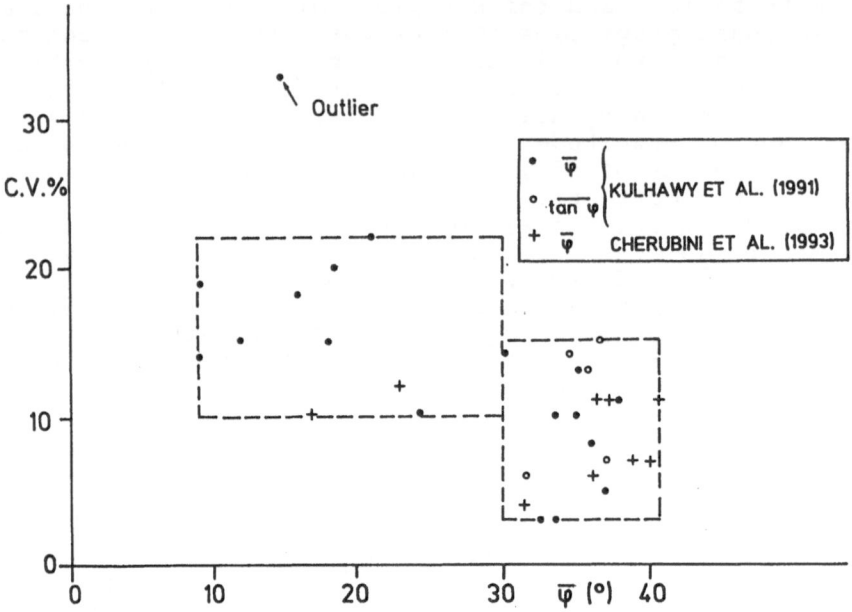

Fig. 2 - The trend of the friction angle variation coefficients as a function of the corresponding mean values.

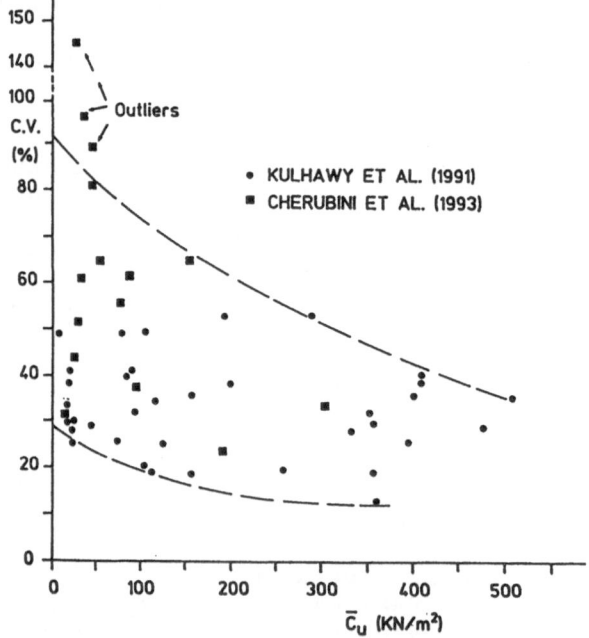

Fig. 3 - The trend of the undrained cohesion variation coefficients as a function of the corresponding mean values.

initial tentative analysis seems to indicate coefficients
of variation ranging from 34 to 47%.
    The two characteristics that describe rock strength
namely, uniaxial compression strength and tensile
strength, exhibit variation coefficients that are medium
to high: in both cases, the highest values amount to 102%.
C.V. shows a marked downward trend as the mean value of
tensile strength rises (fig. 4). Instead, for uniaxial
compressive strength, the tendency for C.V. to drop around
the mean value is only hardly observable.

4 Fluctuation scale

The fluctuation scale measures the distance within which a
soil property shows strong correlations (Vickremesinghe
and Campanella, 1983). δ would seem to be an attribute of
a soil mass. The values of δ are very similar for
different properties in the same soils (Bao et al., 1993).
    Usually, a distinction is made between a fluctuation
scale in the vertical direction and in the horizontal
direction. The lowest and the highest values observed with
the vertical fluctuation scale δv are between 1 and 2 m,
while for the horizontal fluctuation scale δh the
corresponding values range from 5 to over 50 m.

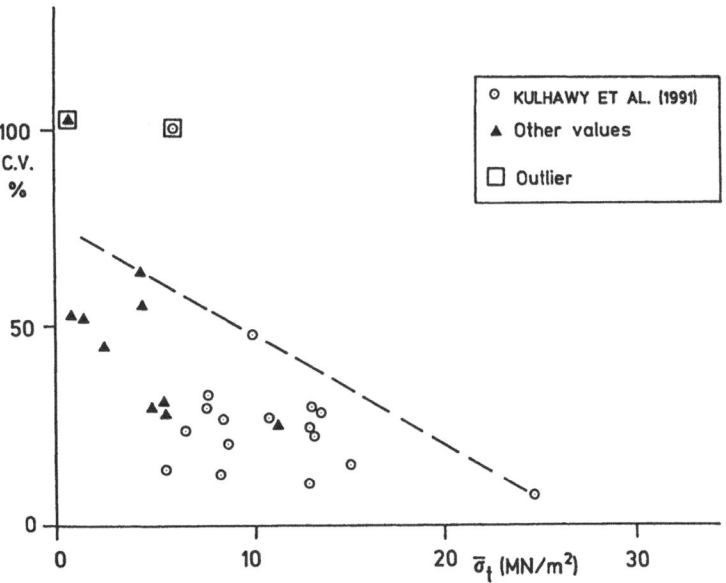

Fig. 4 – The trend of the tensile rock strength
          variation coefficient as a function of the
          corresponding mean values.

The δi's can be evaluated in different ways described in a number of works (see, for example, Vickremesinghe and Campanella, 1983; Vanmarcke, 1977; Keaveny et al., 1989). In the evaluation of this parameter, the results of in situ tests appear to be easier to handle. In particular, tests yielding almost continuous readings, or else readings that are only separated by very short intervals - just a few cm - appear to fit the purpose best.

Doubtless, it is important to have a good understanding of this parameter because it can significantly resolve the working variance in relation to the kind of problem being investigated.

Mostyn and Li (1993) rightly contend that probabilistic models that do not account for the spatial correlation of data should be discarded.

Should the fluctuation scale δ not be considered in the "probabilized" geotechnical model (slope stability, bearing capacity, etc.), this might engender failure probabilities that are often very high, and quite inconsistent with the frequencies observed in actual pratice: the consequence is lack of confidence in the probabilistic model.

Some values of δh and δv, according to the above quoted Authors, are presented in table 2a, b.

5 Further considerations

Other point of interest are the trend of variability of geotechnical data with respect to depth. Precisely in connection with undrained cohesion, we are here discussing the data developed by Bergado and Chang (1989) for the Nong Ngoo Hao clay which is a normally consolidated clay. The mean values of Cu (undrained cohesion) vary with depth across levels about 1 m thick, whereas the coefficients of variation, which in turn are rather variable, do not show any definite upward or downward trend with depth.

Again, with respect to variability of undrained cohesion in the (overconsolidated) London Clay, data reported by Chowdhury (1984) show a trend of coefficients of variation unrelated to depth.

Corcerning the shape of the frequency distribution curves of geotechnical parameters, reference is here made to a synthetic presentation reported by Rethati (1989) and drawn from Corotis et al. (1975). Table 3 shows the most frequent types of distribution for some parameters of this kind.

In his work, Harr (1987) also reports data obtained from several Authors concerning the values of coefficient or correlation between c and $\emptyset$. They are always negative in drained tests, varying fron -0.24 to -0.70. In line with these data are those by Bao et al. (1993) (r(c$\emptyset$)= -

PROPERTY	AUTHORS	δ_h (m)
CONE PENETROMETER RESISTANCE (7 m Depth)	KEAVENY ET AL. (1989)	24.62
CONE PENETROMETER RESISTANCE (9 m Depth)	KEAVENY ET AL. (1989)	66.49
COMPRESSION INDEX C_c	VANMARCKE (1977)	54.86
UNDRAINED SHEAR STRENGTH	VANMARCKE (1977)	46.00
PORE PRESSURE (PIEZOCONE)	VICKREMESINGHE AND CAMPANELLA (1993)	17.75

Table 2a - Literature data relative to horizontal fluctuation scales.

	NORMAL	LOGNORMAL	OTHER TYPES
— Water Content	67 %	33 %	—
— Liquid Limit	67 %	33 %	—
— Plastic Limit	—	100 %	—
— Void Index	67 %	—	33 %
— Consolidation Coeff.	67 %	33 %	—

Table 3 - Types of probability density curves for certain physical characteristics of soils.

PROPERTY	AUTHORS	δ_v (m)
Water Content (%)	Vanmarcke (1977)	1.2
Initial void index e_o	" "	3.05
N_{SPT}	" "	2.44
Undrained Shear Strenght	" "	5.00
" " "	✤ Wu (1974)	0.79
" " "	✤Matsuo and Asaoka (1977)	1.25
" " "	" "	2.50
" " "	" "	1.82
Cone Penetrometer Resistance (Sand)	Alonso and Krizek (1975)	2.20
Friction Penetrometer Resistance (Sand)	" "	1.20
Cone Penetrometer Resistance (Clay)	" "	1.00
Cone Penetration Resistance (Silty Clay)	▲ Gao and Li (1993)	0.40–0.80
Cone Penetration resistance (Muddy Clay)	" "	1.00–1.50
Cone Penetration Resistance (Greyish Clay)	" "	0.60–1.20
Cone Penetration Resistance (Dark Green Clay)	" "	0.50–0.80
Cone Penetration Resistance (Fine Sand)	" "	0.25–0.60

✤ From the article of Quek et al. (1992)

▲ (Shanghai Soils)

Table 2b - Literature   data   relative   to   vertical
          fluctuation scales.

0.4285) which do show a positive, though rather low, correlation between c and the bulk unit weight in the same soil (r(cγ)= =0.2232). Cherubini et al. (1990) find that the correlation coefficient between c and $\emptyset$ for the Matera Clays is equal to -0.61.

6 Conclusions

This brief review allows certain general considerations to be made, namely:

- geotechnical properties, at least those we have investigated, exhibit different degrees of variability;
- a true standardization of the tests will, on the one hand, make it possible to reduce the variance induced by the tests to negligible levels, and will also enable results to be readily compared even when obtained in different laboratories and, maybe, in different countries, should such standardization become universally applicable;
- for some geotechnical parameters, like undrained cohesion, one may well suspect that the observed high variabilities are due to misused statistical procedures;
- knowledge about the fluctuation scale is still inadequate. Instead, this is a parameter of the utmost importance if one is to apply probabilistic methods to geotechnics with success and credibility.

7 References

Alonso, E.E. and Krizek, R.J. (1975) Stochastic formulation of soil properties. **Proc. of the 2nd ICASP.** Aachen, Vol. 4, pp. 9-32.

Bao, C.G. Huang, W.F. and Chang, Q.N. (1993) Reliability analysis of bearing capacity for a gravity wharf foundation in random field theory. **Proc. of the Conference on probabilistic methods in geotechnical engineering.** Canberra, pp. 291-294.

Bergado, D.T. and Chang, J.C. (1989) Application of on inter-active computer aided probabilistic slope stability assessment for embankments os soft Bangkok Clay. **Proc. of The Int. Symp. on Computer and Physical Modelling in Geotechnical Engineering.** Bangkok, pp. 25-52.

Chowdhury, R.N. (1984) Recent developements in landslide studies: Probabilistic Methods. **State of the Report Session VII a International Symposium on Landslides.** Toronto, pp. 209-228.

Cherubini, C. Cucchiararo, L. Germinario, S. and Giasi C.I. (1990) Elaborazione statistica preliminare sulle principali caratteristiche geotecniche delle Argille Azzurre della Città di Matera. **Proc. of 2nd Workshop "Informatica e Scienze della Terra", GIAST.** Sarnano, pp. 63-70.

Cherubini, C. Giasi, C.I. and Rethati, L. (1993) The coefficients of variation of some geotechnical parameters. **Proc. of the Conference on probabilistic methods in geotechnical engineering.** Canberra, pp. 179-183.

Corotis, R.B. Azzouz, A.S. and Krizek, R.J. (1975) Statistical evaluation of soil index properties and constrained modulus. **Proc. of the 2nd ICASP.** Vol. 2, Aachen, pp. 273-293.

Gao, A.Z. and Li J.P. (1993) Reliability analysis on pile bearing capacity. **Proc. of the Conf. on Probability Methods in Geotechnical Engineering.** Camberra, pp. 295-301.

Harr, M.E. (1987) **Reliability based design** in civil engineering. Mc Graw Hill Book Company, N.Y.

Keaveny, J.M. Nadim, F. ang Lacasse G. (1989) Autocorrelation functions for offshore geotechnical data. **5th Int. Conf. on Structural Safety and Reliability ICOSSAR.** pp. 263-270.

Kulhawy, H. Roth, N.J.S. and Grigoriu, N.B. (1991) Some statistical evaluations of geotechnical properties. VI ICASP, Mexico City, pp. 705-712.

Mostyn, G.R. and Li, K.S. (1993) Probabilistic alope analysis. State of play. **Proc. of the Conference on probabilistic methods in geotechnical engineering.** Canberra, pp. 89-109.

Queck, S.T. Chow, Y.K. and Phoon K.K. (1992) Further Contribution to reliability - Based Pile - Settlement analysis. **Journ. of Geotechnical Engineering ASCE.** Vol. 118 n. 5, pp. 726-742.

Rethati, L. (1989) **Probabilistic solution in geotechnics.** Elsevier, Amsterdam.

Wickremesinghe, D. and Campanella R.G. (1993) Scale of fluctuation as a description of soil variability. **Proc. of the Conference on probabilistic methods in geotechnical engineering.** Canberra, pp. 233-239.

Tukey, J. (1977) **Exploratory data analysis.** Addison Wesley Publishing Company.

Vanmarcke, E.H. (1977) Probabilistic modelling of soil profiles. **Journ. of Geot. Eng. Div. ASCE 103 (H).** pp. 1227-1246.

SMALL SCALE VARIABILITY IN CLAY SAMPLES

M. SOULIÉ and E. MASENGO
Geotechnical Division, École Polytechnique, Montréal, Canada

Abstract
In this study of small scale variability of Champlain Sea
Clay's geotechnical properties, two parameters have been
analyzed: the water content and the Swedish Cone resistance.
They were measured on approximately cubical blocks of 0.027m³.
Two blocks have been investigated: the first one gave 2704
water contents and 1352 cone strength values, while the second
one gave 3024 water contents and 1512 cone strength values.
    For each block taken individually, or the two blocks
combined, a geostatical analysis has been made in order to
model the observed spatial variability. The results show in
fact the presence of two types of clay in clearly defined zone
inside each block. A study of the characteristics of the clays
was then carried out. This study has shown that apart from the
structure, the two clay types are similar.
    Finally, by using the variograms which have been obtained,
the variability of the water content of samples with different
sizes, as used in conventional laboratory tests
(consolidation, triaxial tests), has been analyzed and
discussed.
Keywords: Clay, Water content, Shear strength, Variability,
Geostatistics.

1. Introduction

Soil parameters show a large variability in their values. That
variability may be observed either because of some lack of
homogeneity in the soil itself and because of the measurement
errors. Indeed soil is a very non-homogeneous material because
of the way it is formed and the diverse materials from which
it is formed. So great is this non homogeneity that a soil
classified as homogeneous under some soil classification will
show some significant variations in its properties from point
to point. Lumb (1966,1974, 1975) pioneered the research in
this field by developing some models of spatial variability of
soil properties. This work has been followed by many others
Singh and al (1970), Schultze (1971), Vanmarcke (1977), Soulié
(1983), Soulié and al (1990), Recordon and Despond (1977),

81

D. Breysse (ed.), Probabilities and Materials, 81–94.
© 1994 *Kluwer Academic Publishers.*

Réthati (1988) Magnan (1982), Chiasson and al (1993). In all
these studies the main problems were how to model the observed
spatial variability and how to take account of this
variability in practical situations (design of slopes,
foundations, site investigations, etc.). The objective of this
paper will consider only the modeling of the observed spatial
variability. The main difference with the other work is the
scale at which the variability is observed. In most of the
studies, the distance between samples is going from 0.5m to
10m or more. In our work the distance is in the range of cm
and the geotechnical parameters under study are the water
content and cone laboratory shear strength. There are several
reasons for focusing attention on variability at small scale.
First, it is impossible, unless some hypothesis is assumed, to
infer the variability in volume smaller than the volume
defined by the minimum distance between samples. This is not
important if we are interested in the mean or representative
value on large volume. But a model of this variability at
small scale is needed (Cressie 1992), if the objective is to
better understand the observed variability between samples
taken in a same block of soils in laboratory experiments.
Second, even if the objective is to address the variability at
large scale, there is a need to model the variability at the
scale of sample dimension, because the global variance at
large scale depends on the dimensions of the sample, Matheron
(1965), Journel and Huijbregts (1978).

This paper starts with a description of the experimental
procedures. This is followed by a presentation of the results.
Geostatistical approach is then applied to the results and
some conclusions are presented.

2. Experimental tests and procedures

2.1. Soil origin

The soil used in this study was obtained from a test trench
excavated close to Mont St-Hilaire near Montréal. Undisturbed
blocks of clay were recovered below the weathered crest at a
depth varying between 3m and 5m.

This clay is known as the Champlain Sea Clay. The Champlain
Sea occupied the St-Lawrence Valley from approximately 12000
to 15000 years BP. Subsequently marine clays were deposited to
form, in distal regions of the ancient sea, thick massive
clays. In proximal areas the clay can be interlayered. The
normal composition of clay from the Champlain Sea is one
dominated by primary minerals such as quartz, feldspar, mica
and chlorite with the presence of glacial ground amorphous
material and sometimes carbonates.

2.2. Testing and Sampling procedures

Twenty seven cubic blocks (30cm x 30cm x 30cm) were well localized and recovered from the trench. Two blocks have been used in this study.

The water content was taken on cylindrical samples (1cm x Ø1.5cm). As defined in geotechnics, the water content is the ratio between the weight of water and the weight of the dry soil, this ratio being expressed in percentage. The weights were measured with a reproducibility of 0.005g. These samples were taken according to a regular 3 D pattern : 2cm x 2cm in horizontal directions x and y, and 1cm in the vertical direction z.

The cone shear strength was obtained according to the same pattern but every 2cm in the vertical direction, by using a cone laboratory. The penetration of the cone was measured with a precision of 0.5mm, the weight of the cone was 100g and its angle 30°.

3. Tests results

3.1. Water content

For the water content a total of 5728 samples have been measured. The distribution is shown in fig. 1.

The principal statistics are :

mean	w	=	68.77%
minimum	$w_{min.}$	=	55.81%
maximum	$w_{max.}$	=	111.54%
standard-deviation	σ_w	=	6.75%
coefficient of variation		=	9.8%.

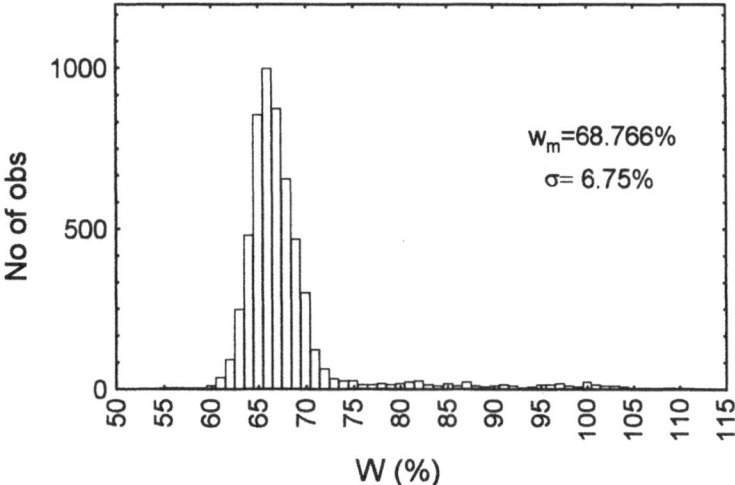

$w_m=68.766\%$

$\sigma= 6.75\%$

Fig.1. Global distribution of water content.

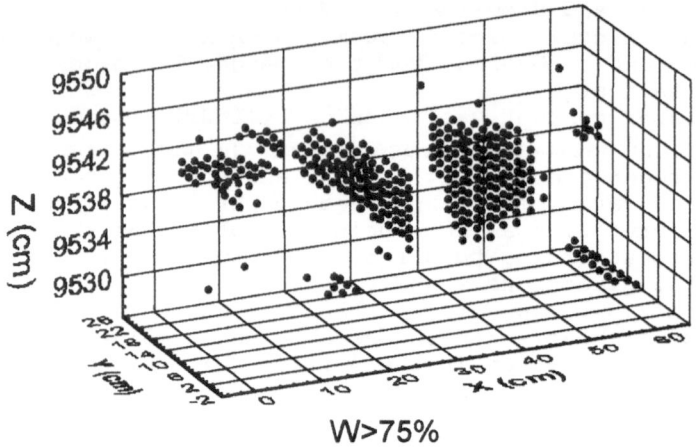

W>75%

Fig.2. Spatial location of water content with
w>75%.

Because of the long tail distribution for the large values
of the water content, the distribution has been split in two
parts with a cut-off value of 75%. The spatial position of the
samples with a value of w > 75% is shown in the figure 2.
The figure shows clearly that these samples are not randomly
distributed into the two blocks. The zones corresponding to w
> 75% are clearly defined. X ray photographs (Fig. 3) confirm
this, where the boundary between the clear zone (w > 75%) and
the dark zone (w < 75%) is sharply defined.

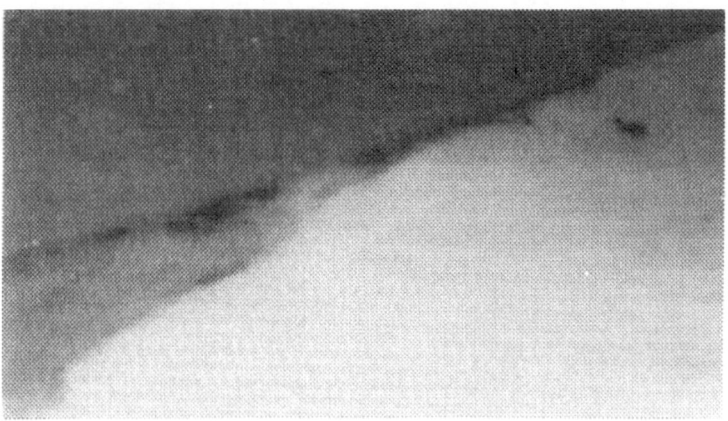

Fig.3. XRay picture of Champlay Sea Clay
(Dark zone W<75%, light zone w>75%).

Mineralogical analysis indicate that these two soils have the same minerals, but with different granulometries. No good reasons have been found to explain the presence of these nodules of soil with a high water content going from 75% to 111%. Because 92% of the samples have a water content < 75%, we decided to characterize the spatial variability of this clay.

The mean value of w is 67% and the variance is 5.24 with a coefficient of variation of 3.4%: this clay can be considered as homogeneous. The distribution is well fitted by a normal distribution (Fig. 4) even if the $\chi 2$ value shows that the difference is significative because of the large number of data (n = 5283) and the assumed independance between the values.

The coefficient of variation is small when we compare it with reported values in the literature, with a range going from 8 to 20 %. (Magnan 1988, Rethati 1988, Lumb 1966, 1974, 1975). Even for "homogeneous" clay Recordon and Despond (1977) obtained values of 11.3 % to 20 %.

The observed variability of the water content can be attributed to measurement error and spatial variability or non-homogeneity. The measurement error of w is the result of the imprecision of three weight measures: the wet sample, the dry sample and the can. The reproducibility of the balance was ±0.005g. The error variance has been estimated to be about 0.15%². Consequently most of the observed variability is due to spatial variability.

The water content as defined in soil mechanics as $w = w_w/w_s$ (where w_w is the weight of water and w_s is the weight of solids) is not an additive variable. Therefore if

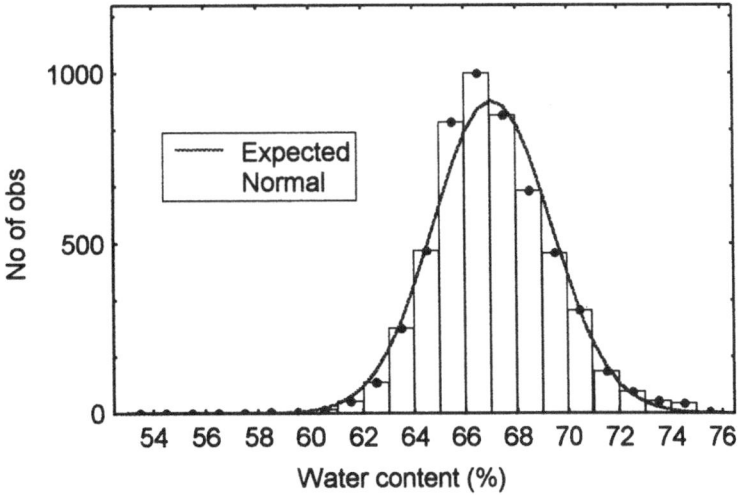

Fig.4. Distribution of water content w<75%.

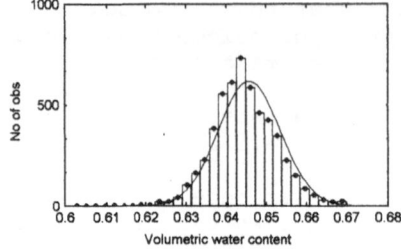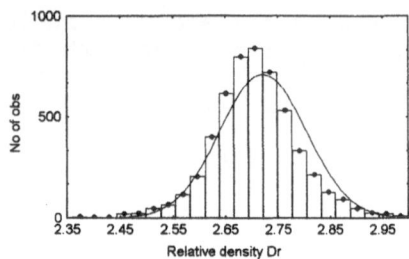

Fig.5. a)Distribution of volumetric water content θ
assuming w<75% and Dr=2.72.   b)Distribution of the
relative density Dr assuming w<75% and θ=0.646

two samples with same volume, have different water contents,
the water content of these samples considered as one is not
the mean value of the two water contents. On the contrary the
volumetric water content θ defined by the volume of water
expressed as a percentage of a given volume of soil is
additive.
    The relation between these two water contents is

$$w = \frac{\theta}{D_r - D_r\theta} \tag{1}$$

where D_r is the relative density of the solid particles.
    The water content is not strictly additive, but in this
study, because of the small variability we are justified to
linearize the equation in the vicinity of the mean value of w
with D_r= 2.72. Therefore it will be acceptable to consider for
the following analyses that w is additive.
    As a consequence the observed variabilility of w can be the
result of the combined variability of θ and D_r. Figure 5
illustrates the distributions of θ assuming D_r=2.72 and of D_r
assuming θ = 0.646. It is interesting to note that the
resulted variability of D_r is not in contradiction with
reported values in the literature (Mitchell 1976).

3.2. Cone Shear Strength

The cone shear strength values (r = 2864) for the two blocks
and the two soils, are distributed according the Figure 6 with
a mean value of 15.6kPa, a variance of 3.5kPa² with a
coefficient of variation of 12%. The discrete set of measured
values is the consequence of the lack of sensitivity of the
test. Nevertheless the relative large number of data (2864)
has allowed to make some correlation of the cone shear
strength values and the water content. To do this, the water
content values have been grouped in class 5% wide and for each
class the mean value Cu has been computed. The results are
plotted in Figure 7. No attempt to fit a specific function

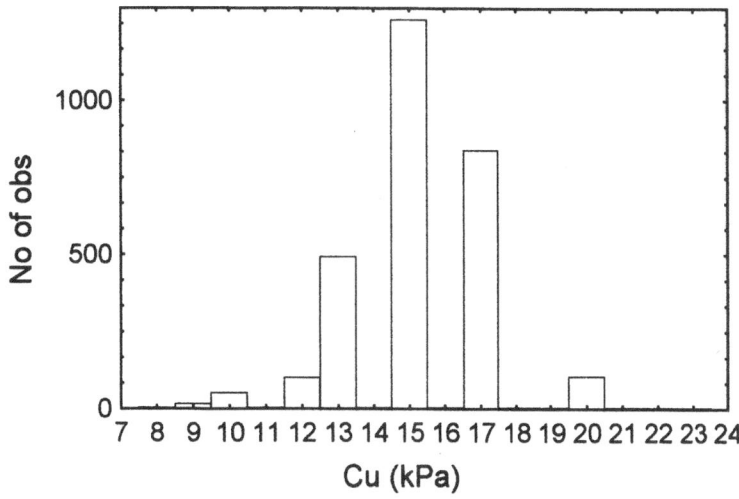

Fig.6. Global distribution of cone shear strength.

has been made, but the figure shows clearly that there is a correlation between the water content and the mean value of the cone shear strength. In this Figure 7, we can see the difference between the two soils in term of correlation, difference which was not noticeable on the global distribution of the shear strength value.

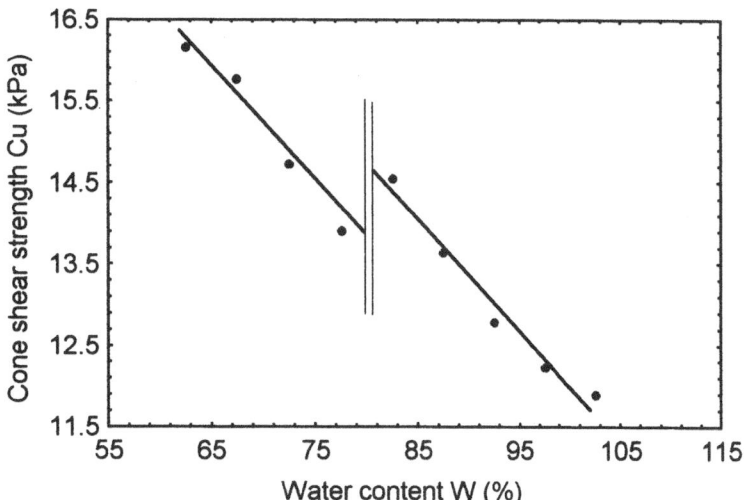

Fig.7. Relation between mean value of cone shear strength Cu and mean value of water content w.

88

4. Modeling spatial variability

4.1. Experimental variograms
Following classical geostatistical approach the observed
variability is analyzed by means of second order moment
approach. Consequently, the variogram is defined as one half
the variance of differences of water content between paired
values separated in space by a distance h. In the case where
the mean value is assumed to be constant the variogram is
expressed by

$$\gamma(h) = \frac{1}{2N_h} E\left[w\left(\vec{x} + \vec{h}\right) - w(\vec{x})\right]^2 \qquad (2)$$

where N_h is the number of pair of values separated by h, $\vec{x}$ is
the vector {x, y, z) of the coordinates of the point value (or
center of the support value).

Using the 5283 values with w < 75%, the Figure 9 shows the
resulting variograms computed in the horizontal and vertical
directions. As it appears it seems that linear variograms,
with anisotropy in the horizontal and vertical directions and
a nugget effect of about $1.5\%^2$ will fit perfectly the observed
value. But as mentioned before the measurement error variance
in$0.15\%^2$ and the nugget value cannot be attributed only to
this error variance.

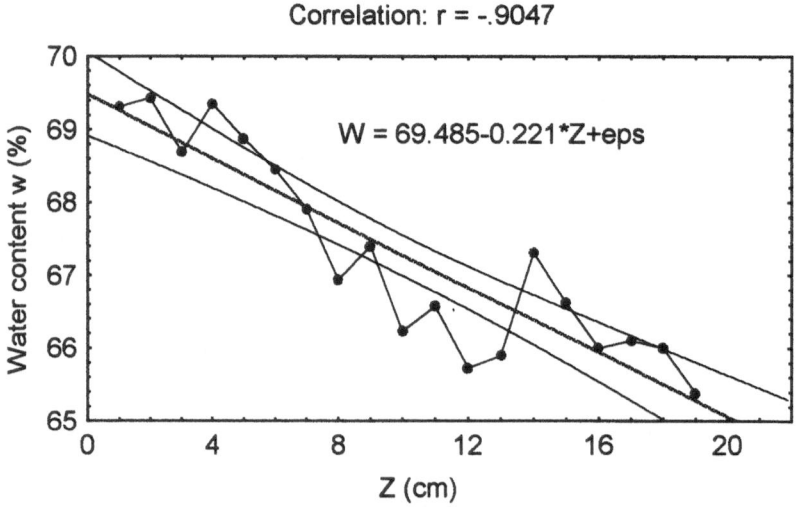

Fig.8. Vertical trend of the horizontal mean value of w
(w<75%0.

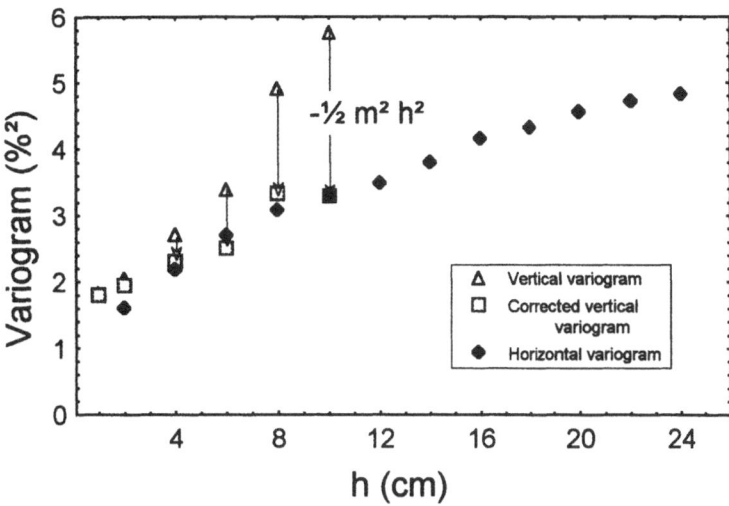

Fig.9. Vertical and horizontal variograms of w (w<75%).

To check if one possible explonation, to the differences between horizontal and vertical variograms, could be the presence of a linear trend in the vertical direction, the mean value of w has been computed for each horizontal slice. The Figure 8 shows clearly, that at the scale of the block size, a trend exists. Assuming a linear trend, the vertical variogram values were decreased by $\frac{1}{2}mh^2$, where m is the linear coefficient of the fitted linear trend. On the figure 9, we see that after correction the variograms are now similar. Therefore the following model has been adapted to describe the spatial variability at the scale of the block size.

$$w(x, y, z) = m(z, \varpi) + \varepsilon_w(x, y, z) \qquad (3)$$

where $m(z, \varpi)$ is the vertical trend, which can be random from one block to another block, and $\varepsilon_w(x, y, z)$ is the fluctuation around this trend with:

$E\{\varepsilon_w(x, y, z)\}=0$

and isotropic variogram $\gamma(h)$.

With a linear trend given by w= 69.485-0.2211z the computed variance of ε_w is 4.126 %² instead of 5.24 %².

4.1.1. Variogram fitting

Fitting a variogram model to experimental values is in some way arbitrary. Indeed the experimental values of the variogram are random variables and therefore the experimental variogram for a given realization can be very different from the true one. Because of this problem of fluctuation, only the value, with a lag less than the half of the maximum size of the black in one direction, will be used.

Another problem is the fact that the values of water content have been obtained on some volume v (1.33 x 1.33 x 1cm) and therefore the computed variogram is a regularized variogram $\gamma_v(h)$ (Matheron 1965, Journel and Huijbregts 1978) and is related to the point variogram by

$$\gamma_v(h) = \bar{\gamma}(v, v_h) - \bar{\gamma}(v, v) \tag{4}$$

where $\bar{\gamma}(v, v_h)$ is the mean value of the variogram $\gamma(h)$, when each extremity of a segment describes independently the two identical volumes v, translated by h. Obtaining the point variogram $\gamma(h)$ from $\gamma_v(h)$ is not an easy task but nevertheless, is possible because for h > the dimensions of v, $\bar{\gamma}(v, v_h) \approx \gamma(h)$ and $\bar{\gamma}(v, v)$ can be numerically estimated.

Different models have been tested. The best results have be obtained with the De Wijs' model (Matheron 1960, Krige 1981):

$$\gamma(h) = 3\alpha \ln(h) \tag{5}$$

where 3α is the absolute dispersion coefficient.

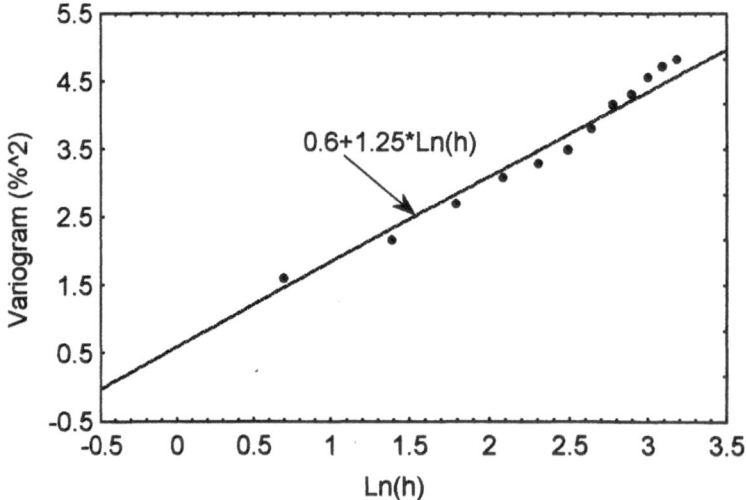

Fig.10. De Wijs' model: estimation of the absolute dispersion parameter.

This "point" variogram is not defined for h = 0 and in fact is not a function but has to be considered in term of distribution. Once regularized on a volume v, $\gamma_v(h)$ is a function and can be used as an ordinary variogram.

To have an estimation of α it is sufficient to plot the experimental values of $\gamma_v(h)$ in function of n(h). For h>λ $\gamma_v(h)$ can be approximated by $N + 3\alpha \ln(h) + 3\alpha\left(\frac{3}{2} - \ln(\lambda)\right)$ where λ is the linear equivalent of v = 2.5 x 1.33 = 3.325cm, and N the nugget effect = 0.15%^{2} (Krige 1981). The value of 3α is 1.25%^{2} and the "point" variogram is:

$$\gamma(h) = 0.15 + 1.25\ln(h) \tag{6}$$

The model variogram $\gamma(h)$, and the numerically computed $\gamma_v(h)$ are plotted in Figure 11.

Experimental variogram does not show well-developed sill. Existence of a sill implies existence of finite population mean and variance, and the De Wijs' model predicts a dispersion variance for volumes of similar shape given by:

$$\sigma^2(v/V) = \bar{\gamma}(V, V) - \bar{\gamma}(v, v) = \alpha\ln\left(\frac{V}{v}\right) \tag{7}$$

To check the validity of this model, observed variance as defined by:

$$\sigma^2_{exp}(v/V) = \sigma^2(v/V) + N + \sigma^2_{T,v} \tag{8}$$

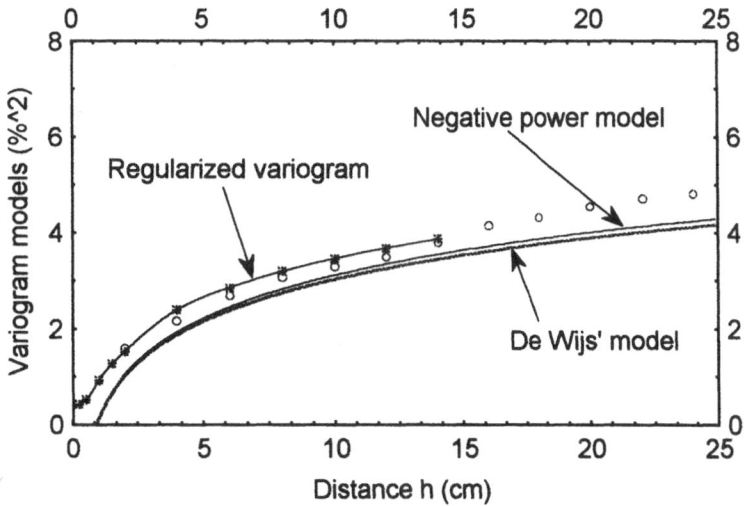

Fig.11. Comparison of variogram models.

where N is the variance error associated with the support v
and $\sigma^2_{T,v}$ is the variance due the trend in the volume V, have
been computed and compared with experimental values for the
following cases:

v   =   sample volume   1.33 x 1.33 x 1 cm
1B   =   1 block   25 x 22 x 18 cm
2B   =   2 blocks   50 x 22 x 18 cm
Oed =   oedometer size   7 x 7 x 1 cm
Tri =   triaxial size   5 x 5 x 12 cm.

The following table presents the results:

Dispersion variance	Model values ($\%^2$)	Experimental values ($\%^2$)
v and 1B	4.92	4.89
v and 2B	5.35	5.24
v and Oed	2.02	*
v and Tri	2.41	*
Oed and 2B	2.20	*
Tri and 2B	1.82	*

Other models have been tested and good fits were obtained
with a negative power-law variogram $\gamma(h) = B\left(1 - h^{-3\alpha_L}\right)$. This
model has been derived by several authors following different
routes (Agterberg 1993, Rose and Srivastava 1993). This model
can be related to the de Wijs' model by assuming that we have
De Wijs' model in the logarithmically transformed variable,

with an absolute dispersion parameter $\alpha_L = \alpha \dfrac{\sigma^2}{\sigma^2_L}$, where σ^2 is

the dispersion variance of w and σ^2_L is the dispersion
variance for Ln(w).In this case the variance of dispersion:

$$\sigma^2(v/V) = m^2\left[\left(\frac{V}{v}\right)^{\alpha_L} - 1\right] \tag{9}$$

The variograms of the two models are not defined for h = 0,
that means that the variance of dispersion increases without
bound, if the volume of the support is becoming smaller and
smaller. This is consistent with the fact that the water
content at the point value is not defined, being 0 when the
point belongs to the grain and infinity when the point belongs
to the water. This de Wijsian model in fact belongs to the
family of multifractals (Agterberg 1993) and can be easily
simulated via a transformed variable (e^w) (Matheron 1960,
Krige 1985). The problem is that the second order
representation cannot be intrinsic at the same time in w and
ln(w). So if it is intrinsic in ln w, this implies that B

parameter is proportional to the square of the experimental mean value of w for the domain V, and will vary from one realisation to another (Matheron 1974).

5. Conclusions

The application of geostatistical methods to this case study provided the following main conclusions:

Water content parameter presents some variability, which is spatially distributed, even in "homogenous" clay.
The observed variability is function of the volume of the sample used for measurement and of the size of the domain. Reported values which will not indicate these volumes are useless.
This variability is well described by one unique parameter, the coefficient of absolute dispersion of De Wijs' model, which is a good example of multifractals at least for this scale.
More research needs to be made on different clays to support the validity of the De Wijs' model.
Similar study using the volumetric water content and the relative density, to characterize the variability, would be useful.

6. References

Agterberg, F.P. (1993) Fractals, Multifractals and Change of Support, in **Geostatistics for the Next Century**. June 3-5, 1993. Montreal, Canada.

Chiasson, P., Lafleur, J., Soulié, M. nd Law, T. (1993) Assessing Natural Clay Variability. **46th Annual Canadian Geotechnical Conference**. Saskatoon, 343-350.

Cressie, N. (1991) **Statistics for Spatial Data**. Wiley Series in Probability and Mathematical Statistics. John Wiley & sons. 1991.

Journel, A. and Huijbregts, C.J. (1978), **Mining Geostatistics**. Academic Press. London.

Krige, D.G. (1985) **Lognormal-de Wisjian Geostatistics for Ore Evaluation**. South African Institute of Mining and Metallurgy. Johannesburg.

Leroueuil, S., Tavenas, F. and Le Bihan, F.P. (1983) Propriétés Caractéristiques des Argiles de l'Est du Canada. **Canadian Geotechnical Journal**, Vol. 20, 681-705.

Lumb, P. (1966) The variability of Natural Soils. **Canadian Geotechnical Journal**, Vol. 3, 74-94.

Lumb, P. (1974) Application of Statistics in Soil Mechanics, in **Soil Mechanics, New Horizons**, Ed. Lee, Butterworths, 44-111.

Lumb, P. (1975) Spatial Variability of Soil Properties. Proc. of the 2nd Int. Conf. on Application of Statistics and Probability in Soil and Structural Engineering, Vol II Aachen.

Magnan, J.P. (1982) **Les Méthodes Statistiques et Probabilitistes en Mécanique des Sols**. Presse des Ponts et Chaussées. Paris, France.

Matheron, G. (1960) Traité de Géostatistique Appliquée. **Mémoires du Bureau de Recherches Géologiques et Minières**, N°14. Édition Technip, Paris.

Matheron, G. (1965) Les Variables Régionalisées et leur Estimation. Ed. Masson, Paris.

Matheron, G. (1974) **Effet Proportionnel et lognormalisé ou : Le Retour du Serpent de Mer**. Note géostatistique N°124. Centre de Morphologie Mathématique, Fontainebleau.

Mitchell, J.K. (1976) **Fundamentals of Soils Behavior. Series in Soil Engineering**. John Wiley & Sons.

Recordon, E. and Despond, J.M. (1977) Dispersion des Caractéristiques des Sols Naturels Considérés comme Homogènes, in **IX ICSMFE**, Tokyo 1977, Proceedings of the Specialty Session N°6.

Réthati, L. (1988) **Probabilistic Solutions in Geotechnics**. Developments in Geotechnical Engineering N°46, Ed. Elsevier.

Rose, C.D. and Srivastava, R.M. (1993) A Fractal Correlation Function for Sampling Patterns, in **Geostatistics for the Next Century**, June 3-5, 1993, Montreal, Canada.

Schultze, E. (1971) Frequency Distributions and Correlations of Soil Properties, in Proc. of 1st Conf. on Application of Statistics and Probability in Soil and Structural Engineering. Hong-Kong.

Singh, A. and Lee, K.L. (1970) Variability in Soil Parameters, in **Eighth Annual Conf. on Geological and Soil Engineering**, Idaho, 1970.

Soulié, M. (1983) Geostatistical Applications in Geotechnics, in **Geostatistics for Natural Ressources Characterization**. Part 2, NATO ASI Series. Reidel Publishing Company. Dordrecht, Holland. 703-730.

Soulié, M., Montes, P. and Silvestri, V. (1990) Modelling Spatial Variability of Soil Parameters. **Canadian Geotechnical Journal**, Vol. 27., 617-630.

Vanmarcke, E.H. (1977) Probabilistics Modeling in Soil Profiles, in **IX ICSMFE**, Tokyo 1977, Proceedings of the Speciality Session N°6.

CLASSIFICATION LOGIC AND CORRELATION BETWEEN SOILS PARAMETERS: APPLICATION TO AN ELASTO-PLASTIC LAW

J.L. FAVRE, A. RAHMA
Ecole Centrale Paris, Paris, France

Abstract
For the soils, the micro structure is given at the scale of the grains whose the characteristics are the micro structure parameters. To predict the macro structure parameters of the soils, it is necessary to have a good and plentiful description of the micro structure. Unfortunately, the data which are available about the grains, in soils mechanics, do not describe the grain but the assembly of the grains and are called physical or identification parameters. To make the best use of these informations we have to find the micro structural mechanical meaning of the parameters and the logic of connection between them and the macro structural mechanical parameters.
This logic is the logic of the resolution of a continuous medium mechanics problem. The identification and physical parameters are shown as micro structure parameters: They are called either the parameters of the **nature** of grains or the parameters of the **assembly** of grains. A lot of correlations between them and the mechanical macro structural parameters of the continuous medium are quoted and explained.
A new application is given for the prediction of the parameters of an elasto-plastic law, using factorial analysis and multiple regression with the data of the data base Modelisol. The models of prediction are tested to find again the curves of three triaxial tests on the same sand.
Keywords: Soils parameters, Discontinuous medium, Elasto-plastic law, Correlations, Data analysis, Prediction.

1 Introduction

·The complexity of the behaviour and the large range of the types of soils led the Civil Engineer to use a lot of parameters to characterise the soils. Moreover, to make up for the small sampling used for a geotechnical project, he tried for a long time with inter and extrapolations, information cross-check, while he constructed many models of the correlations. This multiplication raised the question of the consistency and robustness of these correlations; very few Research Centres tried to propose a rational for the classification and correlations of parameters. Only one of them, in Grenoble then in Paris has developed this question for a long time, Favre (1972, 1980), Simon (1972), Biarez and Favre (1975, 1977), Favre et al. (1989).

D. Breysse (ed.), Probabilities and Materials, 95–108.
© 1994 Kluwer Academic Publishers.

2 Basic problem: The logic of the mechanics of continuous medium applied to a discontinuous medium

In addition to the mechanical parameters of the behaviour models more or less complex, the Civil Engineer uses other parameters called physic and identification parameters (the grain size distribution, the void ratio, the Atterberg's limits, etc...).He creates correlation models between the mechanical parameters on one hand as a search for internal material coherence and between them and physical and identification parameters on the other hand, if not as a prediction of the firsts by the seconds, at least to supplement a too small knowledge on the firsts, to save sampling, to optimise the information.

To understand how the mechanical parameters are connected with the others, it is necessary to look at how they are obtained:

You apply a load to a small volume of the medium.
You obtain an experimental response and you adjust on this response a stress-strain model of whom the parameters are calculated.

This response can be find through the calculation. Then, you have to solve a problem of continuous medium (CM) mechanics but applied to an assembly of grains. You have to write:

The general equations of the bodies equilibrium.
The constitutive equations and interfaces laws of these bodies.
The integration of these laws, with the given boundary conditions

So, for the calculation of the deflection of a beam, you write the equilibrium of the acting and reacting moments and strengths in a section, you write the behaviour (i.e. elastic, linear, isotropic) of the material and you integrate with the geometrical boundaries of the beam using Bernoulli's hypothesis. So, we can take as a basic logic applicable to a discontinuous medium (DM):

general equations (DM) + constitutive equations (DM)
+ boundary conditions (DM) = solutions (CM)   (1)

3 Classification of parameters in soils mechanics

The different mechanical physical and identification parameters can be join to one of the four headings and we are going to explain their specific role in the correlations.

3.1 Solutions: mechanical parameters
The integration of the discontinuous medium behaviour leads to characterise the stacked grains as an imaginary continuous medium with its own mechanical parameters. This behaviour shows clearly characteristics as unreversibility coupled with anisotropy, non linearity, hardening ; these characteristics are to explain through the physical and identification parameters attached to the grains and to their arrangement.

The rheological models have to take into account the stress range, the stress path, and the initial state. According to their level of sophistication, they try to integrate these influences into their formulation itself, to make the mechanical parameters the most intrinsic as possible. It only subsists almost the time the dependence (or not) with the initial state.

3.2 General equations

They govern the static and dynamic equilibrium of the grains between them, in particular with the specific volumetric weight of the grains which presents a low variability ($\gamma s = 27$ kN/m^3).

3.3 Constitutive equations: parameters of nature

The "rheological" meaning of the physical and identification parameters is to be found in the mechanisms implemented in the granular medium. We shall examine rather the ones of the sands.

<u>Mechanisms in the non cohesive media</u>

It has been noticed for a long time, Biarez (1962, 1963) that the number of contacts between grains varies little and that the grains slide over each other rather than that they roll along. So, there is a variation in the geometry of the meshes rather than a multiplication (or reduction) of the number of meshes: mechanically, one is faced with phenomena of grain to grain friction leading to global behaviour, unreversibily created by stress deviators greater than the last which was applied (hardening). So, the reciprocal mobility of grains leads to the behaviour, giving non linearity and unreversibility, rather than the deformability and fragility which apply only to hardened materials.

<u>Parameters of Nature for the sands</u>

   <u>grains rheology</u>

These parameters are not used in practice neither the grain to grain friction. The mineralogy can take into account the different constituents. On the other hand, the granulometry gives an account of the relative mobility.

   <u>granulometric range and size</u>

The size has little effect, except by correlation with a seldom measured parameter, the shape. The granulometric range counts a lot: it governs the grains arrangement.

   <u>shape and surface state</u>

These parameters are difficult to measure; several charts of them exist. Different studies, Cambou (1972), Frossard (1978), Favre (1980), have shown the interdependence between the shape and the size according to the deposition process and the important role of the shape.

   There is little correlations between the physical and rheological parameters

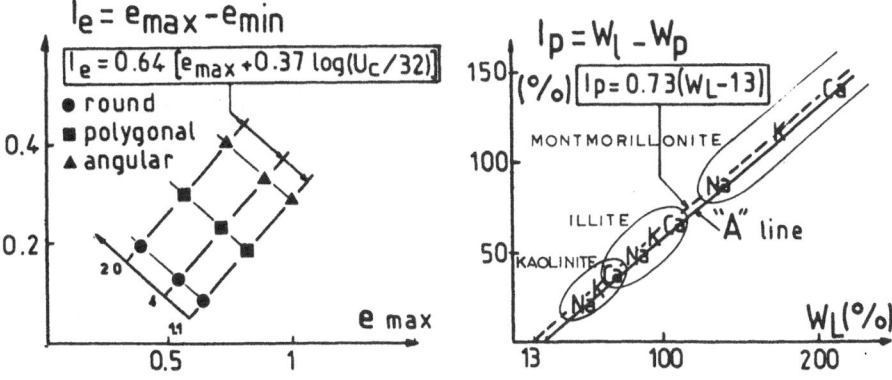

Fig.1. Charts of synthesis nature parameters, Favre et al. (1989)

synthesis nature parameters

We shall call like this, the parameters as e_{min}, e_{max}, γ_{opn}, etc.. which characterise the response of the material to a particular mechanical test and integrate in this way the previous parameters. The studies of different authors, Cambou (1972), Becker et al. (1973), Frossard (1978), Favre (1980), indicate that they vary with the shape and the granulometric range. We can propose a chart for the sands like this one of Casagrandre for the clays, figure 1, but "unfolded" according to the granulometric range which varies for the sands.

The synthesis parameters e_{min} or e_{max} will be, like W_l or W_p for the clays, good parameters for describing the mechanical properties of the sands.

Mechanisms and Nature parameters for the clays

Little is know about the basic mechanisms for the clays despite many more studies than for the sands: with the mechanical connections, we have to add the electrical and chemical connections. These mechanisms are governed by the nature of the clay particle, the kind of particles organisation, the nature of the connected water, and the flaked or/and aggregated structure. Studies were about the role of the physic-chemical parameters as the mineralogy, the cations exchange capacity, the flaked or aggregated state, the %<2 μm, the consolidation stress.

These parameters are not independent and often they explain badly the behaviour. By contrast, the synthesis nature parameters such as W_l or W_p have shown a great explanation potential of the mechanical properties and their correlations between themselves and with the physic-chemical parameters have been the subject of a very important literature gathered by Favre (1980).

3.4 Boundaries conditions: compaction parameters

Role of the density

It acts on the mechanical parameters by two ways:

The initial compaction of the material:

This can take away value: it does not exist a reference density of the material "at rest". So, one tries to connect the mechanical parameters at the previous stress state which has given this "initial" state. That is the over consolidation or the hardening.

The instantaneous compaction:

Beyond this stress state, the material forgets its past and its properties are independent of the initial state but can depend of the instantaneous density while this one varies a lot as with the normally consolidated clays. This dependence can be integrated in the parameter itself : it is the case of the compression index Cc.

This role of the density, clear enough for the clays, is the same for the sands except that these ones get generally an over consolidated behaviour corresponding to high and imprecise enough values of consolidation stress (5 to 50 MPa). So, as a good correlation for the clays has been found between the liquidity index Il and the consolidation stress, σ_c, Zervoyannis (1982) has found for the sands a good correlation between the relative density and the critical consolidation stress due to Seed and Lee, but index linked by the granulometric range.

Parameters of compaction and of mechanical state

A great enough poverty is to note in the description of the boundary conditions of the grains. Apart from the granulometry which we have placed with the parameters of nature, we have only at our disposal a global description of the assembly of the voids in relation to the grains, by the way of the void ratio or the saturation water moisture, and of the quantity of water in relation to the voids, by the way of the degree of saturation.

These two parameters, or any other couple, are sufficient to characterise the ratios of the three phases if one considers the specific volumetric weight of the grains as constant.

Comparing the parameter of global compacity, w, to the two synthesis nature parameters W_l and W_p using the index of consistency $I_c = (W_l-w)/(W_l-W_p)$, you define the more or less great proximity of the material to its liquid or plastic behaviour. It is the same with the relative density $D_r = (e_{max}-e)/(e_{max}-e_{min})$. These parameters, combining the nature with the compacity of grains, will call mechanical state parameters.

The anisotropy of the assembly is little used, the consolidation stress can be an explanation. Finally you get also the ratio of over consolidation.

3.5 The classification of the parameters

We can sum up the classification of the parameters on the figure 2, and give two basic equations of connection between the parameters of the discontinuous medium (DM) and these ones of the continuous medium (CM):

$$\text{Nature (DM)} + \text{Compaction (DM)} \rightarrow \text{Rheology (CM)} \qquad (2)$$

$$\text{Mechanical State (DM)} \rightarrow \text{Rheology (CM)} \qquad (3)$$

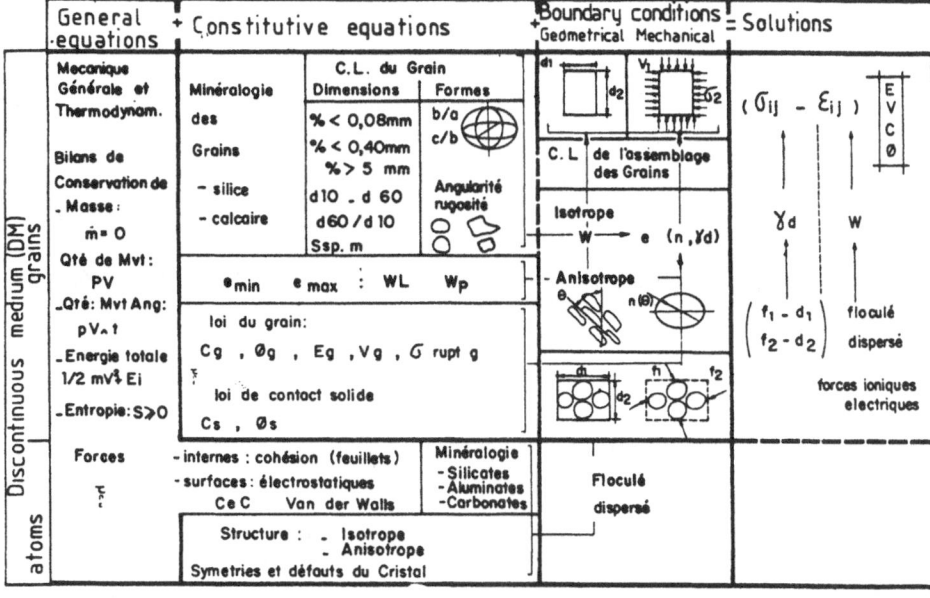

Fig.2. Classification of parameters, Favre (1980)

4 The correlations

The best based correlations are given for the mechanical parameters of the simplest models as the isotropic elasticity, the consolidation and the plasticity with Mohr-Coulomb criterion. Of course, they could be extend over a lot of other models as

Duncan's, Cam-Clay models, etc... We propose following a study with the elasto-plastic model of Hujeux.

4.1 Deformability parameters

For the sands, from a lot of studies, in particular Hardin (1963, 1968), Schultze (1970), Cambou (1972), Al Issa (1973) Favre (1980), we deduct the model

$$E = E_0\sigma^{0.5} \tag{4}$$

where E_0 depends essentially of the mechanical state Dr.

For the normally consolidated clays, among a lot of studies listed in Favre (1980), we can indicate Skempton (1943, 1944), Lumb (1968), Biarez for slurries. These studies concern:

-The compression index Cc; we can adopt a relation close to this one of Skempton,

$$Cc = 0,009(W_1 - 13), \tag{5}$$

which, combined with the relation

$$I_p = 0,73(W_1 - 13) \tag{6}$$

leads to

$$Cc = I_p/82. \tag{7}$$

We can also accept

$$Cs = Cc/4. \tag{8}$$

-The position of the normal consolidation line on the space w, $\log\sigma$. We can deduct that

$$
\begin{aligned}
w &= W_1 \text{ for } \sigma = 6.5 \text{ KPa} \\
&= W_p \text{ for } \sigma = 1 \text{ MPa,}
\end{aligned}
\tag{9}
$$

from which, we can obtain the following relation

$$I_1 = 0.46 (3 - \log\sigma). \tag{10}$$

The oedometric module depends functionally from Cc, σ and e_0

4.2 Rupture parameters

For the sands, Favre has refined the model of the internal friction angle of perfect plasticity.

The angle of peak has been studied by Al Issa (1973) and Frossard (1978) with $K = e_0 tg\varphi_{peak}$. From studies of De Beer (1965), Bishop and Eldin (1953), Bishop (1948), Nash (1953), Favre (1980), we can propose a model close to this one of De Beer for the peak internal friction angle versus the mechanical state Dr with corrections due to the average stress and the granulometric range.

For the clays, the undrained cohesion has been studied a lot. From results of Skempton, Bjerrum, Schofield, Wroth among other people, we can take the relation

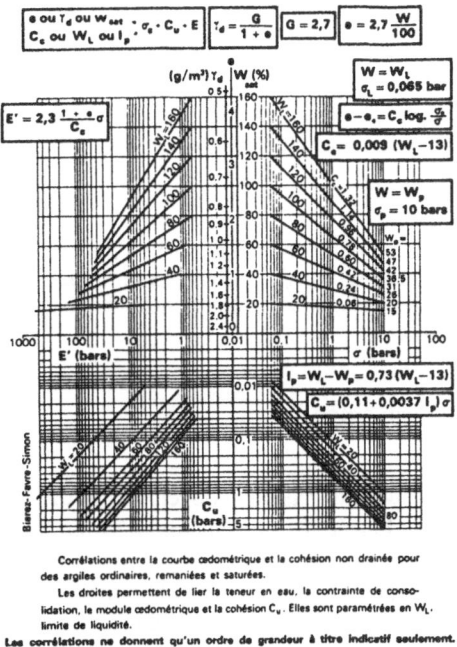

Fig.3. Chart for the disturbed normally consolidated clays

Clayey soils

W_L	20	30	40	50	60	80	100	200
W_p	15	18	20	23	26	31	37	64
I_p	5	12	20	27	34	49	63	136
C_c	0.06	0.15	0.24	0.33	0.42	0.60	0.78	1.68
C_u/σ_c	0.13	0.15	0.18	0.21	0.24	0.29	0.34	0.61
$E/\sigma_c^{(1)}$	56	27	20	16	14	12	11	9
$E/\sigma_c^{(2)}$	41	22	15	11	9	7	6	4
$\phi'r$	30	24	19	15	14	12	10	-
$tg\,\phi'r$	0.58	0.45	0.34	0.27	0.25	0.21	0.18	-
W_{OPN}	10	13.5	17	20.5	24	32	40	-
γd_{OPN}	19.5	18.5	17.5	16.5	15.5	13.5	11.5	-

(1) for $W = W_L$ (2) for $W = W_p$

Sandy soils

$\phi_{pp} = 31.5° + \phi_D + \phi_F + \phi_M + \phi_{Uc}$	
$\phi_F = 1°$ angular	$\phi_{Uc} = 2°$ Uc = 2
$= 0°$ polygonal	$= 3°$ Uc = 3
$= -1°$ round	$= 4°$ Uc = 5
	$= 5°$ Uc = 10
$\phi_D = .5°$ .1 to .315 mm	$= 6°$ Uc = 50
$= 0°$ .315 to 1.0 mm	$\phi_M = 1°$ for each 10%
$= -.5°$ 1.0 to 4.0 mm	of carbonates

Fig.4. Parameters independent
of the compaction

Clayey soils

I_c	0	0.25	0.50	0.75	1	
I_L	1	0.75	0.50	0.25	0	
C_u	1.5 *	5 *	15	40	100	$W_L = 50$
E'	90 *	300 *	1000	3400	11000	
C_u	2.3 *	8 *	25	70	170	$W_L = 100$
E'	62 *	200 *	600	1900	5700	
σ_c	65	23	90	300	1000	

* = for mud these values can be several times greater.

Sandy soils

D_r	0	0.2	0.4	0.6	0.8	1
ϕ peak	32	34	36	39	42	45
$E_0 = \frac{E}{\gamma\sigma}$	4000	5000	6500	7500	9000	11000

(1): Values for σ'_m: every time σ'_m is doubled substract (2 Dr)°.

Correct ϕ as you do for ϕ_{pp}.

Values can vary by 100%: Do not use
for design.

Fig.5. Parameters dependent
on the compaction

of Skempton for the normally consolidated clays,

$$Cu = (0.11 + 0.0037I_p).\sigma \qquad (11)$$

For the residual internal friction angle, among different authors like Skempton (1964), Kenney, Gibson, Grim, we will deduct a relation close to the results of Gibson (1953) for φ' versus W_l.

4.3 Synthesis
We can sum up most results into an chart for the disturbed normally consolidated clays , figure 3, and two types of tables, these for the parameters independent of the density, figure 4, and these for parameters dependent of the density for which one uses the mechanical state I_l and Dr, figure5.

5 Application to the elasto-plastic model

5.1 Cyclade Model
The constitutive model called "Cyclade" has been developed at the Ecole Centrale de Paris and the Institut Français du Pérole. It is an elastoplastic model which employs the concept of critical state. The plastic strains are decomposed in three plane strain mechanisms. The hardening variables are the plastic volumetric strain and an internal variable linked to deviatoric plastic strains of each mechanism, whose initial value r^{el} determines the size of the elastic domain.

We have to predict from the parameters of the discontinuous medium a set of ten parameters which can be classified as follows :
- elastic parameters : E_a, v, n (isotropic non linear elasticity) with :

$$E = E_a(p/p_a)^n \qquad (12)$$

E_a, Young module for $p=p_a=100$ KPa.
v, Poisson's ratio
n, coefficient of non linearity
- size of the elastic domain : r^{el}
- critical state parameters : β, p_{co}, φ with

$$p_c = p_{co} \exp(\beta\varepsilon_v^p) \qquad (13)$$

p_{co}, initial value of the critical pressure p_c, depending on the initial density,
φ, friction angle at perfect plasticity
- plastic deviatoric parameters : a, b
- plastic volumetric parameter : α

These five classes can be divided into two groups:
1- experimental parameters : E_a, v, n, β, p_{co}, φ directly determined from test results (conventional triaxial tests).
2- numerical parameters : r^{el}, a, b, α, found in different equations of the model and determined only by comparing the experimental curves with the computed ones.

5.2 Determination of Model parameters

21 sands were selected from the database MODELISOL, having a large variety of physical properties. Around 150 drained and undrained triaxial tests were available for these 21 sands, tested at different initial densities. The results have been gathered from available papers and entered with a standard form in the database.

The determination of model parameters was made using a curve fitting approach between experimental and numerical stress-strain curves.

The general evolution of these two curves were also taken into account. When several different tests at different initial densities on a same sand were available, this research of an unique set of parameters compatible for all the test results made necessary a long and difficult approach to find the best compromise.

6 Modelling of rheological parameters

We have used mainly two methods : the Principal Components Analysis and the Multiple Regression, with an estimation of the "confidence range" for a probability of 95%.

6-1 Correlation and statistical links

Assuming that previous works (elastic parameters E, ν, n and friction angle φ gave enough elements for the statistical determination of them we have concentrated our study on the specific parameters of the constitutive model Cyclade : r^{el}, a, b, α on one hand and on the critical state parameters β and pc_0 on the other hand.

We have distinguished the parameters depending on the Nature of the granular material r^{el}, a, b, α, β and the parameter pco depending on the mechanical state of the material, usually expressed by the value of critical void ratio e_0.

It could happen that some parameters describing the discontinuous medium would be connected to each others by certain ways. We have tried to use in our analysis the most significant parameters and this selection has first been performed on the basic of statistical independence. They are the following :

* parameters representative of the granulometric range d90, d60, d30, d10, or Cu=d60/d10, Cz=(d30²/d60.d10).
* e_{max}, e_{min} or Ie = e_{max} -e_{min}
* the mineralogy of the grains was usually unknown and we have used the density of the grains γ_s to represent it, even if we are aware that it is not sufficient by it self
* A parameter representative of the mechanical state Dr=$(e_{max}-e_0)/(e_{max}-e_{min})$, which will be introduced in the modelling of pco only.
* Mathematical transformation have been made in some parameters in order to have an homogeneous distribution of all the variables.

6-2 The tools: Principal Components Analysis

At first the results of the Principal Components Analysis figure (5), gave us the strong connections between the parameters of the constitutive model, on one hand, and the parameters describing the granular material on the other hand.

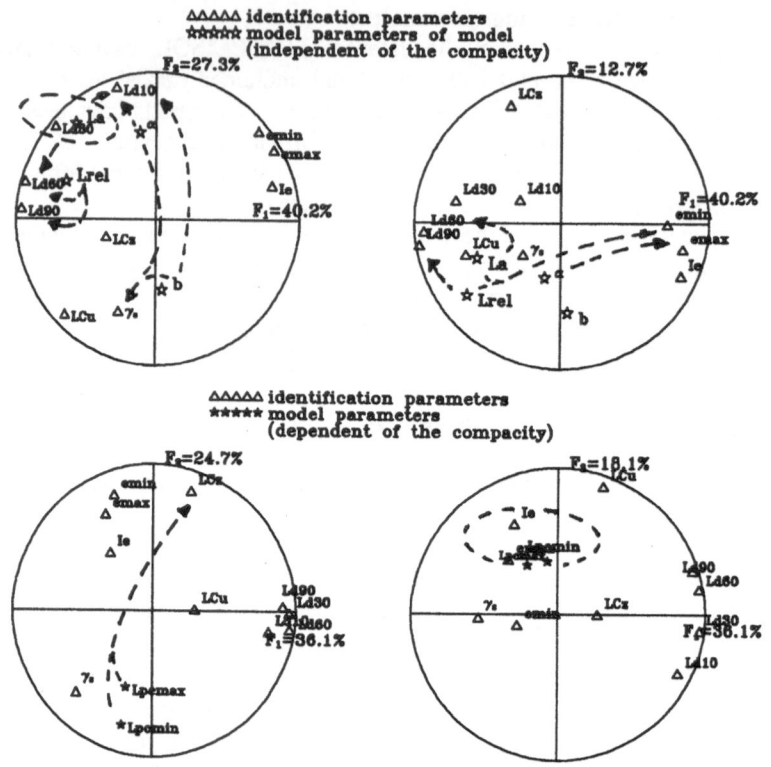

Fig.6. Principal Components Analysis

	calculated value (initial set)	optimized value
$\log r^{el}$	-1.725 ± 0.3 $r^{el}=0.019^*$	-1.7 $r^{el}=0.02$
$\log a$	$-2.93\pm0.25)$ $a=0.00115^*$	-2.82 $a=0.0015$
b	0.256 ± 0.07	0.25
α	2.03 ± 0.5	2.4
$\log pc_{min}$	2.487 ± 0.5	2.57
$\log pc_{max}$	3.103 ± 0.3	3.03

$\log pc$ (Dr=84%)	3.0046^* (pc=1010 kPa)	2.954 (pc=900 kPa)
$\log pc$ (Dr=64%)	2.88^* (pc=760 kPa)	2.875 (pc=750 kPa)
$\log pc$ (Dr=54%)	2.82^* (pc=660 kPa)	2.813 (pc=650 kPa)

* mean value

Fig.7. Validation test

One can conclude that a strong correlation exists between r^{el}, a, b, α, and the granulometric range. This can be understood because r^{el} and a control the initial slope of the stress-strain curve, i.e. the possibility of rearrangement of the granular assembly when submitted to a given loading.

b is also dependent on the granulometric range but mainly on the value of d10, which indicates that the percentage of "fines" plays a great role on b.

The mineralogy of the grains, expressed by the density γ_s, influences the values of b and α. For b this can be related to what has been said above.

For a the explanation has probably to be found in the fact that γ_s is an indirect way of taking into account some characteristics of the grains as the shape and surface state which can influence the dilatancy potential of the assembly.

e_{max} and e_{min} have an significant influence on all these parameters, which can be well understood since they are the results of mechanical tests and take into account in an indirect way most of the characteristics of the granular medium.

β and pco are two parameters which locate the straight line representing the critical state in the (e-logp) diagram. β is the slope and pco is the value of p at critical state for a given void ratio eo of the granular material. Instead of working on β and pco, we decided to determine two values of pc, pc_{max} and pc_{min} corresponding to two different void ratios: e_{max} and e_{min}.

In these conditions we obtain in the (Dr-logp) diagram:

$$\beta_{norm} = (logpc_{max} - logpc_{min}), \text{ because } Dr_{max} = 1 \text{ and } Dr_{min} = 0 \quad (14)$$

$$\text{and } \beta = 2.3 \; \beta_{norm}(1+eo)/Ie \quad (15)$$

β is dependent on the critical void ratio eo, because the mathematical expression of the critical state in Cyclade Model is $pc = pco \; exp(\beta\varepsilon_v^p,)$ which takes into account only the plastic volume change, when the experimental results consider the total volume change. The values of pc_{max} and pc_{min} are of course dependent on e_{max} and e_{min} and also on the granulometric range.

6-3 The results: multiple regression

For all the model parameters we have tried to obtain linear regression models. They are presented below in the form of an mean and standard deviation :

Yi=E(y) $\pm$ tσ, with t=1.96 for Pr=95%, with σ: standard deviation and R: correlation coefficient

$$logr^{el} = 0,33 \; logd60 - 1,2logCz - 1.58 \pm 0.3 \qquad R=0,92 \quad (16)$$

$$loga. = 0,45 \; logd60 - 0.1 \; logCu + 0.9 \; e_{max} - 3.5 \pm 0.2 \qquad R=0.93 \quad (17)$$

$$\alpha = (1,65 \; logd90 - 1,6 \; logd60 + 0,45 \; logd10) +1,75 \; e_{max} - 0.26 \; \gamma_s + 7.2 \pm 0.5$$
$$R=0,92 \quad (18)$$

$$b = 0,013 \; logd10 - 0.4 \; e_{max} + 1.0 \; Ie + 0.08 \; \gamma_s - 1.9 \pm 0.07 \qquad R=0,90 \quad (19)$$

$$\log pc_{min} = -12 \log Cz -4.7\ e_{max} + 9.7\ Ie + 2.2 \pm 0.5 \qquad\qquad R=0.96 \qquad (20)$$

$$\log pc_{max} = -4 \log Cu - 7.5\ e_{min} + 4.75\ Ie + 7.1 \pm 0.3 \qquad\qquad R=0.95 \qquad (21)$$

All the correlation coefficients are greater than 0.9, which indicates that the proposed correlation models are consistent. Some restrictions are still to be considered :
- the size of the sampling remained modest.
- they were uncertainties on the determination of some parameters of the discontinuous medium like the shape of the grains, the mineralogy, the definition of the test used for determining e_{max} and e_{min}.

At the present stage, these correlation models can be considered as very valuable not for the final determination of the model parameters but more likely for the definition of an initial set of parameters, realistic enough to permit an good optimisation process to be undertaken.

6-4 Validation test

At first we have to calculate, for a new given material, the values of the model parameters (r^{el}, a, b, α, pco, β) by using the correlation models which have been determined in the present study and predict "a priori" its behaviour by some simulations using Cyclade Model. The elastic parameters E, ν, n and the friction angle φ have been determined directly from the test results.

Then by comparing experimental and numerical curves, we have to modify gradually the initial set of parameters in order to determine an optimised one which represents at best the behaviour of the material.

The validity of the correlation models can be assessed through two criteria :
* the comparison between initial and final set (mean value and standard deviation).
* the comparison between the different solutions and the experimental curves, as well as the making of the methodology and the procedures which had to be followed.

In figure 6 we present the initial and final values of the parameters and the result after the optimisation process.

7 Conclusion

It turned out that the classification of parameters into parameters of **nature of the discontinuous medium**, parameters of **compaction of the discontinuous medium** and parameters of **rheology of the continuous medium** has been a powerful guide to understand the coherence of the most part of statistical relations of the literature to propose a set of correlations which explains the rheological parameters of the classic behaviour models as elasticity, consolidation and plasticity.

But, this rational is quite powerful for the much more sophisticated models as the elasto-plastic law of Hujeux-Cyclade. In this case, at first, an important thought has to be led on the meaning of the parameters of the model, coming back to the grains mechanisms. Next, the powerful tools of the factorial analysis as Principal Components Analysis and Stepwise Regression have to be used to understand the multiple connections between a lot of variables and samplers. With a such thought, it is possible to construct for the Engineers robust and simple enough prediction models with only two explicative parameters which can be easy to measure on the soils.

Such prediction models have not to replace the tests on samples, but they can give an initial set of parameters very helpful for a preliminary study while one has not yet at one's disposal the results of the site investigations. The above validation test indicates well that one finds again very correctly the experimental response of the material.

Lastly, the probabilistic models are interesting for the uncertainties estimations and for the calculation of the probability of failure of the earthworks.

On this time, some geotechnical data base as MODELISOL which stocks the experimental curves of tests and the parameters of rheological models extracted on them, are in development: The prediction models for sophisticated laws, relying on the logic of parameters classification and connection, are again with good prospects

8 References

Al Issa, M. (1973) Recherche de lois contraintes déformations des milieux pulvérulents. Th.Doct.Ing. Univ. Grenoble.

Becker, E. Chan, C.Y. and Seed, H.B. (1972) Strength and deformation characteristics of rockfill materials in plane strain and triaxial compression tests. Report n° TE 72-3, Dept. Civil Eng., Berkeley.

Biarez, J. (1962) Contribution à l'étude des propriétés mécaniques des sols et des matériaux pulvérulents. Th.Doct.es Sces. Univ. Grenoble.

Biarez, J. and Wiendick, K. (1963) La comparaison qualitative entre l'anisotropie de structure des milieux pulvérulents, C.R. à l'Académie des Sces, 256, p.1217

Biarez, J. Favre, J.L. (1975) Parameters filling and statistical analysis of data in soils mechanics .Proc.2nd I.C.A.S.P. Aachen 1975 Vol 2, pp.249-264.

Biarez, J. and Favre, J.L. (1977). Statistical estimation and extrapolation from observati. Report of Organisers, Spec. Session 6, IX ICSMFE, Tokyo, Vol 3: pp.505-509.

Biarez, J. and Hicher, P.Y. (1987) Simplified Hypotheses on mechanical properties equally applicable to sands and clays. Constitutive equations for granular non-cohesive soils. pp. 19-30. Rotterdam, Balkema.

Biarez, J. and Hicher, P.Y. (1989) An introduction to the study of the relation between the mechanics of discontinuous granular media and the rheological behaviours of continuous equivalent media-application to compaction. Powders and Grains. Conff. 1989. Balkema, Rotterdam.

Bishop, A.W. (1948) A large shear box testing sands and gravels, Proc. 2nd ICSMFE, Rotterdam, Vol.1,ind p.207

Bishop, A.W. and Eldin, A.G. (1953) The effect of stress histoty on the relation between internal friction angle and porosity in sand, Proc. 3th ICSMFE, Zurick, Vol.1, p.100

Cambou, B. (1972) Compressibilité d'un milieu pulvérulent. Influences de la forme et de la dimension des particules sur les propriétés mécaniques d'un milieu pulvérulent, Th. D. Spéc., Grenoble

De Beer, E.E. (1965) Influence of the stress on the shearing strength of sand, Proc. 6th ICSMFE, Montréal, Vol.1, p.165

Favre, J.L. (1972) Pour un traitement par le calcul de probabilité et les statistiques des problèmes de mécanique des sols. Thèse du Doctorat de Spécialité. Grenoble.

Favre, J.L. (1980) Milieu continu et milieu discontinu. Mesure statistique indirecte des paramètres rhéologiques et approche probabiliste de la sécurité. Th.Doct.es Sces. Univ. Paris 6: 2T.

Favre, J.L. Biarez, J. Hicher, P.Y. Rahma, A. (1989) Correlations for granular media, classification logic and connections between classes, in Powders and Grains, A.A. Balkema, Rotterdam, p.201-209

Favre, J.L. Rahma, A. (1990) Prédiction statistique des paramètres d'une loi élastoplastique: Analogie Sable Argile. Approche probabiliste des problèmes en géomécanique. GEOPROBA90. Journée d'Etude, 13 février 1990. Bruxelles

Frossard, E. (1978) Caractérisation pétrographique et proprétés mécaniques des sables. Th.Doct.Ing. Univ. Paris 6.

Gibson, R.E. (1953) Experimental determination of the true cohesion and true angle of internal friction, Proc. 3th ICSMFE, Zurick, Vol.1, p.126

Hardin, B.O. and Richart, F.E. (1963) Elastic wave velocities in granular solis, Proc. ASCE, Vol.89, n°SM1, p.33

Hardin, B.O. and Black, W.L. (1968) Vibration modulus of normally consolidated clay, Proc. ASCE, Vol.94, n°SM2, p.353

Hicher, P.Y. 1987 - Utilisation des corrélations pour une déterminations initiale des paramètres de la loi Hujeux. Rapport EDF.

Hujeux, J.C. (1979) Calcul numérique de problèmes de consolidation élasto-Plastique, Th. DDI. Ecoles Centrale des Arts et Manufactures, Chatenay-Malabry.

Hujeux, J.C. (1985) Une loi de comportement pour le chargement cyclique des sols. Génie Parasismique, Presses de l'Ecole Nationale des Ponts et Chaussées, Paris. pp. 287-302.

Lumb, P. and Holt, J.L. (1968) The undrained shear strength of a soft marine clay from Hong-Kong, Géotechnique, Vol.18, pp.25

Michalski, E. and Rahma, A. (1989). Modelisations du comportement des sols en élastoplasticité: définition des paramètres des modèles Hujeux-cyclade et recherche des valeurs des paramètres pour différents sols. Rapport d'étude pour le MRT. Décision d'aide No.84F1489.

Nash, J.L.K. (1953) The shearing resistance of fine closery graded sand, Proc. 3th ICSMFE, Zurick, Vol.1, p.160

Rahma, A. (1990) Prévision des paramètres de lois élasto-plastiques à partir d'essais de laboratoire. E.C.P & B.R.G.M. GRECO II. Pradet. France. 9-14 DEC. 1990.

Rahma, A. (1991) Modélisation probabiliste des paramètres de loi éléstoplastique Hujeux-Cyclade. Thése à l'Ecole Centrale Paris.

Schultze, E. (1970) Bodenmechanische Probleme bei Sand, Mitteilingen ans Inst. für Verk. Grundbau und B.M. der Tech. Hochschule, V.G.B. Helft 50, Aachen

Simon, J.M. (1972) Propriétés mécaniques des argiles, synthèses des lois usuelles et rôle des vitesses de déformation comparé à celui des températures, Th. D. Spéc., Grenoble

Skempton, A.W. (1943) Notes on the compressibility of clays, n°6 parts 1 and 2, p.119

Zervoyanis, C. (1982) Etude synthétique des propriétés mécaniques des argiles et des sables sur chemin oedométrique et triaxial de révolution. Th.Doct.Ing. Ecoles Centrale des Arts et Manufactures, Chatenay-Malabry.

VARIABILITY OF THE MECHANICAL PROPERTIES OF WOOD: RANDOMNESS AND DETERMINISM

P. CASTERA and P. MORLIER
Laboratoire de Rhéologie du Bois de Bordeaux (CNRS/ INRA/ Université BdxI),
Cestas, France

Abstract
Wood is an extremely variable material, and one reason of this variability lies in its
biological origin. This observation is verified for most wood properties, and at the
different structural levels of the material, from the timber beam to the microscale. This
paper refers to an important data base on the mechanical properties of maritime pine
wood and timber. The results that are presented here only concern clear wood
specimens. The distributional characteristics of the bending stiffness and bending
strength are affected by the biological composition of the population: influence of the
age of the tree and the position in the stem. This result is attributed to the existence of
within tree gradients of wood properties, including the juvenile wood effect. Statistical
relationships between density and modulus of elasticity, or modulus of elasticity and
strength, are also affected by these factors. This might have important consequences
on the prevision of strength properties by non destructive estimators.
Keywords: Weibull distribution, mechanical properties, maritime pine, clear wood,
reconstituted wood design.

1 Introduction

Wood based materials are the most widely used materials throughout the world. It is
clear for everybody that wood is a variable material, depending on the species, soil,
climate, which properties depend on the scale of measurement: clear (defect free)
wood and timber are two different materials (Madsen, 1992) and many authors think
nowadays that wood rheology is helpless for understanding timber mechanics.
Consequently the biological variability of wood is replaced by a technico commercial
variability of timber and one objective is to predict the mechanical performance of a
timber beam by visual or automatic inspection , or non destructive testing.
Nevertheless knowing the variability of clear wood in the tree considered as a wood
factory is of great interest in a fundamental analysis: which part can be explained
(determinism) and which part is due to hasard? Industrial applications of this can be
found in the field of reconstituted wood products (LVL, glulam, pannels). For such
products it is worthwhile knowing whether the variability of the final products may be
reduced by selecting the components on biological or physical criteria.

D. Breysse (ed.), Probabilities and Materials, 109–118.

2 Biological patterns of wood variability

From the biological point of view the main factors that control wood variability can be either classified in environmental or genetic factors, with possible interactions between the two effects. Under the assumption of additivity this leads to the following schematic representation of variability for a given characteristic X:

$$X = \mu + Gi + Ej + GEij + eijk \qquad [1]$$

where μ is the mean value of X for the sampled population, Gi, Ej and GEij represent the deviations due to genetic, environment and their interactions, respectively. eijk is a residual deviation, i.e variations between trees of the same genetic origin growing in the same stand. In this representation X is a local property which is usually measured at a reference position in the stem of the tree, and in most cases X will represent the average density or fibre dimensions at the breast height level (1.30m). For most coniferous species however it is well known that wood structure and properties can differ greatly for one tree according to the position in the stem. Because of its large magnitude compared to all other factors this spatial variability has been the subject of many studies during the last decades. The within tree variability can be described in the following manner:

$$X(r, \theta, z) = P(r, \theta, z) + Q(r, \theta, z) \qquad [2]$$

where (r, θ, z) refers to the spatial position in the stem (r: radial; θ: azimuth; z: vertical).

$P(r, \theta, z)$: within tree gradients.

$Q(r, \theta, z)$: random function accounting for local effects (climatic and sylvicultural mainly).

Illustration on wood density: density is the most commonly used non destructive estimator of wood's mechanical properties, and the effects of environment, sylvicultural treatments and genetics on this characteristic have been widely studied for most plantation grown species. The development of X-ray densitometry in the early sixties to analyse the local variations of wood density in the tree stem (Polge, 1978) has contributed to improve our knowledge on the patterns of variations of this parameter. In figure 1 is presented the distribution of density values obtained from a densitometric profile along a radius of a maritime pine stem. The bimodal form of the distribution is characteristic of heterogeneous woods, which are composed of successive layers of low density material (initial or springwood) and high density material (final or summerwood). From the distribution it is possible to estimate the average values of each kind of layer, the corresponding standard deviations, and the relative final wood content. Figure 2 shows the variations of average initial- and final wood density values when considering different samples in the same tree, or different trees in the same stand. The within- and between tree variations of density have a comparable magnitude, but in the first case the greatest variations occur for final wood density, whereas the between tree variability is mainly due to initial wood variations. This result has already been observed for maritime Pine. Furthermore it seems that initial wood density is under genetic control (Fonseca et al, 1992), i.e Gi in equation [1].

Because r and z in equation [2] are increasing with the age of the tree, there is an indirect effect of age on the distribution of wood properties. An example of variation of final wood density along the radius, i.e $P(r, \theta_0, z_0)$, is given for maritime pine in figure 3. It appears on the figure that the final wood density increases from pith to bark, and this trend is usually attributed to the juvenile wood effect.

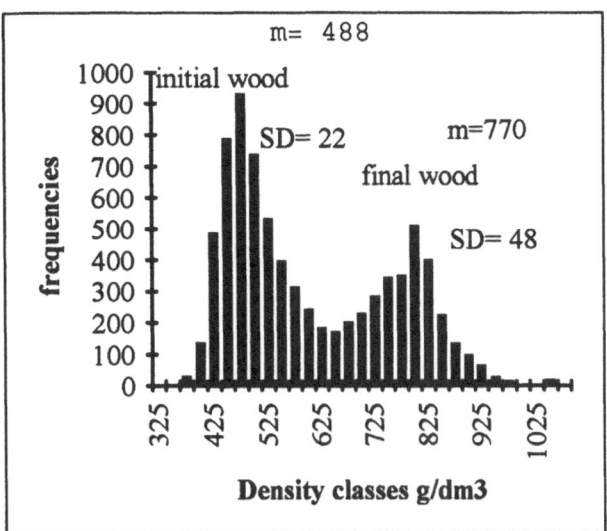

Fig . 1 . distribution of density values obtained from a
densitometric profile on a mature maritime Pine tree
m: average value ; SD: standard deviation.

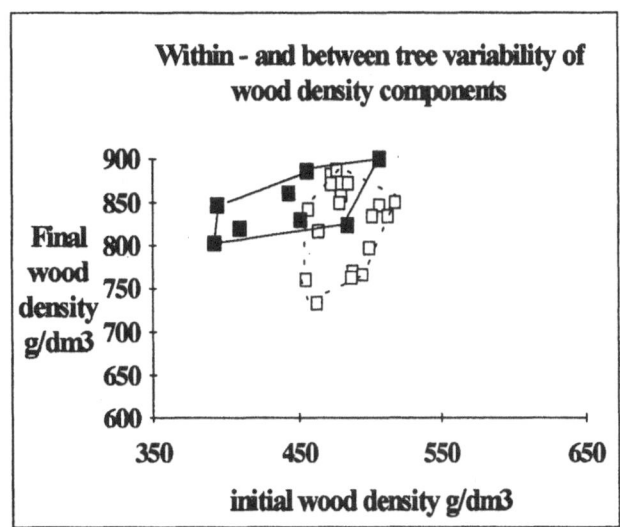

Fig . 2 . Variations of initial and final wood density values:
white squares: for 1 tree (19 samples)
black squares: for 8 trees (one densitometric profile per tree)

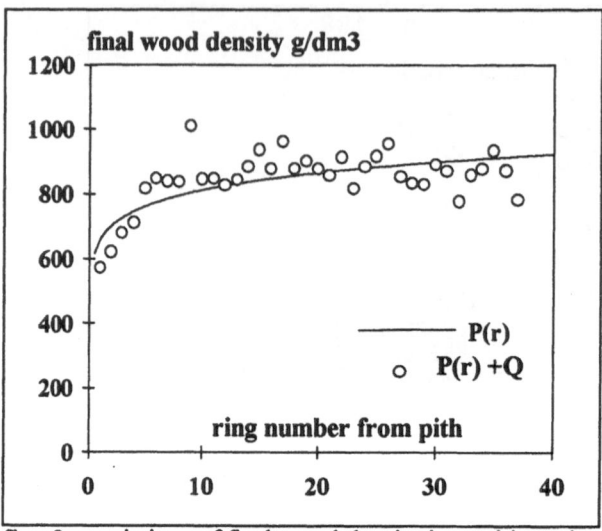

fig . 3 . variations of final wood density in maritime pine from pith to bark at breast height (1.30m).

3 Probabilistic representations of the variability of wood properties

The use of probability functions to analyse and simulate wood variations is of interest for many purposes, such as reliability based design procedures in massive wood or wood based products (glulam, LVL..). The choice of distributions is generally guided by goodness-of-fit analysis (Pellicane, 1985). This analysis has mainly been made for strength properties in bending of timber (MOR), but can also be applied to non destructive estimators of strength, such as density or modulus of elasticity (MOE) (Pellicane, 1993). Data obtained from small, defect free specimens (clear wood), could successfully be used in predicting the characteristics of wood based composite products.

The most common distributions in this field of research are the normal, Log-normal and Weibull distributions. The corresponding probability density functions (pdf's) are given below:

normal:
$$f(x) = \frac{1}{\sigma\sqrt{2\pi}} \exp\left[-\frac{(x-\mu)^2}{2\sigma^2} \right]$$
[3a]

Log-normal:
$$f(x) = \frac{1}{S\sqrt{2\pi}} \exp\left[-\frac{(X-M)^2}{2S^2} \right]$$
[3b]

where: $X = Ln(x)$
M and S are the mean and standard deviation of X

3 parameter Weibull:
$$f(x) = \left(\frac{m}{w}\right)\left[\frac{x-x_0}{w}\right]^{m-1}\left[\exp - \left[\frac{x-x_0}{w}\right]^m \right]$$
[3c]

where: x_0: lower bound of the distribution
m: shape parameter
w: scale parameter

Intrinsec wood properties, such as density or any mechanical property obtained from small, defect free specimens, are usually assumed to be normally distributed. Even for such properties however the validity of the normal distribution depends on how the sample is composed, due to the spatial patterns of wood variations. The use of the Weibull distribution to fit timber datas originates from the theory of brittle fracture that was first proposed by Weibull (1939). Although this theory has proved to be unefficient to explain scale effects and the fracture of wood (Madsen, op.cit), it has two main advantages compared to the normal distribution: the dissymetric form of the distribution, and the existence of a lower bound.

These pdf's have been used to analyse a data base composed of physical and mechanical properties measured on small specimens (wood properties) and full size beams (timber properties). The compression strength parallel to the grain (MCS: maximum crushing strength), the bending stiffness (MOE) and the bending strength (MOR) have been measured on clear samples according to national standards. The MOE and MOR were determined on full size beams (2000x40x100 mm), but the corresponding results are beyond the scope of this paper.

4 Composition of the sample

The parent population is composed of 56 maritime pine trees collected on 8 different areas in the Aquitaine region (Castéra and El Ouadrani, 1992). The trees have been chosen of various ages, from 30 to 80 years old, but similar girth at breast height: 120cm, which is the average harvesting size of maritime pine trees. The population has been divided into 5 classes of age: class 1 (30-40 years old) represents an intensive sylviculture leading to high growth rates and a reduction of harvesting age, class 2 and class 3 (40-50 and 50-60 years old) should be representative of actual sylviculture, whereas class 4 and class 5 (60-70 and 70-80 years old) represent the smallest trees in old traditionnally managed stands (figure 4). The number of trees for each class varies from 10 to 12. Note that the trees of extreme classes are not necessarily representative of the stands in which they have been collected. Furthermore it must be pointed out that the composition of the population does not reflect the variability of maritime pine wood in the Aquitaine forest massive.

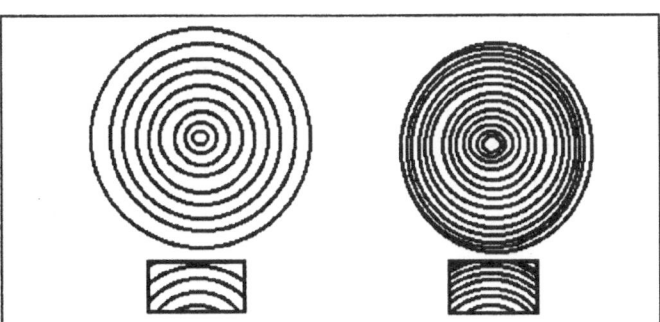

fig . 4 . Comparison between a fast growing tree
anf a traditionnally managed tree

For each tree five 2.50m length logs have been collected, corresponding to increasing height levels. The objective of this data base was to analyse the effects of harvesting age (indirectly growth rate) and position in the stem, on the variability of wood and timber properties, and the quality of the prediction of strength properties by non destructive estimators (density, MOE).Other data on the bending properties of maritime pine wood are available from another base which was differently composed: the sample reflected the variability of products that could be extracted from thinnings and final cuts in different areas. This sample was composed of 45 trees and 5 levels of height. The main results have been reported by C. Délisée (1987).

5 Data analysis

5.1 Characteristics of the sample

Extracted (resin free) density varies in a large range, from 400 to 700 g/dm3. The best prediction of density characteristics (mean and standard deviation) for the whole sample is obtained from the normal distribution, whereas the Weibull distribution provided the best fit according to the Kolmogorov-Smirnoff test for all mechanical characteristics (concerning the maximum crushing strength MCS, equivalent results were obtained from the normal and the Weibull distributions). A significant part of the variability of MOE can be explained by concomitent variations of density, with a correlation coefficient $r= 0.53$ (497 values). The best non destructive estimator of strength properties is the modulus of elasticity. The regression curves for three specimen dimensions are shown in figure 5. the 30x30 mm² (50 values) and 30x50 mm² (262 values) come from a previous data base (G.Guyon, 1987; C.Délisée, op.cit). Although the three samples only deal with defect free specimens, the patterns of variations look different. This result may be partly explained by scale effects. This shows that the prediction equations of strength (MOR) from MOE can differ within one species according to the "parent population". But within one population the residual variability is also very large.

Using two non destructive estimators of strength instead of one (density and MOE) increases slightly the quality of the prediction. In figure 6 the specific bending stiffness MOE/D is used as estimator of the specific bending strength MOR/D. MOE/D also measures the performance of wood as a component in composite structures. It can be seen that the correlation between the two variables is highly significant. This relationship exhibits the influence of structural wood characteristics other than density on its mechanical behaviour (microfibril angle in the cell wall layers).Even in this graphic however residual variability exists, and the distribution of MOR/D cannot be fully predicted by the cdf of MOE/D.

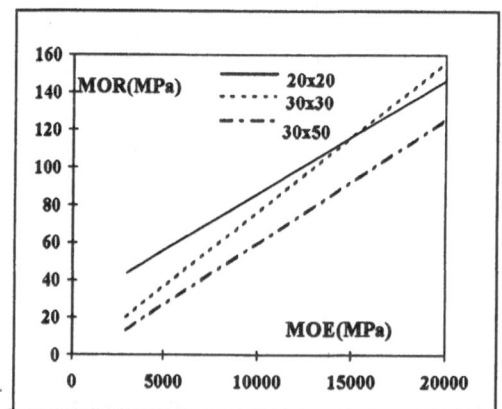

fig.5. relationship between the MOE and MOR
for different specimen dimensions.

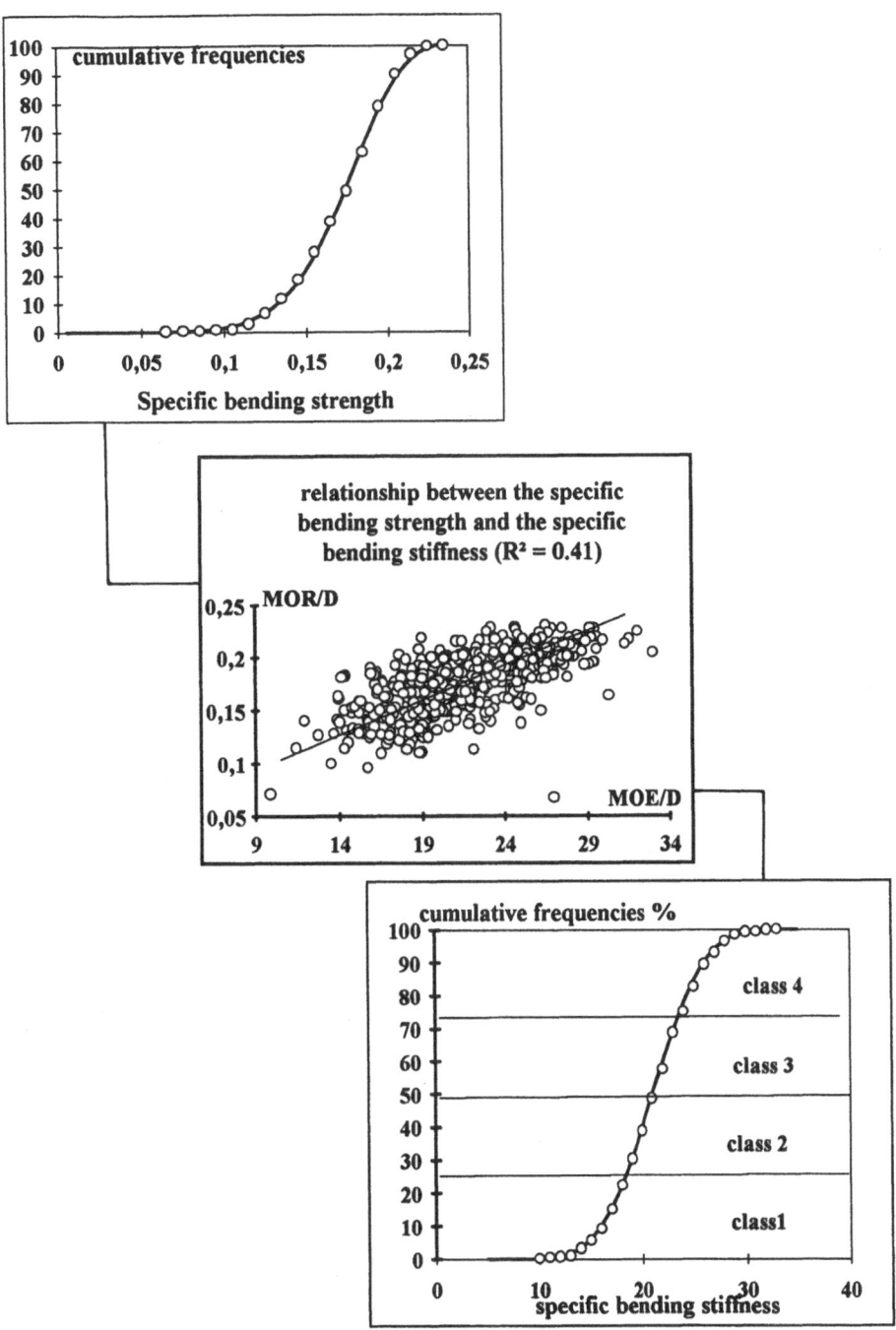

fig . 6 . variability of MOR/D as explained by the variations of MOE/D

5 . 2 Analysis of the residual distributions after grading

The parent population was divided in different classes, using first the density as grading parameter. For each class the experimental distributions of MOR values were fitted to a Weibull distribution. The limited number of values did not enable however to have a correct prediction at the lower tail of the distributions, which is of interest in reliability based design.

It appears that the increase of MOR with density classes does not follow the same patterns for the different fractions of the distribution. The lower fraction is slightly increasing from class 1 (450 to 500 g/dm3) to class 2 (500 to 550 g/dm3), and more drastically increasing from class 2 to class 3 (550 to 600 g/dm3).The same analysis has been made using MOE/D as estimator of MOR/D. The results are presented in figure 7. The variability of MOR/D is greater for the lower classes of specific stiffness.

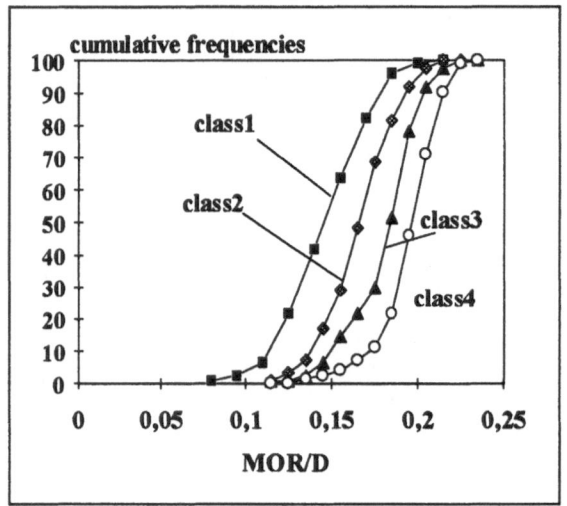

fig . 7 . distribution of MOR/D for different classes of MOE/D
(class1 corresponding to the lower values of the estimator)

5 . 3 Influence of tree age and position in the stem on the variability of wood properties

One important question considering the prevision of wood quality is: do we have to account for the biological patterns of the variability of wood in design procedures for wood and wood based products? An illustration of this is proposed by Tang and Pearson (1992), who studied the consequences of separating juvenile and mature wood in their sample, which have clearly different characteristics, on reliability based design procedures. The important conclusion of the authors was that juvenile wood had an effect on design results when stiffness was the limiting state. This in turn should have consequences in forest management of plantation species. For given dimensions of the trees the proportion of juvenile wood varies with age and position in the stem, and this affects the distributions of wood properties, and the concomitent variations of these characteristics.

Figure 8 for instance shows that the prediction of the bending stiffness by density is different in the buttlog from any other part of the stem.

Similar results obtained on loblolly pine are discussed by Megraw (1985), and the author relates this phenomen to different patterns of variations of microfibril angle in the cell wall layers. This, in turn, affects the variability of MOE/D.

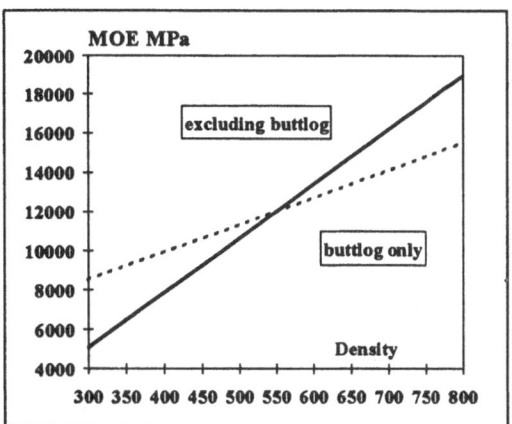

fig . 8 . relationship between MOE and density in maritime pine
after exclusion of the buttlogs

A look at the distributions of MOE at different height levels indicates a decrease of the average stiffness, part of which can be explained by a decrease of density values. A consequence of this is that the distribution parameters of MOE/D are very similar for the different height levels, except for the buttlog and the highest level, which was close to the living crown of the tree. The distribution of MOE/D in the buttlog mainly differed from the other distributions at the lower tail: the lowest values of MOE/D in the parent population were only observed at this position.

On the other hand there is an effect of the age of the tree on the distributions of MOE/D (figure 9). The average specific bending stiffness increases linearly with age, due to a decrease in the proportion of juvenile wood, but the lower tail increases significantly from class 1 to class 2, and then remains constant. The 40 to 50 years old trees look like a transition between the fast growing trees and the traditionally managed trees. Compared to the whole sample, the goodness of fit with a Weibull distribution is not improved when considering separately the different age classes.

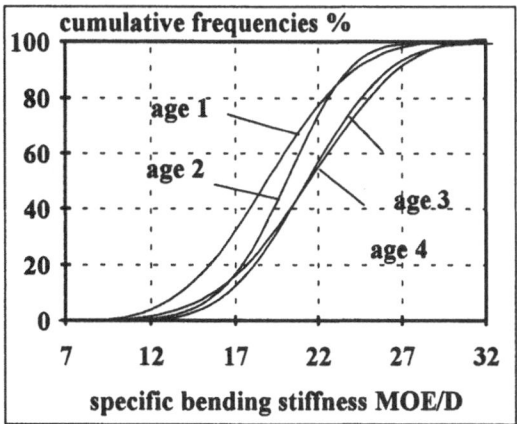

fig . 9 . influence of the age of trees on the
distribution of MOE/D

6 Conclusion

The prevision of the bending stiffness of clear wood by density depends on the biological origin of the material (juvenile or mature wood), and therefore there is an interest in controlling the trends of variations of wood properties in trees. The relationships between MOE and density is similar for the intermediate age classes, but is poor for the younger and older trees. This affects the patterns of variations of MOE/D (performance of the material), which exhibits significant correlations with strength properties.
Projected values of the proportions of juvenile wood at various heights and their variations with the age of the tree indicate that there is no significant effect of height on the proportion of juvenile wood until the living crown, which confirms the fact that the distributions of MOE/D are similar for the different levels. The influence of age is important between 30 to 50 years old, but decreases for older trees. This observation is also in agreement with the observed distributions (figure 9).

Aknowledgments: the Data base was established with the collaboration of CTBA, and the financial support of the MRE.

7 References

Délisée, C (1987) Variabilité des propriétés mécaniques du Pin maritime- effet d'échelle. **Thèse de l'Université de Bordeaux1**, n°90.

El Ouadrani, A. and Castéra, P. (1992) Variation des caractéristiques physiques et mécaniques du bois de Pin maritime (Pinus pinaster Ait.). **IUFRO All-division 5 conference,** Nancy, August 23-28.

Fonseca, F.M.A, Louzada J.L. and Silva, M.E. (1992) Correlations between density components of juvenile and adult wood on Pinus pinaster Ait. **IUFRO All-division 5 conference,** Nancy, August 23-28.

Guyon, G. (1987) Prévisions de la rupture différée du Pin maritime en flexion. **Thèse de l'Université de Bordeaux1**, n°133,.

Madsen, B (1992) Structural behaviour of timber. Timber Engineering Ltd, North Vancouver, Canada.

Megraw, R.A. (1985) Wood quality factors in Loblolly Pine. TAPPI press, Atlanta, USA.

Pellicane, P.J. (1985) Goodness-of-fit analysis for lumber data. **Wood Sci. Technol.,** 19, 117-129.

Pellicane, P.J. (1993) Application of the S_B distribution to the prediction of concomitent wood properties. **Wood Sci. Technol.,** 27, 161-172.

Polge, H. (1978) Fifteen years of wood radiation densitometry. **Wood Sci. Technol.,** 12, 187-196.

Tang, Y. and Pearson, R.G. (1992) Effect of juvenile wood and choice of parametric property distributions on reliability-based beam design. **Wood and Fiber Science,** 24(2), 216-224.

Weibull, W. (1939) A statistical theory of the strength of materials. Ingeniörs Vetenskaps Akademien, Handl., n°151.

STOCHASTIC MODELING OF LUMBER PROPERTIES

DAVID ROSOWSKY
The Johns Hopkins University, Baltimore, MD, USA

Abstract
This paper will present the most recent developments in the statistical modeling of the mechanical properties of structural lumber. In addition to providing a summary of testing programs and model development (primarily in North America), two advanced stochastic modeling approaches will be described. The first is the generation of vectors of correlated lumber properties and the second is the stochastic modeling of the creep mechanism in wood. These recent probabilistic approaches to the modeling of lumber properties have been employed in reliability studies which have formed, and continue to form, the basis of the new load and resistance factor design specification for wood construction in the United States. Similar work is being conducted in Canada, Europe, Australia and New Zealand.
Keywords: Creep, Mechanical Properties, MOR, MOE, Stochastic Models, Structural Lumber, Wood

1 Introduction

The modeling of structural wood as an uncertain material has received considerable attention in North America and Europe in recent years, largely as a result of trends toward probability-based limit states design codes for engineered wood construction. In the past decade, work has focused on the probabilistic modeling of resistance properties, such as strength and stiffness, as well as possible member property correlation. More recently, models have been developed for the viscoelastic behavior of lumber as well as stochastic cumulative damage arising from sustained loading.

Wood is a unique material in that its strength is time-dependent (i.e., dependent on the **rate** and the **duration** of the applied load). Furthermore, as a viscous material, its deformation behavior is a function of time-dependent creep. This poses a unique set of circumstances for a structural reliability analysis in that complete time-dependent characterizations of loads and resistances are required to evaluate performance or safety.

119

D. Breysse (ed.), Probabilities and Materials, 119–129.
© 1994 Kluwer Academic Publishers.

2 Resistance Modeling

A number of experimental programs have been conducted to collect information on the flexural, compressive, and tensile strengths of wood. These tests are generally performed under uniform environmental conditions and using standardized loading methods. Most tests to failure are considered "short-term" tests in which the beams are typically loaded to failure over a period of approximately 5-10 minutes. Other tests have been conducted for longer periods, with some long-term creep (deflection) studies being on the order of 10 years or more.

Some of the earliest statistical models for strength were developed in the U.S. for glued-laminated ("glulam") beams. These members were among the first structural wood elements to be "engineered" and among the first types of members to be considered in duration-of-load (or cumulative damage) studies. Marx and Moody reported on the flexural strength of 103 beams [Marx and Moody, 1981; Moody, 1977] specifically fabricated to represent beams with very low load-carrying capacity, and therefore representing a biased sample set. The modulus of rupture (MOR) results were adjusted to standard conditions of 12 inch (30.5 cm) depth, 12% moisture content, uniform load, and a span-to-depth ratio of 21:1. The data (normalized by the allowable stress) were then fit to a 3-parameter Weibull distribution [Ellingwood et al., 1988]. In a second analysis, an unbiased set of 56 Douglas-fir glulam beams [Sexsmith and Fox, 1978] adjusted to the same standard conditions as the Marx and Moody data was fit using a 2-parameter Weibull distribution [Ellingwood et al., 1988]. Figure 1 shows an example of a data set of 30 glulam beams (AITC designation 24F-V4) adjusted to the above mentioned standard conditions and fit using a 2-parameter Weibull distribution. A 2-P Weibull distribution was also shown to fit the compression parallel-to-grain and tension parallel-to-grain strengths of Douglas-fir glulam beams [Knab and Moody, 1978; Peterson, 1978]. As would be expected, and as has since been shown, the 2-P Weibull distribution is also appropriate for other species as well as for dimension lumber. Statistics describing the modulus of rupture (MOR) and the modulus of elasticity (MOE) for different limit states have been developed and are reported in the literature [i.e., Bodig and Jayne, 1982; Ellingwood, 1981; Forest Products Laboratory, 1987; Gerhards, 1979; Green and Evans, 1987; Littleford, 1978].

Strength and stiffness variability as well as correlation of member properties may be attributed to wood species, grading, size, mill, and environmental conditions during storage and on-site. The coefficient of variation (COV) in the short-term strength (e.g.) accounts for such variability. Correlation **among** member properties may arise from a number of sources [Cramer and Wolfe, 1989; Littleford, 1978; Zahn, 1970]. For example, if an assembly is constructed of lumber of similar grade and species, taken from the same flat, originating at the same mill, and graded using the same technique and personnel, the lumber properties are likely to be correlated. Alternatively, if the lumber comes from a number of different sources, and the grade, species, or origin of the wood is mixed, the strengths may be essentially uncorrelated. If this member-to-member (i.e., for one resistance quantity) correlation were significant, it **could** have the effect of reducing the apparent variability (i.e., COV) in the data. Such correlation may also arise from lumber being stored and transported in the same manner. The degree of this correlation, as well as its effect on reliability of wood systems, has been investigated by Rosowsky and Ellingwood (1990) who conclude that, for most applications, this type of correlation can be neglected.

Correlation **between** member properties refers to the possible correlation of one property (i.e., MOR) on another property (i.e., MOE) for a single member. This correlation has significant implications for the performance and safety of wood assemblies. In a statically indeterminate structure in which load-sharing occurs, stiffer members carry a greater share of the load. The wood industry's current concept of a load-sharing system is a wood floor or roof assembly in which the stronger members carry a larger share of the load. This reflects the understanding that a strong positive correlation exists between strength and stiffness in wood members. A number of different studies have reported correlation coefficients between the flexural MOR and MOE on the order of 0.6-0.9 [Bodig and Jayne, 1982; Curry and Tory, 1976; Ellingwood, 1981; Zahn, 1970], with a consensus value of 0.7 being used most recently in reliability studies which form the basis for new code provisions [Foschi et al., 1989; Rosowsky and Ellingwood, 1990]. A trend toward reduced correlation with increased moisture content has also been observed. Additional studies have suggested similar correlations between tension MOR and MOE and compression MOR and MOE [Curry and Fewell, 1977].

3 Generating Vectors of Correlated Random Lumber Properties

A method for generating correlated random variables is described herein. Specifically, the procedure described is used to simulate sets of correlated MOR and MOE values. As indicated, the procedure may be generalized to vectors of random variables (i.e., $n > 2$). This may be useful for generating correlated sets of lumber specimens, for example.

A procedure for generating correlated random variables (i.e., for use in Monte Carlo simulation) is described by Melchers (1987). The approach to generating correlated non-normal random variables proceeds in two steps: (1) generate appropriately correlated standard normal variates, and (2) transform the standard normal variates into equivalent (correlated) non-normal random variables. A Choleski decomposition scheme is used to generate a vector of correlated standard normal variates, and Rosenblatt transformations are used to transform the normal variates into the corresponding non-normal quantities. The correlation structure in the initial standard normal variates is corrected empirically to account for the transformation, thereby assuring the correct correlation structure in the final, non-normal variables.

The first step is to generate a vector of correlated standard normal variates $\underline{X}$ with a known covariance matrix, $\underline{C}_X$. This is done by premultiplying a vector $\underline{Y}$ of n independent standard normal variates (with associated covariance matrix $\underline{C}_Y$) by the transpose of a transformation matrix $\underline{A}$. This transformation should be orthogonal in order to keep the vector $\underline{Y}$ unchanged under the transformation, $\underline{X} = \underline{A}^T \underline{Y}$. For $\underline{A}$ to be orthogonal, $\underline{A}^T \underline{A} = \underline{I}$ and $\underline{A}^T = \underline{A}^{-1}$.

The transformation matrix can be obtained by considering the covariance matrix $\underline{C}_Y$ such that $\underline{C}_X=\underline{A}^T\underline{C}_Y\underline{A}$. Since the variables in $\underline{Y}$ are independent, the off-diagonal terms of $\underline{C}_Y$ are zero. Further, since the variables in $\underline{Y}$ are standard normal, the standard deviation terms appearing on the diagonal of $\underline{C}_Y$ are unity. Thus, the covariance matrix $\underline{C}_Y$ is by definition the identity matrix $\underline{I}$, and therefore $\underline{C}_X=\underline{A}^T\underline{A}$. By definition, a matrix is symmetric if and only if the matrix transpose is the same as the matrix itself. A symmetric matrix such as $\underline{C}_X$ can be decomposed into a product of a lower triangular matrix $\underline{B}$ and its transpose, using Choleski decomposition: $\underline{C}_X=\underline{B}\underline{B}^T=\underline{A}^T\underline{A}$. Therefore, replacing $\underline{A}^T$ with $\underline{B}$, the original transformation becomes $\underline{X}=\underline{B}\underline{Y}$. The elements of the lower triangular matrix $\underline{B}$ are written using a recursive formula:

$$b_{ij} = \frac{c_{ij} - \sum_{k=1}^{j-1} b_{ik}b_{jk}}{\sqrt{c_{jj} - \sum_{k=1}^{j-1} b_{jk}^2}} \qquad [1]$$

where $\sum_{k=1}^{0}$ is defined as zero. The transformation matrix $\underline{B}$ for the case of two standard normal variates therefore simply becomes,

$$b_{ij} = \begin{bmatrix} b_{11} & 0 \\ b_{21} & b_{22} \end{bmatrix} = \begin{bmatrix} 1 & 0 \\ \rho_{o,12} & \sqrt{1-\rho_{o,12}^2} \end{bmatrix} \qquad [2]$$

where $\rho_{o,12}$ is the correlation coefficient between the standard normals. Therefore, the two correlated standard normal variates can be computed:

$$x_1 = y_1$$
$$x_2 = y_1\rho_{o,12} + y_2\sqrt{1-\rho_{o,12}^2} \qquad [3]$$

Since the transformation described is a mapping onto normal space, the resulting correlation coefficient $\rho_{o,ij}$ is not exactly the desired correlation coefficient ρ_{ij} between the non-normal random variables (i.e., the MOR and MOE). A semi-empirical correction factor, $F=\rho_{o,ij}/\rho_{ij}$, can be used in the transformation to modify the correlation coefficient in the transformation matrix $\underline{B}$. This factor has been developed for various combinations of probability distributions by Der Kiureghian and Liu (1986) and has been shown to be a function of the desired correlation coefficient ρ_{ij} and the moments of the distributions, and is invariant to linear transformations of the random variables. In the case of an assumed MOE distributed lognormally and MOR distributed Weibull (e.g.), the correction factor becomes:

$$F = 1.031 + 0.052\rho_{ij} + 0.011V_E - 0.21V_R + 0.002\rho_{ij}^2 + 0.22V_E^2 +$$
$$0.35V_R^2 + 0.005\rho_{ij}V_E - 0.174\rho_{ij}V_R + 0.009V_EV_R$$

[4]

where V_E is the coefficient of variation in the MOE and V_R is the coefficient of variation in the MOR. Therefore, the value of the correlation coefficient in matrix $\underline{B}$ is computed as $\rho_{o,ij}=F\rho_{ij}$.

Finally, inverse transformations are used to obtain values of the (correlated) random variables. If the correlated standard normal variates are defined as x_1 and x_2 (for the case of correlating pairs of random variables) then the equivalent correlated lognormal and Weibull random variables, E and R respectively, are given by:

$$E = \exp(\xi x_1 + \lambda)$$
$$R = [-\ln(1 - \Phi(x_2))]^{1/k}\alpha$$

[5]

where ξ and λ are the parameters of the lognormal distribution, k and α are the parameters of the Weibull distribution, and $\Phi(\cdot)$ is the standard normal cumulative distribution function.

The effect of correlation on generated pairs of random variables (in this case, values of MOR and MOE) is shown in Figure 2. Here, 200 pairs of random variables corresponding to the MOR and a normalized (i.e., relative) version of the MOE are generated and plotted for values of the correlation coefficient of 0.0, 0.25, 0.5, and 0.95. The specific case shown is for a glulam member in which the mean MOR (in ratio to the nominal design value) is assumed to be 3.17 with a COV of 0.17, and the mean normalized MOE is assumed to be 1.0 with a COV of 0.20. This figure illustrates the effect of increasing correlation while maintaining the relative variability in the two random quantities. Two phenomena are observed with this method. First, if one random variable's uncertainty is dominant (i.e., the COV of one is larger than the other), the choice of distribution for the variable with the lower uncertainty becomes less significant. Second, the procedure fails if the uncertainty in one random variable is very small (i.e., negligible) relative to the other variable. In this extreme case, as one variable becomes essentially more deterministic, the effect of varying the correlation coefficient becomes smaller and a situation of "competing" statistics (i.e., COV and ρ) arises. For the application described herein, however, the COV's are on the same order of magnitude (V_R = 0.20-0.40 typ., and V_E = 0.10-0.20 typ.) and the procedure is successful. Further discussion of this application, including verification of the resulting distributions and correlation structure, is presented in [Rosowsky and Ellingwood, 1990].

4 Modeling Stochastic Creep Behavior

The strength of wood is dependent on the rate and duration of the applied loading. In addition, the strain response of wood increases as the rate of applied stress decreases.

Wood also displays the phenomenon of **creep** in which strains increase while the applied stress is held at a constant value. As a material which exhibits this type of relationship between stress, strain, and time, wood may be considered a viscoelastic material.

Mechanical analog models are convenient representations of the differential equations describing viscoelastic creep behavior. Such models are generally comprised of Maxwell elements (a spring and Newtonian dashpot in series) and Kelvin elements (a spring and Newtonian dashpot in parallel). The Maxwell element is capable of representing initial elastic deformations and time-dependent steady state (or secondary) creep deformation, stress relaxation, and recovery of elastic strain upon removal of the applied stress. The Kelvin element is capable of representing decelerating (or primary) creep. Together, these elements may be used to represent the behavior of a member including creep effects. The simplest model which includes all observed characteristics of viscoelastic materials is the serial combination of a Maxwell element and a Kelvin element. This four-element model, known as the Burger model and shown in Figure 3, is capable of representing elastic deformation, primary and secondary creep, zero-stress creep recovery, and stress relaxation. This type of model has been used in many studies to represent the viscoelastic behavior of wood elements [Hoyle et al., 1985; Fridley, 1992; Fridley at al., 1992; Senft and Suddarth, 1971].

While some work has been done to develop viscoelastic models with stress-dependent parameters [Mukudai, 1983a,b], most models developed for use with wood have stress-independent parameters [Hoyle et al., 1985, 1986; Senft and Suddarth, 1971]. Fridley et al. (1992) fit a stochastic four-element model to the data from tests of nominal 2 × 4 inch (5.08 × 10.16 cm) Douglas-fir lumber for different combinations of load, moisture content, and temperature. To account for hygrothermal effects on the creep behavior, quadratic interpolating functions were developed to describe the mean values for the parameters of each of the four model components as continuous functions of the temperature and moisture content of the wood [Fridley et al., 1992]. The influence of moisture content was found to be roughly an order of magnitude greater than that of temperature. The coefficients of variation of the creep model parameters were found to be essentially invariant with respect to hygrothermal conditions. The means and COV's for each of the parameters (elements) in the four-element model based on the work by Fridley et al. (1992) are shown in Table 1 for three different moisture content conditions. Although these parameters were developed from tests of specific grades of Douglas-fir lumber, current design practice in the United States assumes that **relative creep is comparable across grades and species,** and therefore these models may be assumed to be valid for other types of lumber.

The applications of these stochastic creep models in reliability studies and the development of probability-based design criteria are described in [Philpot and Rosowsky, 1992; Fridley and Rosowsky, 1992].

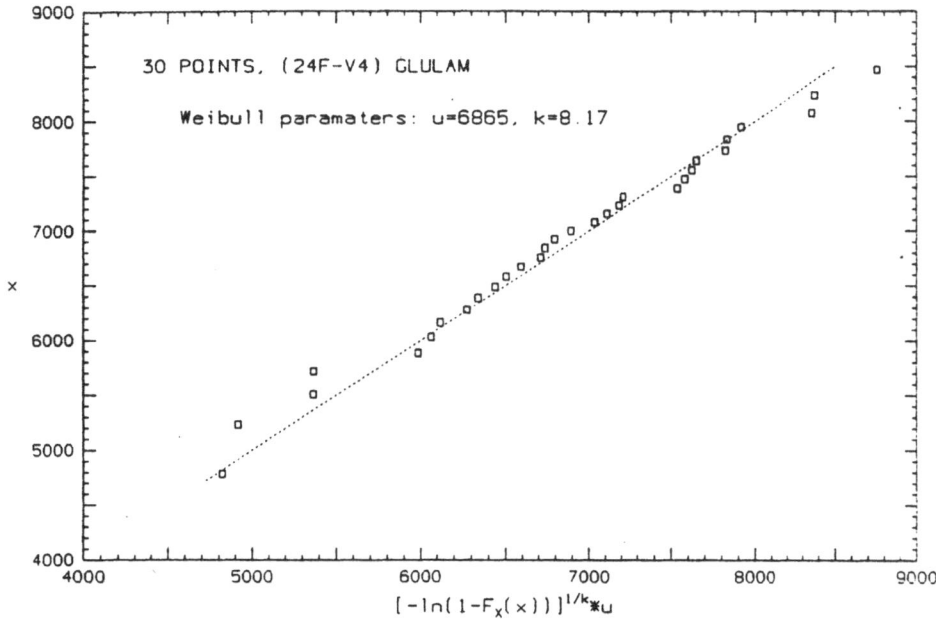

Figure 1: Fitting a 2-P Weibull distribution to flexural modulus of rupture (MOR)

126

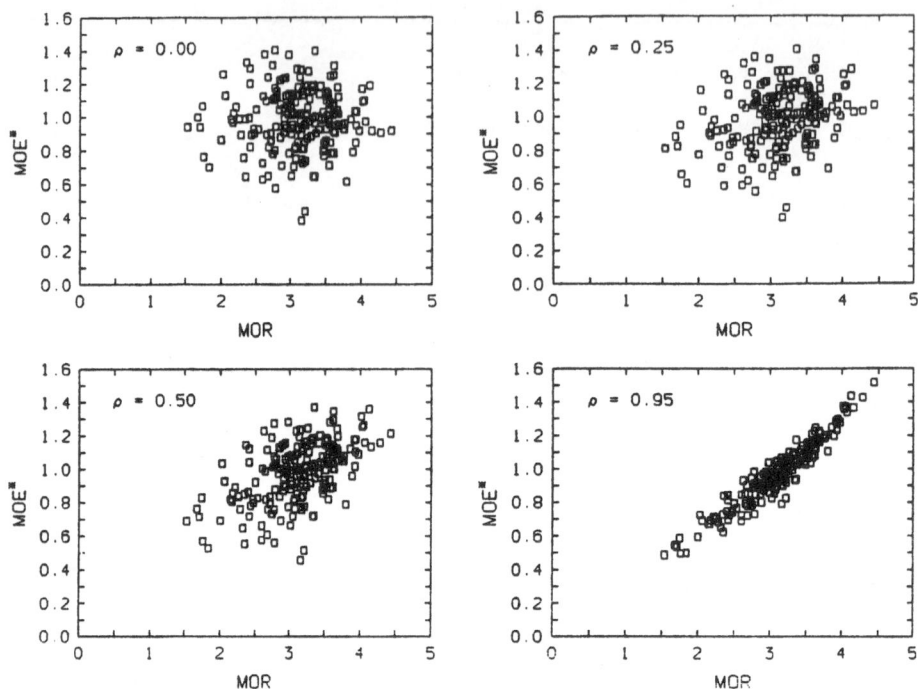

Figure 2: Correlation between MOR and normalized MOE (n=200)

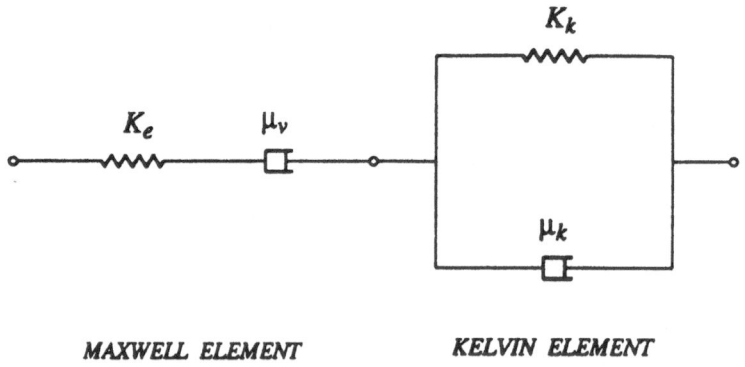

Figure 3: Four-element viscoelastic model

Table 1: Moments for creep model parameters at specific moisture
content (m.c.) conditions (from Fridley, 1992)

CREEP MODEL PARAMETER	15% m.c. (max.) mean	COV	19% m.c. (max.) mean	COV	28% m.c. (max.) mean	COV
K_e						
$(10^6$ psi)	1.68	0.16	1.56	0.16	1.31	0.16
$(10^3$ MPa)	11.58	0.16	10.76	0.16	9.03	0.16
K_k						
$(10^6$ psi)	2.77	0.41	2.17	0.41	1.63	0.41
$(10^3$ MPa)	19.10	0.41	14.96	0.41	11.24	0.41
μ_k						
$(10^{10}$ psi-min)	4.56	0.42	3.36	0.42	1.79	0.42
$(10^7$ MPa-min)	31.44	0.42	23.17	0.42	12.34	0.42
μ_v						
$(10^{14}$ psi-min)	5.56	0.39	4.44	0.39	3.24	0.39
$(10^{11}$ MPa-min)	38.34	0.39	30.61	0.39	22.34	0.39

5 Summary

A summary of recent testing programs and probabilistic model development related to the mechanical properties of structural lumber has been presented. These models have been employed in reliability studies to develop probability-based limit states design procedures for wood structures. As a naturally occurring material, wood properties are highly uncertain and exhibit high variability even within grade. Furthermore, as a material subject to the phenomenon of creep-rupture, these member properties may be dependent on the duration as well as the magnitude of the applied loading. Finally, issues of member strength-stiffness correlation are important in evaluating the in-time performance of multiple-member wood systems. As a consequence, robust probabilistic models for the mechanical properties of structural wood elements are required in order to assess structural performance and safety.

Two advanced stochastic modeling approaches for the mechanical properties of lumber have been described: the generation of correlated lumber properties (i.e., MOR-MOE) and the stochastic modeling of the creep response in wood flexural members. These approaches have been used in recent simulation-based reliability analyses of wood members and systems.

6 References:

Bodig, J. and Jayne, B.A. (1982) **Mechanics of Wood and Wood Composites**, Van Nostrand Reinhold Company, New York.

Cramer, S.M. and Wolfe, R.W. (1989) Load-Distribution Model for Light-Frame Wood Roof Assemblies, **Journal of Structural Engineering**, ASCE, 115(10):2603-2616.

Curry, W.T. and Fewell, A.R. (1977) The Relation Between the Ultimate Tension and Ultimate Compression Strength of Timber and its Modulus of Elasticity, Building Research Establishment, CP 22/77, Princes Risborough Laboratory, Buckinghamshire, England.

Curry, W.T. and Tory, J.R. (1976) The Relation Between the Modulus of Rupture (Ultimate Bending Stress) and the Modulus of Elasticity of Timber, Building Research Establishment, CP 30/76, Princes Risborough Laboratory, Buckinghamshire, England.

Der Kiureghian, A. and Liu, P. (1986) Structural Reliability Under Incomplete Probability Information, **Journal of Engineering Mechanics**, ASCE, 112(1):85-104.

Ellingwood, B. (1981) Reliability of Wood Structural Elements, **Journal of the Structural Division**, ASCE, 107(1):73-87.

Ellingwood, B.R., Hendrickson, E.M. and Murphy, J.F. (1988) Load Duration and Probability Based Design of Wood Structural Members, **Wood and Fiber Science**, 20(2):250-265.

Forest Products Laboratory (1987) **Wood Handbook: Wood as an Engineering Material**, United States Department of Agriculture, Handbook 72.

Foschi, R.O., Folz, B.R. and Yao, F.Z. (1989) Reliability-Based Design of Wood Structures, Structural Research Series Report No. 34, Department of Civil Engineering, University of British Columbia, Vancouver, Canada.

Fridley, K.J. (1992) Designing for Creep in Wood Structures, **Forest Products Journal**, Forest Products Research Society, 42(3):23-28.

Fridley, K.J. and Rosowsky, D.V. (1992) Time-Dependent Load Sharing in Parallel-Member Wood Systems, submitted to **Wood and Fiber Science** (in review).

Fridley, K.J., Tang, R.C. and Soltis, L.A. (1992) Creep Behavior Model for Structural Lumber, **Journal of Structural Engineering**, ASCE, 118(8):2261-2277.

Gerhards, C.C. (1979) Time-Related Effects of Loading on Wood Strength: A Linear Cumulative Damage Theory, **Wood Science**, 11(3):139-144.

Green, D.W. and Evans, J.W. (1987) Mechanical Properties of Visually Graded Lumber, Vols. 1-5, USDA, Forest Service, Forest Products Laboratory, Madison, WI.

Hoyle, R.J., Griffith, M.C. and Itani, R.Y. (1985) Primary Creep in Douglas-fir Beams of Commericial Size and Quality, **Wood and Fiber Science**, 17(3):300-314.

Hoyle, R.J., Itani, R.Y. and Eckard, J.J. (1986) Creep of Douglas-fir Beams Due to Cyclic Humidity Fluctuations, **Wood and Fiber Science**, 18(3):468-477.

Littleford, T.W. (1978) Flexural Properties of Dimension Lumber from Western Canada, Information Report VP-X-179, Western Forest Products Laboratory, Vancouver, British Columbia.

Knab, L.I. and Moody, R. (1978) Glulam Design Criteria for Temporary Structures, **Journal of the Structural Division**, ASCE, 104(9):1485-1494.

Marx, C. and Moody, R. (1981) Strength and Stiffness of Small Glued-Laminated Beams with Different Quantities of Tension Laminations, FPL Report 381, U.S. Forest Products Laboratory, Madison, WI.

Melchers, R.E. (1987) **Structural Reliability: Analysis and Prediction**, Ellis Horwood Limited, distributed by John Wiley and Sons, New York.

Moody, R. (1977) Improved Utilization of Lumber in Glued Laminated Beams, FPL Report 292, U.S. Forest Products Laboratory, Madison, WI.

Mukudai, J. (1983a) Evaluation of Linear and Nonlinear Viscoelastic Bending Loads of Wood as a Function of Prescribed Deflections, **Wood Science and Technology**, 17:203-216.

Mukudai, J. (1983b) Evaluation of Nonlinear Viscoelastic Bending Deflection of Wood, **Wood Science and Technology**, 17:39-54.

Peterson, J. (1978) Tensile Strength of L3 Douglas-fir Glulam Members, **Journal of the Structural Division**, ASCE, 104(1):1-8.

Philpot, T.A. and Rosowsky, D.V. (1992) Effect of Creep on the Performance and Serviceability Reliability of Wood. Part I: Single Members, Structural Engineering Report CE-STR-92-11, Purdue University, West Lafayette, IN.

Rosowsky, D.V. and Ellingwood, B.R. (1990) Stochastic Damage Accumulation and Probabilistic Codified Design for Wood, Civil Engineering Report no. 1990-02-02, The Johns Hopkins University, Baltimore, MD.

Senft, J.F. and Suddarth S.K. (1971) An Analysis of Creep-Inducing Stress in Sitka Spruce, **Wood and Fiber Science**, 2(4):321-327.

Sexsmith, R.G. and Fox, S. (1978) Limit States Design Concepts for Timber Engineering, **Forest Products Journal**, 28(5).

Zahn, J.J. (1970) Strength of Multiple-Member Systems, Research Paper FPL 139, Forest Products Laboratory, USDA, Madison, WI.

EXPERIMENTAL STUDY ON THE SCALE EFFECT ON CONCRETE IN TENSION

P. ROSSI, X. WU, F. LE MAOU, A. BELLOC
Laboratoire Central des Ponts et Chaussées, Paris, France. Geomaterials Joint Research Group.

Abstract
Scale effects in the mechanical behaviour of concrete are important phenomena and are the object of many studies internationally. The Laboratoire Central des Ponts et Chauss,es (LCPC) has been working for several years on the scale effect on the distribution function of the tensile strength of concrete. Very many direct tensile tests have been performed on specimens of different volumes, made of concretes having different formulations. From these experimental investigations has been derived a rather general law that can be used to determine the distribution function (mean and standard deviation) of the tensile strength of the concrete versus the volume of the tensile specimen, for concretes having compressive strengths between 30 and 130 MPa, from knowledge of just the compressive strength obtained on standardized cylinder (in France) and the ratio volume of tensile specimen/volume of coarsest grain of concrete. It should be emphasized that this law is valid only if, on the one hand, the mode of preservation of the direct tensile specimens is identical to that of the compressive specimens, and on the other the aggregates used in the concretes are common aggregates, which eliminates light aggregates and very hard aggregates with a high Young's modulus.
Keywords: Concrete, Scale effect, Tension, random distributions.

1 Introduction

For a few years, the LCPC has been developing a numerical model of the cracking of concrete based on a probabilistic approach (Rossi (1988), Rossi and al. (1987 and 1988)). The main datum that must be known and entered into this model is the distribution function of the tensile strength, which depends on the size of the finite elements of the mesh used. Since, in the model, it is assumed that there is equivalence between the finite element of the mesh and a volume of material, the distribution function of the tensile strength, which depends on the size of the finite elements, can be obtained by performing a large number of direct tensile tests on specimens of different volumes. Although there has been some international research on problems of scale effect in the cracking of concrete (Bazant (1991), Nooru and al. (1991)), there has

D. Breysse (ed.), Probabilities and Materials, 131–139.
© 1994 *Kluwer Academic Publishers.*

not, to my knowledge, been much work on the scale effect on the direct tensile strength of concrete. A first study was performed on this subject at the LCPC in 1985 but published only as an internal report. It consisted of direct tensile testing of cast cylindrical specimens of different diameters and a slenderness ratio of two. Although highly instructive, this study had two main limitations:

The cast specimens, of very small diameter, had wall effect problems (grains too coarse with respect to the diameter of the specimens) that completely perturbed the results, whereas with the specimens of larger diameter there were scale effects that were not sufficiently marked, although significant.

Only one composition of concrete was studied, which made it impossible to draw general conclusions concerning these scale effects.

In 1991, it was therefore decided to conduct a campaign of tests that differed in the following ways from the previous study:

The specimens were cored, not cast.
Three very different formulations of concrete were studied.

It is the results of these two research campaigns that I am going to present.

2. Experimental investigation of the direct tensile behaviour of concrete

2.1 First study

2.1.1 Experimental conditions
In this study, five diameters of cylindrical specimens of slenderness ratio two were tested: 20, 36, 89, 113, and 160 mm. As I have already reported, the specimens were cast. The composition of the concrete studied is given in table 1.

Table 1. Composition of the concrete of the first study, performed in 1985 (per m^3)

Constituents	Weight in kg
Rolled silico-calcareous gravel, 4/12 mm	1100
Rolled silico-calcareous sand, 0/5 mm	700
Cement	350
Water	200

After form removal, the specimens were coated with several layers of resin to prevent drying and so achieve a homogeneous internal moisture content.

The tests were performed between 28 and 40 days after the production of the specimens, on a Zwick tensile press having a capacity of 10 tonnes.

The tests were controlled in loading rate: 0.5 MPa/s.

Steel slabs, of the same diameter as the specimens, were
glued to the specimens, then attached to the tensile press.
It must be reported that, concurrently with the direct tensile tests, compressive tests
were performed on standardized cylindrical specimens (French standard) 160 mm in
diameter and 320 mm high.

The compressive tests were performed on the same days as the
direct tensile tests, and under the same conditions of preservation, i.e. with the
specimens coated with several layers of resin.

2.1.2 Experimental results

Approximately 50 tensile strength values per specimen diameter were kept. Tests in
which there was a failure in the glue or near the ends of the specimens were eliminated.

As for the small-diameter specimens (diameters 20 and 36 mm), it was very quickly
discovered, from the results and from the examination of the fracture facies, that the
structure of the material was perturbed by wall effects during the casting of the
specimens: laitance and bubbling at the surface of the specimens and a large gradient of
fines concentration (low at the "core" of the specimen and large at the perimeter). It
was therefore decided to analyze only the results for the specimens 89, 113, and 160
mm in diameter.

Table 2. Groups the results obtained during this study

Table 2. Results for the tensile tests of the first study

Specimen diameter (mm)	89.0	113.0	160.0
Number of values	46.0	54.0	49.0
Mean tensile strength (MPa)	3.0	2.8	2.5
Standard deviation (MPa)	0.3	0.3	0.2
Mean compressive strength (MPa)		47.5	

It can be seen, in table 2, that there in fact exists a scale effect, but that it is relatively
small: increase of the mean strength, and of the deviation, with the volume of material
loaded.

Statistical tests were performed - Henry's function and the 2, to check that the
experimental distributions obtained follow known theoretical laws. It was found that
these experimental distributions in particular followed two types of laws: the truncated
normal law (no negative values), and Weibull's law. These are laws that are commonly
encountered in investigations of the probabilistic aspects of the failure of
heterogeneous and fragile materials.

3. Second study

A few years later, for the reasons mentioned above, we performed a second experimental investigation on scale effects in direct tensile testing.

3.1 Experimental conditions

In this study, three diameters of specimen were used: 30, 60, and 150 mm. Since the first study had shown that the scale effect is small when the specimen diameters range from 90 to 160 mm, it was decided to concentrate the tests on smaller diameters.

The specimens were cored from large rectangular blocks of concrete having the following dimensions: 102 x 92 x 56 cm. Three types of concrete were tested; their compositions are given in tables 3, 4, and 5. We shall call these concretes concrete 1, concrete 2, and concrete 3, respectively.

All specimens were protected from desiccation by packing, after form removal, in a sheet of plastic paper followed by two layers of self-adhesive aluminium paper (method developed by the LCPC (Attolou and al. (1989)). This wrapping was kept during the test.

The tests were performed at an imposed loading rate of 0.5 MPa/s on a 10-tonne Tinus Olsen press.

Steel helmets were glued to both ends of the specimens, and other steel helmets, fitted with ball-jointed tension rods, were bolted to these helmets. The whole set-up was then placed in the press for the test.

In these tests, the displacement measurement, used for the determination of the Young's modulus, was made using a special extensometer, designed by C. Boulay and al. (1981), that is placed directly on the specimen. This extensometer consists of two aluminium alloy rings, placed at the ends of the specimen (at some distance from the edges to eliminate problems of lateral confinement) using setscrews mounted on elastic blades; these rings constitute the measurement base of three LVDTs placed at 120ø round the larger ring.

Table 3. Composition of concrete 1, per m3.

Constituents	Weight in kg
Crushed silico-calcareous gravel, 5/20 mm	1114
Silico-calcareous sand, 0/4 mm	774
Cement	300
Water	185

Tableau 4. Composition of concrete 2, per m^3

Constituent	Weight in kg
Crushed silico-calcareous gravel, 5/20 mm	1236
Silico-calcareous sand, 0/5 mm	667
Cement	350
Water	158
Superplasticizer	7

Table 5. Composition of concrete 3, per m^3.

Constituent	Weight in kg
Calcareous Boulonnais gravel, 12.5/20 mm	854.0
Calcareous Boulonnais gravel, 4/12.5 mm	411.0
Boulonnais sand, 0/4 mm	326.0
Seine sand	421.0
Cement	42.0
Silica fume from Laudun	112.0
Lomard D superplasticizer	7.6

Concurrently with these direct tensile tests, standardized compressive tests, mentioned above, were performed on the three concretes, at the same age as the tensile tests. The conditions of preservation of the compressive specimens were the same as those of the tensile specimens. The Young's modulus was also determined during these compressive tests, using the same type of extensometer as for the tensile tests.

It should be noted that three compressive tests were performed per type of concrete.

3.2 Experimental results

Since the first tensile study on a given concrete, presented previously, had shown that the distribution function of the tensile strengths followed either a normal law or a Weibull type law, we hypothesized that this is true of all concretes. This hypothesis allowed us, using some statistical laws (based on hypotheses concèrning the values of the means and standard deviations), to reduce the number of specimens to be tested while preserving statistical representativeness. Naturally, the number of specimens depends on the volume of each of them (since the mean and standard deviation depend on this volume). It was thus decided that the minimum number of specimens for each volume would be distributed as follows:

$\emptyset$ 30 mm ----> 12 specimens
$\emptyset$ 60 mm ----> 7 specimens
$\emptyset$ 150 mm ----> 7 specimens

Tables 6, 7, and 8 give the results for concretes 1, 2, and 3, respectively.
In the light of these results, we can make the following observations:

It is found that for the three concretes there exists a scale effect on the direct tensile strength, and that the lower the compressive strength of the concrete the larger this scale effect. If it is assumed that this scale effect is related to the heterogeneity of the material, which is what we think, this result is readily explained by the fact that, the higher the compressive strength, the more homogeneous the concrete (the mechanical characteristics of the matrix become closer to those of the grains). This is found again in the coefficients of dispersion of the tensile strength and Young's modulus, which increase as the volume of the specimen becomes smaller and as the compressive strength of the concrete decreases.

It is also observed that there exists a very slight scale effect on the Young's modulus of the concrete, but one that acts in the reverse direction with respect to the effect on tensile strength. It is thus found that the Young's modulus increases as the specimens become larger. This result may seem astonishing, but can be explained, in our opinion, as follows: because of the different shrinkages existing within the concrete, it contains self-equalizing initial stresses that contribute to "self-prestressing" the material (Acker(acker and al. (1987), Wittmann (1983)). Now, when such a medium is cored, a part of these initial stresses are released in the vicinity of the new surfaces generated by the core drilling. To the extent that, for a given concrete, this zone affected by the core drilling is constant (it is related only to the initial stress field, which depends on the concrete), the smaller the specimen, the relatively larger its small mechanical effect (zone of decreased rigidity) on the whole of the specimen. This is what is observed experimentally. In consequence, we think that the mean value of the Young's modulus is independent of the volume of the specimen, but that on the other hand the coefficient of dispersion of this Young's modulus increases as the volume of material decreases and as the strength of the concrete
increases, as indicated in tables 6, 7, and 8. Finally, we may note that for rather similar specimen dimensions (O = 16, h = 32 mm, and O = 15, h = 30 mm) the Young's modulus in compression is equal to the Young's modulus in tension for the three concretes.

4 Conclusions

We have presented an experimental investigation concerning the scale effect on the tensile strength of concrete and its Young's modulus. The main findings are the following:

i. This scale effect is highly significant (increase of the mean value, and of the

coefficient of dispersion, as the volume of material decreases), and the lower the compressive strength of the concrete, the larger it is.

ii. The Young's modulus seems not to depend on the volume of concrete, but on the other hand the coefficient of dispersion on this modulus decreases as the volume and the compressive strength of the concrete increase.

iii. The conclusions do not concern the concretes in which the grains are weaker than the cement paste (lightweight concretes, and some high-strength concretes).

Table 6. Results for concrete 1

Specimen diameter (mm)	30.00	60.00	150.00
Number of values	15.00	15.00	16.00
Mean value of tensile strength R_t (MPa)	4.80	3.20	2.40
Standard deviation on R_t	1.00	0.60	0.20
*Coefficient of dispersion of R_t	0.21	0.18	0.08
Mean value of Young's modulus in tension E_t (GPa)	35.60	39.00	39.80
Standard deviation on E_t (GPa)	3.30	1.80	0.80
*Coefficient of dispersion of E_t	0.09	0.05	0.02
Mean value of compressive strength (MPa)		35.00	
Mean value of Young's modulus in compression E_c (GPa)		39.60	

* The coefficient of dispersion is the ratio of the standard deviation to the mean.

Table 7. Results for concrete 2.

Specimen diameter (mm)	30.00	60.00	150.00
Number of values	18.00	8.00	9.00
Mean value of tensile strength Rt (MPa)	5.10	4.30	3.30
Standard deviation on Rt (MPa)	1.00	0.70	0.20
Coefficient of dispersion of Rt	0.20	0.16	0.06
Mean value of Young's modulus in tension Et (GPa)	3.10	1.80	1.50
Coefficient of dispersion of Et	0.07	0.04	0.03
Mean value of compressive strength (MPa)		55.80	
Mean value of Young's modulus in compression Ec (GPa)		45.70	

138

Table 8. Results for concrete 3

Specimen diameter (mm)	30.00	60.00	150.00
Number of values	12.00	17.00	7.00
Mean value of tensile strength Rt (MPa)	6.40	6.00	6.00
Standard deviation on Rt (MPa)	0.90	0.70	0.20
Coefficient of dispersion of Rt 0.	0.14	0.12	0.03
Mean value of Young's modulus in tension Et (GPa)	52.30	51.80	53.90
Standard deviation on Et (GPa)	2.70	1.60	1.10
Coefficient of dispersion of Et	0.05	0.03	0.02
Mean value of compressive strength (MPa)		127.50	
Mean value of Young's modulus in compression Ec (GPa)		55.40	

5 References

Acker, P., Boulay, C.and Rossi P. (1987) On the importance of initial stresses in concrete and of the resulting mechanical effects. **Cement and Concrete Research,** vol. 17, pp. 755-764.

Attolou, A., Belloc, A. and Torrenti, J.M. (1989) Methodology for a new protection of concrete against desiccation. Bulletin de Liaison des LPC, n° 164, pp. 85-86 (in French).

Bazant Z. P. (1991) Size effects on fracture and localization: Aperçu of recent advances and their extension to simultaneous fatigue and rate-sensitivity. **Conference on Fracture Process in Concrete Rock and Ceramics,** Noordwijk, the Netherlands, edited by J.G.M. Van Mier, published by E ë FN Spon, London, pp. 417-429.

Boulay, C. and Colson, A. (1981) A concrete extensometer eliminating the influence of transverse strains on the measurement of longitudinal strains. **Materials and structures,** vol. 14, n° 79, pp. 35-38 (in French).

Nooru-Mohamed, M.B. and Van Mier, J.G.M. (1991) Size effects in mixed mode fracture of concrete. **Conference on Fracture Process in Concrete Rock and Ceramics,** Noordwijk, the Netherlands, edited by J.G.M. Van Mier, published by E δ FN Spon, London, pp. 461-471.

Rossi P. (1988) Numerical modelling of the cracking of concrete using a non deterministic approach. **Workshop on Fracture toughness and fracture energy - test methods for concrete and rock,** Sendai, Japan, edited by H. Mihashi, H. Takahashi, and F.H. Wittmann, published by A.A. Balkema, Rotterdam, pp. 301-312.

Rossi, P. and Piau, J.M. (1988) The usefulness of statistical models to describe damage and fracture in concrete. **CNRS-NSF Workshop Strain Localization and Size Effect due to Cracking and Damage**, Cachan, France, edited by J. Mazars and Z.P.Bazant, published by Elsevier Applied Science, London, pp. 91-103.

Rossi, P. and Richer, S. (1987) Numerical modelling of concrete cracking based on a stochastic approach. **Materials and structures**, vol. 20, pp. 334-337.

Wittmann F.H. (1983) Structure of concrete with respect to crack formation. **Fracture Mechanics of Concrete**, edited by F.H. Wittmann, published by Elsevier Sc. Publ., pp. 43-74.

FRACTALS, RENORMALIZATION GROUP THEORY AND SCALING LAWS FOR STRENGTH AND TOUGHNESS OF DISORDERED MATERIALS

Alberto CARPINTERI and Bernardino CHIAIA
Department of Structural Engineering, Politecnico di Torino, 10129 Torino, Italy

Abstract
The problems of the size effects on tensile strength and fracture energy of brittle and disordered materials (concrete, rocks, ceramics, etc) are reconsidered under a new and unifying light cast by Fractal Geometry. In this way we can define new tensile properties, with physical dimensions depending on the fractal dimension of the damaged microstructure, which turn out to be scale-invariant material constants. This represents the so-called renormalization procedure, already proposed in the statistical physics of random processes. Variations in the fractal dimension of fracture surfaces produce variations in the physical dimension of toughness and not, as asserted by some authors, in the measure of toughness.
Keywords: Disorder, Fractals, Renormalization Group, Tensile Strength, Fracture Energy.

1 Introduction

The remarkable observation that a characteristic feature of phase transitions is a discontinuous (catastrophic) change of the macroscopic parameters of the system undergoing a continuous variation in the system state variables, leads us to set the phenomenon of fracture of disordered materials into the wide framework of critical phenomena (Carpinteri 1989b, Herrmann & Roux 1990).

We shall focus on the two main quantities governing the fracture behavior of disordered materials: the ultimate tensile strength σ_u and the fracture energy $\mathcal{G}_F$. The occurrence of criticality in the phenomenon of rupture of solids is evident: both fracture energy and ultimate tensile strength may then be regarded as critical parameters. The former represents the energy required for the separation of a continuum into two parts and is equal to the critical value of the strain energy release rate: this separation is nothing else but a transition from a monolithic phase to a new one, which is characterized by a matter discontinuity. This kind of transition reveals the cohexistence of broken bonds with the sound ones and shows a remarkable physical instability (brittle fracture). On the other hand, the latter parameter, σ_u, is usually defined as the maximum tensile stress that the material can undergo.

A very strong relationship between critical phenomena and fractals can be easily recognized. The physicists have often observed a power-law dependence of the main physical quantities during the experiments on systems exhibiting a phase-transition (Pfeuty & Toulouse 1977). Stanley (1985) first pointed out the deep analogy involving non-integer (fractal) dimensions and scale-invariant phase-transitions, even recalling Renormalization Group Theory: «Since one can introduce fractal dimensions that play the role of critical exponents, and since relations among the dimensions play the role of

141

scaling laws, it is natural to ask: what about Renormalization Group? ».

It should be noted here that any power-law distribution is mathematically equivalent to a fractal distribution. Recalling in fact the basic definitions from Fractal Geometry (Falconer 1990), and more precisely the box-counting dimension, we can say that, ignoring, in a fractal set V, the "objects" of size smaller than δ, which means "measuring at scale δ", the measure of the set is then determined by the following power-law, as δ tends to zero:

$$M_\delta(V) = c\delta^{-\alpha},$$ (1)

where α is the so-called "box-counting dimension" and c is the measure of the set for $\delta = 1$.

Just like the perimeter of a fractal coastline scales with the yardstick length l according to a non-integer exponent α, $P \sim l^{(1-\alpha)}$, the exponents characterizing the power-law scaling of the critical quantities involved in a phase-transition are usually found to have non-integer values. This implies a deep analogy: the common inherent rule is *self-similarity* which is behind this non-standard scaling in both cases. In fact, the scale-invariance of a physical system at a critical point has been clearly recognized either by experimental or theoretical approaches. At the critical point the system shows large fluctuations (regions belonging to one of the two phases) with no characteristic size. Moreover, the hypothesis of scale-invariance implies the existence of a unique model describing the system, which is independent of length scale.

The above scale-invariance forms the basis of the Renormalization Group Theory, which has been successfully applied to the analysis of continuous phase-transitions, and has provided the calculations of the critical exponents. They confirmed to be non-integer numbers. If Fractals could represent the Geometry of disorder, Renormalization Group represents the physical backbone behind it: the two theories are thus so intimately related that we could assert that Fractals are the geometrical aspect of Renormalization Group Theory.

2 Fractal nature of material ligament and size effects on nominal tensile strength

It is well-known that the nominal tensile strength of many materials undergoes very clear size effects. The usual trend is that of a strength decrease with size, and this is more evident for disordered (i.e. macroscopically heterogeneous and/or damaged) materials. Griffith (1921) explained the strength size effect in the case of glass filaments, assuming the existence of inherent microcracks of a size proportional to the filament cross-sectional diameter. Some years later Weibull (1939) gave a purely statistical explanation to the same phenomenon according to the weakest-link-in-a-chain concept. Only recently have the two views been harmonized, enriching the empirical approach of Weibull with the phenomenological assumption of Griffith (Freudenthal 1968, Jayatilaka 1979, Carpinteri 1989a). A *statistical size distribution of self-similarity* may be defined, for which the most dangerous defect proves to be of a size proportional to the structural size (Carpinteri 1986, 1989a). This corresponds to materials presenting a considerable dispersion in the statistical microcrack size distribution (disordered materials). In this case, the power of the LEFM stress singularity, 1/2, turns out to be the slope of the strength versus size decrease in a bilogarithmic diagram. When the statistical dispersion is relatively low (ordered materials) the slope is less than 1/2 and tends to zero for regular distributions (perfectly ordered materials).

Although the above-described view contains the fractal concept of self-similarity, this is circumscribed only to the defect of maximum size, whereas the disordered nature of the material microstructure is completely disregarded. The real nature of the material will be herein described using a more complex fractal model, where the property of self-similarity is extended to the whole defect population. This model represents a more realistic picture of reality and is consistent with the fractal explanation of the fracture energy size effect, which will be proposed in the next section. On the other hand, as will be shown, slope values higher than 1/2 would represent, with both models, a degree of disorder that is so high as to be usually absent in real materials.

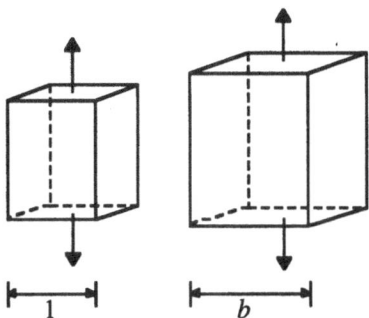

Fig. 1. Geometrically similar bodies loaded in tension.

Let us assume that the reacting section or ligament of a disordered material at peak stress could be represented as a fractal space of dimension $\alpha = 2 - d_\sigma$, with $1 < \alpha \leq 2$ and, therefore, $0 \leq d_\sigma < 1$. The dimensional decrement d_σ may be due to the presence of cracks and voids and hence, generally, to a cross-sectional weakening. Let us consider two bodies, geometrically similar and made up of the same disordered material (Fig. 1). If the ratio of geometrical similitude is equal to b and the *renormalized tensile strength* σ_u^* is assumed to be a material constant and to have the physical dimensions [force] x [length]$^{-(2 - d_\sigma)}$, we have

$$\sigma_u^* = \frac{F_1}{1^{2-d_\sigma}} = \frac{F_2}{b^{2-d_\sigma}},$$ (2)

F_1 and F_2 being the ultimate tensile forces acting on the two bodies respectively. On the other hand, the apparent nominal tensile strengths are respectively

$$\sigma_u^{(1)} = \frac{F_1}{1^2},$$ (3-a)

$$\sigma_u^{(2)} = \frac{F_2}{b^2},$$ (3-b)

where the latter, according to eq. (2), becomes

$$\sigma_u^{(2)} = \sigma_u^{(1)} b^{-d_\sigma}.$$ (4)

We can write the relationship between nominal strengths related to different sizes in logarithmic form

$$\ln \sigma_u = \ln \sigma_u(1) - d_\sigma \ln b. \tag{5}$$

Eq. (5) represents a straight line with slope $-d_\sigma$ in the $\ln \sigma_u$ versus $\ln b$ plane (Fig. 2-a).

An alternative way to explain the decrease of the nominal tensile strength with specimen size is that of considering a sequence of scales of observation. If the total force F transmitted to the specimen is invariant with respect to the scale of observation, we have:

$$F = \sigma_1 A_1 = \sigma_2 A_2 = ... = \sigma_{n-1} A_{n-1} = \sigma_n A_n = \sigma_{n+1} A_{n+1} = ... = \sigma_\infty A_\infty, \tag{6}$$

where the first scale of observation could be the macroscopic one, with $\sigma_1 A_1 = \sigma_u A$, A being the cross-sectional area, and the asymptotic scale of observation could be the microscopic one, with $\sigma_\infty A_\infty = \sigma_u^* A^*$, A^* being the measure of the fractal set representing the damaged ligament.

From the equality between the extreme members of eq. (6) we get

$$\sigma_u = \sigma_u^* \left(\frac{A^*}{A} \right), \tag{7}$$

and therefore

$$\sigma_u = \sigma_u^* \left(\frac{b^{2-d_\sigma}}{b^2} \right), \tag{8}$$

with b equal to the characteristic dimension of the cross section. From eq. (8) we can get a generalization of eq. (5)

$$\ln \sigma_u = \ln \sigma_u^* - d_\sigma \ln b. \tag{9}$$

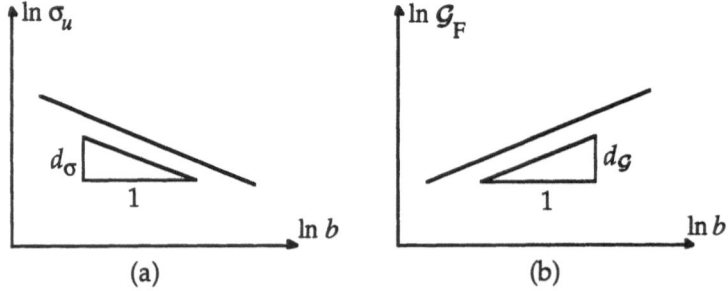

Fig. 2. Size effects on tensile strength (a) and fracture energy (b).

Renormalization group relations analogous to eq. (9) have been proposed by Wilson (1971, 1979) in quantum field theory and statistical mechanics, as well as by Barenblatt

(1979) in the intermediate asymptotics description of turbulence and blasting.

Confirmation of eq. (5) has been provided by several previous investigations, carried out on different metallic and cementitious materials, and with different specimen geometries. These results have been summarized by Carpinteri (1989a). The same trends have been found also in two recent experimental programmes aimed at evaluating the true tensile properties of concrete. A completely new testing apparatus made up of three orthogonally disposed actuators, was utilized at the Politecnico di Torino so that it was possible to perform a true direct tensile experiment on concrete, whereas usually tension and bending are both present. As may be seen from Fig. 3, the slope of the strength decrease proved to be equal to 0.14, thus revealing a material ligament of dimension 1.86, i.e., a fractal set which is very close to a two-dimensional surface. It may be noted that the specimen sizes explored in this investigation ranged over four values of the width: 5, 10, 20, and 40 cm. The fractal nature of the material ligament emerges very clearly at such scales. On the other hand, the property of self-similarity is very likely to vanish or change at higher or lower scales, owing to the limited character of the granulometric curve.

A higher slope of strength decrease was obtained in an experimental research study performed at ISMES-Bergamo, using pre-notched concrete cylinders of diameter 12, 18, 24 and 30 cm, respectively (Fig. 4). The exponent 0.41 is closer to the LEFM limit 0.5, which is valid for disordered materials (high statistical dispersion), and at the same time implies a very weak fractal ligament of dimension 1.59 at peak stress.

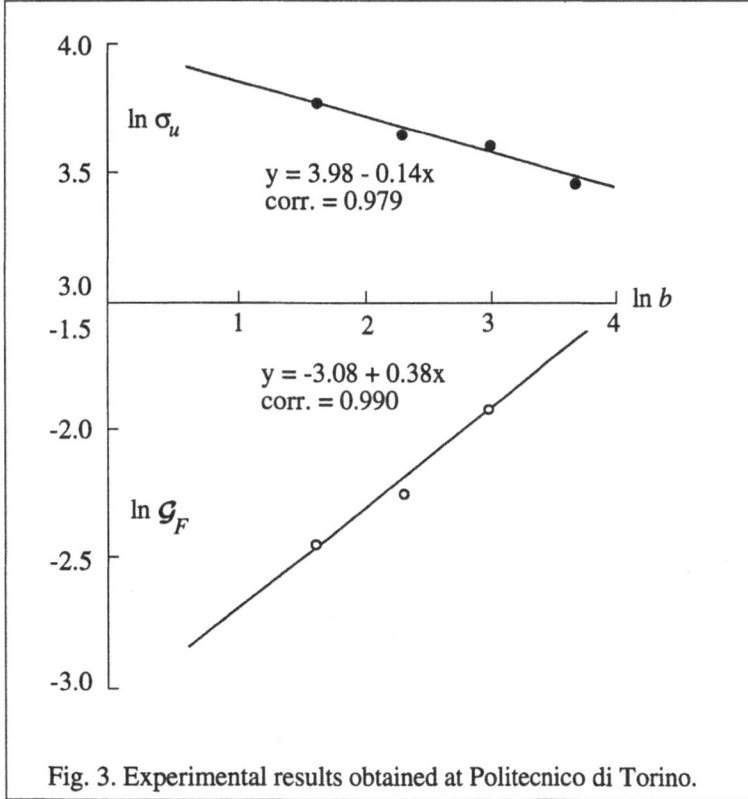

Fig. 3. Experimental results obtained at Politecnico di Torino.

146

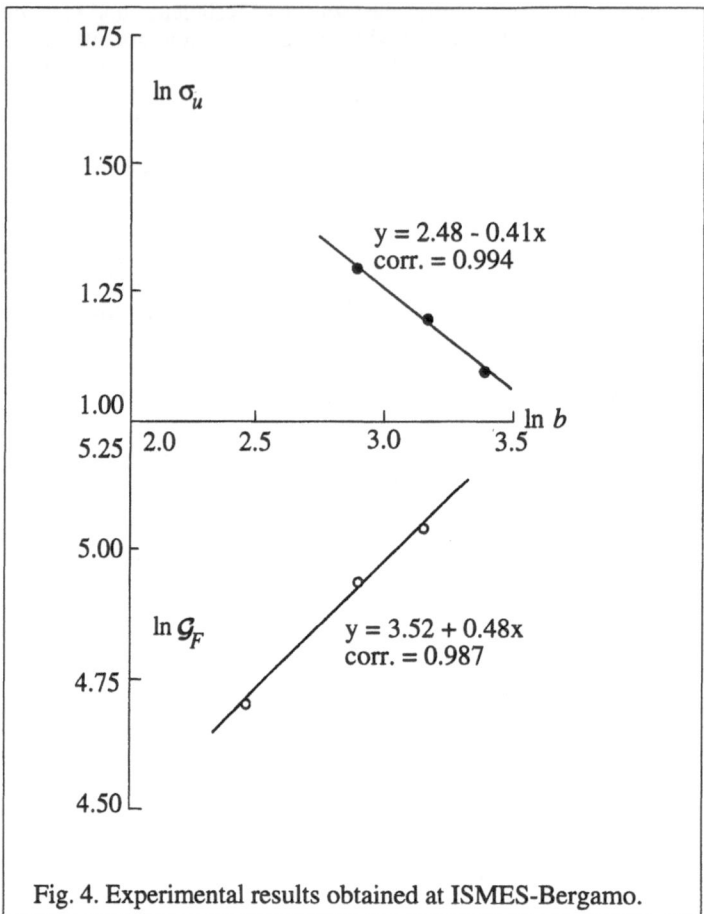

Fig. 4. Experimental results obtained at ISMES-Bergamo.

3 Fractal nature of fracture surface and size effects on fictitious fracture energy

It is well-known how the fracture surfaces of metals (Mandelbrot et al. 1984) and concrete (Saouma et al. 1990) present a fractal nature with a roughness producing a dimensional increment with respect to the number 2. Even in this case we can detect an evident mechanical consequence, considering the size effects on fracture energy $\mathcal{G}_F$. Since Hillerborg's proposal for a concrete fracture test was published as RILEM Recommendation (1985), several researchers have measured a fracture energy $\mathcal{G}_F$ which increases with the specimen sizes and, more specifically, with the size of the uncracked ligament. Such a trend has been systematically found, and in each case the authors of the papers describing these experiments have tried to provide various empirical or phenomenological explanations, without, however, endeavouring to interpret their findings in a larger conceptual framework.

On the other hand, if we wish to understand the experimental observations, it is necessary to abandon the classical thermodynamic concept of surface energy of an ideal solid, and to assume the energy dissipation to be occurring in a fractal space of dimension $\alpha = 2 + d_g$, with $2 \leq \alpha < 3$ and, therefore, $0 \leq d_g < 1$. This represents an attenuation of fracture localization due to material heterogeneity and multiple cracking.

Let us consider two bodies, geometrically similar and made up of the same disordered material (Fig. 1). If the ratio of geometrical similitude is equal to b and the *renormalized fracture energy* $\mathcal{G}_F^*$ is assumed to be a material constant and to have the physical dimensions [force] x [length]$^{-(1+d_\mathcal{G})}$, we obtain

$$\mathcal{G}_F^* = \frac{W_1}{1^{2+d_\mathcal{G}}} = \frac{W_2}{b^{2+d_\mathcal{G}}} , \tag{10}$$

W_1 and W_2 being the energies dissipated in the two bodies respectively.
On the other hand, the apparent fictitious fracture energies are respectively

$$\mathcal{G}_F^{(1)} = \frac{W_1}{1^2} , \tag{11-a}$$

$$\mathcal{G}_F^{(2)} = \frac{W_2}{b^2} , \tag{11-b}$$

where the latter, according to eq. (10), becomes

$$\mathcal{G}_F^{(2)} = \mathcal{G}_F^{(1)} b^{d_\mathcal{G}} . \tag{12}$$

We can write the relationship between fracture energies related to different sizes in logarithmic form

$$\ln \mathcal{G}_F = \ln \mathcal{G}_F(1) + d_\mathcal{G} \ln b . \tag{13}$$

Eq. (13) represents a straight line with slope $d_\mathcal{G}$ in the $\ln \mathcal{G}_F$ versus $\ln b$ plane (Fig. 2-b).

An alternative way of explaining the increase of the fracture energy with specimen size is that of considering a sequence of scales of observation. If the total energy W dissipated by fracture is invariant with respect to the scale of observation, we have

$$W = \mathcal{G}_1 A_1 = \mathcal{G}_2 A_2 = ... = \mathcal{G}_{n-1} A_{n-1} = \mathcal{G}_n A_n = \mathcal{G}_{n+1} A_{n+1} = ... = \mathcal{G}_\infty A_\infty , \tag{14}$$

where the first scale of observation could be the macroscopic one, with $\mathcal{G}_1 A_1 = \mathcal{G}_F A$, A being the cross-sectional area, and the asymptotic scale of observation could be the microscopic one, with $\mathcal{G}_\infty A_\infty = \mathcal{G}_F^* A^*$, A^* being the measure of the fractal set representing the irregular fracture surface.

From the equality between the extreme members of eq. (14) we get

$$\mathcal{G}_F = \mathcal{G}_F^* \left(\frac{A^*}{A} \right), \tag{15}$$

and therefore

$$\mathcal{G}_F = \mathcal{G}_F^* \left(\frac{b^{2+d_\mathcal{G}}}{b^2} \right), \tag{16}$$

with b equal to the characteristic dimension of the cross-section. From eq. (16) we can get a generalization of eq. (13)

$$\ln \mathcal{G}_F = \ln \mathcal{G}_F^* + d_{\mathcal{G}} \ln b. \tag{17}$$

On the other hand, if the pioneering work of Griffith (1921) is revisited considering a fractal crack of projected length $2a$, the fundamental relation of energy balance becomes

$$\frac{\sigma^2}{E} d\left(\pi a^2\right) = 2\mathcal{G}_F^* \, d\left(a^{1+d_{\mathcal{G}}}\right), \tag{18}$$

and therefore

$$\pi \frac{\sigma^2}{E} a \, da = \mathcal{G}_F^*\left(1 + d_{\mathcal{G}}\right) a^{d_{\mathcal{G}}} da. \tag{19}$$

From eq. (19) we get

$$\sigma^2 \pi \, a^{1-d_{\mathcal{G}}} = \left(1 + d_{\mathcal{G}}\right)\mathcal{G}_F^* E, \tag{20}$$

which represents the renormalized critical condition

$$\left(K_{\mathrm{I}}^*\right)^2 = \left(K_{\mathrm{IC}}^*\right)^2. \tag{21}$$

The generalized stress-intensity factor K_{I}^* presents the following physical dimensions:

$$[F][L]^{-\frac{3+d_{\mathcal{G}}}{2}}. \tag{22}$$

When $d_{\mathcal{G}} = 0$, we find again the classical relations of Griffith and Irwin. When $d_{\mathcal{G}} = 1$, as a limit case, we find that the stress singularity vanishes and K_{I}^* assumes the physical dimensions of stress. Similar dimensional transitions have been analysed by one of the authors for strain-hardening materials in the presence of cracks (Carpinteri 1983) and for linear elastic materials in the presence of re-entrant corners (Carpinteri 1987).

The same trends of eq. (13) have been found in recent experimental studies. Tensile testing performed at the Politecnico di Torino has provided a plot slope equal to 0.38 (Fig. 3), which allows a constant (universal) energy parameter to be obtained in the case where the dissipation is considered as occurring in a damaged space of dimension 2.38. Tensile testing at ISMES -Bergamo (Fig. 4) has provided a plot slope very close to the LEFM singularity 1/2, namely 0.48, and thus a fractal dimension 2.48, which implies a very disordered material. It is interesting to note that the range of self-similarity does not extend to the largest size for $\mathcal{G}_F$, just as it does not extend to the smallest size for σ_u. So the ranges of self-similarity for fracture energy and tensile strength do not necessarily coincide.

4 Correlated renormalization for strength and toughness

As regards the dimensional decrement d_σ, as well as the dimensional increment $d_{\mathcal{G}}$, previously defined for the tensile strength and the fracture energy respectively, experimentally they both appear always comprised in the interval [0, 1/2]. The dimensional decrement d_σ tends to the LEFM limit 1/2 only for extremely brittle and disordered materials, as also does the dimensional increment $d_{\mathcal{G}}$. The explanation for the latter bound could arise for Dimensional Analysis reasons.
A generalization of the *brittleness number* defined by one of the authors (Carpinteri 1982, 1985, 1989b) could in fact be the following:

$$s_E^* = \frac{\mathcal{G}_F^*}{\sigma_u^* \, b^{\left(1-d_\sigma-d_{\mathcal{G}}\right)}}. \tag{23}$$

If we postulate that the reversal of the physical roles of toughness and strength is absurd, the exponent of the characteristic linear size b must be positive

$$d_\sigma + d_{\mathcal{G}} < 1. \tag{24}$$

The sum of the dimensional decrement (for material ligament) and the dimensional increment (for fracture surface) must therefore be lower than unity. On the other hand, when for very disordered materials we have $d_\sigma \simeq 1/2$, the upper bound of eq. (24) becomes $d_{\mathcal{G}} < 1/2$.
The above fractal interpretations could be regarded by some as purely mathematical abstractions, if not indeed distortions of reality. The truth is that both classical geometry domains and fractal geometry loci are idealizations of reality. The question that should be answered is the following: which model is closer to a real fracture trajectory in a concrete specimen, a straight line or the von Koch curve? Of course the latter, even though the fractal nature of the fracture trajectory is random and valid only in a limited scale range. This means that, for size scales tending to infinity or, in other words, for very large specimens, tensile strength σ_u and fracture energy $\mathcal{G}_F$ may appear constant by varying the specimen size, whereas, for size scales where random self-similarity holds, the so-called "universal properties" of the system (σ_u^*, $\mathcal{G}_F^*$) are constant, although they are represented by physical quantities with unusual dimensions. The last result represents the target of the so-called "renormalization" procedure, i.e., the determination of physical quantities that are invariant under a change of length scale.

5 References

Barenblatt, G.I. (1979) **Similarity, Self-Similarity and Intermediate Asymptotics**. Consultant Bureau, New York.
Carpinteri, A. (1982) Notch sensitivity in fracture testing of aggregative materials. **Engineering Fracture Mechanics**, 16, 467-481.
Carpinteri, A. (1983) Plastic flow collapse vs. separation collapse (fracture) in elastic-plastic strain-hardening structures. **Materials and Structures**, 16, 85-96.
Carpinteri, A. (1985) Interpretation of the Griffith instability as a bifurcation of the global equilibrium, in **Applications of Fracture Mechanics to Cementitious Composites** (ed. S.P. Shah), Martinus Nijhoff Publishers, Dordrecht, pp. 287-316.
Carpinteri, A. (1986) **Mechanical Damage and Crack Growth in Concrete: Plastic Collapse to Brittle Fracture**. Martinus Nijhoff Publishers, Dordrecht.

150

Carpinteri, A. (1987) Stress-singularity and generalized fracture toughness at the vertex of re-entrant corners. **Engineering Fracture Mechanics**, 26, 143-155.

Carpinteri, A. (1989a) Decrease of apparent tensile and bending strength with specimen size: two different explanations based on fracture mechanics. **International Journal of Solids and Structures**, 25, 407-429.

Carpinteri, A. (1989b) Cusp catastrophe interpretation of fracture instability. **Journal of the Mechanics and Physics of Solids**, 37, 567-582.

Falconer, K. (1990) **Fractal Geometry: Mathematical Foundations and Applications**. John Wiley & Sons, Chichester.

Freudenthal, A.M. (1968) Statistical approach to brittle fracture, in **Fracture** (ed. H. Liebowitz), Academic Press, New York, pp. 591-619.

Griffith, A.A. (1921) The phenomenon of rupture and flow in solids. **Philosophical Transaction of the Royal Society,** London, A221, 163-198.

Herrmann, H.J. and Roux, S. (eds) (1990) **Statistical Models for the Fracture of Disordered Media**. North Holland, Amsterdam.

Jayatilaka, A.S. (1979) **Fracture of Engineering Brittle Materials**. Applied Science, London.

Mandelbrot, B.B. (1982) **The Fractal Geometry of Nature**. W.H. Freeman and Company, New York.

Mandelbrot, B.B., Passoja, D.E. and Paullay, A.J. (1984) Fractal character of fracture surfaces of metals. **Nature**, 308, 721-722.

Pfeuty, P. and Toulouse, G. (1977) **Introduction to the Renormalization Group and Critical Phenomena**. John Wiley & Sons, London.

RILEM Technical Committee 50 (1985) Determination of the fracture energy of mortar and concrete by means of three-point bend tests on notched beams. Draft Recommendation. **Materials and Structures**, 18, 287-290.

Saouma, V.E., Barton, C.C. and Gamaleldin, N.A. (1990) Fractal characterization of fracture surfaces in concrete. **Engineering Fracture Mechanics**, 35, 47-53.

Stanley, H.E. (1985) Fractal concepts for disordered systems: the interplay of physics and geometry, in **Scaling Phenomena in Disordered Systems** (eds. Pynn, R. and Skjeltorp, A.), NATO ASI series B, 133, Plenum Press, New York, pp. 49-69.

Weibull, W. (1939) **A Statistical Theory for the Strength of Materials**. Swedish Royal Institute for Engineering Research, Stockholm.

Wilson, K.G. (1971) Renormalization group and critical phenomena. **Physical Review**, B4, 3174-3205.

Wilson, K.G. (1979) Problems in physics with many scales of length. **Scientific American** (Italian Edition), 23, 140-157.

SIZE EFFECT AND STOCHASTIC NONLOCAL DAMAGE IN QUASI-BRITTLE MATERIALS

I. LAALAI and K. SAB
E.N.P.C.-C.E.R.A.M., Noisy-Le-Grand, France

Abstract
Deterministic nonlocal damage models yield mesh independent finite element calculations. They predict structure size effect which is in accordance with experimental observations on notched specimens made of concrete in 3-point bending and 4-point bending. However, these deterministic models are unable to predict volume size effect exhibited by quasi-brittle materials in direct tension. We investigate in this paper mesh independency and size effect in a nonlocal damage model where disorder is intoduced at the representative volume scale. Numerical simulations on notched and unnotched specimens will show that the proposed stochastic model yields mesh independent calculations and that this model predicts both volume and structure size effects which are consistent with the experimental observations.
Keywords: Nonlocal Damage, Size Effect, Stochastic Model, Finite Element.

1 Introduction

Progressive damage in brittle heterogeneous materials such as concrete, rocks, ceramics, ...etc often produces both strain softening and size effect at macroscopic level. Size effect is a dependence of the mechanical properties of the material on the size of the structure.

Direct tension tests on homothetic specimens made of concrete have been performed by Kadlecek and Spelta (1967), l'Hermite (1973) and Torrent (1977). It has been observed that the average tensile strength σ decreases when the volume V of the specimen increases, and it seems that it follows power law $\sigma = AV^{-\beta}$ where $A>0$ and $\beta>0$ are material constants. This type of size effect is usually called "volume" size effect and it is attributed to the existence of a random distribution of heterogeneities (or defects) inside the material. Hermann and al(1989) have used Monté-Carlo simulations to compute "volume" size effect in direct

151

D. Breysse (ed.), Probabilities and Materials, 151–161.
© 1994 Kluwer Academic Publishers.

tension of a square lattice of elastic beams which obey a
Von-Mises type failure criterion with threshold values
randomly picked according to Weibull law of parameter m. It
has been numerically shown that for $0 < m < 2$ the average

tensile strength σ follows power law $\sigma \approx L^{-\frac{1}{4}}$ where L is the
size of the lattice.

"Structure" size effect concerns homothetic notched
specimens loaded in mode I. It has been experimentally
established by Bazant (1985) and Bazant and Kazemi
(1989) that the nominal strength at failure depends on the
size of the specimen and that it follows a Linear Elastic

Fracture Mechanics (LEFM) criterion in $K_I^c a^{-\frac{1}{2}}$ only for very
large specimens. Here a denotes the length of the notch

and K_I^c is the critical stress intensity factor. Failure in
these materials is the result of progressive microcracking
in a fracture process zone lying ahead of the crack. The
volume of this zone seems to be related to the size of
heterogeneities. It is constant whereas the size of the
structure may be changed. Thus, LEFM applies only to large
structures for which the size of the process zone is
negligible and damage is concentrated in a relatively small
zone around the crack tip. For small enough structures,
LEFM does not apply and we say that "structure" size is
exhibited.

In nonlocal damage models, Pijaudier-Cabot and Bazant
(1987), the width of the process zone is a material
constant introduced in the constitutive equations as
follows. Damage at point $\underline{x}$ is a scalar $D(\underline{x})$ defined by :

$$\Lambda(\underline{x}) = (1 - D(\underline{x}))\Lambda^0(\underline{x}) \tag{1}$$

where $\Lambda(\underline{x})$ and $\Lambda^0(\underline{x})$ denote the current (damaged) and the
initial (undamaged) stiffness matrix at point $\underline{x}$,
respectively. $D(\underline{x})$ varies monotonically from 0 to 1. Its
evolution is described as a function of the average
equivalent strain over the so-called representative volume
at point $\underline{x}$, i.e. a neighbourhood of characteristic length l_c
centered at this point. Thus, strain softening behaviour is
introduced at the representative volume scale. Nonlocal
damage models guarantee that finite element computations on
strain softening materials up to failure remain sound from
a theoretical and computational viewpoint. Damage cannot
localize in regions of arbitrary sizes and the width of the
damage zone (or shear band) is proportional to l_c. Mazars
et al.(1991) have applied a nonlocal damage model to the
prediction of "structure" size effect in cementitious
materials. They have found good agreement with experimental
results. However, their model is not well-adapted to the

prediction of "volume" size effect because the effect of heterogeneity is introduced at the representative volume scale via a deterministic strain softening behaviour.

The main idea of this paper is to introduce disorder at the representative volume scale. Three lengths will be present in the problem: one associated to the size of the structure, L, one related to the width of the localization zone, l_c, and one related to the size of inhomogeneities, d.

$$d \leq l_c \leq L \tag{2}$$

In the model described in section 2, the material behaves in a purely brittle manner at the inhomogeneities scale. It obeys a nonlocal breaking criterion with random threshold values. It will be numerically shown in section 3 that: a) the proposed model yields mesh independent results, b) it produces strain softening at the macro-scale and c) it predicts both "volume" and "structure" size effects which are in agreement with experimental observations. It will be further shown that, as far as size effect is concerned, the deterministic nonlocal damage model of Mazars et al. (1991) and the random network model of Hermann et al. (1989) are special cases of the proposed one.

2 The model

In the model proposed by Bui and Ehrlacher (1981), damage variable D can take only two values: zero and one. Thus, according to (1), the elastic energy density at point $\underline{x}$ is given by:

$$\begin{aligned} w(\underline{x}) &= \tfrac{1}{2}\varepsilon(\underline{x})\Lambda^0(\underline{x})\varepsilon(\underline{x}) && \text{if } D(\underline{x}) = 0 \\ &= \quad 0 && \text{if } D(\underline{x}) = 1 \end{aligned} \tag{3}$$

where ε denotes the strain tensor. The evolution of damage must obey some breaking criterion because damage is described with a discontinuous state variable. Studies on localization instability have shown that a model in which the breaking criterion at point $\underline{x}$ is a function of the strain tensor at this point yields physically unacceptable results. Indeed, the damage zone is restrained to a region of zero volume and failure occurs without dissipation of energy, a consequence of which is a strong mesh dependency in finite elements calculations. See Fedelich (1990) and Laalai (1993).

A possible remedy to this problem is a nonlocal breaking criterion which is taken here as:

154

$$\frac{\overline{w}(\underline{x})}{w^c(\underline{x})} - 1 \leq 0 \qquad (4)$$

where $\overline{w}(\underline{x})$ is the volume average of w over a cube of side l_c centered at point $\underline{x}$, and $w^c(\underline{x})$ is the threshold value at this point.

Let λ denote a vector of control parameters in the sense of displacement or force control in solid mechanics. The problem is to compute the evolution of damage in the solid as λ varies. We numerically solve this problem as follows: the solid is discretized into finite elements, the elastic problem is solved and (4) is computed at the center of each element. As λ progresses, equality in (4) is attained at some "critical" element or at several "critical" elements, simultaneousely. Then, λ is fixed and the "critical" element(s) is (are) removed. This produces an energy redistribution in the solid and (4) might become positive at some elements. In this case, the element(s) for which (4) is maximum is (are) removed and the operation is repeated until (4) becomes negative in all elements (or the structure breaks down). After that, λ can progress until the next rupture occurs.

Several numerical simulations on homogeneous ($\Lambda^0(\underline{x})$ and $w^c(\underline{x})$ are uniform in the solid) notched structures loaded in mode I show that the results are mesh independent if the element size is in the order of l_c. It is further noticed that the width of the damage zone is proportional to l_c and that no "structure" size effect is exhibited. These results are not surprising; they are consistent with the predictions of nonlocal strain softening damage models.

Remark 1: Numerical errors -or if necessary a small random perturbation of the uniform field of threshold values (see Remark 2)- ensure that only stable equilibrium paths are computed at bifurcation points.

So, homogeneous media behave in a purely brittle manner at macroscopic level without any size effect. Our purpose is to study the effect of disorder when it is introduced at a microscopic level. Further on, we shall assume that w^c is a realization of the following stochastic field: the space is divided into regular square (two-dimensional case) or cubic (three-dimensional case) cells of side d where w^c is piecewise-constant. There is no correlation between cells and the value of w^c at cell k, w_k^c, is randomly picked according to the probability function

$$\text{Prob}\left(w_k^c \le t\right) = \left(\frac{t}{w^{max}}\right)^{m/2} \quad 0 \le t \le w^{max} \tag{5}$$

where $m > 0$ and $w^{max} > 0$ are positive parameters. Thus, heterogeneity depends on two parameters: exponent m and the cell's length d. Only the two-dimensional case with $d \le \ell_e$ is adressed in this paper.

Remark 2: For $m = +\infty$, (5) corresponds to Dirac distribution at w^{max}, i.e. a homogeneous material is recovered. For small enough d and large enough m, the above described stochastic field can be considered as a small perturbation of the uniform field w^{max}. It can be used to avoid the computation of unstable equilibrium paths in homogeneous materials. Indeed, for any $m \ne +\infty$, the probability that two cells have the same threshold value is zero. Now, if d is in the order of the finite element size, the probability that two different elements simultaneousely break is zero. Thus, unstable symmetric equilibrium paths will be avoided.

Consider a structure which undergoes increasing loads controlled by one single displacement parameter (λ is now a scalar). The generalized force, F, is defined as the ratio of total work of internal forces to λ. In typical experiments, such as direct tension test or 3-point bending, λ is slowly increased until the specimen breaks down. F is measured and the force-displacement curve is plotted. In order to simulate such failure experiments with the proposed stochastic model, we discretize the structure with a uniform square (or cubic) mesh. The element size must be smaller than d, the size of heterogeneities. Then, we simulate many realizations of w^c and we compute, for each realization, the force-displacement curve. Let $\overline{F_i}$, $\overline{\lambda_i}$ and $\overline{F_{max}}$, $\overline{\lambda_{max}}$ denote the ensemble average of F, λ at the ith rupture and at the maximum force, respectively. The average force displacement curve is defined by linear interpolation between points of coordinates $\left(\overline{F_i}, \overline{\lambda_i}\right)$. The average stress, strain at failure, σ_f, ε_f are defined by $\sigma_f = \overline{F_{max}}/L^{n-1}$ and $\varepsilon_f = \overline{\lambda_{max}}/L^{n-1}$ where L is the size of the structure and n the dimension of space.

3 Numerical results

We have performed numerous simulations in the two-dimensional case on notched and unnotched structures with different values of parameters m and d. The stiffness matrix is isotropic with Poisson's ratio equal to 0.2 in all these simulations.

We find that, alike deterministic nonlocal models, the proposed model is not sensitive to the discretization parameters, i.e. the orientation of the mesh and the element size which has to be smaller than d. Indeed, we find that σ_f, ε_f and the average force displacement curve are mesh independent. Figure 1 shows a typical average curve computed with 3 different element sizes. It concerns direct tension of a square L×L. The ensemble average have been performed on thirty realizations with the following numerical values $m = 1, d = \ell_\circ = 1$ and $L = 24$. This figure shows also that the material is strain softening at the macroscopic level.

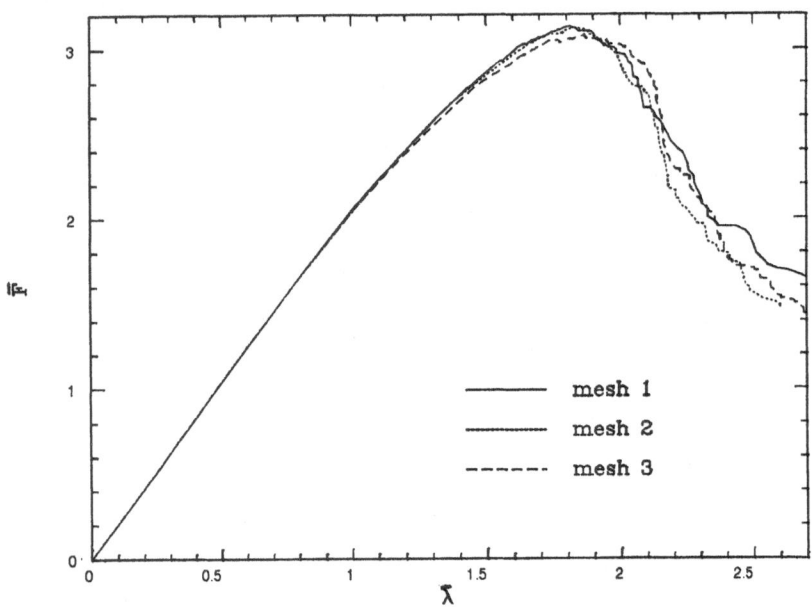

Fig.1 Average force-displacement curve

Remark 3: In order to avoid artificial localizations on boundary, the representative volume at a point near the boundary is extended by periodicity. In this manner, all representative volumes have the same size.

We shall present in subsections 3.1 and 3.2 the "volume" and the "structure" size effects numerically predicted by our model. We have simulated the breakdown of notched and unnotched homothetic structures of size L. Subsection 3.1 concerns a direct tension test on squares L×L and subsection 3.2 concerns a three-point bending test on notched specimens L×0.25L. For each of these tests, let σ_f^r denotes the average maximum strength computed for $L/l_o = 1$, i.e. σ_f^r is the average maximum strength of the representative volume. A straightforward dimensional analysis shows that σ_f/σ_f^r is a function of L/l_o which depends on m and d/l_o (the Poisson's ratio still being 0.2). Thus, we shall assume further on that $l_o = 1, L \geq 1$ and $d \leq 1$ without loss of generality.

3.1 "Volume" size effect
A parametric numerical study on the breakdown of squares L×L undergoing direct tension has shown that, for each value of d, there exists a critical value of m function of d, $m_d \leq 2$, such that: if $m < m_d$, then σ_f/σ_f^r follows power law $L^{-\alpha_d}$. And if $m \geq m_d$, then, there exists a critical length $L_{m,d}$ function of m and d such that σ_f/σ_f^r follows power law $L^{-\alpha_d}$ for $L \leq L_{m,d}$ and remains constant for $L \geq L_{m,d}$ as shown in figure 2.
In other words, $\forall d, \exists\, m_d, \alpha_d$ /

$$\begin{cases} m < m_d \Rightarrow & \sigma_f/\sigma_f^r \approx L^{-\alpha_d} \quad \forall L \geq 1 \\[2ex] m \geq m_d \Rightarrow \exists\, L_{m,d}\, / \begin{cases} \sigma_f/\sigma_f^r \approx L^{-\alpha_d} & \forall\, 1 \leq L \leq L_{m,d} \\ \sigma_f/\sigma_f^r \approx L_{m,d}^{-\alpha_d} & \forall L \geq L_{m,d} \end{cases} \end{cases} \tag{6}$$

Our numerical study shows that exponent α_d is actually a function of d, independent of m, given by the relation

$$\alpha_d = \frac{1}{2\left(\left[\dfrac{1}{d}\right]+1\right)} \tag{7}$$

where $[x]$ denotes the integer part of positive real x.

158

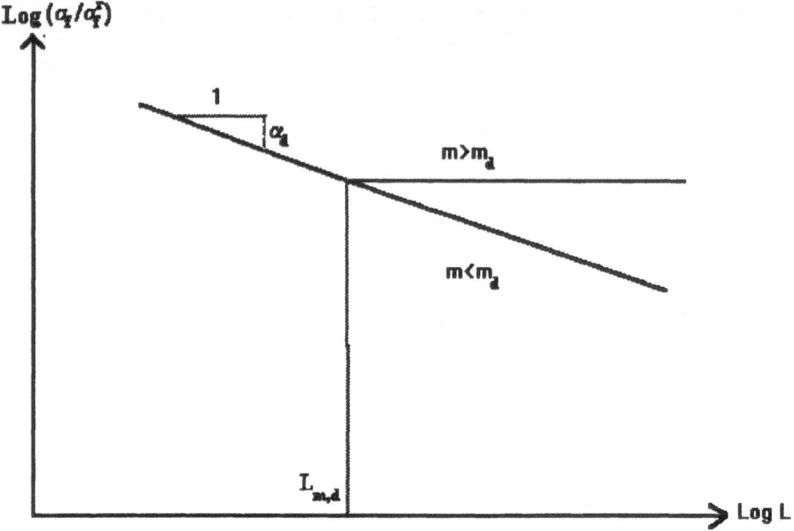

Fig. 2 Volume size effect

The existence of m_d is numerically established as follows: we fix d and we compute $L_{m,d}$ for different values of m $(m \leq 7)$. It is noticed that $L_{m,d}$ increases as m decreases and that it follows a power law in m when m decreases from 7 to 2. m_d is identified as the lower value of m for which $L_{m,d}$ follows this power law. For example, we find that $m_d(d = 1) = 2$ and $m_d(d = 1/3) = 1.21$. Actually, m_d decreases from 2 to 0^+ as d decreases from 1 to 0.

Scaling relationships (6)-(8) also hold for average strain at failure and average force-displacement curve including the post peak response.

For more details on this numerical study, see Sab and Laalai (1993), Laalai (1993), Laalai and Sab (1993).

3.2 Structure size effect
Concerning homothetic notched specimens loaded in mode I, we find that, for $m \geq m_d$, the slope of the curve $Log(\sigma_f/\sigma_f^r) - Log(L)$ is equal to $-\alpha_d$ at the origine and to $-1/2$ at infinity, i.e. an LEFM criterion applies for large structures. See figure 3. We find also that, for $m < m_d$, LEFM does not apply for large structures. This means that cementitious materials should verify $m \geq m_d$.

159

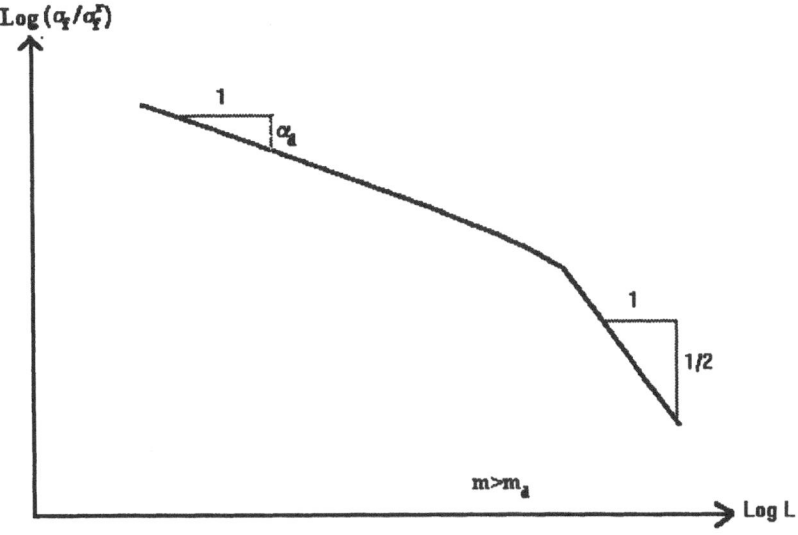

Fig.3 Structure size effect

3.3 Discussion

On the one hand, for $d = 1$, $m < m_1 = 2$, exponent α_1 is equal to $1/4$ and $L_{m,1} = \infty$. We find the "volume" size effect predicted by Hermann et al (1989) with their two-dimensional elastic lattice. This leads us to think that scaling relationships (6)-(8) are "universal" in the sense that they do not depend on the details of the model. Actually, we have proposed in Laalai (1993), Sab and Laalai (1993) and Laalai and Sab (1993) several micromechanics models (including one with elasto-plastic-brittle behaviour at microscopic level) which exhibit the same scaling relationships. On the other hand, for $d \rightarrow 0^+$, $m \neq \infty$, exponent α_d and $L_{m,d}$ go to zero, i.e. there is no "volume" size effect. We find a deterministic strain softening behaviour at the representative volume scale; and the same "structure" size effect predicted by Mazars et al (1991) with their deterministic nonlocal damage model is exhibited. According to Bazant and Pijaudier-Cabot (1989); the characteristic length of concrete is three times the maximum agregate size. We should expect $d/l_c = 1/3$ and $m \geq m_{1/3}$ for concrete. This would mean that concrete exhibits volume size effect up to a critical size beyond it the tensile strength is constant whereas the size of the specimen may be increased. This critical size should be experimentally identified. Indeed, numerical simulations in the three-dimensional case should exhibit the same type of scaling relationships with different values of exponents.

In conclusion, the proposed model yields mesh independent results. It shows that strain softening behaviour at macroscopic scale could be related to disorder at the representative volume scale. Concerning size effect, our model makes a link between random network models and nonlocal damage models.

4 References

Bazant Z.P. (1985), Mechanics of distributed cracking, **Applied Mechanics Review**, vol.39, n°5,675-705.

Bazant Z.P and Kazemi M.T. (1989), Size effect in fracture of ceramics, **Internal Report**, Departement of Civil Engineering, Northwestern University, Evanston,Illinois.

Bazant Z.P. and Pijaudier-Cabot (1989) Measurement of the characteristic length of nonlocal continuum, **Journal of Engineering Mechanics**, ASCE, vol.115, n°4, 755-767.

Bui H.D. and Ehrlacher A. (1981) Propagation of damage in elastic and plastic solids, in **Advances in Fracture Research**, Edited by D. François et al., Pergamon Press, vol.2, 533-552.

Fedelich B. (1990) Trajets d'équilibre des systèmes mécaniques dissipatifs à comportement indépendant du temps physique, **Thesis**, ENPC, Paris

Hermann H.J.,Hansen A. and Roux S. (1989), Fracture of disordered elastic lattices in two dimensions, **Phys. Rev. B**, 637.

Kadlecek V. and Spelta Z. (1967) Effect of size and shape of test specimens on the direct tensile strength of concrete, **Bull. RILEM**,n°36,175-184.

Laalai I., (1993) Effets d'échelle dans les matériaux quasi-fragiles à microstructure aléatoire, **Thesis**, E.N.P.C., Paris.

Laalai I. and Sab K. (1993) Size effect in quasi-brittle materials: a new micromechanics model, in **Mecamat 93, International seminar on micromechanics of materials**, Collection de la D.E.R. d'E.D.F., Eyrolles, Paris.

L'Hermite R. (1973), Influence de la dimension absolue sur la résistance à la flexion, **Annales de l'ITBTP**, 309-310, 39-41.

Mazars J., Pijaudier-Cabot G., and Saouridis C. (1991) Size effect and continuous damage in cementitiuous materials, **Int. J. of Fracture**, 51,159-173.

Pijaudier-Cabot G. and Bazant Z.P. (1987) Nonlocal damage theory, **Journal of Engineering Mechanics**, Vol 113, n°10, 1512-1533.

Sab K. and Laalai I., (1993) Une approche unifiée des effets d'échelle dans les matériaux quasi-fragiles, **C.R. Acad. Sci. Paris**, Série II, 316, n°9, 1187-1192.

Torrent R.J. (1977) A general relation between tensile
  strength and specimen geometry for concrete-like
  materials, **Materials and Structures,** RILEM, Vol 10,
  n°58, 187-196.

ROLE OF DISORDER IN BRITTLE FRACTURE

S. ROUX
Laboratoire de Physique et Mécanique des Milieux
Hétérogènes, Ecole Supérieure de Physique et Chimie
Industrielles

Abstract
We introduce a simple discrete model of brittle fracture
which explicitly takes disorder into account. We report on
one essential property concerning the multifractality of the
stress distribution and draw a few consequences on the local
and the global scales. We analyse the development of the
statistical distribution of stress as a function of the
damage. We study the distribution of stiffness difference
for each elementary fracture. Finally, we introduce a
classification of the scaling effect as a function of the
disorder in the structure.
Keywords: Brittle fracture, Size effects, Scaling, Multi-
fractals, Disorder, Acoustic emission

1 Introduction

In a number of occasions, one is interested in
characterizing and understanding the behavior of
heterogeneous materials on a scale much larger than that of
the elementary constituents. When the local behavior is
linear (elasticity, conductivity, permeability, ...) the
macroscopic behavior will also be linear, and with very few
restrictions, the macroscopic properties will converge under
coarse graining toward a well defined limit, justifying the
powerful concept of equivalent homogeneous medium. However,
when the local properties are non-linear, it may happen that
the macroscopic behavior will be controlled by the
heterogeneities of the medium at all scales. This turns out
to be the case for instance in brittle fracture. Other
examples may also be found in various frameworks: critical
currents in disordered superconductors, clogging of porous
filters by deposition of particles in pores, flow of
threshold fluids in a porous medium, plastic behavior of
heterogeneous solids, etc... We will concentrate on the
particular case of brittle fracture of disordered media. We
will see below that in general, the structure of these
cracks is non trivial, their interactions cannot be
generally neglected, and that scaling concepts are needed to

163

D. Breysse (ed.), Probabilities and Materials, 163–175.
© 1994 Kluwer Academic Publishers.

account for their sample-size dependence.

The problem of incorporating disorder at the local scale in brittle fracture is attracting a lot of effort as can be judged from the recent studies on this subject, (let us mention a few recent books and proceedings giving a comprehensive although partial review : Mazars (1989), van Mier (1991), Herrmann (1990), Charmet (1990)). Despites the variety of models considered, and the basic representation of the local level (networks, finite-element methods, fiber bundles, ...) most of these approaches aim at exploring the macroscopic consequences of fluctuating local properties, in order to study the coarse-grained behavior law of such heterogeneous solids, and the systematic effects associated with a change in scale. In most cases, analytic works are limited to very simple geometries (one dimensional or hierarchical) or extreme hypothesis such as the weakest link concept, and thus, in the case of damage for a two or three dimensional geometry, most information originates from numerical simulations. Therefore, one should be careful in extrapolating their limit of validity. Despite this warning, we will tentatively suggest asymptotic scaling properties using the results of numerical simulations.

It is worth noting that the size effects that emerges in the model, *does not originate from any fractal micro-structure of the medium*. On the contrary, no spatial correlation is initially introduced in the medium. The power-laws and the self-similarity concepts, that we will use to characterise the behavior, result from the collective behavior of the set of micro-cracks and of the damage evolution. Thus the model provides a natural framework to analyse experimental data obtained in the past (see e.g. Carpinteri (1992) and references therein).

2 Model of breakdown

We consider a simple discrete model which captures the essential ingredients of a brittle behavior of a disordered material, and simplify all features which are not expected to play a key role. Let us imagine a regular lattice whose bonds are the basic constituents of the model. Each bond is assumed to have a linear behavior up to a threshold where it fails irreversibly. These elements can either be elastic (linear brittle bars or beams) or electrical circuits (linear fuse). From numerical simulations performed in both frameworks, it turns out that the scaling properties of the mechanical model and those of the electrical analog are indistinguishable (de Arcangelis(1989a), Herrmann (1989), Hansen (1989)). The bonds are thus characterized by two numbers: their stiffness (or conductance), and their threshold stress (or current). Since the elastic modulus of an heterogeneous solid is a property which converges fast toward a scale independent value, we neglect at the local scale the fluctuations of this property, and set the local

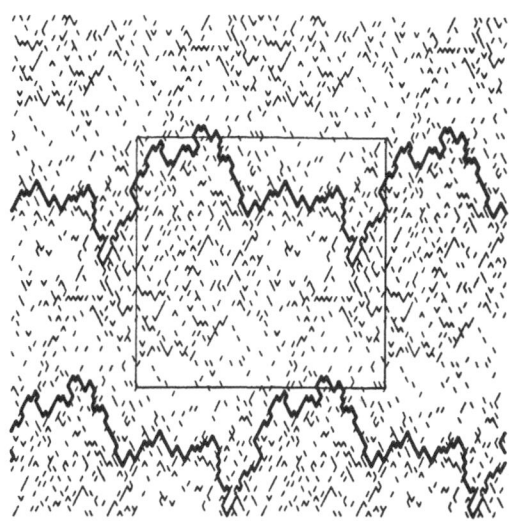

Figure 1: An example of a fuse network just before final breakdown. Doubly periodic boundary conditions are used and two periods in each direction are shown. The square shows one periodic cell of size 50x50. The broken bonds are shown on the dual lattice with lines the thickness of which is proportional to the "opening" of the cracks.

stiffness to a single value chosen to be one. On the contrary, we wish to take into account the effect of a random uncorrelated distribution of thresholds. We thus introduce the probability distribution, p(t), of thresholds t. This function, together with the lattice size, are the only input parameters of the model. The loading is imposed on two opposite borders of the lattice, or through periodic boundary conditions, in such a way that the loading is homogeneous.

The fracture of the lattice is monitored in the following way. At each stage of the fracture process, the behavior of the model is linear from the origin (zero displacement and zero force applied onto the lattice) to a threshold external force. The model makes use of this property to follow the development of the fracture. At each stage, the force applied on the lattice is reset to zero and progressively increased up to the breaking of one single bond. If the forces φ are computed for a unit external force exerted on the latice, the breaking force is determined by the minimum over all bonds of (t/φ). The number of broken bonds is the control parameter of the problem. Figure 1 shows an example of final crack configuration obtained in the simulation.

Using this model, many properties can be studied (de Arcangelis(1989a;1989b), Herrmann (1989), Hansen (1989)). Let us underline one of them and draw some consequences.

166

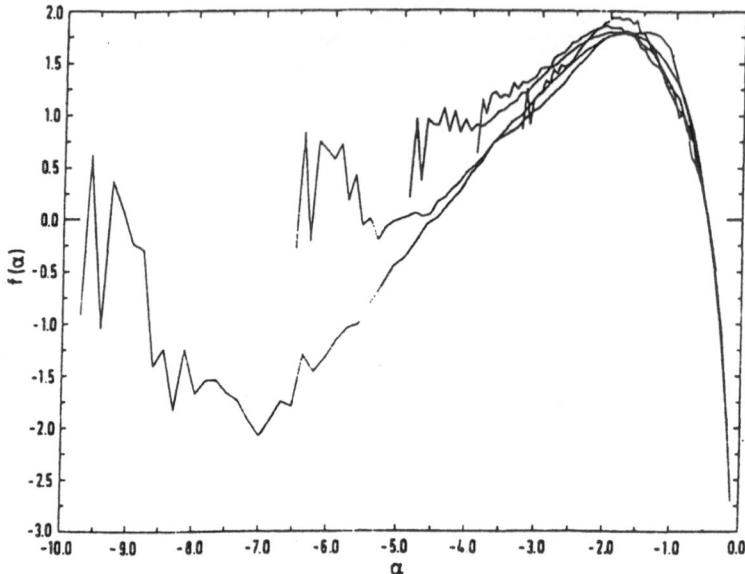

Figure 2: Rescaled distribution of local forces in a beam
network under shear with a uniform distribution of threshold
stress. These distributions collapse onto a single curve
which is the multifractal spectrum.(from Herrmann (1989))

3 One basic property

One of the basic properties observed at the very final stage
of fracture (i.e. when a single bond remains to be broken)
concerns the distribution of local stresses. It has first
been reported by de Arcangelis(1989b) that this
distribution had a *multifractal* character (see Halsey (1986)
for a presentation of multifractality). This means that
this distribution exhibits a size dependence which can be
accounted for by the introduction of reduced variables. Let
$N(\varphi, L)$ be the logarithmically binned histogram of local
force φ in a lattice of size L subjected to a unit external
force applied on the lattice. The use of the scaling index

$$\alpha = \log(\varphi)/\log(L),$$

and of the fractal dimension

$$f = \log(N(\varphi, L))/\log(L)$$

allows to characterize the histogram $N(\varphi, L)$ through the
function $f(\alpha)$, the so-called *multifractal spectrum*, which
is *size independent*. The variable α gives the scaling of
the local force with the system size $\varphi \propto L^{\alpha}$, whereas f is
the fractal dimension of the support of those forces since

$N(\varphi, L) \propto L^f$. One basic consequence is that the correlation length defined on the distribution of forces is equal to the size of the lattice at the final stage. Figure 2 illustrates this size independence for different system sizes (L=4, 8, 16, 32 and 64).

Two extreme cases can be mentioned where this multifractal character is naturally expected. The first case is for a homogeneous system. In this case a simple straight crack will develop throughout the medium, and thus at the very late stage of fracture a single bond remains to be broken. The distribution of force in this case will display a very simple multifractal spectrum containing three particular points: one describing the bulk of the medium (f is equal to the space dimension d) and a trivial α index resulting from the equilibrium $\alpha = 1-d$. A second point describes the singularity of forces at the tip of the crack f=0, and $\alpha = 0$, and, depending on the precise geometry, a third ``cold stress'' singularity may appear (the latter playing in any case a negligible role). The multifractal spectrum will be in this case the convex envelop of those three points (wedge-shaped spectrum).

Another extreme case is to be found when the distribution of thresholds is extremely broad, in the limit where it dominates completely the distribution of local forces in the criterion for fracture. This case can be described as a percolation type problem (with the additional requirement that the complete screening of bonds has to be taken into account, Roux (1988)). It is well known that at the final stage of fracture, i.e. in this particular case at the percolation threshold, the distribution of currents in a random resistor network (or its mechanical equivalent) has a multifractal character, de Arcangelis (1985). We also note that this multifractal character of the current distribution is a very common property which appears in a number of other models. In the case of the Dielectric Breakdown Model (DBM) introduced by Niemeyer(1984), the distribution of the current at the surface of the growing crack is also multifractal. When this model is changed in such a way that breaking may occur at any place in the lattice, with a probability proportional to the local current raised to some power η, again the current distribution in the bulk appears multifractal, Hansen(1990). In this case, the spectrum does

4 Evolution of the distribution during damage

The previously reported property of the local force or current distribution is only observed at the final stage of the process. Therefore, a natural question to ask is what can be established for intermediate stages, which are of greater interest.

In order to get some insight in this question, we use a result known from percolation theory where a similar problem has been treated, Roux (1989).

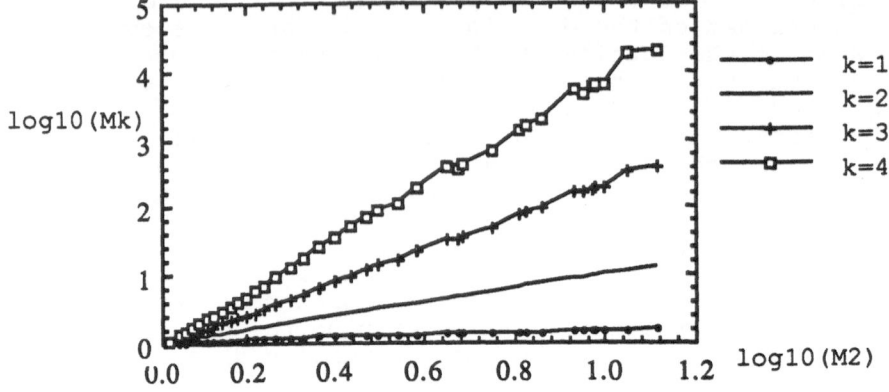

Figure 3: Evolution of the logarithm of the moments M_k for k=1 to 4 as a function of M_2 as the damage evolves. The data has been obtained for 320 lattices of size 40x40. The proportionality of those quantities support the existence of a correlation length which grows as the damage evolves.

The size effect that appears in the multifractality of the stress distribution at the final stage, is expected to appear gradually as the process goes on from the intact state to failure. More precisely, at some intermediate stage, the stress distribution may be described using the same spectrum as at the final stage but at a particular scale, called the correlation length, ξ, which only reaches the system size L at the final stage. More precisely, the moment M_k of order k (which for historical reasons is a pure sum not normalised by the number of elements summed) of the force distribution at the lattice size, $M_k = \Sigma \ |\varphi|^k$ can be related to the moment m_k estimated at the final stage, for the same boundary conditions (say a unit force applied on the lattice).

$$M_k = (L/\xi)^{d+k(1-d)} \ m_k$$

where d is the space dimension. This relation only makes use of the change of number of elements in the sum, and of the change in resulting force felt by a cell of size ξ. Each moment m_k scales with the size of the cell over which they are computed through an exponent $\tau(k)$: $m_k \propto \xi^{\tau(k)}$ so that for a fixed correlation length the scaling of M with L is trivial, whereas the power-law dependence on ξ contains a much richer information.

Unfortunately, ξ is not known a priori, and thus it is difficult to test the form proposed in the above equation.

However, choosing one particular moment, say M_2, as a
reference, it is possible to check a consequence of the
previous equation, i.e. that any moment is a power law of
M_2:

$$[\log(M_k)-(d+k(1-d))\log(L)]/[\tau(k)-d-k(1-d)] =$$
$$[\log(M_2)-(2-d)\log(L)]/[\tau(2)-(2-d)]$$

Moreover choosing a different normalisation (constant
stress, and dividing out each moment by the number of
elements considered) allows to get rid of the L dependence
in the above relation. Therefore using this ensemble, the
log of any moment should be proportional to $\log(M_2)$ as the
damage varies. Figure 3 shows that indeed such a relation
is very closely followed for the moments 1 to 4. Let us
emphasize that the data points of the figure have been
obtained at various stages of the damage process from the
intact stage to the end.

Therefore, this procedure justifies the existence of a
correlation length which gives in practice the size of the
representative volume element (indeed, above this scale, the
distribution of current is homogeneous, whereas below it,
the sample size is too limited to give a full account of the
stress distribution). This intrinsic length scale diverges
at the end of the process. It is difficult to compute
directly this length scale; however, since the logarithm of
M_2 -- and thus of any moment -- is proportional to $\log(\xi)$,
we can obtain an estimate of the variation of ξ simply by
studying this quantity. It is worth being noted that since
M_2 is proportional to the stiffness, the above property
indirectly legitimates the use of the damage parameter, i.e.
the reduction of elastic moduli, as a "good" parameter for
the evolution. However, the relation between the damage and
the density of microcracks is much less trivial than
generally assumed, since a very significant size effect can
be found (see below).

The above mentioned multifractal character of the stress
distribution at the fracture point may appear as a very
academic property, which is difficult to observe
experimentally and whose consequences are not obvious. It
is the purpose of the next two sections to show that this
property is useful in predicting observable features.

5 Acoustic emission

Having noted that the force distribution is multifractal at
the end of the process, we deduce that the distribution of
macroscopic elastic modulus jumps ΔE or compliance jumps
ΔS that would result from the breaking of one single bond
is also multifractal at the late stage of fracture.

Before reaching the final stage of the process, we have
seen above that a correlation length ξ could be introduced
to account for the distribution of forces, and the same

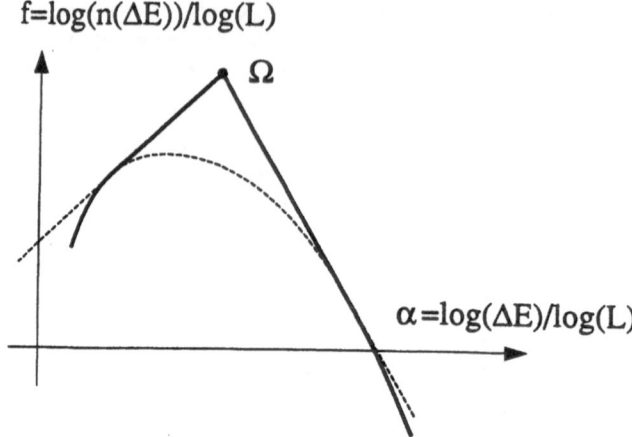

Figure 4: Schematic plot of the construction of the distribution of jumps recorded during the entire fracture process. If the location of the point Ω lies above the spectrum obtained at the end of fracture (shown in dashed line) the histogram will contain two power-laws as shown by the bold line. Otherwise, it is not possible to distinguish between the two spectra.

concept applies to the stiffness or compliance jumps. The initial stage, where the correlation length is equal to the mesh size of the lattice is characterized by a multifractal spectrum reduced to a single point, S. At intermediate stages, $\xi<L$, the spectrum can be obtained from the final spectrum ($\xi=L$) dilated by a factor $\log(\xi)/\log(L)$ from the point S (Roux (1989)).

It is thus possible to use this information to investigate the distribution of a local quantity which is recorded as the fracture process progresses from the initial to the final stage. In particular, the compliance jumps which occur during fracture originate from various stages with different correlation length. If we are interested in the series of jumps which take place during each individual breaking, we can use the form of the spectrum for intermediate stages, $\xi<L$, as well as the correct weighting (specifying the density of events as a function of the correlation length). Under very general conditions, the entire distribution of these jumps regardless of the precise state of the lattice when they were generated can be obtained using a simple geometric construction: One should consider the convex envelope of the multifractal spectra at the final stage and a point Ω in the

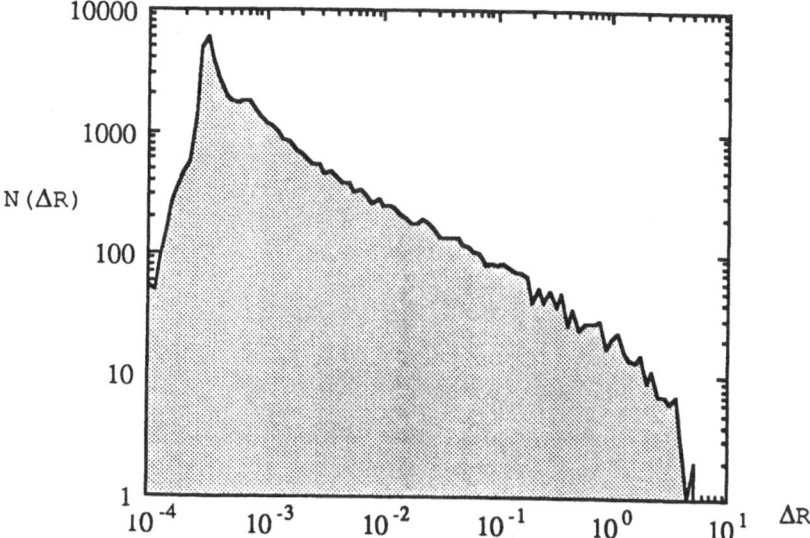

Figure 5: Log-log plot of the histogram of resistance jumps
ΔR recorded during each micro-crack   As predicted by the
argument developped in the text, the histogram essentially
consists of two power-laws for small and large jumps. This
distribution has been obtained for 40x40 fuse lattices such
as shown in Fig.1.

f-α plane. The location of Ω depends on the way the
correlation length reaches L and on the spectrum in the
initial stage. Thus if the point Ω lies below the final
spectrum, the envelope is simply equal to that of the final
stage. On the contrary, if Ω is outside the spectrum, as
shown schematically on figure 4, a wedge, tangent to the
spectrum, appears with a vertex at Ω. As a consequence, for
the latter case, the histogram will contain two power-laws.
Figure 5 shows such a distribution obtained from numerical
simulations on the fuse networks (320 lattices of size 40).
    Figure 6 shows a log-log plot of the histogram of
acoustic emission amplitudes of events which occured during
the fracture of one sample of concrete under a tensile
test. For each of these events, the location of the source
has been computed so as to exclude spurious events. A total
of about 3700 events were recorded and identified. The
experiment has been performed by Y. Berthaud and J. L.
Robert. More details about the experiment can be found in
(Berthaud, 1991). The right hand part of the curve displays
a very clear power-law behavior, which suggests that
acoustic emission may provide a very efficent way of probing
the scaling of the local stress field during fracture.

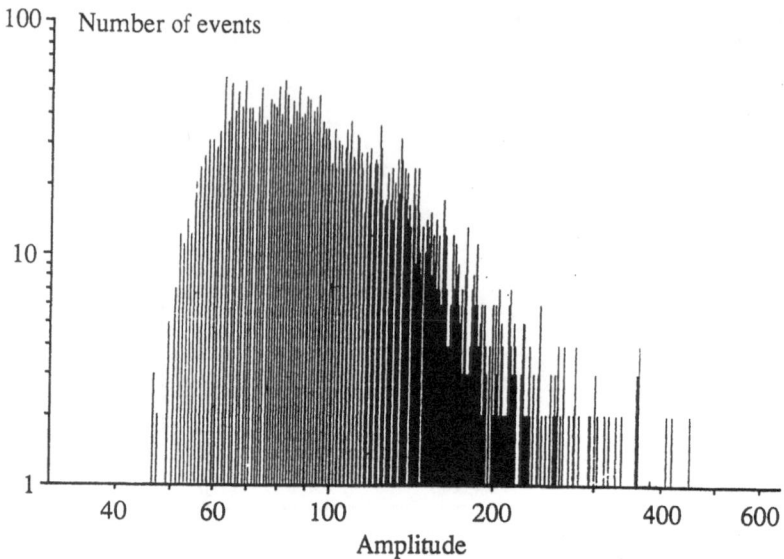

Figure 6: Log-log plot of the histogram of acoustic emission amplitudes recorded during the fracture of a single block of concrete under tension. More than 3700 events have been identified. (Courtesy of Y. Berthaud)

6 Classification of scaling behaviors

The property of having a multifractal distribution of stress or current at the final stage of fracture has also some theoretical implications on the classification of scaling behaviors depending on the distribution of local thresholds (Hansen, 1991). The distribution of local thresholds $p(t)$ is an input of the model. Obviously, this distribution should be set according to some experimental data analysis. It corresponds to an infinite number of degrees of freedom.

However, we have seen that the ``correct'' size-independent variable of the problem was not the stress but the scaling index $\alpha = \log(\varphi)/\log(L)$. In the model, the criterion for choosing the next bond to break is to find the minimum of the ratio t/φ where t is the local threshold. Since α should be used, we are lead to the conclusion that the "correct" variable which represents the threshold is $\alpha_t = \log(t)/\log(L)$, i.e. the scaling index of the threshold. Thus, the relevant representation of the threshold distribution is its multifractal spectrum. The latter is obtained in a similar way as for the force distribution, as the limit for large L of the rescaled log-log histogram of $p(t)$.

This statement seems odd at first sight. Indeed $p(t)$ is not expected to be size dependent, since it is a local material property. Nevertheless, the multifractal spectrum

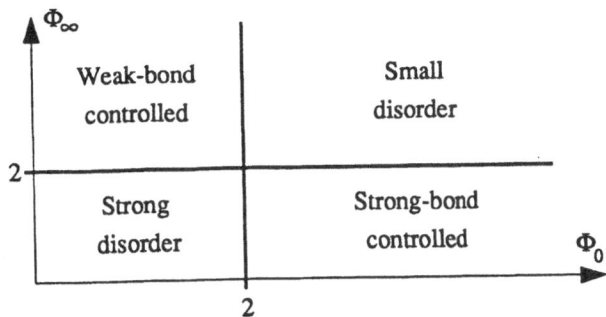

Figure 7: Schematic phase diagram of the scaling behaviors in the fracture process. Depending on the parameters Φ_0 and Φ_∞, different regimes are expected as discussed in the text.

of $p(t)$ is not trivially reduced to a single point. Generally, it contains three points. One which correponds to the ``typical'' threshold: f=d, α=0, and two additional points which indicate how the distribution reaches zero and infinity. More precisely, let us call

$$\Phi_0 = \lim_{t \to 0} \log(t\, p(t))/\log(t)$$
$$\Phi_\infty = - \lim_{t \to \infty} \log(t\, p(t))/\log(t)$$

These two numbers, Φ_0 and Φ_∞, are enough to characterize the scaling of the smallest and the largest element in a sampling of a given size. The spectrum of the distribution also contains two points which indicate this size-effect: (α=-d/Φ_0, f=0) and (α=d/Φ_∞, f=0). If the distribution does not reach zero (resp. infinity), then the limit is not defined, and we take Φ_0=-∞ (Φ_∞=∞ respectively) as a convention.

It is obvious that the spectrum of a distribution contains much less information than the distribution itself (two numbers are enough to characterize it), however, the identification of the proper ``thermodynamic'' variables allows to expect that the scaling properties of the fracture process are uniquely determined by the two scalar numbers defined above. Thus one should be able to draw a ``phase-diagram'' giving the scaling behavior in the Φ_0-Φ_∞ plane (Hansen (1991)).

Let us note that there exist a few simple models where an analytic solution can be obtained (one dimensional case, hierarchical lattices, ..). In all those cases, the above prediction applies. However, in two- or three-dimensional euclidian lattices, it is not possible to map exactly this phase diagram. Thus, we have to rely on numerical simulations in order to identify the scaling properties at any given point. We may also use some perturbation techniques so as to extract the domain of validity of a regime which is well-known (no-disorder case for instance) (Roux (1990)).

174

Figure 7 shows such an attempt to identify various phases, in terms of scaling properties. When Φ_0 and Φ_∞ are large enough, we are close to the situation where no disorder is present (all bonds would have the same threshold). Reducing progressively those parameters allows to obtain an estimate of the validity of this regime. Working along those lines, we have obtained the limits $\Phi_0=2$ and $\Phi_\infty=2$. For lower Φ_0, the fracture process is controlled by the weakest bonds in the system, which induce a sort of "diffuse damage" rather than the initiation of a single dominant crack. For lower Φ_∞, then very strong bonds are numerous enough to stop any propagating crack, and thus diffuse damage will also result. When both Φ_0 and Φ_∞ are smaller than 2, then it is extremely difficult to extract any reliable law.

The results of numerical simulations (de Arcangelis (1989a), Herrmann (1989), Hansen (1989)) seem to indicate that in particular for the weak-bond controlled regime, the scaling properties are insensitive to the value of Φ_0 and Φ_∞, provided they still lie in the same domain.

7 Conclusion

We have reported a few properties of brittle fracture of disordered media. We would like to point out the fact that the use of tools and concepts of statistical physics provides a very natural and powerful guideline to the analysis of such material properties in the case of heterogeneous materials. We have focused our presentation on very general properties, trying to be as distant as possible from a specific model. Some properties reported above have to be checked experimentally very carefully, since it is really through this confrontation that scaling analysis may become a useful tool. Moreover, the numerical modelling is always restricted to a rather small range of sizes due to the rapid increase of computer time required for large system size.

We acknowledge the collaboration of L. de Arcangelis, A. Hansen, H. J. Herrmann, E. L. Hinrichsen, with whom most of the material reported above has been obtained. We are also grateful to Y. Berthaud for providing us with the acoustic emission data. This work is supported by the GRECO "Géomatériaux".

References

Berthaud Y., Ringot E. and Fokwa D. (1991), Cement and Concrete Research, 21, 928
Carpinteri A. (1992), Fractal nature of material microstructure and size effects on apparent mechanical properties, LFM internal report 1/92

Charmet J. C., Roux S. and Guyon E. eds.(1990), Disorder and
    Fracture, Plenum, (New York)
de Arcangelis L., Redner S. and Coniglio A. (1985),
    Anomalous voltage distribution of random resistor
    networks and a new model for the backbone at the
    percolation threshold, Phys. Rev. B 31, 4725
de Arcangelis L., Hansen A., Herrmann H. J. and Roux S.
    (1989a), Scaling laws in fracture, Phys. Rev. B 40, 877
de Arcangelis L. and Herrmann H. J.(1989b), Scaling and
    multiscaling laws in random fuse networks, Phys. Rev. B
    39, 2678
Duxbury P. M., Beale P. D. and Leath P. L. (1986), Size
    effects of electrical breakdown in quenched random
    media, Phys. Rev. Lett. 57, 1052
Duxbury P. M., Leath P. L. and Beale P. D. (1987), Breakdown
    properties of quenched random systems: the random fuse
    network, Phys. Rev. B 36, 367
Halsey T. C., Jenssen M. H., Kadanoff L. P., Procaccia I.
    and Shraiman B. I.(1986), Fractal measures and their
    singularities: the characterization of strange sets,
    Phys. Rev. A 33, 1141
Hansen A., Roux S. and Herrmann H.J. (1989), Rupture of
    central-force lattices, J. Physique 50, 733
Hansen A., Hinrichsen E. L. and Roux S. (1990), Annealed
    model for breakdown processes, Europhys. Lett. 13, 517
Hansen A., Hinrichsen E. L. and Roux S. (1991), Scale-
    Invariant disorder in fracture and related breakdown
    phenomena, Phys. Rev. B 43, 665
Herrmann H. J., Hansen A. and Roux S. (1989), Fracture of
    disordered, elastic lattices in two dimensions, Phys.
    Rev. B 39, 637
Herrmann H. J. and Roux S. eds. (1990), Statistical models
    for the fracture of disordered media, North-Holland,
    (Amsterdam)
Mazars J. and Bazant Z.P. eds. (1989), Cracking and damage:
    strain localisation and size effect, Elsevier, (New
    York)
Niemeyer L., Pietronero L. and Weissman H. J. (1984),
    Fractal dimension of dielectric breakdown, Phys. Rev.
    Lett. 52, 1033
Roux S., Hansen A., Herrmann H. J. and Guyon E. (1988),
    Rupture of heterogeneous media in the limit of infinite
    disorder, J. Stat. Phys. 52, 237
Roux S. and Hansen A. (1989), Off-threshold multifractality
    in percolation, Europhys. Lett. 8, 729
Roux S. and Hansen A. (1990), Early stages of rupture of
    disordered materials, Europhys. Lett. 11, 37
van Mier J.G.M., Rots J.G., and Bakker A. eds. (1991),
    Fracture processes in concrete, rock and ceramics,
    Chapman and Hall, (London)

BLOCKING PROBLEMS IN THE ANALYSIS OF RANDOM FIELDS

L. FARAVELLI
Department of Structural Mechanics, University of Pavia, Pavia, Italy

Abstract
The dependency of structural response measures on the design variables is regarded as a mapping problem. Several mapping models are illustrated, compared and discussed.
A probabilistic model of the mechanical and geometrical properties of continua results from the introduction of random fields and/or, in a discretized continuum, random vectors. The investigation of their effects on the mapping results may require the simulation of some realizations to be incorporated in the data of the numerical (or laboratory) experiments. The association of a different simulated field for each experiment would be too noisy and, hence, blocking techniques must be used.
The paper illustrates them, their optimization and their appropriate use in the successive uncertainty propagation through structural analysis.
Keywords : Blocking, Causal Network, Central Composite Design, Experiment Design, Neural Network, Random Field, Regression Analysis, Response Surface

1 Introduction

The spatial random variability of the mechanical and geometrical properties of some continua can be modelled by random fields and/or random vectors, Augusti et al. (1984), Casciati and Faravelli (1991). For discretized continua, the discretization of any random field still results in a random vector. These probabilistic models can be used in solid mechanics analyses by simple simulation or in the context of some input-output mapping scheme built on the basis of experimental material. Attention is focused on the latter problem and, in particular, on the design of the numerical experiments to be carried out in order to achieve a good analytical approximation on the actual behaviour of the continuum. The association of a different set of simulated fields (vectors) for each experiment would be too noisy

177

D. Breysse (ed.), Probabilities and Materials, 177–195.

and, hence, blocking techniques must be used.

The paper provides an overview of the existing input-output mapping schemes. The possibility of incorporating models of spatial variability is then investigated and, eventually, the operative aspects met in selecting options are discussed.

2 Mapping Models

The basic elements of any structural problems are:
- a vector $\mathbf{X}$ which groups the potentially important input variables;
- a vector $\mathbf{x}$ which groups the actually important variables; it should always be a subset of $\mathbf{X}$;
- a vector $\mathbf{y}$ which groups the response variables of interest;
- a functional relationship $\mathbf{f}(\mathbf{x}, \boldsymbol{\theta}_f)$ which provides values for $\mathbf{y}$ once levels for $\mathbf{x}$ are given; $\mathbf{f}$ may be implicit or explicit, analytical or numerical, fully known, known in form but with unknown parameters $\boldsymbol{\theta}_f$, fully unknown.

The case of interest is the one in which the functional relationship $\mathbf{f}$ must be obtained and the subset $\mathbf{x}$ identified. For this purpose several numerical or laboratory experiments must be carried out; the i-th is characterized by:
- vector $\mathbf{V}_i$ (and, hence $\mathbf{v}_i$) to define the values of the input variables $\mathbf{X}$ ($\mathbf{x}$);
- vector $\mathbf{w}_i$ to collect the measured responses.

Just for sake of notation simplicity, reference is made in the following to a scalar response y and, hence, to scalar formats for f and w. It follows that the unknown structural relationship is:

$$y = f(\mathbf{X}, \boldsymbol{\theta}_f) \tag{1}$$

for which

$$
\begin{aligned}
w_1 &= f(\mathbf{V}_1, \boldsymbol{\theta}_f) \\
w_2 &= f(\mathbf{V}_2, \boldsymbol{\theta}_f) \\
\cdot &= \ldots \\
w_N &= f(\mathbf{V}_N, \boldsymbol{\theta}_f)
\end{aligned}
\tag{2}
$$

N being the number of experiments. This experimental material is not given a priori: indeed, vector $\mathbf{V}$ ($\mathbf{v}$) must be selected following the criteria of experiment design theory which are specific for each analysis purpose, Petersen (1985). Before entering details on this topic, it is worth noting that each experiment is defined by the input vector $\mathbf{V}$ ($\mathbf{v}$) and requires the associated response w be determined in laboratory or by running a computer code. The latter one is regarded as a black box as required in the analysis of complex structural systems, Casciati and Faravelli (1991), where large general-purpose finite-element algorithms are generally adopted.

Using a highly fractionated design on the original space $\mathbf{X}$, the problem of screening variables is the first to be approached. As a result one determines the subset $\mathbf{x}$ of the actually important variables. Examples of this problem in reliability analysis are presented by Faravelli (1992) and Faravelli and Breitung (1993). In those papers the appropriate fractional factorial design is defined. Equations (1) and (2) can now be rewritten

$$y = f(\mathbf{x}, \boldsymbol{\theta}_f) \tag{3}$$

for which

$$
\begin{aligned}
w_1 &= f(\mathbf{v_1}, \boldsymbol{\theta}_f) \\
w_2 &= f(\mathbf{v_2}, \boldsymbol{\theta}_f) \\
. &= ... \\
w_N &= f(\mathbf{v_N}, \boldsymbol{\theta}_f)
\end{aligned}
\tag{4}
$$

Most techniques of experiment analysis do not bother about whether the input (factors) have a qualitative or quantitative nature. In empirical modelling, one is in fact concerned in estimating mean effects and factorial contrasts, as main effects and interactions. However, structural engineering makes primarily use of quantitative models and in this case it is natural to think the response y as a function of the levels of the input variables $\mathbf{x}$. Empirical modelling means now to determine a local interpolation approximation $g(\mathbf{x}, \boldsymbol{\theta}_g)$ of parameters $\boldsymbol{\theta}_g$ to $f(\mathbf{x}, \boldsymbol{\theta}_f)$:

$$y = g(\mathbf{x}, \boldsymbol{\theta}_g) + \varepsilon \tag{5}$$

ε being the model error.

Toward this objective, the main tool is regression analysis. It can be non-parametric, as in the application reported by Casciati and Faravelli (1989), or parametric. In the first case the form of g is regarded as unknown, while in the second case it is assigned and only the parameters are unknown. The simplest case for g is to have a linear nature, so that linear multivariate regression analysis applies. For given levels $\mathbf{v}$ of the variables $\mathbf{x}$, the linear coefficients of the parameters $\boldsymbol{\theta}_g$ are known and can be grouped in matrix $\mathbf{A}$, of size Nxn, with $N > n$. Let $\mathbf{W}$ be the vector which collects the left hand sides in Eq. (4); one simply writes:

$$\mathbf{W} = \mathbf{A}\boldsymbol{\theta}_g \tag{6}$$

The least-squares algorithm provides:

$$\mathbf{A}^T(\mathbf{W} - \mathbf{A}\boldsymbol{\theta}_g) = 0 \tag{7}$$

and, hence, if the columns of $\mathbf{A}$ are linearly independent

$$\boldsymbol{\theta}_g = (\mathbf{A}^T\mathbf{A})^{-1}\mathbf{A}^T\mathbf{W} \tag{8}$$

An estimate s^2 for the variance of ε is then obtained as

$$s^2 = \frac{(\mathbf{W} - \mathbf{A}\boldsymbol{\theta}_g)^T(\mathbf{W} - \mathbf{A}\boldsymbol{\theta}_g)}{N - n} \qquad (9)$$

2.1 Mechanistic model

The mechanisms of some phenomena are understood sufficiently well that useful mathematical models can be directly written down. The mathematical model, then, provides a mapping from the space of the input variables to the one of the response quantities. Most mechanistic models are nonlinear in the parameter: iterative algorithms are therefore necessary to carry out their estimation. Moreover, the problem of choosing experimental conditions for nonlinear models to achieve reliable estimates of the parameters is quite specialistic and can be found in some textbooks. It turns out that a sequential strategy is usually needed, since for this nonlinearity the best selection of the experimental conditions depends on the values of the parameters themself.

Let now the vector $\mathbf{x}$ be a vector of random variables of given joint probability density function (JPDF): the knowledge of the mechanistic model and its parameters makes the estimation of the PDF of y a derived distribution problem which can easily be approximately solved by simulation or by adopting an algorithm of reliability index assessment, Casciati and Faravelli (1991). For instance, let the performance function be:

$$\eta - y \geq 0 \qquad (10)$$

Then, an asymptotic estimation holds for Prob$[y < \eta]$, once the point of maximum log-likelihood function associated with the JPDF of $\mathbf{x}$ under constraint (10) has been identified, Faravelli and Breitung (1993).

Nevertheless several phenomena are too complex or are not sufficiently well understood to permit the mechanistic approach.

2.2 Response surface

Response surface methodology is made by a group of statistical techniques for empirical model building and model exploitation. By careful design and analysis of experiments, it seeks to relate a response variable to the values of a number of input variables. This is generally done by determining the coefficients of a second order polynomial form written in the input space or in a transformation of it. It is the analogous of the analytical second order Taylor expansion.

This functional relationship is fitted on the basis of experimental runs: in each of them the input variables are adjusted to a definite set of levels. Central composite experiment designs provide optimality conditions. It assembles a (fractional) factorial design with a star design and some central points, Box and Draper (1987)

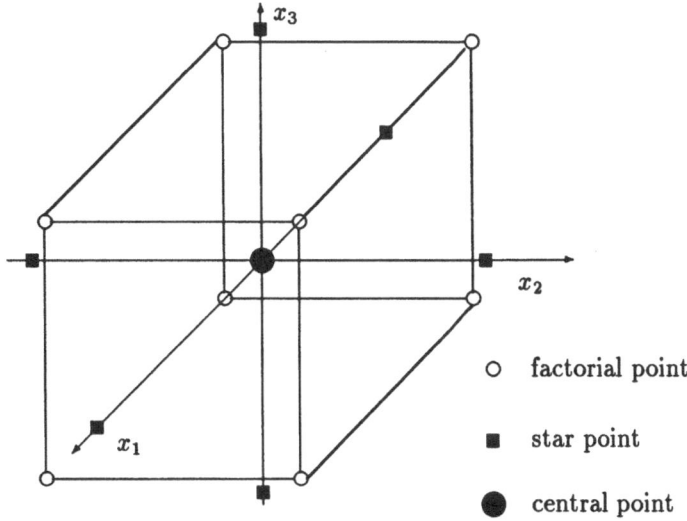

Figure 1: Central composite design in $\mathcal{R}^3$

(see Figure 1). The quadratic form in $\mathbf{x}$ preserves the linearity in the model parameters and multivariate regression analysis (Eq. (8)) can still be used in the analysis. The inherent lack of fit will produce a difference between the calculated response and the observed one: it can be incorporated in the model as an additive zero-mean error term. In a similar way, if repeated runs are made at the same input conditions and the measured response will vary due to basic variability in the experimental material, a pure error term must be incorporated.

Let now the vector $\mathbf{x}$ be a vector of random variables of given joint probability density function: the knowledge of the response surface parameters makes, also in this case, the estimation of the PDF of y a derived distribution problem which can easily be approximately solved by simulation or by adopting an algorithm of reliability index assessment. It is worth noting in particular that the asymptotic estimation for Prob$[y < \eta]$ arising from the knowledge of the point of maximum log-likelihood accounts for a second order relation between y and $\mathbf{x}$.

2.3 Causal probabilistic networks

An alternative idea is that the response of a structural specimen, an existing building or a mathematical model for them can be reached by absorbing and propagating evidence about potential causes or observed effects into a causal network representing the structural behaviour, Faravelli and Gherardini (1992) and Casciati and Faravelli (1993). A causal network is a directed graph $G = (R, E)$, R being a set of nodes and E a set of directed links. The nodes represent conceptual entities

182

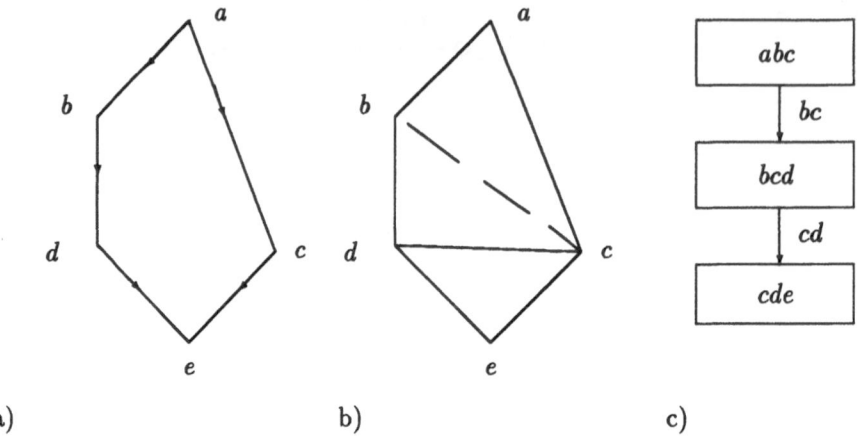

Figure 2: a) Typical causal probabilistic network; b) Moral graph triangulation; c) Junction tree: each item is a clique.

coming from a specific knowledge domain. The links express a causal (in a very broad sense) relationship between these entities. In an object-oriented framework nodes and objects coincide.

If directed links go from entities c and d to entity e, then e (the child) is caused by c and d (the parents), which are denoted as $pa(e)$.

In a probabilistic framework these entities will somehow be associated with random variables. To make the exposition simpler, let these variables be discrete, i.e. they only assume discrete states. A consequence of the causality links is that the probability distribution on each node is a conditional probability distribution specifying, for each possible configuration of states of the parent variables, the probability of the child variable being in each of its states.

The resulting CPN (Causal Probabilistic Network) is a static model for a certain knowledge domain.

It is assumed that the probabilistic structure of the relevant conditional probabilities is known. If one then collects data that completely specify the value of a subset of variables in the network, so that no uncertainty exists about them, new values of the conditional probabilities for the variables that have not been observed can be assessed, by Bayesian schemes of knowledge updating, on the basis of these new pieces of information, Casciati and Faravelli (1990 and 1991).

Straightforward computational procedures may be very inefficient in large networks. A correct and efficient probability updating can be performed by exploiting local relationships between nodes in the graph after concepts as moral graph, clique and junction tree have been introduced (see Figure 2) and implemented, Casciati and Faravelli (1993). Further, the updating algorithm can be described in an object-oriented framework.

The conditional probabilities that characterize the network links between two consequent nodes are the result of present expertise. They can be improved on the basis of a learning strategy as the results of new experiments become available. To provide the capability of learning from a database means modelling the variability in terms of unknown parameters which are estimated and updated on the basis of the accumulated data. Bayesian learning methods for directed graphs are the logical extension of the previous probabilistic treatment.

Assume that the conditional probabilities required in the definition of a probabilistic CPN are unknown quantities. Initial guesses can then be updated as the results of a number of cases accumulate. Spiegelhalter and Lauritzen (1990) suggested, for this purpose, the introduction of supplementary parameter nodes, representing marginal independent random quantities, $\phi_r, r \in R$, whose realization specifies the conditional probabilities tables for the network, Casciati and Faravelli (1993).

The main advantage of a CPN model is that one can exploit his physical understanding of the phenomenon under investigation by isolating simpler and simpler cause-effect relations. On the other side, for large networks with several states for each node, the computational effort may become soon unbearable.

2.4 Neural networks

Neural networks are an implementation of an algorithm inspired by research into the brain. Nevertheless, they have little to do with biology: they just means a technology in which computers learn directly from data. Among others (classification, data compression), function estimation is one of their purposes.

Structural and mathematical differences can be envisaged between a neural network and another. Structural differences appear in the number of layers of data-processing nodes and the types of connections between the nodes. Mathematical differences show up in the equations controlling the network behaviour.

The most popular algorithm is back-propagation: such a network is made from interconnected nodes arranged in at least three layers (see Figure 2). The input layer is passive: it merely receives the data patterns (input vectors) passing into the network. The number of its nodes consequently equals the number of measured data values presented to the network. Unlike the input layer, the output and hidden layers both actively process data.

The output layer produces the network result: a set of continuously variable values (output vectors), one per output node. The hidden layer has no direct connection to input or output: its introduction permits back-propagation to model nonlinear functions of greater complexity. Choosing the number of nodes in the hidden layer almost always involves experimentation.

A single node has many inputs but only one output. Each input is a single data value presented to the node, usually through a connection from a preceding layer. An extra input (the threshold input) acts as a reference level or bias for the processing element. On a serial computer the microprocessor emulates one node at a time. On a parallel computer, each element maps to a physical processor.

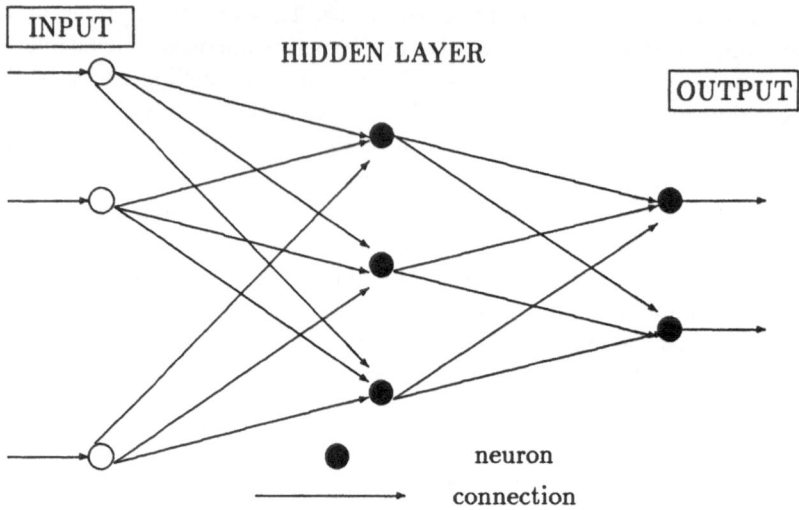

Figure 3: Typical back-propagation neural network.

Associated with each connection there is an adjustable value, the weight. Basically, a node calculates the weighted sum of its inputs, then passes the sum through a function to produce a result. This transfer function is a steadily increasing S-shaped curve which constraints the result within fixed limits. The back-propagation algorithm strength is its ability to change the values of its weights in response to errors. It does this automatically during training, which requires a series of input patterns tagged with their desired output patterns. During training, the network passess each input pattern through the hidden layer to the output layer to generate a result for each output node, which is compared with the actual result. The differences are the output-layer errors, which the network passes back (back-propagation) to the hidden layer using the same weighted connections. Next, each hidden node calculates the weighted sum of the back-propagated errors to find its contribution to the know output errors. After each output and each hidden node finds its error value, it adjust its weights to reduce its error. The equation that changes the weights is designed to minimize the network sum-squared error. The network overall accuracy is improved by the aggregate correction during training. When the network can process input patterns with sufficient accuracy, the weights are saved to preserve what it has learned.

After training, the network should be tested with known data that was not used during training. The network accuracy with patterns outside the training set is called generalization and indicates its reliability in an application. Of course, failure on the testing data is quite possible even for a well-trained network.

After training and testing, the network is ready to process unknown data. Basically, applying a pattern to the input produces a corresponding patterns at the output. The network therefore acts as a model of a function, mapping input patterns to output patterns. It learns this association solely from the training, even if the equation describing the function is nonlinear, unknown or both.

Neural networks are valuable on several counts: they can take data and learn from it, without any prior programmer knowledge of rules (adaptivity: the only prerequisite being a representative sample of the function behaviour); they capture complex interactions among the input variables in a system (nonlinearity); eventually they are highly parallel, in the sense that their numerous identical independent operations can be executed simultaneously by a parallel hardware. On the other sides several weaknesses are slowing the acceptance of neural networks:

- it can be difficult to account for their results: as human experts, neural networks express opinions that they cannot easily explain;

- training methods are imperfectly understood: few definite rules exist for choosing values for training parameters;

- neural networks can consume huge amounts of computer time during training.

3 Incorporating random fields

All the mapping models of previous Section suffer limitations in the number k of variables $\mathbf{x}$. In the presence of stochastic processes and random fields, therefore, it was already recognized that their intensity must be incorporated in vector $\mathbf{x}$, but the temporal and spatial variability cannot, Casciati and Faravelli (1991),
In the presence of a mechanistic model, this temporal and spatial variability will result either in a probabilistic description of the model parameters or in an additive random term ε. In a response surface model the term ε will incorporate the effects of this randomness which does not appear in vector $\mathbf{x}$ but alters the responses computed for the same input levels $\mathbf{v}$.

A different situation arises for casual probabilistic networks, where the presence of stochastic processes and random fields cannot be accounted directly in terms of nodes, but their presence has the effect to alter the structure of the link conditional probabilities. Also there is no matter to add as many neurons to the input layer of a neural network as the global amount of components in the random vector arising from the discretization of stochastic processes and random fields. A new neural network should be conceived, whose input layer is made by state variables which account for all the variables and any randomness.
Thus both the network models require local arrangements in the problem structure which are currently object of investigation.

By contrast, the response surface model just requires that realizations of the spatial and temporal variability be incorporated in the experiment material.
A different set of realizations for each of the N experiments would introduce too much noise. It was therefore proposed, Casciati and Faravelli (1991), that a few realization sets are generated and their use is repeated for blocks of the experimental

material. The association criterion is not free but is governed by the rules of blocking. The experiment design can be arranged in blocks to neutralize the effect of possible inhomogeneities. Orthogonal blocks are defined as the ones whose effects do not affect the usual estimates of the parameters of the second order model. For this purpose each block must be itself a first-order orthogonal design ($\mathbf{v}_b^T \mathbf{v} = 0$) and the fraction of the total sum of squares of each variable contributed by each block must be equal to the fraction of the total observations allotted to the block, Box and Draper (1987).

4 Blocking techniques

The first proposal of using response surface in a probabilistic context, by coupling the regression approach with a second-level reliability approach, dates back ten years, Veneziano et al. (1983). It was also recognized the importance of blocking in studying the randomness introduced by random vectors, stochastic processes and random fields. This gave rise to the need of establishing association criteria of the planned experiments into blocks. The implementation of the whole procedure, as formulated in Faravelli (1989) and Faravelli and Bigi (1990), considered as predominant the number of central points available and regarded as separated entities the star points along a single axis and the subsets in which it is possible to partition the (fractional) factorial by preserving the orthogonality condition ($\mathbf{v}_b^T \mathbf{v} = 0$).

1. Some central experiments are coupled with the experiments of subsets of the (fractional) factorial arising by blocking generation, others are associated with the star points lying along the same axis, the remaining ones, eventually, are runned alone. If the number of central points is lower than the sum of the space dimension and the number of fractional subsets, a single star set and a single fractional subset are joined to form a larger block.

In some special cases (e.g. k, the number of variables, is equal 2) no orthogonal fractional subsets is available. The criterion was to partition the fractional design into two blocks in order to avoid a strict dependency on the results on the single realization associated with all the fractional design.
The algorithm was running well in several applications ranging from crashing problems, Faravelli and Bigi (1990), and fatigue lifetime analyses, Casciati et al. (1993).

A main inconvenience was detected in succesive applications of the algorithm. Some realizations of the stochastic processes may be the cause of large deviations of the response from the average behaviour (for instance, in a cantilever, a very low value of the Young modulus in the fixed section, when the response is in term of the displacement at the free end). Then all the w_i in Eq. (4) of that particular block contain this effect and the regression algorithm is unable to identify the essential of the functional variation of y in the space of $\mathbf{x}$. An improvement was therefore adopted by Breitung and Faravelli, (1993).

2. For the experiments of each block which are not central points, the values w_i are substituted with the values obtained from w_i by subtracting the w in the central point of that block and by adding the average of the w in all the central points considered in the experiment plan.

The studies consequent the previous inconvenience led the author to discover that this particular experiment design was the object of Section 15.3 in the book by Box and Draper, (1987). Adding to the orthogonality condition the constraint that the fraction of the total sum of squares of each variable contributed by each block must be equal to the fraction of the total observations allotted to the block, a different experiment plan must be conceived. The star points along each axis must be grouped in order to have a block of size comparable with the one of the factorial blocks. Moreover the additional constraint requires that the non-dimensional distance α from the central point of each star point be:

$$\alpha = k \frac{1 + \frac{n_{s0}}{n_s}}{1 + \frac{n_{c0}}{n_c}} \tag{11}$$

where n_c is the number of experiments in each subset in which the (fractional) factorial design is partitioned; n_{c0} is the number of central points associated with these subsets; n_s is the number of star points and n_{s0} is the associated number of central points (usually 1). Note that in this way the rotatability of the design is broken, but in all cases the resulting variance contours are adequately close to the spherical contours that the rotatability provides. In summary:

3. Some central experiments are coupled with the experiments of subsets of the (fractional) factorial arising by blocking generation, a further one is associated with the single block of the star points.

The last blocking technique make use of a very limited number of central points and this could cause inaccuracy in the estimate of the pure error. In particular for $k = 2$, 2 further central points will be added to the 2 required by star and factorial blocks. Eventually, one can think to couple device 2. with criterion 3.:

4. The criterion in 3. is used but, for the experiments of each block which are not central points, the values w_i are substituted with the values obtained from w_i by subtracting the w in the central point of that block and by adding the average of the w in all the central points considered in the experiment plan.

5 A numerical example

The four blocking techniques illustrated in the previous section were tested in a very simple example which has all the features of a complex structure to be analysed by a finite element approach: a horizontal cantilever divided into five elements of random length (Gaussian distribution with mean 1 m and coefficient of variation

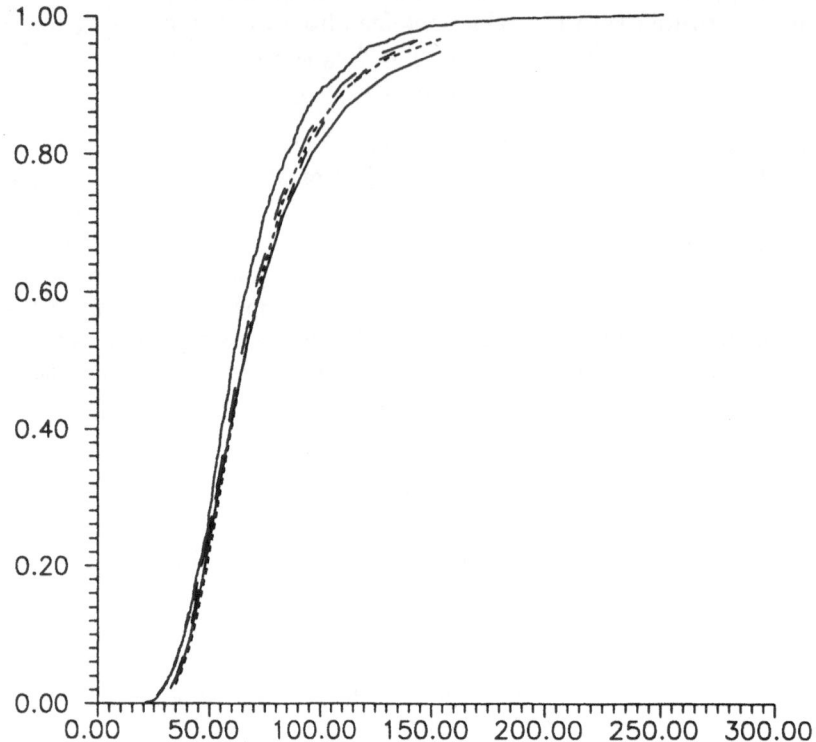

Figure 4: Problem B - Comparison, in a linear scale, between the simulation result and different response surface approximations. The optimal transformation is introduced. The solid line is for technique 1.; the short dashed line is for technique 1. with non-orthogonal blocks for the factorial design; the medium dashed line is for technique 2.; the long dashed line is for technique 3.

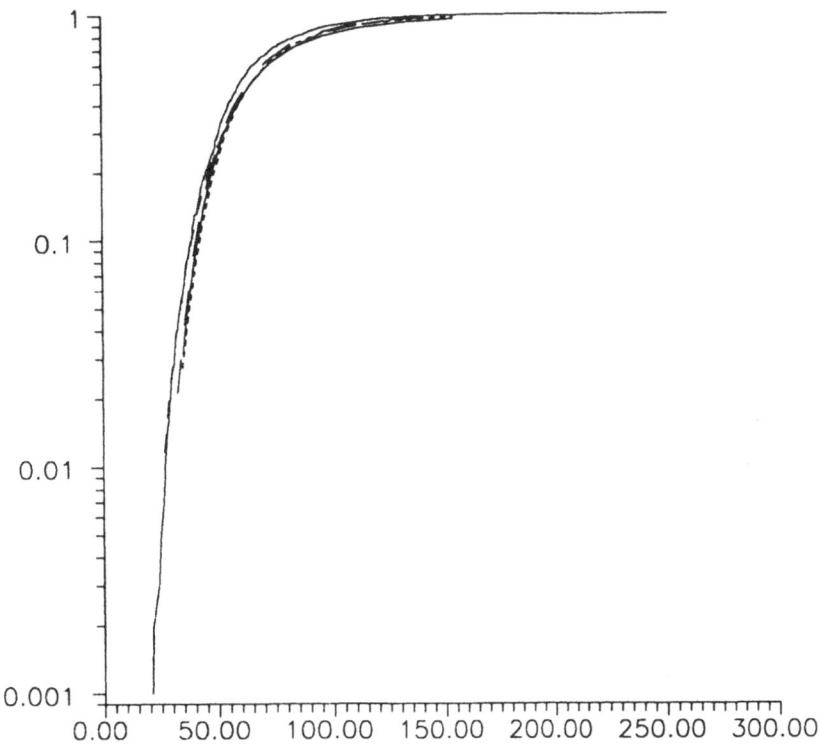

Figure 5: Problem B - Comparison, in a logarithmic scale, between the simulation result and different response surface approximations. The optimal transformation is introduced. The solid line is for technique 1.; the short dashed line is for technique 1. with non-orthogonal blocks for the factorial design; the medium dashed line is for technique 2.; the long dashed line is for technique 3.

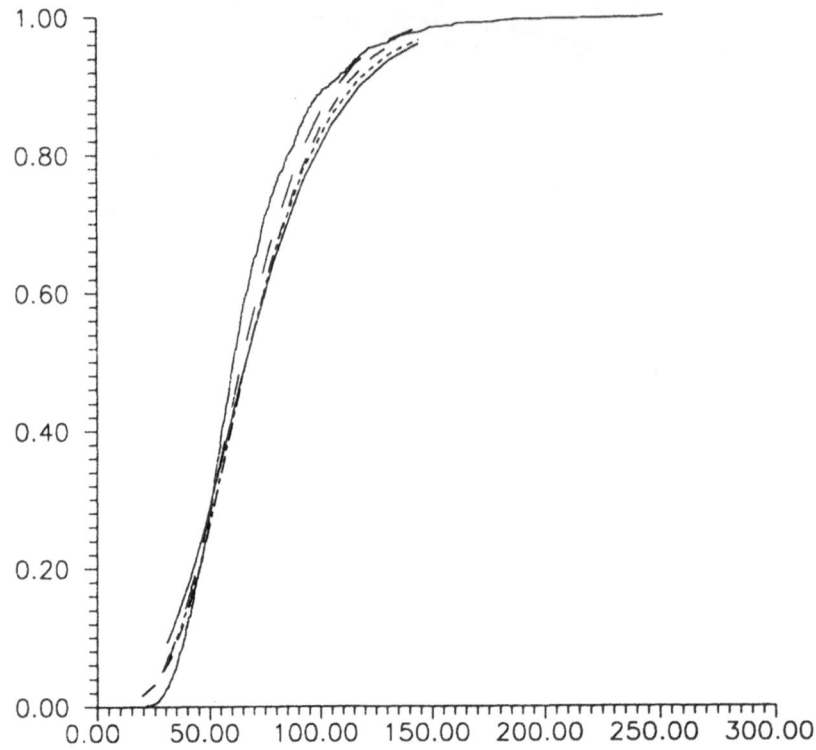

Figure 6: Problem B - Comparison, in a linear scale, between the simulation result and different response surface approximations. The response surface is built in the original space. The solid line is for technique 1.; the short dashed line is for technique 1. with non-orthogonal blocks for the factorial design; the medium dashed line is for technique 2.; the long dashed line is for technique 3.

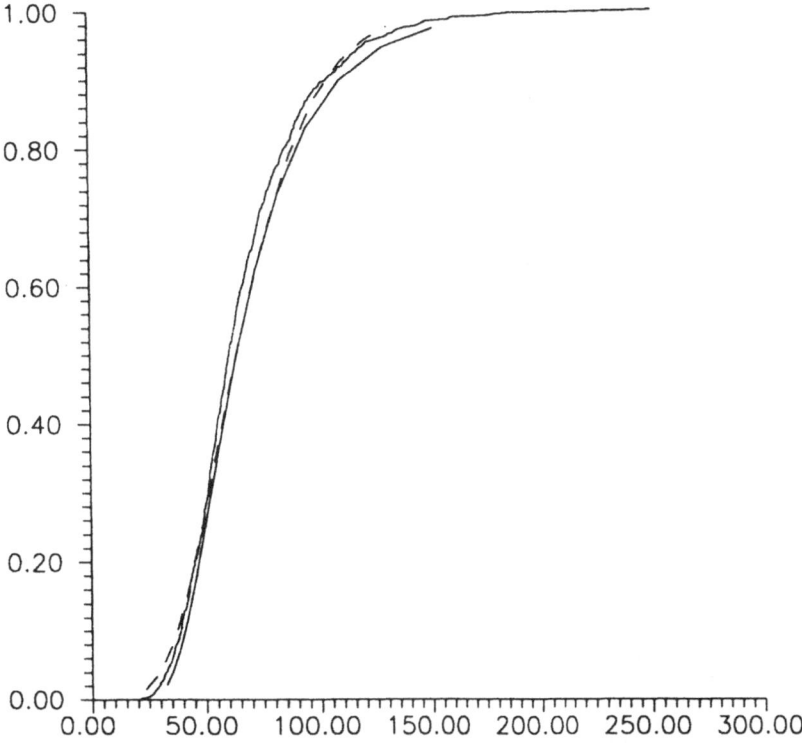

Figure 7: Problem B - Comparison, in a linear scale, between the simulation result and different response surface approximations. The optimal transformation is introduced. The solid line is for technique 3.; the dashed line is for technique 4.

0.05) with a vertical load of 1 t at the end.

The beam section is 20 cm by 5 cm and the Young modulus is 2100 t/cm². Two different problems are studied:

[A.] The height of the beam section is regarded as a stochastic process. Its mean is random (Gaussian variable of mean 20 cm and coefficient of variation 0.10) and the deviations from the mean are a Gaussian white noise with zero mean and standard deviation 2 cm. The Young modulus is deterministic.

[B.] The height of the beam section is defined as in problem A., but also the Young modulus is regarded as a stochastic process. Its mean is random (Gaussian variable of mean 2100 t/cm² and coefficient of variation 0.05) and the deviations from the mean are a Gaussian white noise with zero mean and standard deviation 105 t/cm².

The response of interest is the end vertical displacement which is given in millimiters in the following figures. The strong nonlinearity on the section height results in a large unsimmetry of the probabilistic distribution of the response.

Before the results are illustrated two further aspects must be clarified: a simulation of size 1000 is used for comparison among the different techniques; the second order polynomial response surface is not generally written in terms of the response y but of its optimal transformation Y. The latter one is found automatically by the computer code resulting from the research activity on this subject. Nevertheless, comparisons are also made for response surfaces written directly on y. The results are summarized in Figures 4 to 7 for problem B and in Figures 8 to 9 for problem A.

The sequence of Figures 4 to 7, for the problem with three random variables ($k = 3$) and two stochastic processes, makes evident the advantage of techniques 2. and 3. over technique 1., which, by contrast, is not affected to much by errors in forming the orthogonal blocks. Moreover, technique 4. provides a further improvement to the technique 3..

For the problem with two random variables ($k = 2$) and one stochastic processes, the advantage of technique 2. is still evident but technique 3. looses precision over technique 1. due to the use of just two blocks in the experiment design. Nevertheless, technique 4. still provides a further improvement to the technique 3..

6 Conclusions

The link between structural response measures and the associated values of the design variables is regarded as a mapping problem. Classical and non-classical (or alternative) regression analysis schemes have been considered for approaching it. Among them, response surface turns out the algorithm which in principle is most suitable for incorporating the random spatial distribution of material properties.

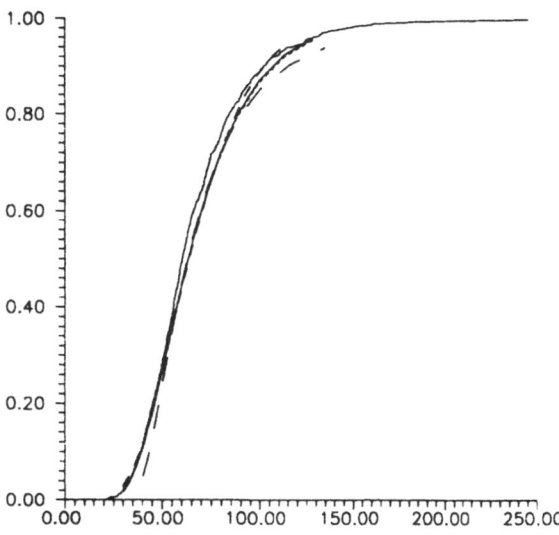

Figure 8: Problem A - Comparison, in a linear scale, between the simulation result and different response surface approximations. The optimal transformation is introduced. The solid line is for technique 1.; the medium dashed line is for technique 2.; the long dashed line is for technique 3.

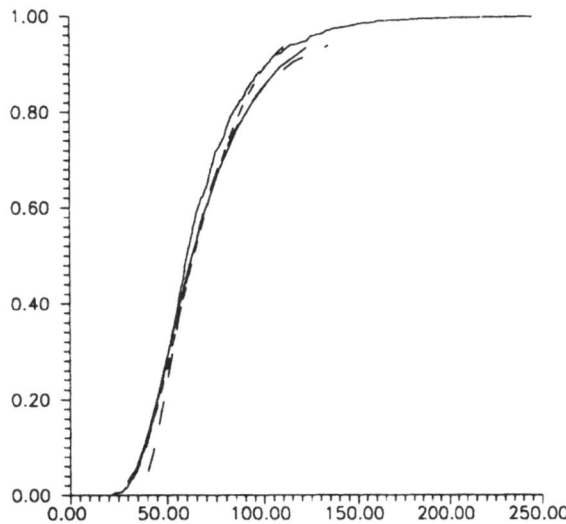

Figure 9: Problem A - Comparison, in a linear scale, between the simulation result and different response surface approximations. The optimal transformation is introduced. The short dashed line is for technique 2.; the long dashed line is for technique 3 and the solid line is for technique 4.

Nevertheless, the actual implementation of coupling random fields and response surface requires subtle and sophisticated considerations of blocking theory. Some alternative operative approaches were proposed and their performance was compared toward a simple numerical example of stochastic finite element analysis. The effect of design replication on the conclusions reached in this paper will be the object of further successive developments.

Acknowledgement - This research was supported by a grant from the Innovative Technology Committee of the Italian Research Council (CNR - 93.03398.CT11).

References

Augusti G., Baratta A. and Casciati F. (1984), **Probabilistic Methods in Structural Engineering**, Chapman & Hall, London

Box G.E.P. and Draper N.R. (1987), **Empirical Model-Building and Response Surfaces**, John Wiley, New York

Breitung K. and Faravelli L. (1993), Log-likelihood Maximization and Response Surface in Reliability Assessment, **Nonlinear Dynamics**

Casciati F. and Faravelli L. (1989), Biaxial Damage Assessment, **Proc. ASCE Structures 89**, 497-506

Casciati F. and Faravelli L. (1990), Expert Systems and Seismic Vulnerability, **Proceedings 9th ECEE**, Moscow, Vol I: 23-32

Casciati F. and Faravelli L. (1991), **Fragility Analysis of Complex Structural Systems**, Research Studies Press, Taunton

Casciati F. and Faravelli L. (1992), A Knowledge-Based System for Seismic Vulnerability Assessment of Masonry Buildings, **Microcomputers in Civil Engineering**, 6, 291-301

Casciati F. and Faravelli L. (1993), Seismic Prevention for Monumental Heritage by Artificial Intelligence Diagnosis, in **Conservation of Stone and Other Materials**, (Thiel M.J. ed.), Chapman & Hall, London, vol.2, 439-444

Casciati F., Colombi P. and Faravelli L. (1993), Lifetime Prediction of Fatigue Sensitive Structural Elements, **Structural Safety**, 12, 105-111

Faravelli L. (1989), A Response Surface Approach for Reliability Analysys, **Journal of Engineering Mechanics, ASCE**, 115 (12), 2763-2781

Faravelli L. (1992), Structural Reliability via Response Surface, in **Nonlinear Stochastic Mechanics** (Bellomo N. and Casciati F. eds.), Springer Verlag, 213-224

Faravelli L. and Bigi D. (1990), Stochastic Finite Elements for Crash Problems, **Structural Safety**, 7 (3-4), 113-130

Faravelli L. and Gherardini P. (1992), Expert System Modules for System Control, in Wen Y.K. (ed.), **Intelligent Structures II**, Elsevier, 128-143

Nexpert Object (1987), **Reference Manual** (Rel. 2.0), Neuron Data, Palo Alto, CA (USA)

Petersen R.G (1985), **Design and Analysis of Experiments**, Marcel Dekker inc., New York

Spieghlhalter D.J. and Lauritzen S.L. (1990), Sequential Updating of Conditional Probabilities on Directed Graphical Structures, **Networks**, 20, 579-605

Veneziano D., Casciati F. and Faravelli L. (1983), Method of Seismic Fragility for Complicated Systems, **Proc. 2nd CSNI Specialist Meeting on Probabilistic Methods in Seismic Risk Assessment for Nuclear Power Plants**, Livermore, 67-88

EXPERIMENTAL DESIGNS FOR AN EXPERIMENTAL MODELLING OF MATERIAL OR STRUCTURAL TESTS

G. REGNIER
ENSAM - LTVP, Paris, France
B. SOULIER
ENS - LMT, Cachan, France

Abstract
The experimental study of material or structural behaviours often consists of a multiparametric analysis. To describe the physical phenomenon in a limited region, it is valuable to identify the coefficients of a multilinear model with a regression. But the precision on these coefficients depends on the arrangement of the test matrix which represents the parameter values taken for each test. The optimal matrix form is reached by an orthogonal test matrix, but these orthogonal designs are restricting and sometimes impossible to realise because of physical constraints. Nevertheless, according to a criterion, optimal designs can be defined by respecting certain constraints : D-optimal designs are presented.
    The influence of identified noise parameters can be studied and minimised by a specific arrangement of test matrix : a product design. In this way, the time can be taking into account.
    Finally, a precise parametrical study of numerical or finite element models can be performed with experimental designs where the code runs are considered as tests.
    Keywords : Linear Regression, Experimental Modelling, Orthogonal Design, D-optimal Design.

1. Material and structural tests

The performance of industrial parts often depends on good material and structural studies. Several inner and outer parameters can have an influence on the material or the structural behaviour : thermal and mechanical charges, atmospheric conditions, chemical environment, light or radioactive sources. The quality of the structural parts is generally linked to a multiparametric study in which several interactions between these parameters have to be taken into account.

1.1 Studied experimental problems
This paper discusses experimental problems where (Fig.1) :

D. Breysse (ed.), Probabilities and Materials, 197–208.
© 1994 *Kluwer Academic Publishers.*

The responses are clearly identified (problem outputs).
More than two controlled and independent parameters are
studied (problem inputs).
A complete model which can predict the physical
behaviour, does not exist.

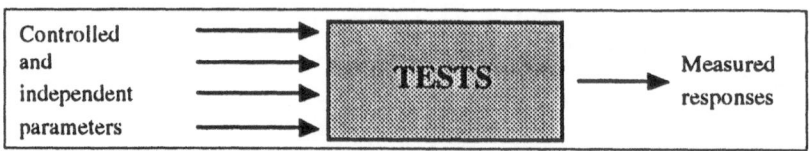

Fig.1. Form of studied experimental problems

Thus, the tests realised in order to update analytical or
numerical models are not taken into consideration.

1.2 Experimental data modelling

Choosing experimental test points has a considerable
interest for engineers and researchers to learn as much as
possible about experiments (Schimmerling, 1992). It is
still more important in the case of designs with a small
test number when experiment costs are high (e.g. material
or structural tests).
Let an experimental problem in which :

The response depends on p controlled and statistically
independent input parameters.
The parameter values are chosen in a limited
p-dimensional region in which it is desirable and
feasible to perform tests.

To get an image of the response surface and to determine
the influence of parameters in a given region, it is
advantageous to build from the experimental results a
multilinear model. The model coefficients are estimated
with a least-square calculation. However, the accuracy of
these coefficients depends on the experimental test points.
It is linked to parameter variance, parameter correlation
and response variance in the studied region.
The experimental points are very often chosen with
empiricism, without precise strategy. The purpose of this
paper is to discuss the choice of experimental points for
the coefficient identification of a linear model, that is
to say to study the supposed influent parameters.

2. Multi-linear regression

2.1 Calculation of regression coefficients

A linear model is an equation which expresses a response y
function of k parameters represented by k continuum
variables x1,...,xk (or discrete variables with whole
values for qualitative parameters) under the following
form:

$$y = c + \sum_{i=1}^{k} a_i . x_i + e \tag{1}$$

where c and the a_i are the k+1 coefficients to estimate, e
the error between the model and the response. This error is
supposed to be a random Gaussian variable.

Let the response y measured on a n-test-sample, the
model form is shown in fig.2.

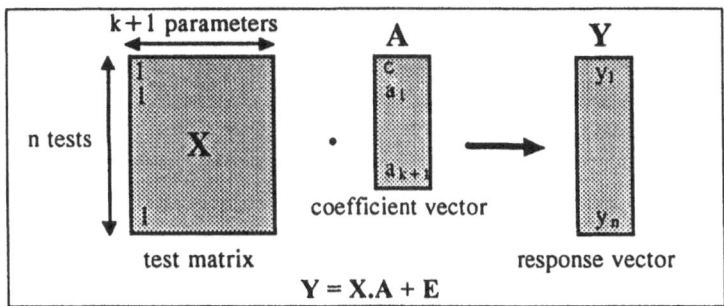

Fig.2.   Matrix representation of the problem (the ith line
of the matrix groups the parameter values taken in
the ith test, E is the error vector)

Let the ordinary error indicator E^tE : the sum of the
square differences between the estimated responses and the
measured responses. To minimise E^tE, we must calculate :

$$\frac{\partial}{\partial A} E'E = 0 = \frac{\partial}{\partial A} \left[(Y - XA)'(Y - XA) \right] \tag{2}$$

We obtain the estimated coefficient vector $\hat{A}$:

$$\hat{A} = (X'X)^{-1} X'Y \tag{3}$$

2.2 Experimental design principle

With the Gauss-Markov hypothesis (statistically independent
parameters and constant variance on the experimental
studied region), the variance-covariance matrix is equal
to:

$$\text{cov} (\hat{A}) = (X'X)^{-1} \sigma^2 \tag{4}$$

where the diagonal terms are the coefficient variances
(cov(ai,ai)=var(ai)) and the non-diagonal terms the
correlation between parameters:

$$[\text{cov}(ai,aj)] = \begin{pmatrix} \text{var}(a1) & . & . & . & \text{cov}(a1,ak) \\ & . & \text{var}(a2) & . & . & . \\ . & & . & . & . & . \\ . & & & . & . & . \\ . & & & & . & . \\ \text{cov}(ak,a1) & . & & . & . & \text{var}(ak) \end{pmatrix} . \sigma^2 \qquad (5)$$

For a given response variance, the coefficient precision depends on the matrix terms : the lower they are, the better is the precision. This rule is the basis of experimental design building which consists of choosing the experimental points to minimise the terms of $(X^tX)^{-1}$

2.3 Choice of experimental points

The optimal design occurs when $(X^tX)^{-1}$ is diagonal (matrix of Hadamart type), these designs are called orthogonal designs. Alternatively, several criteria can minimise the terms of the $(X^tX)^{-1}$ matrix, for example :

Minimisation of the $(X^t X)^{-1}$ trace
Maximisation of the $(X^t X)$ determinant (called D-optimal designs

3. Orthogonal designs

3.1 Building of test matrix

The building of the test matrix consists of a matrix choice in a set of orthogonal tables. Several methods have been developed to build these designs in an easier manner : Box and al. (1978), Taguchi and al.(1987). Nevertheless, the building of the test matrix remains often difficult, because it is an optimisation between :

The number of tests directly linked to the experimental costs.
The number of studied parameters and interactions.
The number of values for each parameters (which depends on the supposed linearity or non-linearity of each parameter)
The value of response variance.
The desirable precision on the parameter coefficients

At the moment, this work needs the intervention of an experimental design specialist who is able to limit the error risk for model coefficient calculation. These errors are generally due to several departure hypotheses.

3.2 An example of orthogonal design : local surface shrinkage determination on injected polymer plates

The experimental study of shrinkage for injected polymer parts is very important to understand and predict the part deformation, in order to reduce the high costs of mould prototype design.

The prediction of the processing condition influence (Fig.3) was performed on injected plates (Régnier and al., 1993). A plate mould was engraved with 30μm deep lines. The shrinkage at different locations on the plates was calculated by comparing the distances between the engravings on the mould and the reproduced engravings on the plates. The engraving distances were measured on translation stages under a microscope (Régnier and al., 1992).

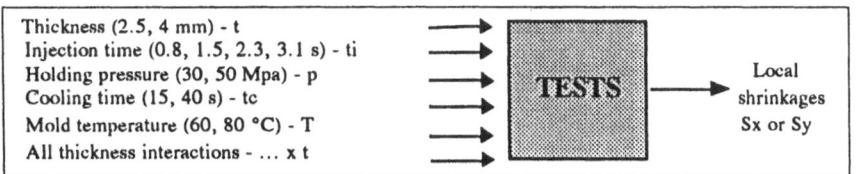

Fig.3. Processing conditions influence on the shrinkage

A sixteen-orthogonal design was performed twice (Fig.4) in order to identify a linear model (6) :

$$Sx(\%) = 1.335 + 0.02\ ti - 0.063\ ti^2 - 0.117\ p - 0.113\ tc + 0.018\ T$$
$$- 0.005\ ti.t - 0.042\ ti^2.t - 0.064\ p.t - 0.024\ tc.t - 0.02\ T.t \qquad (6)$$

Test N°	TINJ (s)	PM (Mpa)	EP (mm)	TREF (s)	TM (°C)	Rx (%)	
1	0.8	31.1	2.5	15	60	1.171	1.173
2	1.5	31.1	2.5	40	60	1.031	1.04
3	2.3	31.1	2.5	40	80	1.137	1.106
4	3.1	31.1	2.5	15	80	1.268	1.337
5	0.8	50.5	2.5	40	80	0.964	0.973
6	1.5	50.5	2.5	15	80	1.186	1.177
7	2.3	50.5	2.5	15	60	1.113	1.123
8	3.1	50.5	2.5	40	60	0.95	0.93
9	0.8	31.1	4	15	80	1.744	1.725
10	1.5	31.1	4	40	80	1.603	1.586
11	2.3	31.1	4	40	60	1.577	1.573
12	3.1	31.1	4	15	60	1.802	1.793
13	0.8	50.5	4	40	60	1.129	1.123
14	1.5	50.5	4	15	60	1.454	1.512
15	2.3	50.5	4	15	80	1.489	1.522
16	3.1	50.5	4	40	80	1.146	1.117

Fig.4. Test matrix and shrinkage measurement
(at a certain location on the plate)

According to the experts, the model form is always chosen before the tests .In order to reduce the number of

tests certain terms should be neglected (quadratic terms
and interactions) . In fact, the experiment design is
appropriate because eleven coefficients could be precisely
(see variance analysis in the next chapter) identified with
only sixteen tests.

To compare the parameter influence, it is better to
normalise the parameter variation : -1 for each minimum
value and +1 for each maximum value. The mean effect of
some parameters or interactions can be seen and compared on
the graphs of the figure 5.

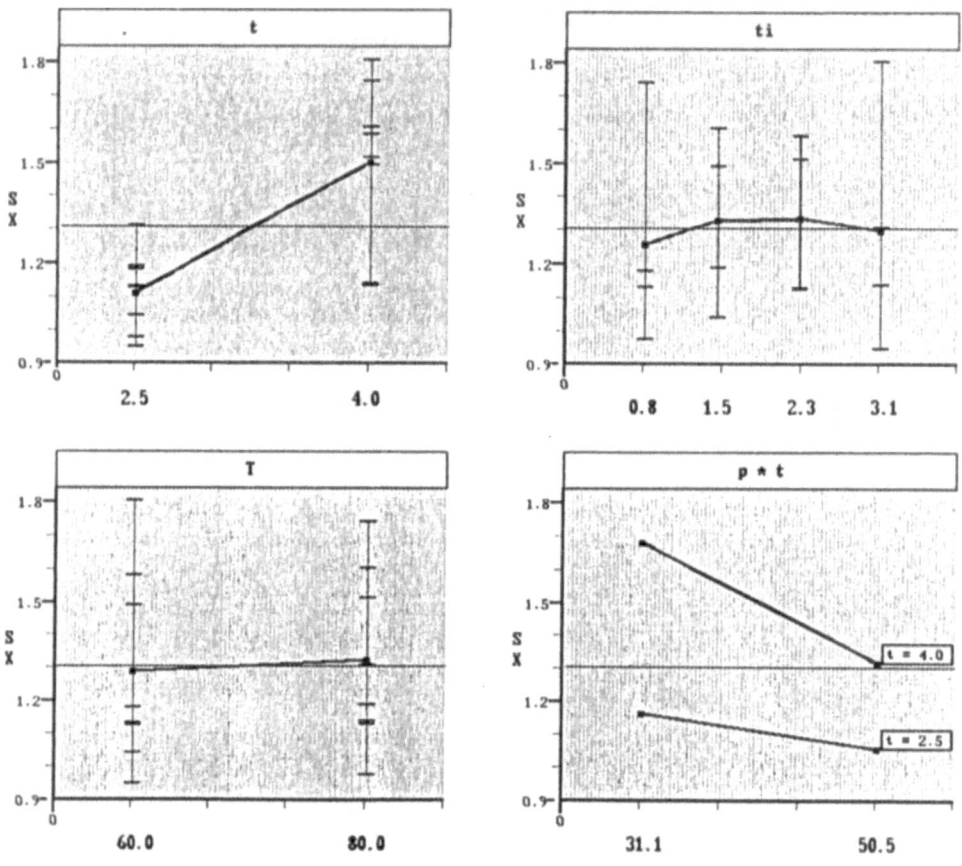

Fig. 5. Mean parameters effects

Variance analysis

The variance analysis is very important to validate the
hypotheses made for the initial model form and to evaluate
the quality of the design.

First of all, the repeatability error (which
characterises the variability of the measurements done
under the same conditions : in the same laboratory, on the
same test system, with the same operator) determinated, or
better the reproducibility error (Repeatability and

reproducibility are defined in the international standard ISO 5725). So, the repeatability variance can be compared to the residual variance (linked to the differences between model prediction values and measured values) : the supposed form model is appropriate if the two variances are close to each other (definition of adequation - fig.6)

Moreover, the comparison between the repeatability variance and the variance resulting from the parameter variation gives the parameter influence probability (Fig.6).

ANAVAR : MODEL-SX

Mean response				1.2992	Number of tests		32
Residual squared adjustement				0.993			
Standard deviation of residues				0.0221	Model degree of freedom		14
Standard deviation of repeatability error				0.0198	Number of repetitions		2
Fisher ratio		3.19	Probability	0.265	Adequation		YES

Factors	Squares sum	d.o.f	Standard d.	Fisher ratio	Probability	Conclusion
p	0.441	1	0.664	1130.0	0.0	HIGHLY.SIGNIFICANT
T	0.0107	1	0.104	27.4	8.22E-5	HIGHLY.SIGNIFICANT
tc	0.406	1	0.637	1040.0	0.0	HIGHLY.SIGNIFICANT
t	1.21	1	1.1	3080.0	1.28E-7	HIGHLY.SIGNIFICANT
ti	0.0319	3	0.103	27.1	1.89E-6	HIGHLY.SIGNIFICANT
ti * t	0.0113	3	0.0612	9.57	7.42E-4	HIGHLY.SIGNIFICANT
tc * t	0.0185	1	0.136	47.3	3.81E-6	HIGHLY.SIGNIFICANT
T * t	0.0131	1	0.115	33.5	2.79E-5	HIGHLY.SIGNIFICANT
p * t	0.133	1	0.365	340.0	0.0	HIGHLY.SIGNIFICANT
Residues	0.00877	18	0.0221			
Total	2.28	31	0.271			

Fig.6. Variance analysis
(specific code developed at Dassault Aviation)

4. D-optimal designs

The use of orthogonal designs is limited and sometimes impossible when :

There are constraints between parameters,
Old tests must be taken into account.

The strategy consists of adding and deleting one or more tests to the initial tests to achieve an increase of the $[X^t X]$ matrix determinant. Several algorithms are well known such as the Fedorov's algorithm (1972) or the DETMAX algorithm (Mitchel, 1974).

The D-optimal design method is described by an academic example : the study of the breaking force for composite plates bolted on metal plates (fig.7).

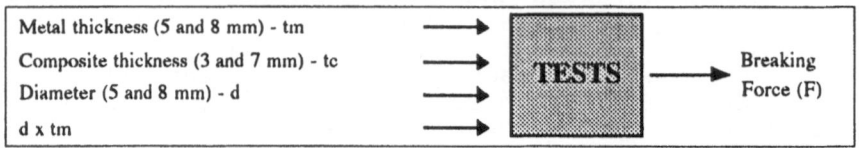

Metal thickness (5 and 8 mm) - tm	→		
Composite thickness (3 and 7 mm) - tc	→	TESTS	Breaking
Diameter (5 and 8 mm) - d	→		Force (F)
d x tm	→		

Fig.7. Test definition of composite bolted plates.

A linear experimental model of the assembly breaking force built with an orthogonal design (7) is compared to a D-optimal one (8). The two designs are built with eight tests (fig.8). This is sufficient if the influence of each parameter are supposed linear (moreover, all interactions could be determined without correlation for the orthogonal design). The D-optimal design respects a technological constraint : "tm/tc > 0.5"

Orthogonal design :
$$F(N) = 77618 + 13895\,d + 14697\,tc + 8768\,tm + 2830\,d.tm \tag{7}$$
D-optimal design
$$F(N) = 75533 + 13405\,d + 16845\,tc + 6683\,tm + 2340\,d.tm \tag{8}$$

Orthogonal design

d	tc	tm	F
5.0	3.0	5.0	45073
5.0	3.0	8.0	58906
5.0	7.0	5.0	70496
5.0	7.0	8.0	80415
8.0	3.0	5.0	58683
8.0	3.0	8.0	89023
8.0	7.0	5.0	101146
8.0	7.0	8.0	117200

$\det(X^{t}X) = 32768$

D-optimal design

d	tc	tm	F
5.0	3.0	5.0	45073
5.0	5.0	8.0	69371
5.0	7.0	5.0	70496
5.0	7.0	8.0	80415
8.0	3.0	5.0	58683
8.0	5.0	8.0	95566
8.0	7.0	5.0	101146
8.0	7.0	8.0	117200

$\det(X^{t}X) = 20480$

1/8	0	0	0	0
0	1/8	0	0	0
0	0	1/8	0	0
0	0	0	1/8	0
0	0	0	0	1/8

11/80	0	-1/20	1/80	0
0	1/8	0	0	0
-1/20	0	1/5	-1/20	0
1/80	0	-1/20	11/80	0
0	0	0	0	1/8

Covariance matrix

Fig. 8 Comparison of orthogonal and D-optimal design

The comparison of the two covariance matrix (fig.8) shows the loss of precision on model coefficient estimation for D-optimal design :

The increase of the diagonal term value corresponding to the coefficient variances of the model (except for the diameter and the interaction, for that matter the

variations of diameter and interaction coefficients are
low in the two models).
Low values for the non-diagonal terms of D-optimal
design : the correlations between factors occur but remain
low.
The simulation on the orthogonal design with both models
leads to a residual standard deviation equal to 6450 N for
the first model (7) and 8860 N for the second model (8). On
the orthogonal design domain, the model obtained by an
orthogonal design is better than the model obtained by a D-
optimal design. But the residual standard deviation of the
D-optimal design on its smaller domain is equal to 5800 N.

These results show that D-optimal designs can be an
alternative method with a limited accuracy loss when
orthogonal designs cannot be used.

Moreover, the confidence interval on each coefficient
estimation can be calculated : the response repeatability
variance and the coefficient variance (covariance matrix)
are known.

5. Other types of experimental design applications

5.1 Orthogonal product design

This particular method (Taguchi and al., 1989) makes it
possible to minimise the influence of identified noise
parameters. This method consists of repeating the principal
design for all configurations of the noise parameters just
controlled for the tests. Of course, the test number
increases quickly. In fact, all the interactions between
noise parameters and others parameters are calculated. The
optimisation can be achieved by choosing robust
configurations of principal parameters by adapting
interactions values of noise parameters (Shoemaker and al.,
1991).

This method can also be applied for time analysis in
which the time is considered as a noise parameter in order
to know the evolution of time influence for each parameter
or principal interaction (Pillet, 1992).

5.2 Numerical experimental design

Many scientific phenomena or structural behaviours are now
investigated by complex finite element models. In general,
the exploitation of these models is not optimal. A computer
experiment is a number of code runs with various parameter
inputs. The experimental design technique makes it possible
to carry out a parametrical analysis in a limited region
(Sacks and al., 1989 and 1990). The response could be
predicted all over the test region with a Bayesian approach
(Currin and al., 1988), but although the output of a
computer is deterministic. An image of the response surface
can be quickly obtained with a multilinear model. This

simplified model can be assimilated to a power expansion in the studied region.

The finite element model analysis of a bolted C/Sic plate is shortly presented (Fig.9).

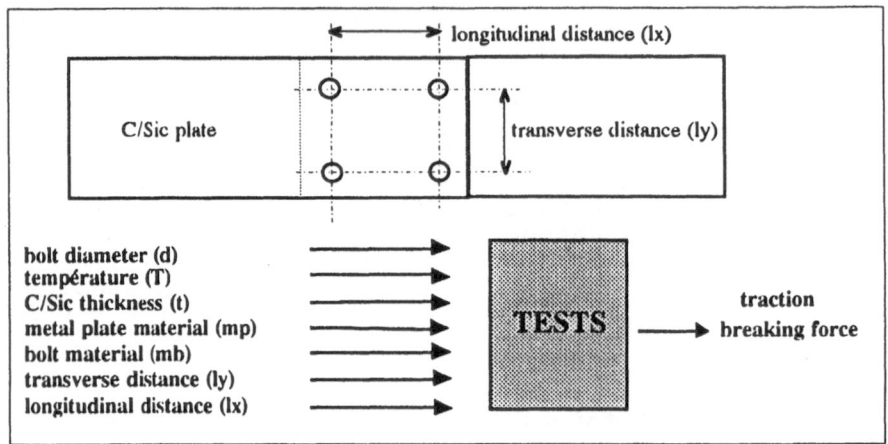

Fig.9.   Inputs and output of a finite element module

According to an orthogonal design, thirty-two code runs were performed for 4 temperature values (140, 260, 380, 500°C). For each temperature, the coefficients of a linear model have been identified and the parameter influences has been determined (9).

$$F_{140}(N) = 23486 + 492\ ly - 490\ lx - 5458\ t - 291mb - 313\ mp$$
$$+7883\ d - 221d^2 + 1194\ d.t - 1273\ d^2.t - 404\ lx.ly \tag{9}$$

In fact, a product design (4 x 32 tests) was realised to know the influence evolution of all parameters by varying temperature (Fig.10).

Conclusion

For material tests, when an experimental and multiparameter study has to be done, several benefits can be obtained by using experimental designs:

A multilinear model can be directly identify from tests.
Interactions between parameters can be studied.
For a given number of tests, a better precision for model coefficients can be obtained.
When a repeatability study has been done, the variance analysis give a probability for each influence and indicates if the supposed model is adequate.

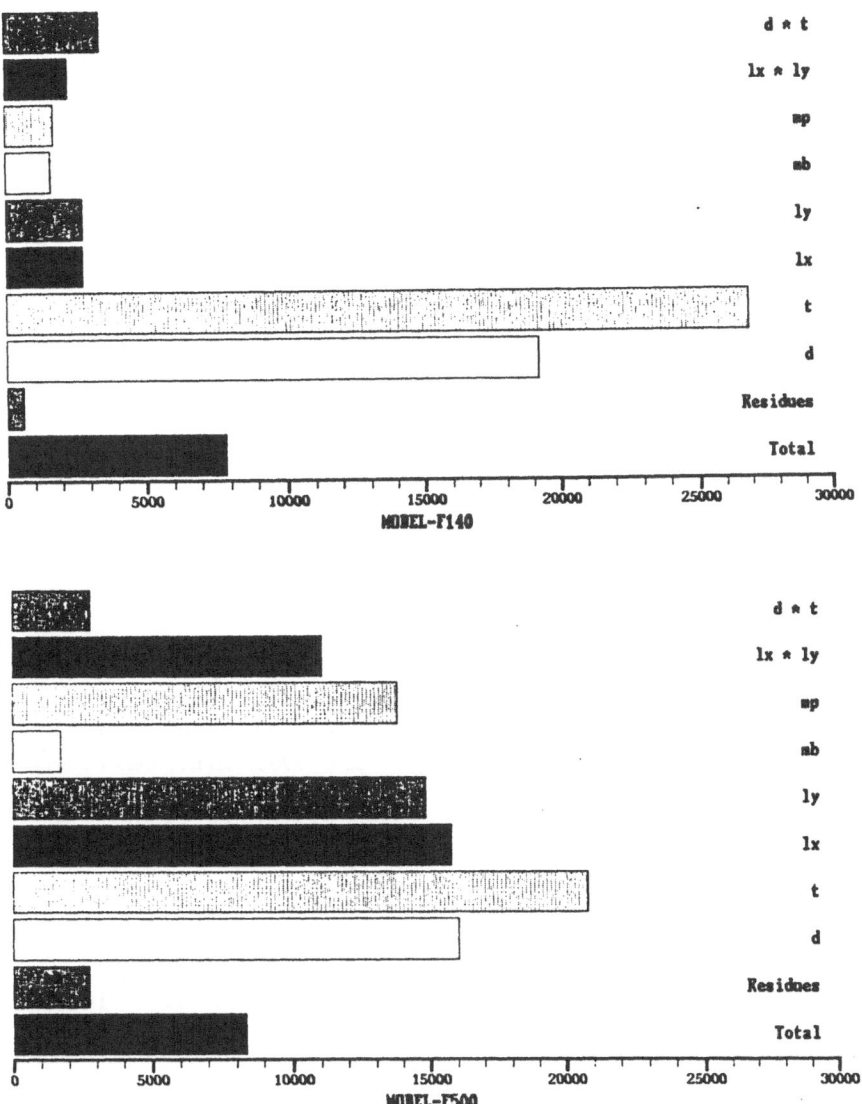

Fig.10. Comparison of the parameter influences for 140°C
         and 500°C, with the calculated induced variances by
         the parameter variations

The research of the best linear model can be performed, according to the tests and to a criterion : the minimisation of the residual variance for example.

By using product designs, the minimisation of the noise parameter influences can be performed. With the same approach, the variation of the principal parameter influences can by analysed when parameters such as time or temperature are varying.

Finally, for finite element models, precise parameter study in a defined region can take advantage of this working method.

References

Box G., Hunter W.G. and Hunter J.H. (1978). Statistics for experimenters. **Wiley Interscience**.

Currin C., Mitchell T., Morris M., Ylvisaker D. (1988) A bayesian approach to the design and analysis of computer experiments. Oak Ridge National Laboratory - Martin Marietta - **ORNL 6498**.

Fedorov V.V. (1972). Theory of optimal experiments. **Academic Press**, New York.

Mitchell T.J. (1974). An algoritm for the construction of D-optimal experimental design. **Technometrics**, vol 16-2, 203-210.

Pillet M. (1992). Introduction à la méthode des plans d'expériences par la méthodologie TAGUCHI. **Editions d'Organisations**, Paris.

Régnier G. and Trotignon J.P. (1992). Local orthotropic shrinkage measurement in injected moulded polymer plates. **Polymer Testing**, accepted 21 nov. 1992.

Régnier G. and Trotignon J.P. (1993). Local orthotropic shrinkage prediction in injected moulded polymer plates. **PPS 9**, 59-60.

Sacks J, Schiller S.B., Welch W.J. (1989). Designs for computer experiments. **Technometrics**, vol 31, 41-47.

Sacks J, Welch W.J., Mitchell T.J., Wynn H.P. (1990). Design and analysis of computer experiments. **Statistical Science**, vol 4, 409-435.

Schimmerling C.(1992). Design of experiment for the improvment of testprocedure : application to engine coolant. **Workshop DEINDE** (DEsign INDustrial of Experiments), Politechnico di Torino, 11-12 nov.1992.

Shoemaker A.C., Tsui K.L., Jeff Wu C.F.(1991). Economical experimentation methods for robust designs. **Technometrics**, vol 33, n°4.

Taguchi G. and Konishi S. (1987) Orthogonal arrays and linear graph. **American Supplier Institute Press**.

Taguchi G., Elsayed A., Lisiang T. (1989) Quality engineering in production system. **Mc Draw-Hill International**.

REPORT ON THE FIRST ROUND TABLE SESSION: TOOLS

LUC DAVENNE* and JEAN-PIERRE VILOTTE**
* Laboratoire de Mécanique et de Technologie, Ecole Normale Supérieure de Cachan,
94235 Cachan Cedex, France
** Département Terre Atmosphère Océan, Ecole Normale Supérieure,
75231 Paris Cedex 05, France

Abstract
This paper is the report on the round table discussion about the tools. The authors made a subjective arrangement of the discussion into several themes and kept only the main ideas of the speaker talkings. They hope they didn't change the meaning of the people tellings in rewriting some phrases.

1 Introduction

Most of the discussion was concerned with the physical properties and the response of non linear and non homogeneous materials under typical engineering situations. The common underlying problem is to relate the macro-scale mechanical response to the micro-structure and to the physical processes which take place at that scale. Different aspects have been discussed:
The first one deals with potential tools for describing the non homogeneous space distributions of the physical and the geometrical variables at the different scales.
The second one is more concerned on how and at which level such an information should be introduced in order to predict a macroscopic response. When is a probabilistic approach really needed?
The third one is on the use of computer simulation when dealing with stochastic variables or fields.
And the last part deals with the actual use of probabilistic models in engineering compared to the use of deterministic models.

2 Fractals as a descriptive tool

Fractals and multi-fractals seem to be a tool to describe the disorder. However, one have to bare in mind that fractal distributions are practically limited within some range defined by a lower and upper cut-off. Therefore the use of a fractal description must include those upper and lower cut-off. [Bolle].
Fractals are often associated with probabilistic models while it seems that they are more related to a deterministic point of view. [Breysse].

209

D. Breysse (ed.), Probabilities and Materials. 209–212
© 1994 *Kluwer Academic Publishers.*

Fractal dimensions are in general measures of the disorder and they can be deterministic or not. [Carpinteri].

There is often a confusion about fractals and probabilistic models. I do see why fractals are sometime related to probabilistic models. It appears to me that fractals are mainly a description of some geometrical invariance which might be observed as a result of a deterministic or a probabilistic process. They are indeed often but not always associated with randomness. In the nature, many observations in space, but also in time, do exhibit fractal statistics which may be analysed by particular tools. Those tools are nowadays well developed for time series, as encountered in meteorology or fluid dynamics. [Vilotte].

Fractals provide useful statistical tools to analyse observed variability. It seems clear that we do not, and never will be able to describe nature in all its complexity. When dealing with problems involving a wide spectrum of scales, it is of importance to be able to assess some scaling laws in order to extrapolate experimental measures. [Soulié].

In a lot of physical phenomena we do have a loss of information. In that situation we do have tools to measure this. The fractal statistics are in that sense one of those tools. Those measurements tell us that information is lacking and that there is a kind of disorder. From this point, each of us can have different approaches. [Casciati]

We ear more and more about fractals: the fractal geometry of fracture, roughness or related topics. Most of the work on fractals appears to be essentially descriptive, providing numbers like fractal dimensions, and to have its best applications on image processing: compression and storage, movie production. Applications in engineering or mathematical applications are still scarce. There is no fractal calculus in the sense of formulating differential or integral operators, boundary value problems. We still do not know how to write elasticity equations for a medium which is fractal or which have fractal boundaries. How to characterise operators such as divergence, gradient, and others for such a medium ? Very few projects in that direction meat the standards of physical description such as invariance. Can we formulate a fractal calculus ? [Chudnovski]

There is a beginning of stochastic boundary value problems, for example the work of Panagiotopoulos who treated elastic structures with fractal boundaries. [Ostoja-Starzewski]

Fractals are a good tool to describe things, they were not introduced to understand things. If we describe things, then we are able to understand things and then we will go to a predictive level. [Franceskonis]

When we speek about fractals we often think about self-similarity. It exist other scale anisotropic invariances, the self-affinity, that is useful in the describing of the fracture surfaces. [Chudnowski]

3 Micro to macro-structure response

We are here mainly concerned with the homogenisation of non linear materials. Most of the work going on in that direction is based on a deterministic approach since we have no tools for a probabilistic approach in the case of non linear behaviours. The real question would be in fact whether it is necessary or not to use probabilistic tools in homogenisation. Good constitutive behaviour approximations of composite materials are indeed obtained using just variational principles and a self-consistent approach, or some modified versions. At the macroscopic scale where you do have a lot of averaging, the material will usually have a deterministic behaviour. The picture is rather different when the overall behaviour of the material is governed by very local properties or rare events such as in the case of fracture. For this case, very local defects shall control the initiation of the cracks, and small parts of the micro-structure shall drive the overall behaviour. In such a case, global averaging procedures such as the classical homogenisation will be misleading. Indeed you may have cracks or a network of cracks for which the volume fraction of cracks is zero and where a simple global averaging will not work. Probabilistic approach should be used when the physics is controlled by very rare events. [Jeulin]

Homogenisation which is somehow an averaging technique is not well adapted for problems like crack propagation which are sensible to local fluctuations and defects. Furthermore those fluctuations may strongly change in time as the crack and the damage processes evolve. In such a case we do need some statistical tools to describe those fluctuations in space and their time evolution from witch we can derive some macroscopic description.[Vilotte]

In fracture, events are at various scales, starting from the macro going to the micro-scale. They are so connected that we cannot break the chain. We start with uncertainty at some micro level and it cascades up to the macroscopic level because of the strongly non linear and unstable phenomena. Like in the critical phenomena, it has a very high sensitivity to the fluctuations at all scales. In such cases, other methods than probabilities are helpless. But the models become very complex with a lot of adjustable parameters, each of them is random, this means that each of them is a base of distribution. This represent such an amount of information that it rises the question on how predictive the result is. [Chudnovski]

4 Computer Simulation

Laboratory experiments are often very difficult and time consuming. Getting a few points on a curve can take several years. It is therefore quite fascinating and tempting for material scientist to use computer simulations. Some of those simulations start to build a model and then to introduce some disorder. In doing so, one has to be quite cautious and pay attention to validate the model and make it robust. In particular, disorder has to be introduced in the basis of physical arguments and observations. Problems with disorder can easily be simulated by putting a specific noise at a certain level. The main problem is to identify the real source of noise and to check how this problem is sensitive to other noise parameters. Otherwise it may confirm the old theory that the probabilistic methods can proof anything. [Casciati]

Numerical simulations using random variables or random fields are intrinsically difficult and need to be implemented carefully. In particular one have to pay attention to the discretisation and to the relative scale of the discretisation versus the micro-structure. One problem in the discretisation of random fields or of random variables, is that neither the discretised histograms nor the distributions of the correlation function are known. This also depend on whether the discretised variables are additive or not. In the first case, the change of the distribution as a function of the support is well known. It is the result of simple convolutions. In the later case, no simple solution exist and one needs some models. This is the case for variables such as the elastic moduli. Here, fractals or multi-fractals can help to introduce some scale effects. [Jeulin]

5 Probabilistic approaches

There is apparently an unequal treatment between the deterministic and the statistical models. Well posed deterministic laws such as the Newton's laws in mechanics, the electromagnetic laws, the diffusion laws, which all have been for a long time part of our educational training, are nowadays commonly accepted. Although we know how to collect statistical data from experiments, how to treat them and fit some physical response, statistics still appears in engineering as secondary with respect to the more classical. The picture is quite different in physics, at least if we recall quantum mechanics. For problems in which defect distributions at a small scale is of primary importance like in the case of the fracture phenomena, the macroscopic results do possess uncertainties. On a philosophical basis, I would say that a challenging question rises here: can we formulate variational principles which would allow to formulate deterministic and statistical mechanics under the same roof? Can we formulate a variational approach in statistical continuum mechanics? [Chudnovski]

Today, the statistical methods are still in their majority just descriptive. There is a scepticism in the community toward probabilistic methods because there is a lack of examples that allow us to predict something. [Chudnowski]

There is different way of doing statistical mechanics. It can be only statistical analysis or it can be a probabilistic mechanical analysis. I think that probabilistic methods are coming naturally. They are already often advocated when dealing with extreme events. In offshore engineering, all codes which lead to offshore design are based on probabilistic methods, this is the same in aerospace engineering. There is also a beginning in the Eurocodes for civil engineering. [Schueller]

In engineering, especially geotechnical engineering, people are used to think in terms of probability or statistics. In mining engineering, it is a standard to describe slope stability, and decisions are based on probabilistic models. [Einstein]

Other fields like mining or petroleum industry have a lot of experience in describing their phenomena using probabilistic approaches. In those fields they use stochastic approaches, and numerical models in order to describe flow in complex media, or to interpolate data. [Jeulin]

MORPHOLOGICAL RANDOM MEDIA FOR MICROMECHANICS

D. JEULIN
Centre de Geostatistique, Ecole des Mines de Paris, Fontainebleau.

Abstract
As a consequence of microstructural heterogeneities, fluctuations in the
mechanical properties of materials are observed in experiments. This requires
a probabilistic approach to relate the microstructure to the overall materials
properties, and to predict scale effects in the fluctuations of properties.
In this presentation, the problem of the strength of materials, is specifically
addressed, and various models of random structures developed for fracture
statistics are introduced. In fact, any fracture criterion is sensitive to
microstructural heterogeneities, such as flaws with low strength, or defects
inducing a local stress concentration. There is a large effect of small scale
heterogeneities for fracture phenomena. The approach combines the selection
of appropriate fracture criteria and random structure models.
Keywords: Fracture Statistics, Change of Scale, Random Structures, Weakest
Link, Crack Arrest.

1 Introduction
Prediction of the overall behavior of heterogeneous media from their structure
is of importance in many fields of applied sciences: in materials science,
this is the way to design and produce materials with customized
microstructures, as far as the final use properties are concerned. When
considering the physical properties of heterogeneous media and their
fluctuations at various scales, it is necessary to introduce appropriate
techniques and models. This problem is encountered in many situations, such
as flows in porous media, elastic properties, strength of composite materials,
etc. It is solved by different methods, using in any case models of random
media.

The macroscopic (or effective) properties (like an overall elasticity
modulus) can be estimated from partial knowledge of the microstructure (lower
scale) by variational methods. In that case, approximations of average
properties of infinite random media are obtained, using homogenization
techniques as in Matheron (1967) or in Sanchez Palencia and Zaoui (1987).

The statistical distribution of the properties of finite domains can be
obtained theoretically for physical properties sensitive to local
heterogeneities, like strength of materials. Scale effects, like the change
of mean properties and of their variance with the size of specimens, can be
predicted and compared to experimental data.

In this paper, some recent models for fracture statistics models are
reviewed. They enable us to predict fracture probabilities of materials under
various loadings and on different scales. The approach is the following: 1)

213

D. Breysse (ed.), Probabilities and Materials, 213–224.
© 1994 *Kluwer Academic Publishers.*

Choice of local (i.e. point-wise) and of macroscopic fracture criteria; the first type accounts for crack initiation, and the second for the fracture of a specimen. 2) Random structure models, defined on a point scale, for which the calculation of the fracture probability and of scaling laws is possible; this usually involves simplifications, such as the use of the stress field seen by an equivalent homogeneous medium, using the so-called local approach (Pineau, 1981).

The proposed random function models with a point support, and appropriate change of supports, are introduced below.

2 Choice of a fracture criterion

The first step required to develop fracture statistics models is the choice of **local and of global fracture criteria.** A local criterion is sensitive to the fracture initiation, while a global or macroscopic criterion accounts for the fracture of a domain.

2.1 Local fracture criteria

Various local criteria can be used: the fracture is initiated at points in the structure where some intrinsic property of the material is reached, as the result of the applied load. Usually, this property is the **critical stress** $\sigma_c(x)$, or more generally the **critical stress intensity factor** $K_{Ic}(x)$ for the tensile fracture in linear elastic brittle materials (Jeulin 1991, 1992b). When there is **competition between several fracture mechanisms,** as for cleavage and intergranular fracture in rocks and in metals, multivariate criteria and multivariate random function models can be used (Jeulin 1991, 1992b). The **local fracture energy** $\gamma(x)$ corresponding to the creation of a fracture surface is used in (Jeulin 1992a).

2.2 Global fracture criteria

The following macroscopic fracture criteria, involving different fracture assumptions, were proposed (Jeulin 1991, 1992b):

The **weakest link model** is well suited for the brittle fracture of materials, as in the cleavage of steel at low temperature (Pineau 1981); it corresponds to a sudden propagation of a crack after its initiation.

Models with a **damage threshold** generalize the previous one; they are valid for a fracture with several potential sites for crack initiation.

Models with a **Griffith crack arrest** criterion compare, for each step of a crack path, the local fracture energy $\gamma(x)$ to the stored energy $G(x)$ due to the deformation of the material.

Formally, the first type of criterion uses a **change of support** of the information by the operator $\wedge$ (infimum); the second family is connected to a change of support by convolution; finally, the last criterion involves a change of support by the operator $\vee$ (supremum).

In the present approach, an **equivalent homogeneous medium** with a random critical stress is used. This simplification, which separates the applied field and the critical field, enables us to obtain closed form results without any simulation. This is justified for media with a single component, like polycrystals in metals or in rocks. However, this approach cannot account for small scale stress fluctuations induced by the microstructure when the components have different mechanical behaviors. The three above-mentioned macroscopic fracture criteria give the models introduced in the following sections.

3 Fracture statistics models for brittle materials
Based on the weakest link assumption, these models assume the fracture of a part, as soon as for a single point x_0, we have $\sigma(x_0) > \sigma_c(x_0)$. It corresponds to the immediate propagation of a crack after its nucleation. To estimate the probability of fracture with this assumption, it is necessary to know the probability distribution of the minimum of the values $(\sigma_c(x) - \sigma(x))$ over the loaded domain. This can be done for some random structure models when using the deterministic field $\sigma(x)$ seen by an equivalent homogeneous medium as in the local approach introduced by Pineau (1981).

For a stationary random function $\sigma_c(x)$, the probability of no fracture of a specimen B under the deterministic stress field $\sigma(x)$ is given by:

$$P\{\text{no fracture of B}\} = P(\sigma) = P\{x \in H_{\sigma c}(\sigma)\} \tag{1}$$

In Eq. (1), $H_{\sigma c}(\sigma)$ is the set of implantations of the specimen B for which the minimum of the values $(\sigma_c(x) - \sigma(x))$ remains positive. In general, the probability law (1) is not available. However it can be calculated in a closed form for specific models of microstructures, as developed in (Jeulin 1990, 1991, 1992b, 1993). For instance, the **Boolean random varieties** describe structures with different geometrical defects: points or grains, fibers, strata. They can be constructed in two steps as follows in the three dimensional space: start with a sequential set of Poisson varieties $V_k(t)_i$ (for $0 \le t \le T$) with intensity $\theta(t)$ (points for k=0, lines for k=1, or flats for k=2), as defined by Matheron (1975); consider independent primary random functions with compact support $Z'_t(x) \le \sigma_m$, where σ_m is the homogeneous strength of a matrix. The Boolean random variety of dimension k $Z(x)$ is obtained from:

$$Z(x) = \wedge \{ Z'_t(x-y); \ y \in V_k(t)_i; \ i \in \mathbb{N}; \ 0 \le t \le T \} \tag{2}$$

With this construction, various types of defects in a homogeneous matrix are obtained: random grain defects (negative of Fig. 1 in the plane), fiber defects (negative of Fig. 2 in the plane), and strata defects (inducing fibers by plane sections). For these models, the probability of fracture (1) is expressed as:

$$P\{\text{no fracture of B}\} = \exp -\int_0^T \bar{\mu}(H_{Z'_t}(\sigma - \sigma_m)^c) \ \theta(dt) \tag{3}$$

In Eq. (3), the average measure $\bar{\mu}$, taken over the realizations of Z'_t, depends on k; it is the volume of the set H^c (complementary of H) for grains, $(\pi/4)S$ (S being the surface area) for fibers, and the integral of mean curvature M for strata.

By construction, these models allow correlations between the critical

216

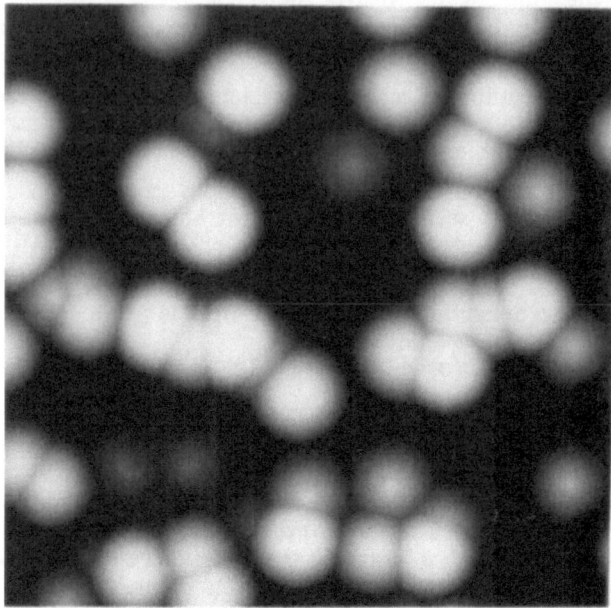

Fig. 1 Simulation of a Boolean random function in the plane (512x512x8 bits), using the supremum of conical primary random functions with radius 45 pixels.

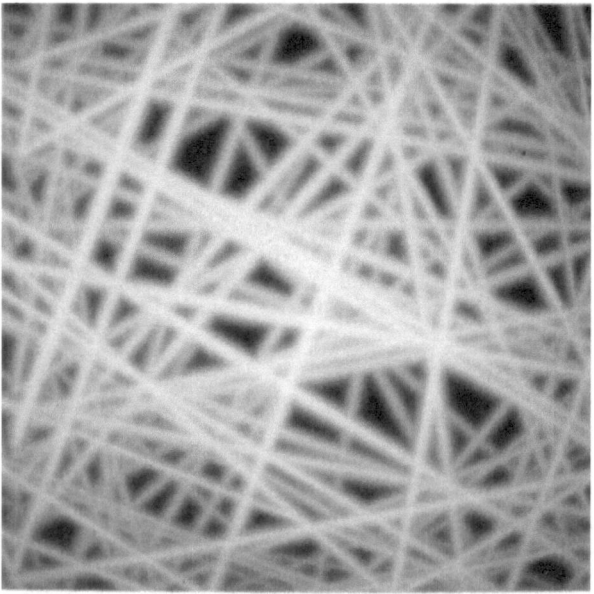

Fig. 2: Simulation of Boolean random fibers in the plane (512x512x8 bits), using the supremum of conical primary random functions with radius 15 to 25 pixels.

stresses observed at different locations (even on any distance for the random fibers or strata!). In addition, on a mesoscale, they predict through the general equation (3) size, shape (of the specimen) and microstructural effects on the fracture probability, that can be compared with experimental data. In the mentioned references, the effect of various stress fields (homogeneous, or at the crack tip of elastic-plastic materials) was studied.

A particular model of this type gives the well-known Weibull distribution, often used in applications, as a probability of fracture under a homogeneous stress field. Obtained with $\sigma_m = + \infty$, a power law intensity θ for $\sigma > \sigma_0$, and point defects, it gives:

$$P\{\text{no fracture of } B\} = \exp - V(B) \left(\frac{\sigma - \sigma_0}{\sigma_u} \right)^m \qquad (4)$$

Other examples include the Pareto distribution, a sigmoïdal intensity (used for defects on carbon fibers by Baxevanakis et al. (1992)), and the critical stress distribution induced from the size distribution of defects estimated from image analysis on polished sections (Berdin et al. 1991, 1993).

Other models, derived from the **Dead Leaves** random functions, are proposed in Jeulin (1990, 1991).

For the weakest link models, the **size effect** is the decrease of the median strength with the volume of the specimen, since the probability to observe a critical defect increases with the size. Its analytical shape depends generally on the choice of the model and on the statistical properties of the defects (size, shape, critical stress). However, a cracked specimen of an elastic-plastic material nearly follows a Weibull distribution with the power 4, which gives no influence of the distribution of the defects strengths on the size effect. For a fixed population of defects and a homogeneous stress field, the size effect increases in the order: strata, fibers, grains.

4 Fracture statistics models with a damage threshold

It is possible to generalize the weakest link criterion in various ways, as shown below (Jeulin 1991, 1992b, 1993).

4.1 Critical volume fraction of defects

It can be assumed that a specimen B under the stress field $\sigma(x)$ will not fail as long as the domain $D \subset B$ where $\sigma_c(x) < \sigma(x)$ is such that $V(D)/V(B) < p_c$, where the critical volume fraction p_c is a parameter of the model, indirectly accounting for percolation effects. This assumption can be used for damaging materials (neglecting the redistribution of the stresses during the process, or the growth of microcracks initiated on the sites with low critical stress) or for ductile fracture with cavities growing from inclusions.

Asymptotic results, valid for samples larger than the microstructure, are obtained from the bivariate distribution of the random function $\sigma_c(x)$. It enables us to calculate the mean m and the variance D^2 of the random variable $V(D)/V(B)$, that is asymptotically normal for large samples B. Therefore:

$$P\{\text{no fracture of B}\} = \frac{1}{\sqrt{\pi}} \int_{-\infty}^{p_c - m} \exp(-y^2) dy \qquad (5)$$

For a given stress field $\sigma(x)$, $P\{\text{fracture of B}\}$ decreases with the threshold p_c, and is equal to $1/2$ for $m = p_c$. In the case of a homogeneous stress field, the median fracture strength σ_M is given by $F(\sigma_M) = p_c$, where F is the probability distribution of the random function $\sigma_c(x)$; there is no size effect for the median stress, and a convergence towards a deterministic behavior. It depends on the microstructure through F, and on p_c.

Similar models with two fracture modes in competition were built from a mosaïc structure simulating a polycrystal based on a Voronoï tessellation (Jeulin 1992b).

4.2 Critical density of defects

For defects like points, lines, flats, the critical volume fraction should be replaced by a critical density of defects θ_c (for instance in numbers) (Jeulin 1991). If $N(\sigma)$ is the random number of defects in B where $\sigma_c(x) < \sigma(x)$, it is assumed that the fracture occurs when $N(\sigma) > n$, with $n = \theta_c V(B)$; this enables the nucleation of a given density of microcracks before the failure of the specimen B. The fracture probability can be deduced from the discrete distribution of $N(\sigma)$: for small size specimens,

$$P\{\text{no fracture of B}\} = p_0 \qquad (6)$$

Equation (6) is a restatement of the weakest link assumption, limited here to the small scale. When $n \geq 1$, the fracture probability is obtained as:

$$P\{\text{no fracture of B}\} = p_0 + p_1 + \ldots + p_n \qquad (7)$$

In the case of a distribution of defects according to a Poisson process, as for the Boolean varieties, $N(\sigma)$ is a Poisson random variable with parameter $\theta' = \int_0^T \bar{\mu}(H_{Z_t},(\sigma - \sigma_m)^c)\, \theta(dt)$. Equation (3) corresponds to $N(\sigma) = 0$. For a uniform stress field, the fracture strength of specimens B with increasing sizes becomes a constant: there is no size effect, as in section 4.1. On intermediate scales, it appears that large specimens are less sensitive to the most severe defects.

This type of approach was used recently in the case of two dimensional woven composites (Baxevanakis et al. 1993): the critical density model was used in the identification of defects from mechanical tests, and for the simulation of damage by finite element calculations.

5 Fracture statistics model with a crack arrest criterion

In this part, microstructural information along the crack path is used to estimate the probability of fracture of random media, as a function of microgeometrical characteristics, namely the spatial distribution of the specific fracture energy $\gamma(x)$. The approach is limited to crack propagation in two dimensional media. Some specific random media are considered, for which the probability of fracture can be theoretically calculated by application of the Griffith's crack arrest criterion. The obtained results differ from the weakest link approach mentioned above.

5.1. Crack propagation and the Griffith's criterion
Brittle random media, with a homogeneous constitutive law (namely elastic with Youngs' modulus E), but with a random fracture energy $\Gamma(x)$ are considered. Following Chudnovsky and Kunin (1987, 1992), a potential fracture path P(s,d) connecting the source s to the destination d is assumed to satisfy, for every point x on the path:

$$2\Gamma(x) \leq G(x) \tag{8}$$

The energy release rate $G(x)$ depends on the location of the crack front x, and also on the overall crack path from s to x. A candidate crack path P must satisfy:

$$\underset{V}{} \{ 2\Gamma(x) - G(x); \ x \in P \} \leq 0 \tag{9}$$

For a random function $\Gamma(x)$, the probability of fracture can be calculated from equation (9). In Chudnovsky and Kunin (1987, 1992), the potential crack paths P are assumed to be realizations of a diffusion stochastic process independent on Γ. This involves crack paths which do not depend locally on the underlying microstructure.

Using a minimal fracture energy criterion, the crack paths P would be the shortest paths obtained from the generalization of the Fermat's principle (Jeulin 1988, 1992a, 1993; Jeulin et al. 1992). For a given loading condition, the fracture will occur along this path if the criterion (9) is satisfied. This is easy to obtain on simulations if the function $G(x)$ is approximated by the case of a straight propagation of the crack. It is possible to limit the study to the extension of a linear crack with length 2a, along a segment (or ligament) of length 2b (according to a crack advance in mode I). This approach enables us to estimate the probability of fracture of a specimen containing a crack in this configuration, according to the loading conditions and to its random microstructure. Below, two models of random functions for $\Gamma(x)$ are compared: the Poisson mosaïc, and the Boolean random functions (Jeulin 1992a). The following two different conditions of loading of the crack are examined, as in Chudnovsky and Kunin (1992):

A **uniform stress** σ orthogonal to the crack is applied at infinity; in that case, the energy release rate $G(x)$ increases along the crack path, resulting in an unstable crack propagation for a homogeneous medium:

$$G(a+x) = \pi \ \sigma^2 \ (a+x)/ \ E \tag{10}$$

A **concentrated load** (also noted σ for convenience) at the tip of the crack; for this configuration, $G(x)$ decreases while the crack propagates, resulting in a stable crack propagation (which may end in a crack arrest):

$$G(a+x) = k \ \sigma^2/ \ [E \ (a+x)] \tag{11}$$

5.2. Probability of fracture and scale effect for the Poisson Mosaïc.

The Poisson mosaïc is a particular model, which might simulate a random polycrystal with local changes of fracture energy, as was the case of the Voronoï mosaïc used in section 4. It is built in two steps:

A **Poisson tessellation** with parameter λ delimits the grain boundaries; they are made of Poisson lines in the plane for a two-dimensional medium or of Poisson planes in the three-dimensional space (but generating Poisson lines in planar sections);

For grains of the tessellation, the fracture energy γ are independent realizations of a **random variable** Γ, with the cumulative distribution function $F(\gamma) = P\{ \Gamma < \gamma \}$; this is also the distribution function of the random function $\Gamma(x)$ built by the random mosaïc.

For the unstable crack propagation (10) (or for any loading where G increases with x), the following probability of fracture $P(a+b)$ is obtained:

$$P(a+b) = F(G(a)/2) \exp[-\lambda \int_{2a}^{2(a+b)} (1-F(G(u)/2)) \, du] \qquad (12)$$

P increases (and converges toward 1), when the crack length 2a or the applied stress σ increases. It also increases for lower λ corresponding to a coarser microstructure: with this model, small grains improve the ability to resist the crack growth, as a result of a higher probability to meet grains with a large fracture energy γ along the same crack path.

For the **stable crack propagation** (11) (or for any loading where G decrease with x),

$$P(a+b) = F(G(a+b)/2) \exp[-\lambda \int_{2a}^{2(a+b)} (1-F(G(u)/2)) \, du] \qquad (13)$$

Considering now samples with a similar geometry (with a constant ratio a/b), **scale effects** depending on the distribution $F(\gamma)$ are observed. If the range of the distribution is finite ($F(\gamma) = 1$ for $\gamma \geq \gamma_c$), the probability of fracture becomes equal to 1 beyond the critical fracture length $2a_c$ given by $2a_c = 4 \gamma_c E / (\pi \sigma^2)$. If the distribution $F(\gamma)$ possesses a **tail**, such as

$1-F(\gamma) \approx \gamma^{\alpha}$ when γ becomes infinite, the scale effects strongly depend on the positive coefficient α:

If $\alpha = 1$, the asymptotic probability of fracture becomes independent of the specimen size and there is no scale effect:

$$P(a+b) = y^{\lambda c} \text{ with } y = a / (a+b) \text{ and } c = E / (2\pi \sigma^2) \qquad (14)$$

If $\alpha \neq 1$, the large scale behavior of the probability of fracture becomes:

$$P(a+b) = \exp [- \lambda (2c)^{\alpha} a^{1-\alpha}(1- y^{\alpha-1}) / (\alpha-1)] \qquad (15)$$

When $\alpha < 1$, $P(a+b)$ converges to 0 for increasing sizes: the growth of $F(\gamma)$ towards 1 is so slow that the crack is stopped with a probability 1 by grains with large fracture energy.

When $\alpha > 1$, $P(a+b)$ converges to 1 for increasing sizes, as for a distribution with a finite range.

From equations (14) and (15) scale effects for the change of the median strength σ_M with the size of the specimen are obtained:

if $\alpha = 1$, $\sigma_M \simeq \sqrt{-\text{Logy}}$, and there is no scale effect;

if $\alpha \neq 1$, $\sigma_M \simeq d^{\frac{1-\alpha}{2\alpha}} |1 - y^{\alpha-1}|^{1/(2\alpha)}$, where d is the size of the specimen; therefore σ_M increases with d for $\alpha < 1$ and σ_M decreases with d for $\alpha > 1$.

For the stable crack propagation (11), it is easy to deduce from equation (13) that P(a+b) converges towards 0, whatever the distribution F, contrary to the other loading condition. The crack is stopped almost surely for specimens with increasing sizes.

5.3. Probability of fracture, and scale effect for the Boolean Mosaïc

The Boolean mosaïc is a particular Boolean random function (Jeulin 1990, 1991), obtained as follows:
- a material with a constant fracture energy γ_0 is considered (for simplification in the notations, assume that $\gamma_0 = 0$); it is equivalent to examining the case of cracks with a length larger than $2a_0$, that should be propagating in the homogeneous material according to the criteria given by equations (8) and (9).
- on every point of a Poisson point process (with intensity $\theta(u)$) there is implanted a random function $\Gamma'(u)$. The value $\Gamma(x)$ of the fracture energy at point x is given by the supremum over all the "primary" random functions Γ' covering x, instead of the infimum for the model used in section 3 (see Fig. 1). For the general model, any dependence of Γ' on u is allowed. In the present case, we consider a fracture energy ($\gamma = u$) constant inside realizations of a random grain X' (as for the Boolean random set (Matheron 1967)) and a measure θ such that $\int \theta(u)du$ remains finite. As for the Poisson mosaïc, either two-dimensional Boolean random functions or two-dimensional sections of a three-dimensional model are considered here. . The main morphological difference with the previous model is that, if the **Poisson mosaïc** is well suited for simulating a **polycrystal**, the **Boolean random function** with convex grains X' is a good simulation of a **matrix with a constant fracture energy γ_0** containing **reinforcing inclusions** with a larger fracture energy γ. The probability of fracture deduced from the criterion (9) is obtained for a given crack path by:

$$P \{\text{fracture} \} = \exp [-\mu_0] \qquad (16)$$

In equation (16), the coefficient μ_0 depends on the loading conditions through G(x), on the measure $\theta(\gamma)$, and on the random set X', that we assume to be a convex set. This equation is obtained as a particular case of the properties of the Boolean random functions and is detailed in (Jeulin 1992a).
- In the case of the **unstable crack propagation** (10), the probability of fracture converges toward 1 when increasing separately the crack length 2a or the applied stress σ. This is similar to a result obtained in subsection 5.2. Here, the effect of the size of the inclusions is the following: the

probability of fracture decreases with the size of the reinforcing inclusions X'. This is due to the fact that, with this size, the overall area fraction of higher strength material increases. The **effect of the measure** $\theta(\gamma)$ is the following. If $\theta(\gamma) = 0$ when $\gamma \geq \gamma_c$, the probability of fracture reaches 1 for large specimens when the crack length $2a$ is beyond the critical fracture length $2a_c$ given by $2a_c = 4 \gamma_c E / (\pi \sigma^2)$. For other measures, the asymptotic behavior of large size specimens depend on the **tail of the measure** θ for large γ. When $\theta(\gamma) = \theta \gamma^{-\alpha}$ with $\alpha > 1$, the cumulative distribution of the random function Γ is:

$$F(\gamma) = P \{ \Gamma(x) < \gamma \} = \exp [- \frac{\theta}{\alpha - 1} \gamma^{1-\alpha}] \tag{20}$$

For large γ, $F(\gamma)$ admits a power law tail of the distribution, since

$1 - F(\gamma) \simeq \frac{\theta}{\alpha - 1} \gamma^{1-\alpha}$. This is close to the Poisson mosaïc model

with similar distributions, leading to equations (14) and (15).
- If $\alpha = 2$, $\mu_0 = 4\theta c [A(X')/a - 2 L(X') \text{Log}(y)/\pi]$. It results:

$$P(a+b) = y^{8\theta cL(X')/\pi} \exp [- 4\theta c(A(X')/a] \tag{21}$$

with $y = a / (a+b)$ and $c = E / (2\pi \sigma^2)$

Equation (21) can be compared to equation (14), corresponding to a similar situation for a different model: the fracture probability slightly increases with the size of similar specimens, and converges towards a constant depending only on the geometry (through y) and on the microstructure (through θ and L(X')). For this specific intensity θ, there is no size effect on the fracture statistics.
- If $\alpha \neq 2$, we obtain:

$$\mu_0 = 4 \left[\frac{c}{a} \right]^{\alpha-1} \frac{1}{\alpha-1} \left[A(X') + 2 L(X') a /\pi \frac{y^{\alpha-2}-1}{2-\alpha} \right] \tag{22}$$

When $\alpha < 2$, μ_0 diverges for increasing sizes of the specimen, so that the probability of fracture converges to 0. When $\alpha > 2$, μ_0 converges to 0, and the probability of fracture of large specimens converges to 1. This is consistent with the results obtained for a Poisson mosaïc and a similar distribution function (after replacement of α in the previous case by by α-1 in the present case). The same scale effect for the median strength σ_M as in subsection 5.2 is obtained.
- For the **stable crack propagation** (11), the coefficient μ_0 in equation (16) diverges with the size of the specimen, so that the probability of fracture becomes 0 for large samples, as for the other model.

5.4. Conclusion
It was possible to derive a theoretical probability of fracture due to

unstable or stable straight crack propagation in two types of random media, namely the Poisson mosaïc and Boolean random functions. Resulting size effects are similar for the two models, and are strongly dependent on the tail of the cumulative distribution function of the fracture energy γ $(1 - F(\gamma) \simeq \gamma^{-\alpha})$: for a very slow growth of $F(\gamma)(\alpha<1)$, the probability for large specimens to encounter a locally large fracture energy is so important that the probability of fracture decreases to 0. When $\alpha = 1$, the probability of fracture of large specimens is a constant depending on the geometry and on the microstructure, but not on the size. When $\alpha > 1$, in any large specimen, a long crack is subject to propagation until fracture.

6) Conclusion

The models introduced above present some simplifications coming from the basic assumptions recalled in section 2. Their main advantages are the following:

-Various morphologies of microstructural defects (inclusions in a matrix, polycrystals, fiber and strata,...), inducing correlations on various ranges, can be described and simulated.

-Exact theoretical results, coherent at different scales, are available.

-Depending on few parameters (two, three, or more), they can be tested from data on various scales: on the macroscopic scale, by means of the experimental distributions obtained from mechanical tests on various specimens and geometries; on a microscopic scale, by means of image analysis measurements after localization of the defects (Berdin et al. 1991, 1993).

-Various scaling laws are obtained, according to the chosen fracture criteria, or to the appropriate random structure models, reflecting the situations occurring with experimental data: the overall strength of specimens may decrease with their size (weakest link), may be size invariant (damage threshold), or may even increase (crack arrest). However, one must be aware of the fact that different combinations of fracture criteria and microscopic models can result in the same size effects, as underlined in (Jeulin 1991). Therefore it is unwise to draw definite conclusions solely on size effects, without any indication on the microstructure and on the micro mechanisms of fracture.

-The models can be easily introduced in post-processor calculations in a finite element code (Berdin et al. 1991, 1993). New extensions combine the simulation of population of defects after their identification from the models and mechanical tests, and the simulation of the progression of damage in composites by finite element calculation (Baxevanakis et al. 1993); this is a way to account for stress redistribution in microcracking processes, which is difficult to handle by analytical calculations.

The use of these models is not restricted to the simulation of critical fracture criteria, as proposed in this paper. In fact they are able to simulate other physical random media with a microstructure, including the distribution of multivariate or tensors properties, that can be used for other purposes such as homogenization calculations, as proposed in (Jeulin 1988).

References

Baxevanakis C., Jeulin D., Valentin D. (1993) Fracture statistics of single

224

fibre composites specimens, **Composites Sciences and Technology,** 48, pp. 47-56.

Baxevanakis C., Boussuge M., Jeulin D., Munier E., Renard J. (1993) Simulation of the development of fracture in composite materials with random defects, **Proc. of the International Seminar on Micromechanics of Materials,** MECAMAT'93, Fontainebleau, 6-8 July 1993, Eyrolles, Paris, pp. 460-471.

Berdin C., Baptiste D., Jeulin D., Cailletaud G. (1991) Failure models for brittle materials, in J. G. M. van Mier et al. (eds), **Fracture Processes in Concrete, Rock and Ceramics,** E. et F.N. Spon, London, pp. 83-92.

Berdin C. (1993) Etude expérimentale et numérique de la rupture des matériaux fragiles, **Thesis,** Ecole des Mines de Paris.

Chudnovsky A., and Kunin B. (1987) A probabilistic model of brittle crack formation, **J. Appl. Phys.** 62 (10), 4124 .

Chudnovsky A., and Kunin B. (1992) Statistical Fracture Mechanics, in M. Mareschal et B.L. Holian (ed), **Microscopic Simulations of Complex Hydrodynamic Phenomena,** Plenum press, New York, pp. 345-360.

Jeulin D. (1988) On image analysis and micromechanics, **Revue Phys. Appl.,** 23, pp 549-556.

Jeulin, D. (1990) Random fields models for fracture statistics, Actes du 32ème Colloque de Métallurgie, INSTN, **Ed. de la Revue de Métallurgie,** n°4, pp. 99-13.

Jeulin, D. (1991) Modèles Morphologiques de Structures Aléatoires et de Changement d'Echelle. Thèse de Doctorat d'Etat ès Sciences Physiques, University of Caen.

Jeulin, D., Vincent, L., Serpe G. (1992) Propagation algorithms on graphs for physical applications **J. Visual Comm. Image Represent.** 3, 2, 161 .

Jeulin D. (1992a) Some Crack Propagation Models in Random Media, communication to the Symposium on the Macroscopic Behavior of the Heterogeneous Materials from the Microstructure, ASME, Anaheim, CA, Nov 8-13, 1992. **AMD Vo. 147,** pp. 161-170.

Jeulin D. (1992b) Morphological Models for Fracture Statistics, Communication to the CMDS7 Conference, Paderborn (14-19 June 1992), **Materials Science Forum,** Vol.123-125 (1993), K. H. Anthony and H. J. Wagner (ed), Transtech Publications, pp. 505-513.

Jeulin D. (1993) Random Functions and Fracture Statistics Models, In A. Soares (ed), **Geostatistics Troia '92,** Kluwer Academic Publ., Dordrecht (Quantitative Geology and Geostatistics 5) Vol. 1, pp. 225-236.

Jeulin D. (1993) Damage simulation in heterogeneous materials from geodesic propagations, Engineering computations, vol. 10, pp 81-91.

Matheron G. (1967) **Eléments pour une théorie des milieux poreux,** Masson, Paris.

Matheron G. (1975) **Random sets and integral Geometry,** J. Wiley.

Pineau A. (1981) Review of fracture micromechanisms and a local approach to predicting crack resistance in low strength steels, **Proc. of 5th Int. Conf. on Fracture,** Cannes; D. François (ed), vol. 2, pp 533-577.

Sanchez Palencia E., Zaoui A. (ed) (1987) **Homogenization Techniques for Composite Media,** Lecture Notes in Physics vol. 272, Springer Verlag.

Serra J. (1982) **Image analysis and Mathematical Morphology,** Academic Press, London.

EFFECTIVE ELASTIC PROPERTIES OF SOLIDS WITH RANDOMLY LOCATED DEFECTS

M. KACHANOV, I. TSUKROV and B. SHAFIRO
Tufts University, Medford, MA, 02155 USA

Abstract
Effective moduli of solids with cavities of various shapes are analyzed. The analysis is based on the elastic potentials of solids with defects. The structure of the potential dictates the proper measures of density of defects. It also establishes the overall anisotropy in the case of non-random orientation.
Keywords: Effective Properties, Cavities, Cracks, Anisotropy.

1 Introduction

The problem of effective elastic properties of solids with defects is a classical one, and a number of various approximate schemes exist.

The simplest (and rigorous) scheme is the approximation of non-interacting defects. Even this scheme, however, gave rise to an ambiguity: it is frequently identified with the linearized approximation of small defect density ("dilute limit"). As discussed by Kachanov (1992, 1993), these two approximations are actually different, and linearization with respect to the defect density only reduces the range of applicability of the approximation of non-interacting defects. Moreover, for *cracks* the approximation of non-interacting cracks remains accurate at high crack densities whereas the linearized expression loses accuracy. Therefore, the mentioned linearization appears to be an unnecessary construction.

For *interacting* defects, a rich menu of approximate schemes exists:

- self-consistent scheme (SCS)
- generalized SCS (GSCS)
- differential scheme (DS)
- Mori-Tanaka's scheme (MTS)
- models of second order in defect density.

It seems that this list can easily be expanded: for example, a hybrid of SCS and DS in which the defects are introduced not in one step (as in SCS) or in infinitely many steps (as in DS), but, for example, in two or three steps, can be developed; another obvious possibility is a differential version of the GSCS.

A reader who wants to actually *use* the theory, being confronted with this choice, may well be puzzled.

The question: "which of the schemes is better?" has no universal answer - the problem is uncertain, unless the statistics of *mutual positions* of defects is specified.

225

D. Breysse (ed.), Probabilities and Materials, 225–240.

A more meaningful question is: which scheme is better provided the *mutual positions of defects are random* ?

Note that solutions for the *periodic* arrangements provide no general guidance. For example, consider a doubly periodic array of parallel cracks, with ℓ and h being the horizontal and vertical spacings between cracks. Then, letting h be sufficiently large, while $\ell \to 0$, creates an array that has an infinitesimal crack density but reduces the stiffness to zero. Letting, on the other hand, ℓ be sufficiently large, while $h \to 0$ leads to an opposite example of an array of infinite density that, nevertheless, produces only an infinitesimal impact on the effective stiffness. (This example demonstrates that neither upper not lower universal bounds can be established for cracked materials. Note, also, that analysis of "ordered" crack arrays requires a different crack density parameter - a fourth rank tensor Ω, see Kachanov, 1992).

Some insight into the problem can be gained through *stress superpositions*. Consider the case when the defects are traction free cavities. Then the problem can be represented, in a usual way, as a superposition of several problems containing one defect each (Fig. 1). In each sub-problem, the traction on a defect consists of two parts:

$t = n \cdot \sigma$ (induced by the remotely applied σ) and the interaction traction Δt generated in all other sub-problems along the site of the considered defect. (We assume that the RVE constitutes a sufficiently large, statistically representative part of a solid subjected to σ at infinity; then replacement of tractions (generally, variable) on the boundary of the RVE by the ones projected on it by σ will affect only a "boundary layer" and will produce only a small error in calculation of the overall strain in RVE. See Beran, 1968 and Hashin, 1983 for a discussion of related matters).

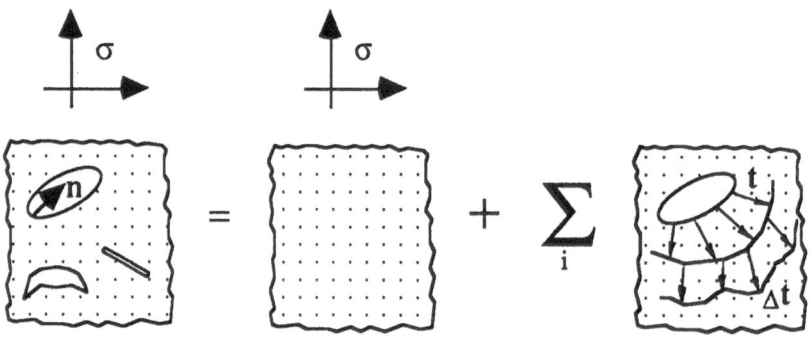

Fig. 1. Stress superposition for a solid with cavities or cracks.

Intuitively, it appears reasonable to assume that, for holes with *random mutual positions*, the interaction tractions Δt reflect the average "stress environment" in the solid phase, i. e. $\Delta t = n \cdot \sigma^s$, where σ^s is the average stress in the solid phase. This assumption constitutes the scheme of Mori and Tanaka (1973), MTS (formulated by them for a more general case of inclusions). MTS received recently a theoretical support in the works of Tandon and Weng (1986), Weng (1992) and Benveniste (1990) and an experimental support in the works of Ferrary and Filipponi (1991) and Biolzi *et al* (1993). Note that MTS follows from a more general and sophisticated method of effective field due to Kanaun and Levin (see Kanaun, 1983), which takes the statistics of mutual positions into account: if this statistics is set to be random, then MTS follows.

In the case of cavities or cracks, the average stress in the solid phase σ^s is given exactly by the following simple formula

$$\sigma^s = \frac{1}{1-p}\sigma \tag{1.1}$$

where p is porosity (relative volume fraction of cavities). Thus, porosity raises all the stress components σ_{ij} by the same ratio. For *cracks*, $p=0$ and $\sigma^s = \sigma$, thus reflecting the fact that the competing interaction effects of shielding and amplification cancel each other, if mutual position of cracks are random. As a result, MTS coincides with the approximation of non-interacting cracks. Extensive computer experiments on interacting cracks in both isotropic (Kachanov, 1992) and anisotropic (Mauge and Kachanov, 1992) matrices confirm that this approximation remains accurate at high crack densities.

2 Elastic potential for the isotropic solid with one cavity

2.1. General relations
The analysis is done in the framework of linear elasticity; non-linear effects caused by closing of holes by compressive stresses (stiffness increasing with compression, different moduli in compression and tension, etc.) are not covered. In the case of compressive loading, this restriction translates into a restriction on the magnitude of stresses at which the linear elastic results are valid (see Walsh, 1965 for estimates of stresses needed to close a hole and produce non-linear effects).

Our analysis is based on results for one cavity. The starting point is the observation that the total strain in a solid subjected to a remotely applied stress σ and containing a cavity is given by a sum

$$\varepsilon = \mathbf{M}^O : \sigma + \Delta\varepsilon \tag{2.1}$$

where $\mathbf{M}^O$ is the compliance tensor of the matrix; a colon denotes contraction over two indices. The additional strain due to introduction of a cavity is

$$\Delta\varepsilon = -\frac{1}{2V}\int_\Gamma (\mathbf{un} + \mathbf{nu})\, d\Gamma \tag{2.2}$$

where $\mathbf{u}$ and $\mathbf{n}$ denote displacements of the cavity boundary Γ and a unit normal to Γ (directed inwards the cavity), V is the (total, including the cavity) reference volume and $\mathbf{un}$, $\mathbf{nu}$ denote dyadic (tensor) product of two vectors. In the 2-D case, the volume V changes to a representative area A.

The representation (1.2) directly follows from application of the divergence theorem to a strained solid (see footnote in Hill, 1963) when the solid contains a cavity. It has been used in the literature since 1970's (see, for example, Vavakin and Salganik, 1975).

The strain $\Delta\varepsilon$ is a linear function of the applied stress:

$$\Delta\varepsilon = \mathbf{H} : \sigma \tag{2.3}$$

where the fourth rank tensor **H** can be called a *cavity compliance tensor*. Since it enters the analysis through the elastic potential (see (2.4)), it possesses the usual symmetries of the elastic compliance tensor.

Equation (2.3) is simply a statement that the response of the system is linear elastic and needs no derivation. The actual construction of **H** was done in the literature for several cavity shapes, by utilizing elasticity solutions. For 2-D elliptical holes, the **H**-tensor was constructed by Kachanov (1993), for 2-D circular holes - by Nemat-Nasser and Hori (1993). Kachanov *at al.* (1994) constructed **H** for a number of complex shapes (polygonal in 2-D and ellipsoidal in 3-D) and, also, identified the class of cavities that can be characterized by a *second* rank tensor (rather than the fourth rank **H**). For a *crack*, **H** = n**Bn** where **n** is a unit normal to the crack and **B** is the COD tensor of a crack relating the average COD to the vector of applied uniform traction (**B** was constructed for elliptical cracks in the isotropic matrix by Kachanov, 1992 and for cracks in 2-D anisotropic matrix by Mauge and Kachanov, 1992).

Our analysis is based on the elastic potential of a solid with cavities. The potential in stresses (complementary energy density) of a solid with one cavity can be represented as a sum of two terms:

$$f(\sigma) = \tfrac{1}{2}\,\sigma : \varepsilon(\sigma) = \tfrac{1}{2}\,\sigma : \mathbf{M}^0 : \sigma + \tfrac{1}{2}\,\sigma : \mathbf{H} : \sigma \equiv f_0 + \Delta f \tag{2.4}$$

where f_0 is the potential in the absence of a cavity. The potential-based approach covers cavities of various shapes in a unified way and dictates the choice of proper parameters of cavity density.

Referring the reader to Kachanov *et al* (1994) for details, we present here the potential Δf for cavities of various shapes. These representations utilize the results of 2-D elasticity and Eshelby's solution for an ellipsoidal inclusion.

Notations are as follows. E_0 and v_0 denote, in the 3-D case (and in the case of 2-D plane stress) Young's modulus and Poisson's ratio of the matrix; in the 2-D plane strain, E_0 and v_0 are to be understood as $E_0/(1-v_0^2)$ and $v_0/(1-v_0)$. A (or V) are the reference area (or volume), S (or V_{cav}) is the area (volume) of a hole.

Expressions in large braces in the formulas (2.5-2.15) are parameters that characterize the considered defect. We emphasize that these parameters are not introduced arbitrarily but emerge naturally, as terms in the elastic potential.

2.2. Potentials of solids with various 2-D holes.

CRACK

$$\Delta f = \frac{\pi}{E_0}\,\sigma \cdot \sigma : \left\{ \frac{l^2}{A}\,\mathbf{nn} \right\} \tag{2.5}$$

The crack thus is characterized by a symmetric second rank tensor $(l^2/A)\,\mathbf{nn}$.

CIRCULAR HOLE

$$\Delta f = \frac{1}{2\,E_0} \left\{ \frac{\pi\,a^2}{A} \right\} \left[4tr\,(\sigma \cdot \sigma) - (tr\,\sigma)^2 \right] \tag{2.6}$$

The hole is characterized by its relative area $(\pi\,a^2/A)$.

ELLIPTICAL HOLE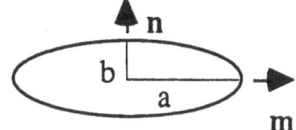

$$\Delta f = \frac{1}{2\,E_0} \left[4tr\,(\sigma\cdot\sigma) - (tr\,\sigma)^2 \right] \left\{ \frac{\pi\,a\,b}{A} \right\} + \frac{\pi}{E_0}\,\sigma\cdot\sigma : \frac{1}{A} \left\{ a^2\,\mathbf{nn} + b^2\,\mathbf{mm} - ab\mathbf{I} \right\} \tag{2.7}$$

The hole is characterized by two parameters: its relative area S/A and the second rank symmetric tensor $a^2\,\mathbf{nn} + b^2\,\mathbf{mm} - ab\,\mathbf{I}$. The potential (2.7) consists of the isotropic (proportional to S/A) and anisotropic (orientation-dependent) parts. In the limiting case of a crack, no degeneracies arise and (2.5) is recovered.

HOLES OF THE REGULAR POLYGONAL TYPE

(A) Shapes producing **anisotropic** response: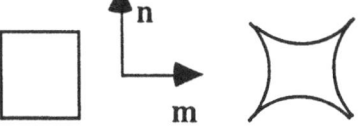

$$\Delta f = \frac{1}{E_0} \left[h_3\,tr\,(\sigma\cdot\sigma) - h_2\,(tr\,\sigma)^2 \right] \left\{ \frac{S}{A} \right\}$$

$$+ \frac{1}{E_0}(h_1 + h_2 - h_3)\,\sigma : \left\{ \frac{S}{A}\,(\mathbf{mmmm} + \mathbf{nnnn}) \right\} : \sigma \tag{2.8}$$

The hole is characterized by its relative area S/A, three scalar "shape factors" h_1, h_2, h_3 and a fourth rank tensor $(S/A)(h_1 + h_2 - h_3)(\mathbf{mmmm} + \mathbf{nnnn})$. For a square approximated by three terms conformal mapping onto a unit circle (almost indistinguishable from the ideal square) $h_1 = 1.530$, $h_2 = 0.334$, $h_3 = 2.660$. For a four-cusps hypotrochoid, $h_1 = 17/8$, $h_2 = 1/8$, $h_3 = 9/2$. The potential consists of the isotropic part and the anisotropic part (containing the fourth rank tensor).

(B) Shapes producing the **isotropic** response:

$$\Delta f = \frac{1}{E_0} \left[h_3\,tr\,(\sigma\cdot\sigma) - h_2\,(tr\,\sigma)^2 \right] \left\{ \frac{S}{A} \right\} \tag{2.9}$$

The hole is characterized by its relative area S/A and two "shape factors" h_3, h_2. For a triangle approximated by three terms conformal mapping onto a unit circle (almost

230

indistinguishable from the ideal triangle) $h_1=2.082$, $h_2=0.502$, $h_3= h_1 + h_2= 2.584$. For a three-cusps hypotrochoid, $h_1 =7/2$, $h_2 = 1/2$, $h_3 = h_1 + h_2 = 4$. The potential is isotropic, i.e. independent of the hole orientation.

2.3 Potentials of solids with various 3-D cavities.

PENNY-SHAPED CRACK

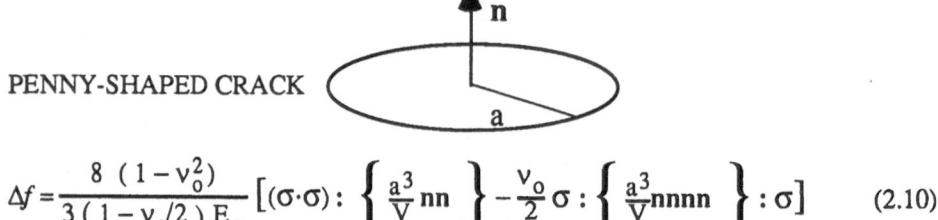

$$\Delta f =\frac{8\ (1-\nu_0^2)}{3\ (1-\nu_0/2)\ E_0}\left[(\sigma\cdot\sigma):\left\{\frac{a^3}{V}\,nn\right\}-\frac{\nu_0}{2}\,\sigma:\left\{\frac{a^3}{V}nnnn\right\}:\sigma\right] \qquad (2.10)$$

The crack is characterized by a second rank symmetric tensor $(a^3/V)nn$ and a fourth rank fully symmetric tensor $(a^3/V)nnnn$; the latter enters with the small multiplier $\nu_0/2$ and plays, therefore, only a minor role (Kachanov 1980, 1992).

ELLIPTICAL CRACK

$$\Delta f =\frac{\pi}{2}\,(\sigma\cdot\sigma):\left\{\frac{a\,b}{V}\frac{\eta+\zeta}{2}\,nn\right\}+\frac{\pi}{2}\,\sigma:\left\{\frac{a\,b}{V}[\xi-\frac{\eta+\zeta}{2}]nnnn\right.$$

$$\left.+\frac{a\,b}{V}\frac{\eta-\zeta}{2}\,n(ss\text{-}tt)n\right\}:\sigma \qquad (2.11)$$

The crack is characterized by the second and fourth rank tensors. The coefficients ξ, η and ζ have the dimension of length and are expressed in terms of elliptic integrals.

SPHERE

$$\Delta f =\frac{3\,(1-\nu_0)}{4\,(7-5\,\nu_0)\,E_0}\left[10\,(1+\nu_0)\,tr\,(\sigma\cdot\sigma)-(1+5\,\nu_0)\,(tr\,\sigma)^2\right]\left\{\frac{V_{cav}}{V}\right\} \qquad (2.12)$$

The cavity is characterized by its relative volume. The potential is isotropic.

SLIGHTLY DEFORMED SPHERE

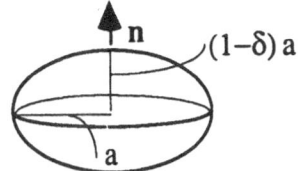

$$\Delta f = \frac{3\,(1-\nu_0)}{4\,(7-5\,\nu_0)\,E_0}\left[\,10\,(1+\nu_0)\,tr\,(\sigma\cdot\sigma) - (1+5\,\nu_0)\,(tr\,\sigma)^2\right]\left\{\frac{V_{cav}}{V}\right\}$$

$$+\frac{3\,(1-\nu_0^2)}{7\,(7-5\,\nu_0)^2\,E_0}\left[-10(5-7\,\nu_0)\,tr\,(\sigma\cdot\sigma) + (17-35\,\nu_0)\,(tr\,\sigma)^2\right]\left\{\frac{V_{cav}}{V}\,\delta\right\}$$

$$+\frac{3\,(1-\nu_0^2)}{7\,(7-5\,\nu_0)^2\,E_0}\left[30(5-7\nu_0)\,\sigma\cdot\sigma - 3(17-35\nu_0)(tr\,\sigma)\sigma\right]:\left\{\frac{V_{cav}}{V}\,\delta\,nn\right\}$$

$$(2.13)$$

The cavity is characterized by (1)two scalars: its relative volume V_{cav}/V and $V_{cav}\delta/V$, where δ is the amount of distortion and (2)the symmetric second rank "shape distortion tensor" $(V_{cav}\delta/V)nn$.

SLIGHTLY "INFLATED"
PENNY-SHAPED CRACK

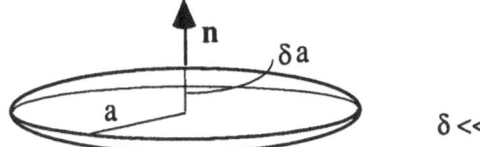

$$\delta \ll 1$$

$$\Delta f = \frac{8\,(1-\nu_0^2)}{3\,(1-\nu_0/2)\,E_0}\left[\,(\sigma\cdot\sigma):\left\{\frac{a^3}{V}nn\right\} - \frac{\nu_0}{2}\sigma:\left\{\frac{a^3}{V}nnnn\right\}:\sigma\,\right]$$

$$+\frac{1}{2\,E_0}\left[\,(1+\nu_0)\,tr\,(\sigma\cdot\sigma) - \nu_0\,(tr\,\sigma)^2\right]\left\{\frac{V_{cav}}{V}\right\}$$

$$-\frac{1}{2\,E_0}\,(1+\nu_0)\left[\,(1-2\nu_0)\,(tr\,\sigma)\,\sigma + 2\,\sigma\cdot\sigma\right]:\left\{\frac{V_{cav}}{V}nn\right\}$$

$$+\frac{1}{2\,E_0}\,(2+5\,\nu_0 - 6\,\nu_0^2)\,\sigma:\left\{\frac{V_{cav}}{V}nnnn\right\}:\sigma$$

$$(2.14)$$

The cavity is characterized by two tensorial *crack* parameters (see the expression 2.10 for an uninflated crack). The effect of "inflation" is characterized by (1) scalar V_{cav}/V, (2) two "porosity tensors": symmetric second rank tensor $(V_{cav}/V)\mathbf{nn}$ and fully symmetric fourth rank tensor $(V_{cav}/V)\mathbf{nnnn}$.

NEEDLE

$$\Delta f = \frac{1}{2E_0} \left[4(1-v_0^2)\, tr\,(\sigma \cdot \sigma) - (1-2v_0^2)\,(tr\,\sigma)^2 \right] \left\{ \frac{V_{cav}}{V} \right\}$$

$$+ \frac{1}{E_0}(1-2v_0)(1+v_0)\left[(tr\,\sigma)\sigma - 2\sigma\cdot\sigma \right] : \left\{ \frac{V_{cav}}{V}\mathbf{nn} \right\}$$

$$- \frac{1}{E_0}v_0(1+v_0)\,\sigma : \left\{ \frac{V_{cav}}{V}\mathbf{nnnn} \right\} : \sigma \tag{2.15}$$

The cavity is characterized by (1)its relative volume V_{cav}/V and (2)two "porosity tensors": the symmetric second rank tensor $(V_{cav}/V)\mathbf{nn}$ and the fully symmetric fourth rank tensor $(V_{cav}/V)\mathbf{nnnn}$.

ELLIPSOID OF REVOLUTION AND GENERAL ELLIPSOID
For these more general cases, the formulas are rather lengthy and contain components of Eshelby's tensor (see Kachanov et al, 1994 for details).

3 Effective properties of solids with non-interacting defects. Proper parameters of defect density

The effective properties for solids with non-interacting defects were first derived by Bristow (1960) for cracks and by Eshelby (1957) for spherical pores.
Our formulation in elastic potentials yields the effective properties for non-interacting defects of all shapes considered in the preceding section. Arbitrary non-random orientational distributions (anisotropy of the effective properties), as well as *mixtures* of defects of diverse shapes, are covered in a unified way.
In the approximation of non-interacting defects, each defect is placed into the externally applied stress σ and does not experience any influence of other defects.
Thus, the potential Δf is obtained by summation (over all defects) of the terms in large braces in (2.5 - 2.15). These sums constitute *proper parameters of defect density*.

We call the density parameters "proper" if the effective moduli can be expressed as unique functions of them. Such functions are *universal*: they cover, in a unified way, all orientational distributions of defects. We emphasize that the choice of the proper parameters is not arbitrary, but is *dictated* by the structure of the potential Δf.

We consider now several types of defects and the corresponding density parameters.

(A) CRACKS, 2-D CASE.
The elastic potential takes the form

$$\Delta f = \frac{\pi}{E_0} \sigma \cdot \sigma : \alpha \tag{3.1}$$

so that the crack density tensor

$$\alpha = (1/A) \sum_i (l^2 \, \mathbf{nn})^i \tag{3.2}$$

(introduced by Vakulenko and Kachanov in 1970's, see Kachanov, 1980, 1992 for references and details) is the proper parameter of crack density. Its linear invariant $tr\,\alpha = (1/A)\sum l^2 \equiv \rho$ is the conventional scalar crack density (introduced by Bristow, 1960). Since α is a symmetric second rank tensor, the effective properties are always *orthotropic*, for any orientational distribution of cracks.

The effective moduli for any orientational distribution of cracks immediately follow from (3.1). For example, Young's modulus for randomly oriented cracks $E=E_0(1+\pi\rho)^{-1}$; for parallel cracks, $E=E_0(1+2\pi\rho)^{-1}$.

(B) CRACKS, 3-D CASE

$$\Delta f = \frac{8\,(1-v_0^2)}{3\,(1-v_0/2)\,E_0}\,[\,(\sigma\cdot\sigma): \frac{1}{V}\sum_i (a^3\,\mathbf{nn})^i - \frac{v_0}{2}\sigma : \frac{1}{V}\sum_i (a^3\,\mathbf{nnnn})^i : \sigma\,] \tag{3.3}$$

The structure of Δf shows that, in addition to the crack density tensor α, the fourth rank tensor $(1/V)\sum_i (a^3\,\mathbf{nnnn})^i$ emerges as a second density parameter. Since, however, the corresponding term enters Δf with a relatively small multiplier $v_0/2$, this tensor plays a minor (as compared to α) role. Orthotropy holds with high degree of accuracy (see Kachanov 1980,1992). Arbitrary orientational distributions of cracks are covered by (3.3) in a unified way.

(C) 2-D ELLIPTICAL HOLES
In this case (considered in detail by Kachanov, 1993) the potential has the form

$$\Delta f = \frac{1}{2\,E_0}\,[\,4tr\,(\sigma\cdot\sigma) - (tr\,\sigma)^2\,]\,p + \frac{1}{E_0}\sigma\cdot\sigma : (\beta - p\mathbf{I}\,) \tag{3.4}$$

so that the proper parameters of density are:

• porosity $p = (1/A)\,\pi \sum_i (ab)^i$

• hole density tensor $\beta = (1/A)\,\pi \sum_i (a^2\,\mathbf{nn} + b^2\,\mathbf{mm})^i$.

The case of circular holes is recovered at $\beta = p\mathbf{I}$. In the case of cracks, $p = 0$ and $(1/\pi)\beta$ coincides with the crack density tensor α.

Note that, even in the case of randomly oriented holes (isotropic moduli), porosity alone is insufficient for the characterization of effective properties - a second scalar parameter, "eccentricity"

$$q = tr\beta = (1/A)\Sigma(a-b)^2 \tag{3.5}$$

is needed. The entire range of elliptical shapes is covered in a unified way; no degeneracies (or a need in special transition to limit) arise in the limiting case of cracks ($p \to 0$). Since β is a symmetric second rank tensor, the effective properties are orthotropic for any orientational distribution (similarly to the case of cracks).

(D) CAVITIES OF OTHER SHAPES

The potential Δf and the effective moduli for cavities of the shapes considered in Section 3 can be obtained in a similar way, see Kachanov et al. (1994) for details.

4. Interacting defects

For interacting defects, the problem of effective properties was considered in literature for a number of cavity shapes in the framework of several approximate schemes. The case of cracks was considered by O'Connell and Budiansky (1974) and Budiansky and O'Connell (1976) in the framework of SCS and by Vavakin and Salganik (1975) in the framework of DS; the latter authors also considered 2-D circular holes. For the latest references, we mention the work of Zimmerman (1986, 1991) which analyzed cavity compressibilities; of Zhao and Weng (1990) on 2-D elliptical inclusions and Zhao et al (1989) on 3-D ellipsoidal inclusions (in the framework of MTS), of Berryman (1980) and Thorpe and Sen (1985) on randomly oriented 2-D elliptical inclusions (SCS), and the work of Day et al (1992) and Jasiuk et al (1992, 1994) on numerical simulations of 2-D holes in a network of elastic springs.

We consider the case when the mutual positions of defects are random (otherwise, the problem of effective properties is uncertain - it depends on the specific defect pattern). Then, as discussed in section 1, the scheme of Mori-Tanaka (MTS) is appropriate. In the framework of MTS, a representative defect is placed in the average (over the solid phase) stress σ^s, given by (1.1). Then the effective properties are obtained from the ones for non-interacting defects by a simple adjustment: replacing $\varepsilon(\sigma)$ in (2.4) by $\varepsilon(\sigma^s) = (1-p)^{-1} \varepsilon(\sigma)$, we obtain the potential as

$$\Delta f = \frac{1}{1-p} \, \Delta f_{\text{non-int}} \tag{4.1}$$

and the effective moduli for all types of defects considered in Section 3 immediately follow. For example, Young's modulus for randomly oriented elliptical holes is

$$\frac{E}{E_o} = \frac{1}{1 + (1-p)^{-1}(3p+q)} \tag{4.2}$$

where the multiplier $(1-p)^{-1}$ accounts for interactions.

5. Mixture of interacting cavities of diverse shapes

Such mixtures are of practical interest; for example, a mixture of cavities and cracks can be considered as a simple model of a microcracked porous rock.

Analysis in terms of elastic potentials allows one to consider mixtures of cavities of diverse types. Indeed, for non-interacting cavities, the elastic potential is a sum:

$$\Delta f_{\text{non-int}} = \sum_J \Delta f^{(J)}_{\text{non-int}} \qquad (5.1)$$

where $\Delta f^{(J)}$ denotes Δf for the cavities of the type (J) considered separately.

For interacting cavities, considered in the approximation of Mori-Tanaka's scheme,

$$\Delta f_{\text{int}} = \frac{1}{1-p} \sum_J \Delta f^{(J)}_{\text{non-int}} \qquad (5.2)$$

Since p entering (5.2) is the *overall* porosity (due to holes of all types), Δf cannot be represented as a sum of $\Delta f^{(J)}$. This produces an *asymmetry of the interaction effect* between different types of holes: in a mixture of more "elongated" and more circular holes, the more "circular" ones produce a higher impact on the more "elongated" ones than vice versa (they change the average stress environment for the more "elongated" ones to a larger extent than vice versa).

For example, in the mixture of elliptical holes and cracks, *cracks do not affect holes* (as far as the effective moduli are concerned) since do not contribute into the overall porosity p and thus they do not change the average stress environment for holes, whereas *holes do affect cracks*. It is seen from the fact that in the sum

$$\Delta f = \frac{1}{1-p} \Delta f_{\text{holes}} + \frac{1}{1-p} \Delta f_{\text{cracks}} \qquad (5.3)$$

the first term (due to holes) contains no information on cracks whereas the second term (due to cracks) *does* contain information on holes, through the parameter p.

The choice of proper parameters of density for mixtures of interacting defects is, again, dictated by the elastic potential. For example, for a 2-D mixture of interacting polygonal holes and cracks, the proper parameters are :

• porosity p

• fourth rank (fully symmetric) tensor

$$\gamma = \sum_J (h_1 + h_2 - h_3)^{(J)} \gamma^{(J)} \qquad (5.4)$$

• crack density tensor α .

We consider now two examples of mixtures.

2-D MIXTURE OF TRIANGULAR HOLES AND RANDOMLY ORIENTED CRACKS (Fig. 2).

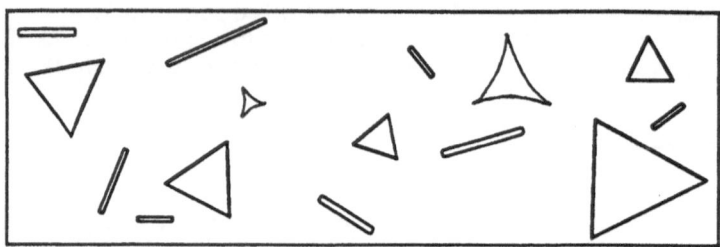

Fig. 2. Mixture of triangular holes and randomly oriented cracks.

In this case, using (2.9) and (2.5) we obtain the following potential:

$$\Delta f = \frac{p}{1-p} \frac{1}{E_0} [h_3 tr\,(\sigma \cdot \sigma) - h_2 (tr\,\sigma)^2] + \rho \frac{1}{1-p} \frac{1}{E_0} (\pi/2)\,tr\,(\sigma \cdot \sigma) \qquad (5.5)$$

where ρ is the conventional scalar crack density. The Young's modulus is:

$$E/E_0 = [1 + \frac{2(h_3 - h_2)p}{1-p} + \pi\rho \frac{1}{1-p}]^{-1} \qquad (5.6)$$

where for (almost ideal) triangles, $h_3 = 2.584$, $h_2 = 0.502$; for three-cusps hypotrochoid $h_3 = 4$, $h_2 = 1/2$.

3-D MIXTURE OF SPHERICAL CAVITIES AND RANDOMLY ORIENTED PENNY-SHAPED CRACKS

The potential is isotropic and is expressed in terms of two scalar density parameters, porosity p and crack density ρ:

$$\Delta f = \frac{1}{1-p} (\Delta f_{non-int}^{spheres} + \Delta f_{non-int}^{cracks}) = \frac{p}{1-p} \frac{3(1-v_0)}{4(7-5v_0)E_0} [10\,(1+v_0)\,tr\,(\sigma \cdot \sigma)$$

$$- (1+5v_0)\,(tr\,\sigma)^2] + \frac{\rho}{1-p} \frac{8\,(1-v_0^2)}{45\,(1-v_0/2)\,E_0} [(5-v_0)tr\,(\sigma \cdot \sigma) - \frac{v_0}{2}\,(tr\,\sigma)^2] \quad (5.7)$$

The effective Young's modulus is:

$$E = E_0 \left\{ 1 + \frac{p}{1-p} \frac{3(1-v_0)(9+5v_0)}{2(7-5v_0)} + \frac{\rho}{1-p} \frac{8\,(1-v_0^2)(10-3v_0)}{45(1-v_0/2)} \right\}^{-1} \quad (5.8)$$

The approximation of *non-interacting* defects is recovered from (5.5) - (5.8) by omitting the terms $(1 - p)^{-1}$, which accounts for interactions.

6 Comparison with other approximate schemes

Aside from the approximation of non-interacting defects and the approximation of the "average stress environment" (MTS), the most frequently used approximations are the self-consistent scheme (SCS) and the differential scheme (DS).

The SCS and DS belong to the class of effective *matrix* schemes: they model interactions by placing the defects into a matrix with altered elastic moduli.

The SCS places non-interacting defects into the matrix having *effective* moduli. The SCS results can be obtained in a straightforward way if the formulation in potentials is used. Indeed, equating $f_0 + \Delta f$ to the potential f of the effective matrix, we specify Δf by placing the defects into the *effective* matrix. Then the results of Section 3 are readily utilized. For example, for 2-D randomly oriented elliptical holes,

$$E/E_0 = 1 - 3p - q \tag{6.1}$$

For a mixture of triangular holes and randomly oriented cracks,

$$E/E_0 = 1 - [\pi\rho + 2p \ (h_3 - h_2)] \tag{6.2}$$

The DS differs from SCS in being incremental: the density of defects is increased in small steps, and the reference matrix is re-calculated at each step. For example, in the 2-D case of randomly oriented elliptical holes,

$$E/E_0 = \exp (- 3p - q) \tag{6.3}$$

For a mixture of triangular holes and randomly oriented cracks, the DS result is

$$E/E_0 = \exp [\pi\rho + 2p \ (h_3 - h_2)] \tag{6.4}$$

Fig. 3 compares SCS and DS with MTS for elliptical holes. The difference between MTS and other schemes depends on the ellipses eccentricity. Note that DS is quite close to MTS for circles, but differs substantially for cracks. As the holes become more elongated, the difference between MTS and the approximation of non-interacting holes gradually vanishes; the two schemes coincide for *cracks*.

It appears that DS and, in particular, SCS, substantially overestimate the effective compliance. This is confirmed by extensive computer experiments on cracks, in both isotropic (Kachanov, 1992) and anisotropic (Mauge and Kachanov, 1992) matrices.

It seems that the underlying reason for this overestimation is that placement of a representative defect into the effective *matrix* (rather than into the effective *stress*) may distort the actual mechanics of interactions (illustrated in Fig. 1). In particular, the effective matrix schemes predict that the interactions *always soften* the effective response (as compared to the approximation of non-interacting defects). This ignores the stiffening impact of the shielding mode of interactions, which may partially (or fully, for cracks) cancel the softening impact of the amplifying mode. We also note that SCS and DS are insensitive to the average "stress environment" in the matrix, which seems physically unreasonable (Kachanov et al.,1994).

238

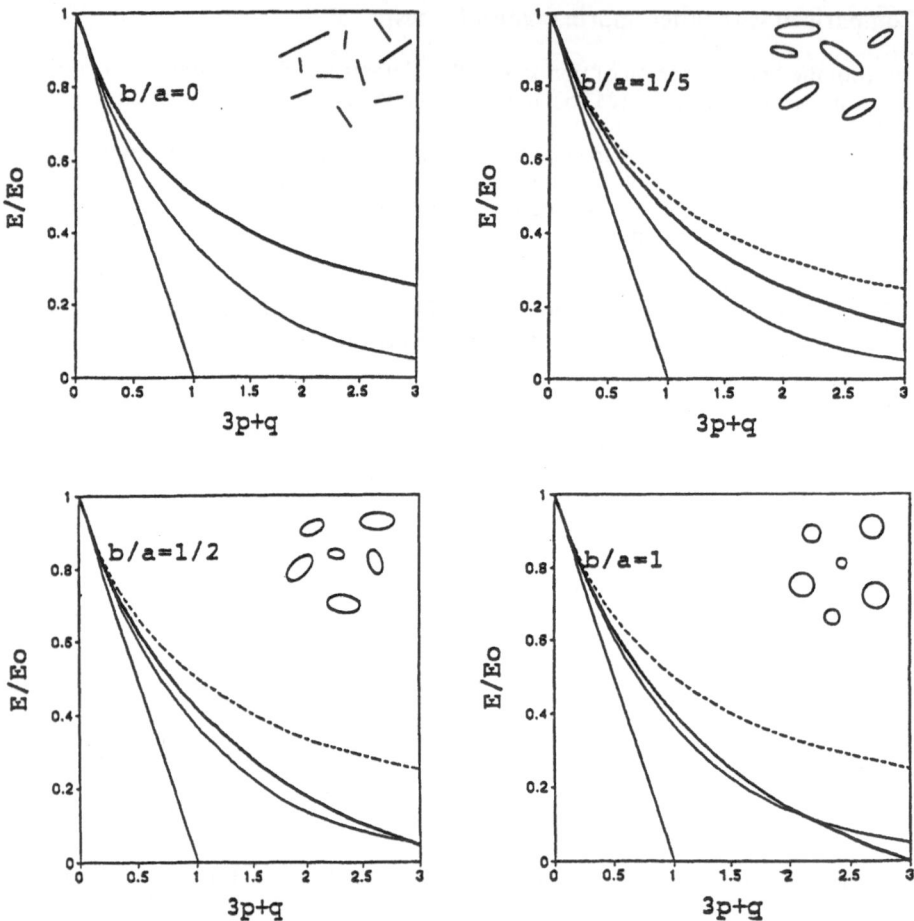

Fig. 3. Comparison of MTS (heavy solid line) with SCS(solid straight line) and DS (solid line) for randomly oriented elliptical holes of various eccentricity *b/a*. Dashed line corresponds to the approximation of non-interacting holes.

Acknowledgment

This research was supported by the U.S.Department of Energy and by the Air Force Office of Scientific Research through grants to Tufts University.

References

Benveniste, Y. (1986) On the Mori-Tanaka method for cracked solids, **Mech. Res. Comm.**, **13**(4), 193-201.

Benveniste, Y., Some remarks on three micromechanical models in composite media, **J. Appl. Mech.**, **57**, 474-476 (1990).

Beran, M.J. (1968) **Statistical Continuum Theories**, Wiley, N.Y.

Berryman, G.J. (1980) Long-wavelength propagation in composite elastic media (parts I and II), **J. Acoust. Soc. Am.**, **68**(6), 1809-1831.

Biolzi, L., Castellani, L. and Pitacco, I. (1993) On the mechanical response of short-fiber reinforced polymer composites, **J. Mater. Sci.**, in press.

Bristow, J.R. (1960) Microcracks and the static and dynamic elastic constants of annealed and heavily cold-worked metals, **British J. Appl. Phys.**, **11**, 81-85.

Budiansky, B. and O'Connell, R.J. (1976) Elastic moduli of a cracked solid, **Int. J. Solids & Struct.**, **12**, 81-97.

Day, A.R., Snyder, K.A., Garboczi, E.J. and Thorpe, M.F. (1992) The effective moduli of a sheet containing circular holes, **J.Mech.Phys.Solids**, **40** (5), 1031-1051.

Eshelby, J.D. (1957) Elastic inclusions and inhomogeneities. In **Progress in Solid Mechanics**, (Sneddon, I and Hill, R., eds.), Vol. II, North-Holland, Amsterdam.

Ferrari, M. and Filipponi, M. (1991) Appraisal of current homogenizing techniques for the elastic response of porous and reinforced glass, **J. Am. Ceram. Soc.**, **74**, 229-231.

Hashin, Z. (1983) Analysis of composite materials - a survey, **J.Appl.Mech.**, **50**, 481-505.

Hill, R. (1963) Elastic properties of reinforced solids: some theoretical principles, **J.Mech.Phys.Solids**, **11**, 357-372.

Jasiuk, I., Chen, J. and Thorpe, M.F. (1992) Effective elastic properties of two-dimensional composites: material with polygonal holes, In **Processing, Fabrication and Manufacturing of Composite Materials**, (T.S. Srivatsan and E.L. Lavernia, eds.), ASME, 61-73.

Jasiuk, I., Chen, J. and Thorpe, M.F. (1994) Elastic moduli of two-dimensional materials with polygonal and elliptical holes, **Appl. Mech. Reviews**, January issue, in press.

Kachanov, M. (1980) Continuum model of medium with cracks, **J.Eng.Mech.Div.**, **106**(EM5), 1039-1051.

Kachanov, M. (1992) Effective elastic properties of cracked solids: critical review of some basic concepts, **Appl. Mech. Reviews**, **45**(8), 304-335.

Kachanov, M. (1993) Elastic solids with many cracks and related problems, in **Advances in Applied Mechanics**, vol. **30** (Hutchinson, J.W. and Wu,T.Y., eds), Academic Press, 259-445.

Kachanov, M., Tsukrov, I. and Shafiro, B. (1994) Effective moduli of a solid with cavities of various shapes, **Appl. Mech. Reviews**, January issue, in press.

Kanaun, S. (1983) Elastic medium with random field of inhomogeneities, Chapter 7 in book: Kunin I.A., **Elastic Media with Microstructure**, vol. II, 165-228, Springer.

Mauge, C. and Kachanov, M. (1992) Interacting arbitrarily oriented cracks in anisotropic matrix. Stress intensity factors and effective moduli. **Int.J.Fracture**, **58**, R69-R74.

Mori, T. and Tanaka, K. (1973) Average stress in matrix and average elastic energy of materials with misfitting inclusions, **Acta Met.**, **21**, 571-574.

Nemat-Nasser, S. and Hori, M. (1993) **Micromechanics: overall properties of heterogeneous materials**. Elsevier, Amsterdam.

O'Connell, R.J. and Budiansky, B. (1974) Seismic velocities in dry and saturated cracked solids, **J. Geophys. Res.**, **79**, 5412-5422.

Tandon, G.P. and Weng, G.J. (1986) Average stress in the matrix and effective moduli of randomly oriented composites, **Compos.Sci.&Techn.**, **24**, 111-132.

Thorpe, M. F. and Sen, P.N. (1985) Elastic moduli of two-dimensional composite continua with elliptical inclusions, **J.Acoust.Soc.Am.**, **77**(5), 1674-1680 .

Vakulenko, A. and Kachanov, M. (1971) Continuum theory of medium with cracks, **Mech. of Solids**, **6**(4), 145-151, Plenum Publ.Co. (English transl. of **Izvestia AN SSSR, Mekhanika Tverdogo Tela**).

Vavakin, A.S. and Salganik, R.L. (1975) Effective characteristics of nonhomogeneous media with isolated inhomogeneities, **Mech. of Solids**, **10**, 58-66, Plenum Publ.Co. (English transl. of **Izvestia AN SSSR, Mekhanika Tverdogo Tela**).

Walsh, J.B. (1965) The effect of cracks on the compressibility of rocks, **J.Geophys. Res.**, **70**(2), 381-389.

Weng, G. J. (1990) The theoretical connection between Mori-Tanaka's theory and the Hashin-Strikman bounds, **Int.J.Eng.Sci.**, **28**(11), 1111-1120.

Zhao, Y.H. and Weng, G.J. (1990) Effective elastic moduli of ribbon-reinforced composites, **J.Appl.Mech.**, **57**, 158-167.

Zhao, Y.H., Tandon, G.P. and Weng, G.J. (1989) Elastic moduli of porous materials, **Acta Mechanica**, **76**, 105-130.

Zimmerman, R.W. (1986) Compressibility of two-dimensional cavities of various shapes, **J.Appl.Mech.**, **53**, 500-504.

Zimmerman, R.W. (1991) **Compressibility of Sandstones**. Elsevier, Amsterdam.

EXPERIMENTAL CHARACTERIZATION, MICROMECHANICAL SIMULATION AND SPATIO-STOCHASTIC APPROACH OF CONCRETE BEHAVIOURS BELOW THE REPRESENTATIVE VOLUME

C. HUET
Swiss Federal Institute of Technology, Lausanne, Switzerland

Abstract
°The implications of granular based microstructure of concrete for the significance of experimental testing results are examined. The universal basic conditions - derived by the author in 1980 - for making use of the effective properties concept are recalled together with a spatio-stochastic approach based on statistical laboratory testing procedures. Some problems arising when the concept of representative volume cannot be used are studied. The concept of apparent properties introduced by the author in 1990 is described. The dependency of these apparent properties upon various kinds of boundary conditions and upon the size of the specimens are studied . Hierarchic ordering - obtained through a variational approach - of various kinds of boundary conditions are presented. Corresponding micromechanical numerical simulations are presented. A - described in details elsewhere - integrated approach (S.C.I.M.) associating these techniques with physical and computer aided microstructural observations is briefly presented.

Some implications of these results for size of specimens optimization and machine design in material testing, for defect sensitivity analyses and for homogenization techniques of materials with distributed heterogeneity levels are derived.
Keywords: Concrete, Representative volume, Apparent properties, Micromechanics, Spatio-stochastic approach.

1. Introduction

In current structural design and quality assessment, concrete is most often considered as a homogeneous material, although its heterogeneous nature - obtained by manufacturing - is obvious. This involves spurious effects like size effects, scatter of results, sensibility to local defects, non-linearities in the behaviours, that are frequently observed in the laboratory although they contradict the basic assumptions of continuum mechanics normally used in structural calculations. This makes one question the representativity of the specimens behaviour as providing models for the behaviour of the material in structural elements.

For a long time, the consequences of heterogeneity for the properties of concrete have been the subject of numerous studies. Many of these are recalled in various contributions to the International Conference held in Lausanne in March 1993 about Micromechanics of Concrete and Cementitious Composites, see Huet (1993a).

We present here, in its present state of development, an outline of the approach we have developed in order to cope with these consequences.

The notation used is given in Table 1.

D. Breysse (ed.), Probabilities and Materials, 241–260.

Table 1. Nomenclature

D_O	Domain occupied in space by a material body.
∂D_O	Boundary of D_O.
Γ	Internal interface of a heterogeneous body.
$\Gamma \rightarrow \Gamma^+$	In one medium, with outward normal n^+ ; $a = a^+$ on Γ^+.
$\Gamma \rightarrow \Gamma^-$	In one medium, with outward normal n^- ; $a = a^-$ on Γ^-.
$[a]_-^+ = a^+ - a^-$	Jump bracket of a.
ρ	Mass density.
x	Position of a material point at time t.
v	Velocity of a material point.
σ	Stress tensor.
ε	Strain tensor.
d	Strain-rate tensor.
u	Internal energy per unit mass.
s	Entropy per unit mass.
q	Heat current vector.
T	Absolute temperature.
g	Gradient of the reciprocal temperature.
$f = \rho b$	External force density per unit volume.
r	Non-mechanical external energy supply per unit volume.
b	External force density per unit mass.
ξ	Rate of internal production of entropy per unit volume.
$E[a] \Leftrightarrow \bar{a}$	Mathematical expectation or ensemble average or stochastic average of the variable a.
$<a>$	Spatial average of the variable a on a domain D.
$\dot{a}$	Material time derivative.
a'	Local fluctuation $a - <a>$ of a around its volume average.
$\times$	Tensor product (dyadic).
$\cdot$	Once contracted tensor product.
$:$	Twice contracted tensor product.
a^T	Transpose of the tensor a.
$\text{Sym } a$	Symmetric part $\frac{1}{2}\left(a + a^T\right)$ of the tensor a.
$F_\varepsilon^\sim, F_\sigma^\sim$	Potential energy and complementary energy functionals, respectively.

2. Some motivations and historical notes

In order to get a better appreciation of what is the situation at present time for experimentalists, perhaps it may be useful to remember what it was about thirty years ago. In fact our own concern in the field dates back to 1960 and was at that time beginning to take explicitly the experimental scatter into account beyond the now classical tools of statistical analysis and planning and to evaluate the representativity of laboratory results by comparison with the behaviour of the material in a real structure. For us, these first questionings about the effects of heterogeneity on the significance of experimental results and the design of specimens came out when we were dealing with road bituminous concretes, see Huet (1963), (1965), that are often tested through (molded or sawn) specimens of small size and are rather often used in quite thin layers, Fig 2.1, that may involve very strong wall effects. More specifically we were designing specimens and procedures for the development of testing methods for complex moduli and fatigue, see Huet (1963), (1965), that are now of current use in many laboratories, especially in Europe, see for instance Francken (1973), (1977), (1991), Ugé et al. (1977), Moutier (1989), Moutier et al. (1991), Assef-Vaziri (1987), Chauvin (1991), Huet (1991), AFNOR (1992). But at that time, the statistical theory of heterogeneous media was not yet developed. Due to this lack of theoretical results, there was no mean to cope with - for instance - the possible implications of the already very well known wall effects on the behaviour of a pavement structure. For instance, the basic paper by Eimer (1967) and the book by Beran (1968) came several years later. Thus, it is only through qualitative reasoning and practical considerations about equipment limitations of the time that we arrived at the solution of a sawn specimen with rather limited dimensions, Fig. 2.2. For coarse aggregates, larger specimens where designed later by C.R.R. in Brussels with our cooperation. Larger specimens for granulates coarser than 20 mm in diameter are now recommended by AFNOR (1992).

Nevertheless we would like to point out here the pioneering works performed by Pierre Dantu at L.C.P.C. in Paris during the fifties and early sixties. In addition to being the inventor of the moiré method for continuous extensometry, Dantu (1958) provided one of the first - if not the first - experimental results that was made available about microstresses distribution in concrete, studied through photoelastic coatings, Fig 2.3. In the same (1958) paper he gave - through surface, line and volume averages - several definitions of statistical physical variables like stress and strain tensors, but without discussing their possible equivalence, saying only that they can be obtained one from the other by integration. He gave also experimental results about the differences in strain and stress distribution between molded and sawn specimens and, in part II, a description of his moiré method, that might be used also for the evaluation of volume averages through the formulas of section 4 below, and is at the root of more recent methods like, for instance, laser holographic interferometry.

At that time, we were not aware of other pioneer works by Kröner (1958) that we discovered very much later, and it is only at the end of our studies about bituminous concrete in the early sixties that the basic works by Hill (1963), Hashin and Shtrikman (1961), (1963), Dantu and Mandel (1963) involving variational approaches for bounding overall properties were published. In his paper with Mandel, Dantu provided the results of a systematic experimental program that was probably the first attempt to derive experimental verification for these bounds.

Some aspects of these experimental tests may seem nowadays rather questionnable. For instance, Dantu's observation that the Hashin bounds were violated in some cases might probably be attributed to the fact that the model specimens of Dantu - as they can still be seen in the 1963's paper - were probably not really isotropic in the bulk, contrarily to the requirements of Hashin's theory. But to our knowledge, no further verification about this experimental violation has been made since that time. This would be of some importance since the Hashin method has provided roots for much more recent extensions of the bounding theory to non-linear problems, see Willis and Talbot (1991), Willis (1993).

Having been - from 1960 to 1963 - a working neighbour of Dantu at LCPC, we got strong impressions of his micromechanical studies and results . Thus we are pleased to acknowledge here his work, which is perhaps not very well known to people who entered this field much more recently.

Since after 1965 we moved to other fields and Institutes, we remained inactive in this field during more than ten years, except for an attempt, Huet (1967), of extending the Hill bounds to materials with capillary stresses, a work that was privately communicated to Mandel and discussed with him, but remained unpublished.

244

2.1   2.2

Fig. 2.1. Microstructure of a three layers - made of three different bituminous concretes - of the surface course of a road pavement The thicknesses from bottom to top are 9, 3 and 3 cm respectively .

Fig. 2.2. Sawn trapezoidal specimen as designed for fatigue and complex modulus testing, from Huet (1963). The dimensions are 25, 2.5 and 2.5 to 5.5 cm.

When we came back to the field in the late seventies, very much progress had been achieved on the theoretical side for the periodic and random cases. For the latter case, that was of most interest for us, we were specially impressed by the works of Ekkehart Kröner and his systematic theory, as explained in his (1972) monograph with further developments in Kröner (1977), (1979), etc. In fact, this systematic theory, mainly based on the use of Green functions and of a statistical hierarchy of correlation functions of increasing orders, provides an approach general enough to cope with the effective mechanical and physical properties of every family of random materials provided we restrict to the linear domain. Morover, the use of variational theorems made possible the derivation of an infinite set of narrower and nar

Fig. 2.3. Photoelastic isochromes for the measurements of microstresses on a sawn concrete specimen under load, from Dantu (1958). The original picture is in color.

rower bounds - of increasing orders in Kröner's terminology - depending on the amount of statistical information injected in the problem, see for instance Kröner (1982).

This character of not being attached to a specific material but to be a rather general tool made this systematic theory very attractive for us, and we first tried to see how its domain of application could be extended to other types of behaviours than the linear elastic or similar ones.

3. The problem of inelastic and climate depending behaviours: the universal assimilation conditions for the existence of an effective medium

Kröner's systematic theory was mainly developed for isothermal problems of the kind of linear elasticity for which the starting point was the Lamé equations. Also Kröner was making an explicit use of what he named the Hill (1963) condition, equating the mechanically defined and energetically defined effective moduli, see for instance Huet (1990). This energetic definition was needed in order to make possible the use of the variational theorems providing the set of bounds of Hill for the zeroth and first orders, of Hashin for the second order and of Kröner for higher orders. The Hill condition requires that the covariance of the stress and strain is vanishing. Mandel, in Dantu and Mandel (1963), stated that this condition was always fulfilled in statistically homogeneous materials, but at the time his proof seemed to us not fully convincing, as involving perhaps unjustified implicit assumptions or approximations. On the other hand, Kröner (1972) demonstrated that they were not fulfilled in several situation, this even in the framework of isothermal linear elasticity.

But our experimental studies dealt with dissipative and highly temperature or moisture dependent materials like bituminous mixes - and later wood - or with chemical reaction, like cement concrete. Thus the question arised: was it still possible to a priori define the concept of an equivalent homogeneous medium - the effective medium - for such materials.

Starting from that part of the set of equations of a general thermo-mechanical problem which is independent of the material under consideration - the universal balance equation for mass, momentum, energy and entropy - we derived, Huet (1981), (1982) what we named the universal assimilation conditions to a general effective continuum. They are recalled in Tables 3.1 and 3.2. As can be seen, they have forms generalizing the Hill condition, that can be recovered from them in the case of linear elasticity.

TABLE 3.1 . Assimilation conditions
to a general effective continuum.

$$\overline{\text{div } \rho\,'v\,'} = 0 \tag{3.1}$$

$$\overline{\rho}\,\overline{v\,'.\text{ grad }v\,'} + \overline{\rho\,'\dot{v}\,'} = \overline{\rho\,'b\,'} \tag{3.2}$$

$$\overline{\rho}\,\overline{v\,'.\text{ grad }u\,'} + \overline{\rho\,'\dot{u}\,'} = \overline{\sigma\,':d\,'} \tag{3.3}$$

$$\overline{\rho}\,\overline{v\,'.\text{ grad }s\,'} + \overline{\rho\,'\dot{s}\,'} = \overline{\left(\dfrac{1}{T}\right)r\,'} - \overline{\text{div}\left[\dfrac{1}{T}\,q\,'\right]} \tag{3.4}$$

TABLE 3.2 . Whole set of sufficient assimilation
conditions to a general effective continuum.

$$\overline{\rho'v'} = \overline{\rho'\dot{v}'} = \overline{\rho'b'} = \overline{\rho'\dot{u}'} = \overline{\rho'\dot{s}'} = 0 \tag{3.5}$$

$$\overline{v'.\text{ grad }v'} = \overline{v'.\text{ grad }u'} = \overline{v'.\text{ grad }s'} = 0 \tag{3.6}$$

$$\overline{\sigma':d'} = \overline{\left(\dfrac{1}{T}\right)r'} = \overline{\left(\dfrac{1}{T}\right)q'} = 0 \tag{3.7}$$

When they are fulfilled, it is possible to write the local universal balance equations in terms of stochastic averages just as if the material were homogeneous. Thus, this defines the domain of situations for which the heterogeneous character of the material can be ignored, as done in the current engineering practice. In the definition of this domain both extrinsic and intrinsic features of the material system may be involved. This has to be checked in each particular case. Then, their content of information can be used in subsequent derivations. In particular, from that can be stated, Huet (1988), that the Clausius-Duhem inequality, Truesdell (1969), keeps its classical form in terms of the same stochastic averages of the physical variables.

As elucidated by Beran (1968), these stochastic averages are locally defined on corresponding points of a set of random realizations of the structure under consideration. Of course, this set can be a set of specimens.

4. Experimental determination of constitutive equations for the effective medium

Since - fortunately for having a compliant world - the universal balance equations are not enough to obtain a well posed problem of thermomechanics, we need constitutive equations for the effective medium to be considered equivalent to the real heterogeneous medium we are dealing with. These constitutive equations can be provided by experiment only. But the latter requires specimens of finite volume, and thus the consideration of volume averages since the internal local states are most often inaccessible to measurements, by contrast with the external surface states like in Dantu's 1958 experiments.

In his systematic theory, Kröner and most other authors switch from stochastic averages to volume averages through the use of an ergodicity assumption. The latter requires statistical stationarity and the whole statistical information being contained in one single realization. This implies infinite volumes in uniform conditions, this being once again inaccessible to experiment.

This leads in the practice to the concept of representative volume, with dimensions such that the obtained overall properties are the same as for specimens of larger sizes. This implies dimensions large enough by comparison with the size of the coarsest heterogeneity. When this is achieved, it is furtherly assumed that the overall properties are independent of the boundary conditions. This allows one to make use indifferently - in analytical derivations - of two special cases of boundary conditions for which the Hill condition expressed in terms of volume averages is fulfilled: the two classical kinematic uniform boundary conditions (ε_0 - KUBC) with displacement vector of the form $\varepsilon_0 \cdot x$ imposed on the whole boundary, and the static uniform boundary condition (σ_0 - KUBC) with traction vector $\sigma_0 \cdot n$ imposed on the whole boundary.

Performing a set of testing programs on a specimen with representative volume and measuring simultaneously the external surface displacements and tractions allows one - at least in principle - to determine the constitutive equations through the resolution of a set of inverse problems.

This is obtained - at least for the case with perfect internal interfaces - through the universal relationships, Eqs (4.1) and (4.2), that exist between the volume averages of ε and σ and the surface displacements and tractions that are accessible to direct measurements for the cases of perfect interfaces:

$$< \varepsilon > = \frac{1}{V} \int_D \text{sym grad } \xi \, dV = \frac{1}{V} \int_{\partial D} \text{sym } (\xi \times n) \, d\Sigma + \frac{1}{V} \int_\Gamma \text{sym } [\xi]_-^+ \times n^+ d\Gamma \tag{4.1}$$

$$< \sigma > = \frac{1}{V} \int_{\partial D} \text{sym } (x \times P) \, d\Sigma + \frac{1}{V} \int_D x \times (f - \rho \dot{v}) \, dV +$$

$$+ \frac{1}{V} \int_\Gamma \text{sym} \left(x \times [\sigma \cdot n^+]_-^+ \right) d\Gamma \tag{4.2}$$

where the symbol $[a]_-^+$ denotes the discontinuity bracket across the interface :

$$[a]_-^+ = a^+ - a^- \tag{4.3}$$

If the ergodic assumption was true at the specimen level, performing the test on one single specimen would be enough. Furthermore, performing the test on several other specimens would yield the same results without any scatter. Every experimentalist knows that this is not what happens in the laboratory.

5. Apparent properties and spatio-stochastic approaches for specimens not having the representative volume

As said in the introduction, it is a daily observation in laboratories that test results exhibit experimental scatter. This means that the ergodicity assumption is too strong to fit the experimental practice. Thus two questions arise: what is the real status of the material property parameters that we derive from tests results involving specimens of given finite volume, and how are they possibly related to the effective properties for cases of structures with dimensions much wider than the specimen ones. For instance, these questions are particularly crucial for dam concrete with granules diameters that can be larger than 10 cm and for which representative volumes have dimensions larger than one meter, for which no machine is large and strong enough, and for which boring cores of large diameters is very expensive or simply impossible.

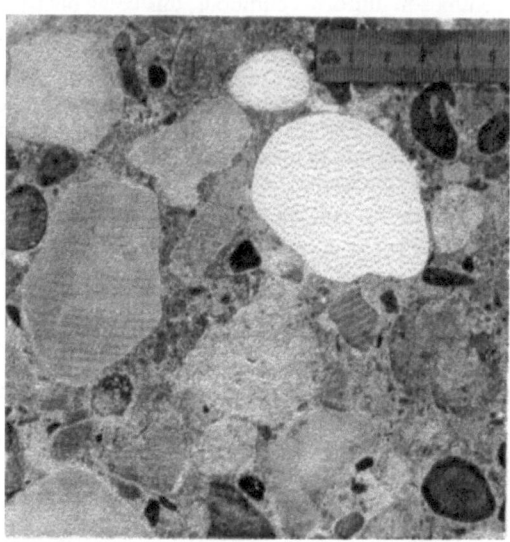

Fig. 5.1. Cross-section of a prismatic dam concrete specimen with dimensions 30, 15 and 15 cm. The dimensions of the coarsest granules here are about 7 cm in diameter. There exist dam concretes with diameters of coarsest granules exceeding 15 cm.

This led us, Huet (1990), (1991b), to introduce the concept of apparent properties in place of effective properties. The former are still defined in uniform boundary conditions of the kind defined in Section 4. This because the Hill condition is still valid for such boundary conditions even when the ergodic assumption is relaxed. This allows one to identify once again the mechanical definition - through a Hooke law written between volume averages - and the energetic definition through the averaged free energy or free enthalpy, both written in terms of strain - respectively stress - averages. From this, variational arguments can still be used providing various bounds.

But here, by contrast with the case with the representative volume, the results depend on the boundary conditions to which they relate. Thus, in the definitions of the material coefficients, the kind of boundary conditions must be specified.

Moreover, for this case we make use of the spatio-stochastic approach that we introduced in Huet (1981), (1984), (1988). From this, see Huet (1990), (1991), we may define safely kinematic statistical apparent modulus and compliance tensors C_ε^{app} and S_ε^{app} through

$$\overline{< \sigma >_\alpha} = C_\varepsilon^{app} : \overline{< \varepsilon >_\alpha} \quad ; \quad \varepsilon_0 \text{- KUBC} \tag{5.1}$$

for a set of specimens D_α being each submitted to ε_0 - KUBC.

From:

$$\overline{<\sigma>_\alpha} = \overline{C_{\varepsilon\alpha}^{app} : <\varepsilon>_\alpha} = \overline{C_{\varepsilon\alpha}^{app} : \varepsilon_0} = \overline{C_{\varepsilon\alpha}^{app}} : \varepsilon_0 = \overline{C_{\varepsilon\alpha}^{app}} : \overline{<\varepsilon>_\alpha} \tag{5.2}$$

this yields :

$$C_\varepsilon^{app} = \overline{C_{\varepsilon\alpha}^{app}} \quad ; \quad S_\varepsilon^{app} = \left(\overline{C_{\varepsilon\alpha}^{app}}\right)^{-1} = C_\varepsilon^{app^{-1}} \tag{5.3}$$

Here, as in the laboratory practice, the N specimens D_α have all the same dimensions. In addition, they must form an exhaustive partition of the initial large specimen D_0.

We can also define static statistical apparent and compliance and modulus tensors S_σ^{app} and C_σ^{app} through:

$$\overline{<\varepsilon>_\alpha} = S_\sigma^{app} : \overline{<\sigma>_\alpha} \tag{5.4}$$

for the set of specimens D_α each submitted to σ_0 - KUBC. From :

$$\overline{<\varepsilon_\alpha>} = \overline{S_{\sigma\alpha}^{app} : <\sigma>_\alpha} = \overline{S_{\sigma\alpha}^{app} : \sigma_0} , = \overline{S_{\sigma\alpha}^{app}} : \sigma_0$$

$$= \overline{S_{\sigma\alpha}^{app}} : \overline{<\sigma>_\alpha} \quad ; \quad \sigma_0\text{-SUBC} \tag{5.5}$$

this yields :

$$S_\sigma^{app} = \overline{S_\sigma^{app}} \quad ; \quad C_\sigma^{app} = \left(\overline{S_{\sigma\alpha}^{app}}\right)^{-1} = S_\sigma^{app^{-1}} \tag{5.6}$$

Iterate applications of the classical minimum theorems of elasticity - using the potential energy and complementary functionals $\tilde{F_\varepsilon}$, $\tilde{F_\sigma}$ respectively - yield then the two following hierarchies of bounds, Eqs. (5.7) and (5.8). They relate the statistical apparent properties defined on various partitions of a given initial specimen D_0 into a set of coarse specimens on the one hand, and of small specimens on the other hand:

$$\left(\overline{C^{-1}}\right)^{-1} \leq C_{\sigma f}^{app} \leq C_{\sigma c}^{app} \leq C_{\sigma o}^{app} \leq C_{\varepsilon o}^{app} \leq C_{\varepsilon c}^{app} \leq C_{\varepsilon f}^{app} \leq \overline{C} \tag{5.7}$$

$$\left(\overline{S^{-1}}\right)^{-1} \leq S_{\varepsilon f}^{app} \leq S_{\varepsilon c}^{app} \leq S_{\varepsilon o}^{app} \leq S_{\sigma o}^{app} \leq S_{\sigma c}^{app} \leq S_{\sigma f}^{app} \leq \overline{S} \tag{5.8}$$

where the subscript o relates to the single initial specimen D_O, subscript c to coarser specimens, and subscript f to finer specimens.

When the initial specimen D_O has the representative volume, $C_{\sigma_O}^{app}$ and $C_{\varepsilon_O}^{app}$ merge into the effective modulus C^{eff}, and similarly $S_{\varepsilon_O}^{app}$ and $S_{\sigma_O}^{app}$ into the effective compliance S^{eff}.

It should be pointed out that the inequalities (5.7) and (5.8) are valid for the statistical apparent properties and not for the individual ones.

Furthermore it can be conjectured that they remain valid for representative stochastic sampling in place of exhaustive sampling.

In Hazanov and Huet (1993), we have recently defined apparent properties for mixed boundary conditions and derived the following additional inequalities for one single specimen D_α with or without the representative volume:

$$C_\sigma^{app} \le C_m^{app} \le C_\varepsilon^{app} \tag{5.9}$$
$$S_\varepsilon^{app} \le S_m^{app} \le S_\sigma^{app} \tag{5.10}$$

where the subscript m relates to the mixed boundary conditions. The extension to an exhaustive partition or a representative sampling has still to be studied.

The results (5.9) and (5.10) have two practical consequences:
- unspecified mixed boundary conditions do not provide bounds for the effective modulus;
- some specific mixed boundary conditions may provide apparent modulus equal to the effective modulus, at least for some components.

Since most testing machines provide rather poorly known mixed boundary conditions, this may have important consequences for the interpretation and use of the results.

It turns out that the classical Hill bounds - for which the present approach gives a new proof - encompass all the other bounds. This means that apparent properties may provide results that are equivalent to higher order bounds in Kröner's sense. This can be attributed to the fact that the coarser the subspecimens, the richer is the statistical information included and made useful in the testing program.

6. Numerical simulations

In order to check the validity of the above theoretical results, a set of numerical simulations have been performed on new versions of the so-called numerical concrete of Wittmann, Roelfstra and Sadouki (1985), Sadouki (1987), Roelfstra (1989). These new versions were specially developed in order to allow at will the sharing of one given initial specimen into smaller ones without modifying the internal microstructure nor the mesh, both in 2D and 3D. Illustrative examples are given on Figures 6.1 to 6.9 for kinematic, static and mixed boundary conditions. This numerical approach has also been used for the study and optimization of the cutting procedure to be used in the experimental programm, Figs. 6.5 and 6.6. More details about these numerical results can be found in Amieur et al. (1990), (1991), (1992), Guidoum and Navi (1993), Guidoum et al. (1992a), (1993b).

Fig. 6.1. Specially developed finite element meshes allowing subdivisions of specimens into parts without change of the mesh; from Amieur et al. (1993).

252

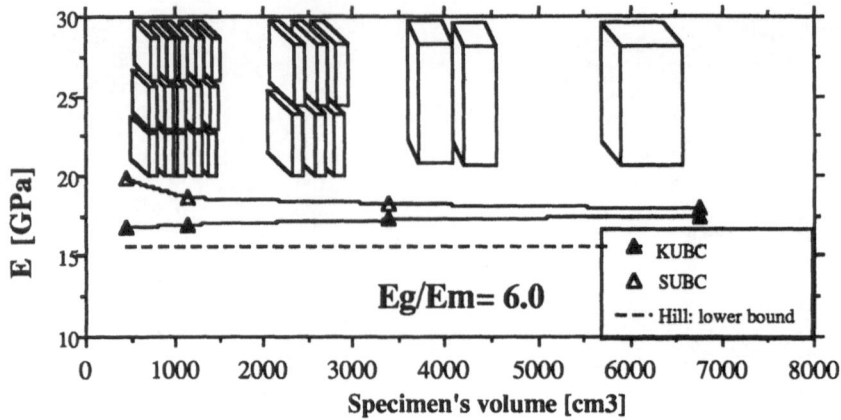

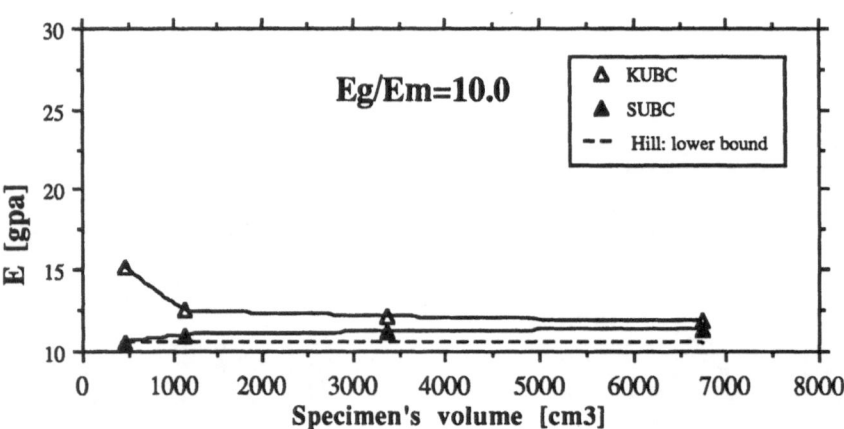

Fig. 6.2. Evolution of the apparent moduli in strain boundary condition (upper curves) and stress boundary conditions (lower curves)for two values of the modulus of the mortar, the modulus of the grains being unchanged; from Amieur et al. (1991).

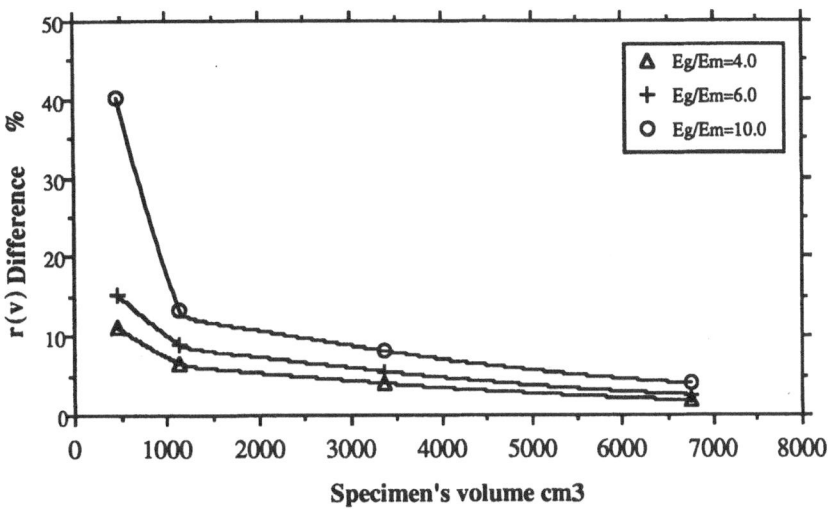

Fig. 6.3. Evolution of the ratio of the difference between E_ε^{app} and E_σ^{app} to the effective modulus E^{eff}, as a function of the specimen's volume (sizes of specimens) for several ratio of the rigidities Eg/Em; from Amieur et al. (1993).

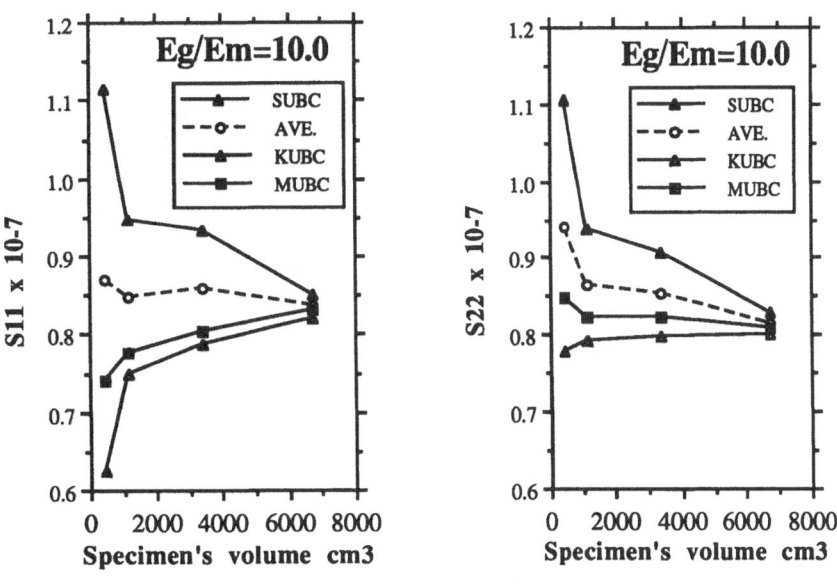

(a)   (b)

Fig. 6.4. Comparison of the compliance components S_{11} and S_{22} for 3 types of boundary conditions: KUBC, SUBC, and MUBC; from Amieur et al.(1993).

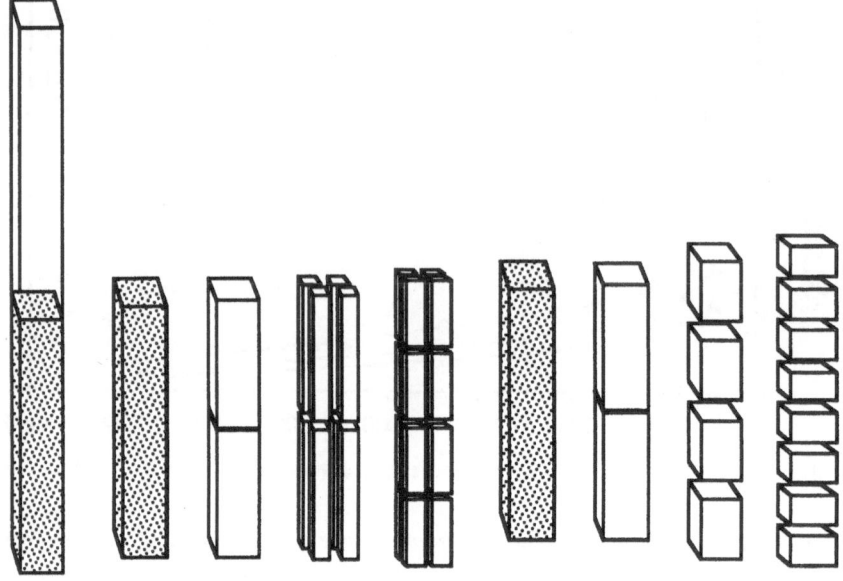

(a) Variant I   (b) Variant II

Fig.6.5. Two types of cutting procedure; from Amieur et al. (1993).

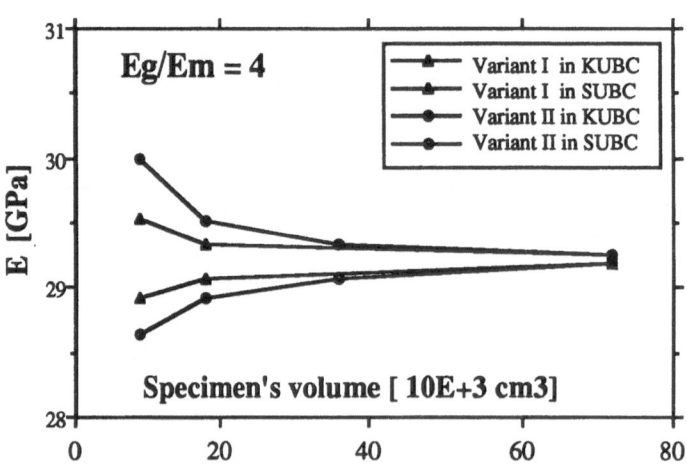

Fig.6.6. Influence of the cutting procedure on the difference between E_σ^{app} and E_ε^{app} for $Eg/Em=4$; from Amieur et al. (1993).

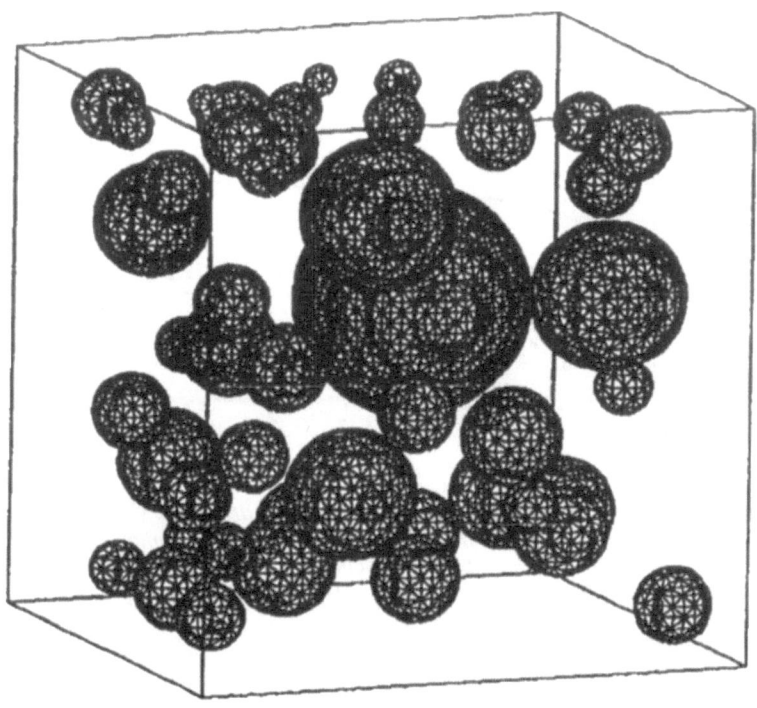

Fig. 6.7 . Finite element mesh of a 3 D numerical model of concrete. Only the grains
are represented. There are 60048 matrix elements ,
22491 inclusion elements and 21793 nodes, from Guidoum and Navi (1993a).

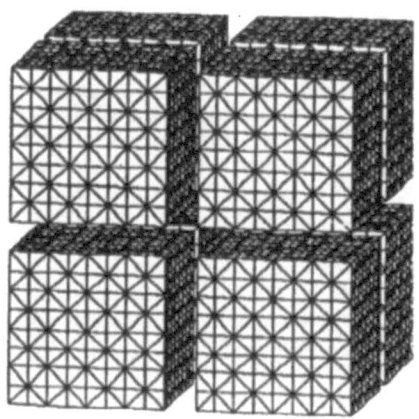

Fig. 6.8. Subdivision of the above specimen into 8 smaller pieces.

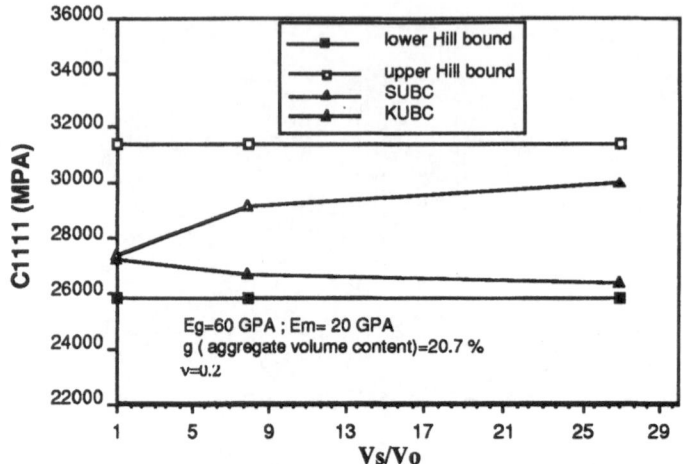

Fig. 6.9. Apparent values of C1111 in kinematic (upper curce) and static (lower curve) uniform boundary conditions as function of the relative volume of the specimen for the initial specimen of Figure 6.6 to the volume of the sub-specimens obtained by successive subdivisions forming exhaustive partitions; from Guidoum and Navi (1993a).

7. Discussion and conclusion

From the approach presented in this paper, it turns out that even using the simplest tools of the general theory of statistical continuum mechanics can provide a better understanding of the significance of the laboratory testing practice for heterogeneous materials like concrete. On the other hand, from the procedures used in laboratory practice, it is possible to elaborate new concepts that avoid the restrictive assumptions made in the classical theoretical analyses.

The results presented in this paper deal first with the problem of effective properties, the use of which corresponds to the wish of the structural engineer to deal with simple homogeneous media only when performing structural calculations on complicated structures. We have seen how the effective property concept is in fact a trick allowing to get rid of the heterogeneous character of the material. We have also seen how the validity domain of this concept can be determined in each particular situation through assimilation conditions that are of universal character, that is, applicable to every kind of material and behaviour.

It was also shown in this paper that the spatio-stochastic approach that we introduced first in 1981 allows one to give apparent estimates bounding the effective properties provided appropriate uniform boundary conditions are used. But since the laboratory testing equipments are in mixed boundary conditions, for which various kinds of results can be obtained, the study has to be continued for those cases. The fact the two classical kinds - static and kinematics - of uniform boundary conditions provide bounds for apparent properties for all kinds of mixed boundary conditions seems to be an important result from this view-point of the laboratory practice .

Numerical simulations have shown that , even for specimens having almost the representative volume, the expected hierarchic ordering between the static and kinematic boundary conditions is obtained : the kinematic boundary conditions gives always higher moduli - in the sense of order relationships for tensors of the fourth rank - than the static one. In addition, static and kinematic uniform boundary conditions give almost same apparent properties for specimens considered as being near the representative volume. This, combined with the above results for mixed boundary conditions, confirms that the effective properties can really be considered as independent of the boundary conditions -as usually assumed- since all kinds of boundary conditions are thus bracketed by merging lower and upper bounds.

This does not hold when the specimens do not have the representative volume. The results are dependent - not only on the specimens dimensions (size-effects) as normally assumed - but also on the boundary conditions, a fact that is most often not perceived. This means in particular that the designers and builders of testing equipments should take more care in the definition of the boundary conditions imposed by their machines, and in providing enough information about the corresponding characteristics. For this purpose, and for the purpose of optimizing the size and shape of the specimens in terms of the involved heterogeneities and other parameters - including economic ones - numerical simulations in 3D and even in 2D turn out to be helpful. An experimental program using these optimizing tools is presently under progress on concrete, see Amieur et al. (1993).

Up to now, the presented approach has been used mainly for thermoelastic properties. It can be transferred without difficulties to other physical properties given by instantaneous linear current-gradient laws, like thermal conductivity for instance. It seems that these results can be easily extended to those non-linear and/or dissipative behaviours that can be linearized in the form of a tangent linear law, written in rate or incremental forms.

For still more complicated behaviours, the corresponding theory has still to be developed, and only a complete micromechanical approach through analytical derivations or numerical simulation can be used.

In particular, even the case of linear viscoelasticity with wide spectrum cannot be dealt with in the general case because of the lack of variational results - in the form of minimum theorems - for the derivation of bounds. Nevertheless our recent results on this latter topic, Huet (1988b), (1992), (1993b), allows one to think that extensions to this case will be possible in the future for some cases at least, and in particular for semi-elastic heterogeneous media, for which only one constituent is viscoelastic. Nevertheless, the results that we obtained on complex moduli with our coworkers in Chetoui et al. (1986) and also in Huet et al (1991b) might be used here for more general cases.

In structures of small thickness and involving thus strong wall-effects and high gradients, the situation is still more involved, and the use of an explicit - numerical or analytical - micromechanical approach can no more be avoided, even at the level of structural calculations. The same holds true for properties that relate to the behaviour of singularities or defects like cracks, local defects, etc.

This is the reason for which, in recent years, we worked in order to develop a micromechanical approach in the form of the SCIM system: System for Concrete Integrated Micromechanics. The word integrated corresponds to the organization of the physical and numerical simulation tools into a completely linked chain of units involving in particular macroscopic and microscopic observations, image analysis and processing, direct meshing, final computing and various kinds of exploitation units. A detailed presentation of SCIM is given in Huet (1993c), Navi and Pignat (1993), Sunderland and Tolou (1993), Guidoum and Navi (1993), Guidoum et al. (1993), Amieur et al. (1991),(1993), Huet et al. (1991a), (1991b).

For the derivation of 3 D overall constitutive equations in the case of inelastic behaviours, it makes use of the dissipative internal variables approach that we developed in (Huet 1988b), (1993d), Huet et al (1991a) from the Biot (1954), (1955), and Mandel (1979) theories through our analysis of the local state principle, Huet (1979), (1982b),.

But since defects or singularities in heterogeneous materials provide high scatter in the results, the probabilistic and statistical tools have still to be applied to the results of these micromechanical computations, that have to be performed on a large enough number of random realizations of the microstructure.

Also the theoretical results that have been obtained for the first statistical moments (averages) have to be extended to higher orders ones - and specially to the second ones (variances and covariances) that play the role of scatter and correlation parameters very much often used by experimentalists in the laboratories.

Acknowledgments

Important supports to this research from E.P.F.L., the Swiss National Foundation for Scientific research under contracts Nr. 21-27962.89 and 20.32206.91, the Foundation for

258

Research of the Swiss Cement Industry and the Federal Office for Water Economy are gratefully acknowledged. We express also our gratitude to our collaborators in L.M.C. who agreed
to develop their own work along some of the lines described in this paper.

References

A.F.N.O.R. (1992). Mesure des caractéristiques rhéologiques des mélanges hydrocarbonés.
Partie 2 : Détermination du module complexe par flexion sinusoïdale. French Standard NF
P 98-260-2,Paris.

ASSEF-VAZIRI A.H. (1987). Essai de fatigue par flexion alternée sur éprouvettes trapézoï-
dales. Report no 13/83 for the Roads Federal Office, E.P.F.L., Laboratoire des Voies de
Circulation (LAVOC), Lausanne.

AMIEUR M., NAVI P., WANG J., HUET C. (1990) Simulation numérique des effets
d'échelle dans les solides hétérogènes. Proc. 25th Colloquium of the Groupe Français de
Rhéologie. Rhéologie des matériaux du Génie Civil, Grenoble, 28-30 novembre, pp. 445-
457.

AMIEUR M., HAZANOV S., HUET C.(1991). Fractal modelling of concrete, 11th Int.
Conference on Structural Mechanics in Reactor Technology, Tokyo, Japan, HO4/2 pp. 109-
114.

AMIEUR M., HAZANOV S. , HUET C. (1993). Numerical and experimental study of size
and boundary conditions effects on the apparent properties of specimens not having the representative volume. In C. HUET (Ed.). Micromechanics of Concrete and Cementitious
Composites. Presses Polytechnique et Universitaire Romandes, Lausanne, pp. 181 - 202.

BERAN M.J. (1968). Statistical continuum theories. Wiley (Interscience), New York.

BIOT M.A. (1954). Theory of stress-strain relations in anisotropic viscoelasticity and relaxation phenomena. J. Appl. Physic, 13, 11, 1385-1391.

BIOT M.A. (1955). Variational Principles of Thermodynamics with Application to
Viscoelasticity. Phys. Review, 97, 6, pp. 1453-1469.

CHAUVIN J.J. (1990). L'essai de module complexe utilisé pour la formulation des enrobés.
Proc. RILEM Symposium on Mechanical tests for bituminous mixes. Chapman & Hall, pp.
367-381.

CHETOUI S., NAVI P., HUET C. (1986). Recherches sur l'évaluation des propriétés macro-
scopiques des matériaux hétérogènes viscoélastiques anisotropes, In C. Huet, D. Bourgoin,
S. Richemond, Rhéologie des Matériaux Anisotropes, Français de Rhéologie, Cepadues,
Toulouse, 307-326.

DANTU P. (1958). I. Etude des contraintes dans les milieux hétérogènes. Application au béton. II. Utilisation des réseaux pour l'étude des déformations. Annales I.T.B.T.P., Paris, no
121.

DANTU P., MANDEL J. (1963). Contribution à l'étude théorique et expérimentale du coefficient d'élasticité d'un milieu hétérogène, mais statistiquement homogène. Annales des Ponts
et Chaussées, Paris, 133, 2, 115-146.

EIMER, C.Z. (1968). The boundary effect in elastic multiphase bodies. Archiwum
Mechaniki:Stosowanej, 1, 20, 87-93.

FRANCKEN L. (1973). Module complexe des enrobés bitumineux, Influence des caractéristiques des bitumes et de la composition des mélanges. Research report 164, Centre de
Recherches Routières, Brussels.

FRANCKEN L. (1977).Module complexe des mélanges bitumineux. Bull. Liaison Labo.
Ponts et Chaussées, Special Issue V, pp.181-198.

FRANCKEN L.(1991). Propriétés mécaniques des enrobés : influence de la composition.
Proc. Journées d' Etudes AFREM, November 28-29, Saint-Rémy les Ch., pp.185-196.

GUIDOUM A., NAVI P. (1993a). Numerical simulation of thermo-mechanical behaviour of
concrete through a 3-D granular cohesive model, In C. HUET (Ed.), Micromechanics of
Concrete and Cementitious Composites. Presses Polytechnique et Universitaire Romandes,
Lausanne, pp. 213 - 228.

GUIDOUM A., NAVI P., HUET C. (1993b). 3-D Numerical study of heat transfer and ther-

Conference on Numerical Methods for Thermal Problems. Pineridge Press, Swansea, pp. 917-927.

GUIDOUM A., NAVI P., HUET C. (1993c). Numerical evaluation of microstructural effects on the shrinkage properties of concrete through a three-dimensional model of cohesive granular material. 5th RILEM International Symposium on Creep and Shrinkage of Concrete, Barcelone, 6-9 septembre (In press).

HASHIN Z., SHTRIKMAN S. (1961). Note on a variational approach to the theory of composite elastic materials. J. Franklin Institute, v. 271, 4.

HASHIN Z., SHTRIKMAN S. (1963). A variational approach to the theory of the elastic behaviour of multiphase materials. J. Mech. Phys. Solids, v. 11, 127-140.

HAZANOV S., HUET C. (1993). Bounds to Mixed Boundary Conditions Effects on the Overall Properties of Elastic Heterogeneous Bodies not having the Representative Volume (submitted for publication).

HILL R. (1963). Elastic properties of reinforced solids : some theoretical principles. J. Mech. Phys. Solids, v. 11, 357-372.

HUET C. (1963). Comportement viscoélastique d'un matériau hydrocarboné, C.R. Ac. Sc. Paris, t 257 , p. 1438-1442.

HUET C. (1965) Etude, par une méthode d'impédance, du comportement viscoélastique des matériaux hydrocarbonés, Annales des Ponts et Chaussées, 6, p. 3-69.

HUET C. (1967). Unpublished result.

HUET C. (1981). Remarques sur l'assimilation d'un matériau hétérogène à un milieu continu équivalent. In C. Huet and A. Zaoui. Comportements rhéologiques et structure des matériaux. Presses ENPC, Paris, 231-245.

HUET C. (1982a). Universal conditions for assimilation of a heterogeneous material to an effective medium. Mechanics Research Communications, 9 (3), p. 165-170.

HUET C. (1982b). Topics in thermodynamics of rheological behaviours, Rheologica Acta, 21, p. 360-365.

HUET C. (1984). On the definition and experimental determination of effective constitutive equations for heterogeneous materials. Mechanics Research Communications, 11 (3), p. 195-200.

HUET C. (1988 a). Some basic principles in the thermorheology of heterogeneous materials. In H. Giesekus and M.F. Hibberd (Eds.), Progress and Trends in Rheology, Supplement to Rheologica Acta, Vol. 26, pp. 1-8.

HUET C. (1988b). Definition of the out equilibrium entropy through the use of markoffian variables and the example of elasto-visco-plastic mechanical models. Kestin-Lehmann Symposium on "What is the correct form of the Gibbs equations for inelastic solids", Bochum.

HUET C. (1988c). Bounds for the viscoelastic properties of heterogeneous materials through new variational theorems. In Uhlherr P.H.T. (Ed.), Xth Int. Congress on Rheology. Australian Soc. Rheology, Sydney, 1, pp. 422-424.

HUET C. (1990). Application of variational concepts to size effects in elastic heterogeneous bodies. J. of the Mechanics and Physics of Solids, vol 38, pp. 813-841.

HUET C.(1991a).The role of mechanical tests for characterization, design and quality control of bituminous mixes. In H.W Fritz and E. Eustacchio (Eds.). Mechanical Tests for Bituminous Mixes, Characterization, Design and Quality Control, Proc. of the 4th International Symposium RILEM TC 101-BAT, Budapest, 23-25 October 1990, v II, 1991, p. 202-204.

HUET C. (1991b). Hierarchies and bounds for size effects in heterogenenous bodies. In G. Maugin (Ed.), Proceedings of the Sixth Symposium on Continuum Models and Discrete Systems, Dijon, Longmans, London, pp. 127 - 134.

HUET C. (1992).Minimum theorems for viscoelasticity. Eur. J. Mech., A/Solids, 11, No 5, 653-684, 1992.

HUET C., Ed. (1993a). Micromechanics of Concrete and Cementitious Materials. Presses Polytechniques et Universitaire Romande, Lausanne.

260

HUET, C. (1993b). An integrated approach of Concrete Micromechanics. In C. Huet (Ed.). *Micromechanics of Concrete and Cementitious Composites*. Presses Polytechniques et Universitaires Romandes, Lausanne, pp. 117-146.

HUET C. (1993c). Dissipative minimum theorems for viscoelasticity (submitted for publication).

HUET C. (1993d). Some basic tools and pending problems in the development of constitutive equations for the delayed behaviour of Concrete, 5th RILEM International Symposium on Creep and Shrinkage of concrete, Barcelone, 6-9 septembre, (In press).

HUET C., ZAOUI A., Eds. (1981). Rheological behaviour and structure of materials. Presses ENPC, Paris.

HUET C., HADJ MOHAMED B., HAZANOV S. (1991a). Internal variables technique for elasto-visco-plastic mechanical models of heterogeneous materials. 11th International Conference on Structural Mechanics in Reactor Technology, Tokyo, Japan, pp. 18-23.

HUET C., NAVI P., and ROELFSTRA, P. (1991b). A new homogeneisation technique based on Hill modification theorem, In G. Maugin (Ed.), Proceedings of the Sixth Symposium on Continuum Models and Discrete Systems, Dijon, Longmans, London, pp. 135 - 143.

KRÖNER E. (1958).Berechnung der elastischen Konstanten des Vielkristalls aus den Konstanten des Einkristalls. Z. Phys. 151, 504.

KRÖNER E. (1972). Statistical continuum mechanics. Springer-Verlag, Berlin and New York.

KRÖNER E. (1977). Bounds for effective elastic moduli of disordered materials. J. Mech. Phys. Solids 25, 137.

KRÖNER E. (1981). Linear properties of random media: the systematic theory. In C. Huet and A. Zaoui (Eds.). Rheological Behaviour and Structure of Materials, Presses ENPC Paris, pp. 15-40.

KRÖNER E. (1986). Statistical modelling. In J. Zarka et Gittus (Eds.), Modelling Small Deformation of Polycrystals. Elsevier, No 1926, 229.

MANDEL J. (1979). Variables cachées, puissance dissipée, dissipativité normale. Sciences et techniques de l'Armement, Paris 53, 4, pp. 525-538.

NAVI P., PIGNAT C. (1993). Tri-dimensional simulation of microstructural development of cement paste during hydration. In C. HUET (Ed.). Micromechanics of Concrete and Cementitious Composites. Presses Polytechniques et Universitaires Romandes, Lausanne, pp. 147 - 158.

ROELFSTRA P.-E. (1989). A Numerical Approach to investigate the Properties of Concrete: Numerical Concrete, Doctoral Thesis No 788, Lausanne EPFL.

SADOUKI H. (1987). Simulation et Analyse Numérique du Comportement Mécanique des Structures Composites , Thèse no. 676, Swiss Federal Institute of Technology, Lausanne.

SUNDERLAND H., TOLOU A., HUET C. (1993). Multilevel numerical microscopy and tridimensional reconstruction of concrete microstructure. In C. HUET (Ed.) Micromechanics of Concrete and Cementitious Composites. Presses Polytechniques et Universitaires Romandes, Lausanne, pp. 171 - 180.

TRUESDELL C. (1969), Rational Thermodynamics, Mac Graw Hill, New-York.

WANG J., HUET C. (1993). A Numerical Model for studying the Influences of Pre-existing Microcracks and Granular Character o n the Fracture Mechanics of Concrete. In C. HUET (Ed.) Micromechanics of Concrete and Cementitious Composites. Presses Polytechniques et Universitaires Romandes, Lausanne, pp. 229 - 240.

WITTMANN F. H., Ed. (1982). Fundamental Research on Creep and Shrinkage of Concrete. Martinus Nijhoff Publishers, The Hague.

WITTMANN F. H., ROEFLSTRA P.E., SADOUKI H. (1985). Simulation and analysis of composite structures. Mat. Sc. Eng., 68, pp. 235-248.

WITTMANN F. H., STEIGER T., SADOUKI H. (1993) Experimental and Numerical Study of Effective Properties of Composite Materials. In C. HUET (Ed.), Micromechanics of Concrete and Cementitious Composites. Presses Polytechniques et Universitaires Romandes, Lausanne, pp. 61 - 86.

MICROMECHANICAL MODEL FOR RANDOMLY PACKED GRANULES

CHING S. CHANG
University of Massachusetts, Amherst, Massachusetts, USA

Abstract
During deformation of a granular material, both forces and moments are transmitted
in the medium through the inter-particle contacts, caused by the relative movements
of particles. Treating the material as a medium of Cosserat type, this paper describes
the framework of a general constitutive model that takes into account the load carrying mechanisms at inter-particle contacts. The moduli are derived as an explicit
function of the properties of inter-particle contacts. Characterizing the heterogeneous
strain field by a distribution function, mechanical behavior of the random granular
medium can be more realistically modelled.
Keywords: Random media, Micromechanics, Granular mechanics.

1 Introduction

Owing to the particulate nature, sands carry the applied load through resistance at
inter-particle contacts. To account for this load carrying mechanism in the stress-
strain modelling, it is necessary to consider the contact behavior between particles. A
number of studies have been attempted along this line of approach. For example,
work can be found in Duffy (1959), Duffy and Mindlin (1957), and Deresiewicz
(1958) for regular packing of spheres, and Digby (1981), Walton (1987), Jenkins
(1988), Chang (1988) and Bathurst and Rothenburg (1988) for random packing of
spheres. A more general formulation is to include the mechanics of particle rotation
in the derivation of the constitutive law. Along this line, work can be found in
Kanatani (1979), Chang and Liao (1990), Chang and Ma (1990, 1992).

   In this paper, we first outline the general formulation of the constitutive law that
leads to the moduli explicitly in terms of contact properties. Closed-form of elastic
moduli for packings of equal size spheres are derived under the condition of uniform
strain. A method of homogenization considering heterogeneous strain field is also
described towards a better modelling for the mechanical response of granular material.

D. Breysse (ed.), Probabilities and Materials, 261–271.
© 1994 *Kluwer Academic Publishers.*

2 Constitutive Theory

Granular material can be viewed at two different scales, namely, 1) representative unit, and 2) micro-element. A representative unit is defined as an assembly that contains a large number of particles to be representative of the granular material. At micro-scale, each particle is a micro-element of the system. A schematic representation of the two scales of granular material is shown in Fig. 1.

Microstructure of each inter-particle contact can be described by a contact normal vector n_i^{nm} and a contact area A^{nm}. For each micro-element, the microstructure of a particle (i.e., micro-element) is described by the particle volume V^n and by a set of vectors that provide the connecting information with its neighboring particles, $b^n=\{r_i^{nm}, r_i^{mn}, n_i^{nm}, A^{nm}; m=1,2\ldots,M^n\}$. The value of M^n represents the total number of neighboring particles. The vector r_i^{nm} joins the centroid of particle n to the contact point with particle m. For equal-sized spherical particles, the set of connecting vectors is reduced to $b^n=\{r_i^{nm}; m=1,2\ldots,M^n\}$.

Microstructure of the granular media is characterized as an ensemble of micro-elements. The geometry of the granular material can be described by a fabric space $B=\{b^n; n=1,2,\ldots,N\}$.

REPRESENTATIVE UNIT

MICRO-ELEMENT

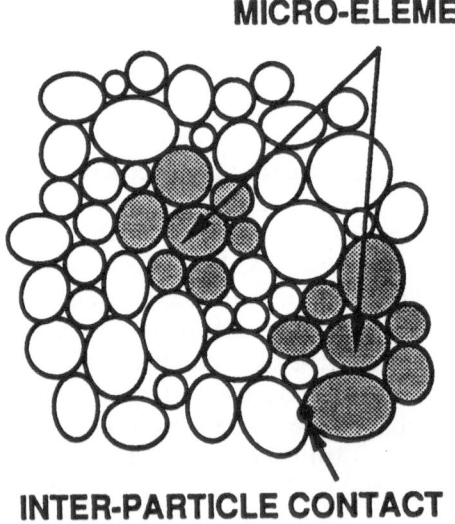

INTER-PARTICLE CONTACT

Fig. 1. Schematic figure for two scales of granular material.

We envisage a simple conceptual model in which the constituent particles are treated as rigid particles. The particles are viewed to be connected at contact points by imaginary springs. Two types of springs are used, namely, the rotational spring and the stretch spring. The rotation springs, transmitting contact moments, representing the contact resistance to the relative rotation of two particles. The stretch springs, transmitting contact forces, representing the contact resistance to the relative translation of two particles.

2.1 Two-particle kinematics

During deformation of packing, particles undergo two modes of movement for a particle: translation, Δu_i, and rotation, $\Delta \omega_i$. Based on the kinematics of two rigid particles of convex shape, the relative displacement $\Delta \delta_i$ and the relative rotation $\Delta \theta_i$ between particle 'n' and particle 'm' at the contact point 'c' are given by

$$\{ \Delta D_i^{nm} \} = [T_{ij}^n] \{ \Delta U_j^n \} - [T_{ij}^m] \{ \Delta U_j^m \} \tag{1}$$

where $\{\Delta D_i^{nm}\}$ is the generalized relative contact-displacement vector, $\{\Delta U_i^n\}$ is the generalized particle displacement vector, $[T_{ij}^n]$ is the transformation matrix.

$$\{\Delta D_i^{nm}\} = \begin{pmatrix} \Delta \delta_i^{nm} \\ \Delta \theta_i^{nm} \end{pmatrix}; \quad \{\Delta U_i^n\} = \begin{pmatrix} \Delta u_i^n \\ \Delta \omega_i^n \end{pmatrix}; \quad [T_{ij}^n] = \begin{bmatrix} \delta_{ij} & e_{ijk} r_k^n \\ 0 & \delta_{ij} \end{bmatrix} \tag{2}$$

The quantity e_{ijk} = the permutation symbols used in tensor representation for cross product of vectors. δ_{ij} is the kronecker delta. r_n is the vector joining centroid of the nth particle to the contact point with particle m.

2.2 Inter-particle contact law

The generalized contact-displacement vector $\{\Delta D_j^{nm}\}$ relates to the generalized contact-force vector $\{\Delta F_i^{nm}\}$ through a generalized contact stiffness tensor $[K_{ij}^{nm}]$, defined in the following:

$$\{\Delta F_i^{nm}\} = [K_{ij}^{nm}] \{\Delta D_j^{nm}\} \tag{3}$$

where

$$\{\Delta F_i^{nm}\} = \begin{pmatrix} \Delta f_i^{nm} \\ \Delta m_i^{nm} \end{pmatrix}; \quad [K_{ij}^{nm}] = \begin{bmatrix} k_{ij}^{nm} & 0 \\ 0 & g_{ij}^{nm} \end{bmatrix} \tag{4}$$

The stiffness tensor for stretch spring k_{ij} is expressed by the normal and shear spring constants, k_n, k_r,

$$k_{ij} = k_n n_i n_j + k_r (s_i s_j + t_i t_j) \tag{5}$$

The stiffness tensor for rotational spring g_{ij} is expressed by the torsional and rolling spring constants, g_n, g_r,

$$g_{ij} = g_n n_i n_j + g_r (s_i s_j + t_i t_j) \tag{6}$$

The basic unit vectors, $\mathbf{n}$, $\mathbf{s}$ and $\mathbf{t}$ of the local coordinate system, are constructed at each contact. The vector $\mathbf{n}$ is outwardly normal to the contact plane. The other two orthogonal vectors $\mathbf{s}$ and $\mathbf{t}$ are on the contact plane. These spring constants can be elasto-plastic as discussed in Chang et. al. (1992a, 1992b). Here, we limit our discussion to the elastic range.

2.3 Local strain of a micro-element

We now treat the discrete system as an equivalent continuum. The displacement and rotation of discrete particles are viewed as continuous vector fields. The gradients of displacement and rotation fields at particle n with respect to its neighboring particles represent the measures of strain.

Considering particle rotation, it has been shown (Chang and Liao, 1990, Chang and Ma, 1990) that deformation strain can be more generically defined for granular soil in terms of two variables: the displacement gradient $\Delta u_{i,j}$ and the particle rotation gradient $\Delta\omega_{i,j}$ in the following manner:

$$\{\Delta E_{ij}^n\} = [\nabla_{ijk}]\{\Delta U_k^n\} \tag{7}$$

where

$$\{\Delta E_{ij}^n\} = \begin{pmatrix} \Delta\varepsilon_{ij}^n \\ \Delta\gamma_{ij}^n \end{pmatrix}; \quad [\nabla_{ijk}] = \begin{bmatrix} \partial_j\delta_{ik} & -e_{ijk} \\ 0 & \partial_j\delta_{ik} \end{bmatrix} \tag{8}$$

∂_j in the Gradient operator $[\nabla_{ijk}]$ is defined as $\dfrac{\partial}{\partial x_j}$.

The symmetrical part of the strain $\Delta\varepsilon_{(ij)}$ is equal to the symmetrical part of displacement gradient, representing the usual strain associated with particle n. The skew symmetric part of the strain $\Delta\varepsilon_{[ij]}$ represents the net spin of particles (i.e., the difference between rigid body rotation of the particle group and the average rotation of individual particles).

Based on a kinematic assumption of affinity (i.e., linear displacement field), we can furnish a convenient kinematic relationship between the local strain and the relative displacement $\Delta\delta_i^{nm}$ at the contact between the center particle 'n' and its neighboring particle 'm', given by (Chang and Ma, 1990)

$$\{\Delta D_i^{nm}\} = [T_{ik}^n]\{\Delta E_{jk}^n\} \ L_j^{nm} \tag{9}$$

where, L_j^{nm} is the branch vector joining the centroids of particle 'n' and particle 'm'.

2.4 Local stress of a micro-element

Stress associated with a particle is measured from the contact force and contact moment transmitted to the particle from its neighboring particles. The relationship between the contact forces and the stress of the particle can be defined by employing the theorem of mean stress (Chang and Liao, 1990, Christoffersen et. al., 1981).

$$\{\Delta S_{jk}^n\} = \frac{1}{2V^n}[T_{ik}^n]\sum_m \{\Delta F_i^{nm}\} L_j^{nm} \tag{10}$$

where

$$\{\Delta S_{ij}^n\} = \begin{pmatrix} \Delta\sigma_{ij}^n \\ \Delta\mu_{ij}^n \end{pmatrix} \tag{11}$$

where V^n is the volume associated with the n-th particle. Summation of the associated volume over all particles is equal to the total volume of the representative unit, i.e., $V = \sum_m V^n$.

2.5 Local constitutive law

Based on the relationships: stress versus contact forces (Eq. 10), contact law (Eq. 3), and strain versus contact displacement at contact (Eq. 9), the constitutive equation for the particle group can thus be derived. For simplicity, we neglect the couple stress and polar strain, the constitutive relationship is given in the following incremental form:

$$\Delta\sigma_{ij}^n = C_{ijkl}^n \ \Delta\varepsilon_{kl}^n \tag{12}$$

where C_{ijkl}^n is the local stiffness tensor for the particle group associated with the n-th particle, given by

$$C_{ijkl}^n = \frac{1}{2V^n}\sum_m L_i^{nm} K_{jk}^{nm} L_l^{nm} \tag{13}$$

3 Constitutive tensor for a representative volume

Here, we consider a given volume to be representative of the granular solid and the boundaries of the said volume are subject to displacements or tractions compatible with the applied strain or stress. Under such conditions, Hill (1967) has shown that, the applied stress and strain can be expressed as the volume averages of their corresponding quantities at local level.

Thus the overall stress and strain for the representative unit, denoted by $\Delta\varepsilon_{ij}$ and $\Delta\sigma_{ij}$, are regarded as volume averages of the local stress and local strain at the micro-element level, such that

$$\Delta\sigma_{ij} = \frac{1}{V}\sum_{n} V^n \Delta\sigma_{ij}^n \qquad (14)$$

$$\Delta\varepsilon_{ij} = \frac{1}{V}\sum_{n} V^n \Delta\varepsilon_{ij}^n \qquad (15)$$

Corresponding to the overall stress and strain, we seek the overall stiffness tensor for the representative unit in the general expression of the overall stress-strain relationship:

$$\Delta\sigma_{ij} = C_{ijkl} \Delta\varepsilon_{kl} \qquad (16)$$

It is noted that, due to the heterogeneous strain field, the overall stiffness tensor C_{ijkl} is not equal to the volume averaged stiffness tensor, $\overline{C}_{ijkl}$, defined by

$$\overline{C}_{ijkl} = \frac{1}{V}\sum_{n} V^n C_{ijkl}^n \qquad (17)$$

However, the assumption of uniform strain is reasonable for the condition where small amount of sliding occurs, i.e. in the elastic range (Chang and Misra, 1990b). Under the assumption of uniform strain, the overall stiffness tensor C_{ijkl} is the same as the volume average of local stiffness tensor, given by

$$C_{ijkl} = \overline{C}_{ijkl} = \frac{1}{2V} \sum_{n}\sum_{m} L_i^{nm} K_{jk}^{nm} L_l^{nm} \qquad (18)$$

3.1 Constitutive matrix for isotropic microstructure
For a suitably large representative volume with a large number of contacts, the summation can be expressed in a integral form with a directional distribution density function $\xi(\alpha, \beta)$. For equal size granules, the integral form is given in the following:

$$C_{ijkl} = \frac{2r^2N}{V}\int_0^{2\pi}\int_0^{\pi} n_i(\alpha, \beta)\, K_{jk}(\alpha, \beta)\, n_l(\alpha, \beta)\, \xi(\alpha, \beta)\, \sin\alpha\, d\alpha\, d\beta \qquad (19)$$

For arbitrary anisotropic condition, the function $\xi(\alpha, \beta)$ can be expanded into a spherical harmonics expansion. Discussion on packing structure and mechanical properties of granules can be found in Chang and Misra (1990a). For isotropic case, the function is equal to $1/4\pi$. Thus the forth rank constitutive tensor can be derived, expressed in Voigt notation as follows:

$$\begin{Bmatrix} \Delta\sigma_{xx} \\ \Delta\sigma_{yy} \\ \Delta\sigma_{zz} \\ \Delta\sigma_{(xy)} \\ \Delta\sigma_{(yz)} \\ \Delta\sigma_{(zx)} \\ \Delta\sigma_{[xy]} \\ \Delta\sigma_{[yz]} \\ \Delta\sigma_{[zx]} \end{Bmatrix} = \begin{bmatrix} \lambda+2\mu & \lambda & \lambda & 0 & 0 & 0 & 0 & 0 & 0 \\ \lambda & \lambda+2\mu & \lambda & 0 & 0 & 0 & 0 & 0 & 0 \\ \lambda & \lambda & \lambda+2\mu & 0 & 0 & 0 & 0 & 0 & 0 \\ 0 & 0 & 0 & 2\mu & 0 & 0 & 0 & 0 & 0 \\ 0 & 0 & 0 & 0 & 2\mu & 0 & 0 & 0 & 0 \\ 0 & 0 & 0 & 0 & 0 & 2\mu & 0 & 0 & 0 \\ 0 & 0 & 0 & 0 & 0 & 0 & Z & 0 & 0 \\ 0 & 0 & 0 & 0 & 0 & 0 & 0 & Z & 0 \\ 0 & 0 & 0 & 0 & 0 & 0 & 0 & 0 & Z \end{bmatrix} \begin{Bmatrix} \Delta\varepsilon_{xx} \\ \Delta\varepsilon_{yy} \\ \Delta\varepsilon_{zz} \\ \Delta\varepsilon_{(xy)} \\ \Delta\varepsilon_{(yz)} \\ \Delta\varepsilon_{(zx)} \\ \Delta\varepsilon_{[xy]} \\ \Delta\varepsilon_{[yz]} \\ \Delta\varepsilon_{[zx]} \end{Bmatrix} \qquad (20)$$

The form of the constitutive matrix is found to resemble that for isotropic elastic medium. In addition to the relationship between the symmetric stress $\Delta\sigma_{(ij)}$ and symmetric strain $\Delta\varepsilon_{(ij)}$, this constitutive matrix gives relationship between asymmetric stress $\Delta\sigma_{[ij]}$ and asymmetric strain $\Delta\varepsilon_{[ij]}$, represented by a constant Z termed as spin modulus. The spin modulus is the resistance against particle spin which would transmit shear force through inter-particle contact and cause asymmetric shear stress of the packing.

The constitutive constants for the granular material can be expressed explicitly in terms of contact stiffness. The Lame constant for the packing is

$$\lambda = \frac{2}{5\upsilon r} (k_n - k_r) \qquad (21)$$

where υ is a dimensionless packing parameter, defined as a function of the total contact number per unit volume, which can be in turn related to void ratio and coordination number.

$$\upsilon = \frac{3V}{r^3 N_c} = \frac{4\pi(1+e)}{\bar{n}} \qquad (22)$$

The packing parameter ranges from 0.1 for very dense packing to 1.0 for very loose packing.

The shear modulus expressed by contact stiffness is given by

$$G = \frac{1}{5\upsilon r} (2k_n + 3k_r) \qquad (23)$$

and the spin modulus is given by

268

$$Z=\frac{1}{\upsilon r}k_r \qquad (24)$$

The corresponding Young's modulus, Poisson's ratio and bulk modulus can be expressed by contact stiffness as

$$E=\frac{2k_n}{\upsilon r}\left(\frac{2k_n+3k_r}{4k_n+k_r}\right) \qquad (25)$$

$$\nu=\frac{k_n-k_s}{4k_n+k_s} \qquad (26)$$

$$B=\frac{2}{3\upsilon r}k_n \qquad (27)$$

3.2 Homogenization

To derive the overall stiffness tensor C_{ijkl} considering a heterogeneous strain field, it is essential to account for the nature of the randomly distributed local stiffness C_{ijkl}^n and local strain Δe_{ij}^n of the micro-elements in the granular system. The problem falls into the category of homogenization theories (Sanchez-Palencia, 1987). Self-consistent methods, similar to that used in the study of poly-crystalline behavior (Mura, 1985, Hutchinson, 1970, Nemat-Nasser and Mahrabadi, 1984), have been applied to study random packings of spheres (Chang et. al., 1992c). An alternative method has been investigated which deals directly with the strain fluctuation (Chang et. al. 1992a).

To facilitate a rational averaging process, we introduce a distribution tensor of strain, which relates the overall strain to the local strain of a micro-element, given by,

$$\Delta e_{ij}^n=H_{ijkl}^n\Delta e_{kl} \qquad (28)$$

where the dimensionless tensor, H_{ijkl}^n, associated with each micro-element, describes indirectly the degree of heterogeneity of the material. From Eq. 15, the following condition must be satisfied:

$$\frac{1}{V}\sum_n H_{ijkl}^n V^n=I_{ijkl} \qquad (29)$$

where I_{ijkl} is a fourth rank identity tensor defined in terms of Kronecker delta δ_{ij} as

$$I_{ijkl}=\frac{1}{2}(\delta_{ik}\delta_{jl}+\delta_{il}\delta_{jk}) \qquad (30)$$

Thus, H_{ijkl}^n represents a distribution function in tensorial form.

With the distribution of strain (Eq. 28), using Eqs. 12 and 14, the stiffness tensor C_{ijkl} can be derived as a 'weighted' volume average of micro-element stiffness tensor C_{ijkl}^n, given by

$$C_{ijkl} = \frac{1}{V} \sum_n C_{ijmn}^n H_{mnkl}^n V^n \tag{31}$$

Since the representative unit consists of a large number of randomly arranged particles, the heterogeneous system can be viewed as a statistical homogeneous system. The strain distribution tensor H_{ijkl}^n is a function of the stiffness fluctuation at the n-th micro-element represented by a dimensionless tensor v_{ijkl}^n. This tensor is defined as the deviation of stiffness divided by the average stiffness of the packing, given by

$$v_{ijkl}^n = \overline{C}_{ijmn}^{-1} (C_{mnkl}^n - \overline{C}_{mnkl}) \tag{32}$$

Based on the results of numerical simulation, it was found that a simple and suitable form for the distribution tensor is a generalized tensorial form of exponential type, i.e.,

$$H_{ijkl}^n = A_{ijmn} \, e^{-\alpha v_{mnkl}^n} \tag{33}$$

The exponential term in Eq. 33 is defined as follows:

$$e^{-\alpha v_{mnkl}^n} = I_{mnkl} - \alpha \, v_{mnkl}^n + \frac{\alpha^2}{2!} v_{mnpq}^n v_{pqkl}^n - \cdots \cdots \tag{34}$$

The constant tensor A_{ijkl} in Eq. 33 must satisfy the condition of identity (i.e., Eq. 29). Thus it can be obtained by

$$A_{ijkl}^{-1} = \frac{1}{V} \sum_n e^{-\alpha v_{ijkl}^n} V^n \tag{35}$$

where the coefficient α is a constant governing the degree of scatter in strain distribution. As $\alpha = 0$, there is no strain variation; the condition is equivalent to the Voigt assumption of uniform strain. As the coefficient 'α' increases, the strains become more scattered while the stresses become more uniform. The coefficient α on the neighborhood of 0.5 has been found to best represent the strain and stress fluctuations of materials with random structure, based on a study of computer simulation (Chang, 1993). The overall stiffness tensor predicted with the consideration of heterogeneous strain field tend to be 10%-20% lower than that predicted with uniform strain assumption.

4 Summary and conclusion

This paper describes the framework of a general constitutive model that takes into account the effects of microstructure and the contact behavior. Elastic moduli are derived as an explicit function of the properties of inter-particle contacts. Characterizing the heterogeneous strain field by a distribution function, mechanical behavior of the random granular medium can be more realistically modelled.

5 Acknowledgment

The research work reported in this paper has been supported by grants from the Division of Mechanical and Structural Systems under the U.S. National Science Foundation and from the Division of Aerospace Studies under the U.S. Air Force Office of Scientific Research.

6 References

Bathurst, R.J. and Rothenberg, L. (1988) Micromechanical Aspects of Isotropic Granular Assemblies with Linear Contact Interactions. **Journal of Applied Mechanics, ASME,** Vol. 55, 17-23.

Chang, C.S. (1988) Micromechanics Modelling of Constitutive Equation for Granular Material. **Micromechanics of Granular Materials** (eds J. T. Jenkins and M. Satake), Elsevier Science Puallishers, 271-278.

Chang, C.S. (1993) Micromechanical Modelling of Deformation and Failure for Granulates with Frictional Contacts. **Mechanics of Materials,** Vol. 16, Elsevier Science Publishers, 13-24.

Chang, C.S., Chang, Y. and Kabir, M. (1992a) Micromechanics Modelling for the Stress-Strain Behavior of Granular Soil - I: Theory. **Journal of Geotechnical Engineering, ASCE,** Vol. 118, No. 12, 1959-1974.

Chang, C.S., Kabir, M. and Chang, Y. (1992b) Micromechanics Modelling for the Stress-Strain Behavior of Granular Soil - II: Evaluation. **Journal of Geotechnical Engineering, ASCE,** Vol. 118, No. 12, 1975-1994.

Chang, C.S. and Liao, C. (1990) Constitutive Relations for Particulate Medium with the Effect of Particle Rotation. **International Journal of Solids and Structures,** Vol. 26, No. 4, 437-453.

Chang, C.S. and Ma Lun (1990) Modelling of Discrete Granulates as Micropolar Continuum. **Journal of the Engineering Mechanics Division, ASCE,** Vol. 116, No. 12, 2703-2721.

Chang, C.S. and Ma, L. (1992) Elastic Material Constants for Isotropic Granular Solids with Particle Rotation. **International Journal of Solids and Structures, Pergamon Press,** Vol. 29, No. 8, 1001-1018.

Chang, C.S. and Misra, A. (1990a) Packing Structure and Mechanical Properties of Granulates. **Journal of the Engineering Mechanics Division, ASCE,** Vol. 116, No. 5, 1077-1093.

Chang, C.S. and Misra, A. (1990b) Application of Uniform Strain Theory to Hetero
    geneous Granular Solids. **Journal of the Engineering Mechanics Division,
    ASCE,** Vol. 116, No. 10, 2310-2328.

Chang, C.S., Misra, A. and Acheampon, K. (1992c) Elastoplastic Deformation of
    Granulates with Frictional Contacts. **Journal of the Engineering Mechanics
    Division, ASCE,** Vol. 117, No. 8.

Christoffersen, J., Mehrabadi, M.M. and Nemat-Nasser, S. (1981) A Micromechanical
    Description of Granular Material Behavior. **Journal of Applied Mechanics,** Vol.
    48, No. 2, 339-344.

Deresiewicz, H. (1958) Stress-Strain Relations for a Simple Model of a Granular
    Medium. **Journal of Applied Mechanics, ASME,** 402-406.

Digby, P.J. (1981) The Effective Elastic Moduli of Porous Granular Rock. **Journal of
    Applied Mechanics, ASME,** Vol. 48, No. 4, 803-808.

Duffy, J. (1959) A Differential Stress-Strain Relation for the Hexagonal Close Packed
    Array. **Journal of Applied Mechanics, Trans ASME,** 88-94.

Duffy, J. and Mindlin, R.D. (1957) Stress-Strain Relations and Vibrations of Granular
    Media. **Journal of Applied Mechanics, ASME,** 585-593.

Hill, R. (1967) The Essential Structure of Constitutive Laws for Metal Composites
    and Polycryctals. **Journal of Mechanics and Physics of Solids,** Vol. 15, No. 2,
    79-95.

Hutchinson, J.W. (1970) Elastic-Plastic Behavior of Polycrystalline Metals and
    Composites. **Proc. of Royal Soc. of London,** Vol. A319, 247-272.

Jenkins, J.T. (1988) Volume Change in Small Strain Axisymmetric Deformations of a
    Granular Material. **Micromechanics of Granular Materials** (eds M. Satake and
    J.T. Jenkins), Elsevier, Amsterdam, The Netherlands, 143-152.

Kanatani, K. (1979) A Micro-polar Continuum Theory for the Flow of Granular
    Material. **Int. J. Engng. Sci.,** 17, 419-432.

Mura, S. (1985) Micromechanics of Defects in Solids. **Martinus Nirjhoff,** The
    Haque, Netherlands.

Nemat-Nasser, S. and Mahrabadi, M. M. (1984) Micromechanically Based Rate
    Constitutive Descriptions for Granular Materials. **Mechanics of Engineering
    Materials** (eds. C. S. Desai and R. H. Gallagher), John Wiley, N. Y. pp.451-464.

Sanchez-Palencia, E. and Zaoui, A. eds. (1987) **Homogenization Techniques for
    Composite Media,** Lecture Notes in Physics, Vol. 272.

Walton, K. (1987) The Effective Elastic Moduli of a Random Packing of Spheres.
    Journal of Mechanics and Physics of Solids, Vol. 35, No.3, 213-226.

A VORONOI CELL FINITE ELEMENT MODEL FOR RANDOM HETEROGENEOUS MEDIA

SOMNATH GHOSH
SURESH MOORTHY, The Ohio State University, Columbus, USA

Abstract
A finite element model has been developed for analysis of heterogeneous media, in which second phase inclusions are arbitrarily dispersed within a matrix. A mesh generator based on Dirichlet tessellation, discretizes the heterogeneous domain, accounting for the arbitrariness in location, shape and size of the second phase. This results in a network of convex "Voronoi" polygons which form the elements in a finite element mesh. An assumed stress hybrid formulation has been implemented for accommodating arbitrary multi-sided elements in the finite element model. Composite element formulations have been developed to incorporate the effect of second phase within each element. Numerical examples have been conducted for small strain elasto-plasticity to validate the model.
Keywords: Dirichlet Tessellation, Voronoi cell finite element, Elasto-plasticity.

1 Introduction

The last three decades have seen a tremendous surge in the advancement of science and technology for heterogeneous materials. Notable among this class of materials are metal/alloy systems with second phase in the form of precipitates *and* composite materials containing a dispersion of fibers, whiskers or particulates in the matrix. Vastly improved properties have resulted in an increased utilization of some of these materials in industries. It is evident that development of robust analytical/numerical models are indispensable for predicting the evolution of state variables such as displacements, strains, stresses, strain energy and material properties such as yield strength and strain hardening in actual microstructures of heterogeneous materials. Within the framework of small deformation linear elasticity theory for composite materials, a number of analytical micromechanical models have evolved. These models predict effective constitutive response at the macroscopic level from characteristics of microstructural behavior. Notable among them are the models based on : (i)

D. Breysse (ed.), Probabilities and Materials, 273–284.
© 1994 *Kluwer Academic Publishers.*

variational approach using extremum principles by Hashin and Strikman (1963), (ii) probabilistic approach by Chen and Acrivos (1978), (iii) self-consistent schemes Hill (1965), Budiansky (1965), and (iv) the generalized self consistent model Christensen and Lo (1979). Extensions of linear elastic models to the elastic-plastic domain for small strains have been accomplished by the use of Mori-Tanaka mean-stress theory in Tandon and Weng (1988) and Teply and Dvorak (1988) . Though these analytical models are reasonably effective in predicting equivalent material properties for relatively simple geometries and low volume fraction of second phase inclusions, they suffer from some serious limitations. Arbitrary distribution of shapes, sizes and location of the second phase cannot be treated with these models. Constitutive response of the constituent phases are also somewhat restricted and predictions with large property mismatches are not reliable. *Unit Cell* models, e.g. Tvergaard (1990), are becoming increasingly popular among researchers who use computational methods. These models generate the material response through detailed discretization of a representative volume element (RVE) of the composite microstructure. Macroscopic periodicity conditions are assumed on the unit cells. Most of these models also make assumptions on local periodicity, thereby making the unit cells very simple. While periodic spatial distribution is often useful to predict optimum properties, the fact remains that real microstructures are seldom periodic. Additionally, these models utilize conventional mesh generators using coordinate transformation methods, discrete transfinite mapping methods or quadtree/octree approaches. For arbitrary or random microstructures, morphological incompatibility of the generated mesh with the physical domain may arise with these generators. Tremendous efforts are required to adequately represent the microstructural morphology. This leads to prohibitively large computational models that pose efficiency problems.

Motivation of the present research is derived from the want of computational models for actual heterogeneous materials with arbitrary microstructures. It is aimed at establishing a direct correlation between techniques of quantitative metallography for composite microstructures and their stress/deformation analysis. The finite element model evolves naturally from a composite microstructure by *Dirichlet Tessellation* of the domain. Tessellation methods have been used by Spitzig, Kelly and Richmond (1985) in conjunction with quantitative characterization of micrographs obtained by automatic image analysis systems. In this paper, the tessellated two-dimensional microstructural domain is directly cast into a finite element model without any further discretization. Tessellation of a representative material element (RME) generates multi-sided convex "Voronoi"polygons based on domain morphology. Ghosh and Mukhopadhyay (1993) have introduced a method where these Voronoi polygons containing one second phase inclusion at most, can be treated as composite elements in a FEM formulation. In this paper, the above work has been extended to plane strain/stress elasto-plasticity problems. Also alternate methods have been introduced to formulate composite elements dealing with element heterogeneity.

2 Mesh Generation With Dirichlet Tessellation

A two dimensional mesh generator is devised for plane sections of multiphase materials which are assumed to consist of unidirectional fibers or particulates. Based on information regarding the boundary of the domain, locations, shapes and sizes of the inclusions, discretization takes place automatically by *Dirichlet tessellation* of a domain to yield a network of convex "Voronoi polygons", containing one inclusion each, at most. Dirichlet tessellation is defined as the subdivision of a plane, determined by a set of points. Each point has associated with it, a region of the plane that is closest to it than to any other. Mathematically speaking, let $P_1(x_1), P_2(x_2), \ldots, P_n(x_n)$ be a set of n distinct random points in plane. Then the interior of the Voronoi polygon associated with the i-th labeled point P_i is the region D_i defined as

$$D_i = \{x : \mid x - x_i \mid < \mid x - x_j \mid, \forall \ j \neq i\} \tag{1}$$

The aggregate of all such regions D_i constitute the Dirichlet tessellation in plane. Each region may be perceived of as the intersection of open half planes bounded by the perpendicular bi-sectors of lines joining the point P_i with each of its neighbors P_j. It is easy to see that this property renders the Voronoi polygons convex. Generation of tessellation within a restricted "window" has been discussed in Green and Sibson (1978). For this case, tessellation for a finite number of points $P_1, P_2, \ldots, P_n$ inside a window W to be made up of regions or *tiles* given as

$$D_i = \{x \in W : \mid x - x_i \mid < \mid x - x_j \mid, \forall \ j \neq i, P_j \in W\} \tag{2}$$

Based on these principles, a robust mesh generator to create convex elements within a composite domain, has been developed by Ghosh and Mukhopadhyay (1991). Generating points in tessellation have been replaced by second phase inclusions with finite size. The effects of non-convex boundaries, multiply connected domains, and shapes and sizes of inclusions have been taken into account in the discretization. An example of discretizing a connecting.rod section, infested with circular inclusions is presented in figure 1. The particles are distributed with a controlled particle density of 28,628 particles per square meter, mean radius of 0.0008 m and standard deviation of 0.12 x 10^{-3} mm corresponding to a normal distribution.

3 Finite Element Formulation With Voronoi Polygons

Voronoi cells make rather unconventional elements due to the fact that different elements can have different number of sides. Application of the displacement finite element method to an element with n nodes runs into difficulty when $n > 4$, because it is impossible to ensure interelement displacement compatibility with n-term polynomial representations. Additionally, rank deficiencies in the stiffness matrix may result. To avoid these difficulties and represent Voronoi cells as conforming elements, the assumed stress hybrid method introduced by Pian (1964), is invoked. In

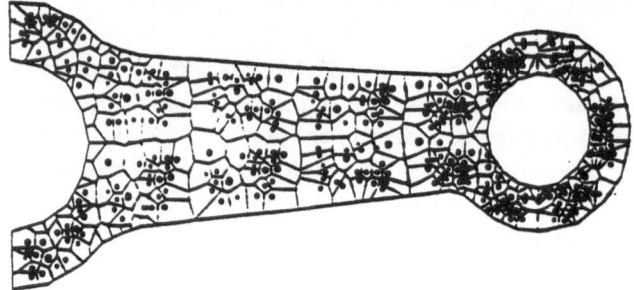

Figure 1: Discretization of a connecting rod section

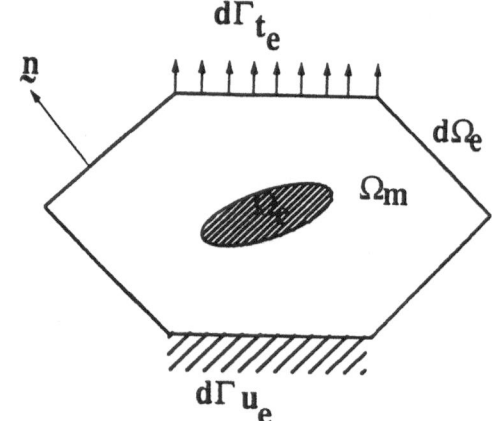

Figure 2: A typical Voronoi element with an inclusion

this method. element stiffness matrices are derived by assuming compatible displacement fields along interelement boundaries and an equilibriated stress distribution in the interior of each element. The element formulation is based on the principle of minimum complementary energy.

3.1 Elasticity
The complementary energy functional for a Voronoi element is of the form :

$$\Pi_e = -\int_{\Omega_e} B(\boldsymbol{\sigma})d\Omega + \int_{\partial\Omega_e} \boldsymbol{\sigma}.\mathbf{n}.\mathbf{u}d\partial\Omega + \int_{\Gamma_{t_e}} \bar{\mathbf{t}}.\mathbf{u}d\Gamma \tag{3}$$

where $\boldsymbol{\sigma}$ is the equilibriated stress field in the element domain Ω_e, $\mathbf{u}$ is the compatible displacement field on the element boundary $\partial\Omega_e$ with an outward normal $\mathbf{n}$, $\bar{\mathbf{t}}$ is prescribed traction field on Γ_{t_e}, the element boundary that coincides with global traction boundary. For linear elastic problems, the complementary energy density $B(\boldsymbol{\sigma})$ takes the form

$$B(\boldsymbol{\sigma}) = \frac{1}{2}\{\boldsymbol{\sigma}\}^T[\mathbf{S}]\{\boldsymbol{\sigma}\} \tag{4}$$

where S_{ijkl} are components of the elastic compliance tensor. In applying the finite element method, the assumed stress field need not be continuous across the interelement boundaries but equilibrium must be maintained with surface tractions. In the application of variational principles, the equilibrating stress field is expressed as a polynomial in the interior of the element as :

$$\{\sigma\} = [P]\{\beta\} \text{ in } \Omega_e \tag{5}$$

where $\{\sigma\}$ is a column vector of stress components, $\{\beta\}$ is a column of m undetermined stress coefficients $\beta_1, \beta_2 \cdots \beta_m$ and $[P]$ is a m x m matrix containing functions of coordinates x,y corresponding to the chosen polynomial. The prescribed boundary displacements $\{u\}$ can be interpolated from generalized displacement $\{q\}$ at the nodes, in the form

$$\{u\} = [L]\{q\} \text{ in } \partial\Omega_e \tag{6}$$

where elements of the matrix $[L]$ are functions of boundary coordinates. Stationarity of Π_e, i.e. $\frac{\partial \Pi_e}{\partial \beta_i} = 0$, yields

$$[H]\{\beta\} = [G]\{q\} \tag{7}$$

where,

$$[H] = \int_{\Omega_e} [P]^T [S][P] d\Omega$$

$$[G] = \int_{\partial\Omega_e} [P]^T \{n\}[L] d\Omega$$

Substitution of $\{\beta\}$ in the expression for complementary strain energy $\Pi = (\sum_e \Pi_e)$ for the entire domain and setting the first variation $\delta\Pi = 0$ gives

$$[K]\{q\} = [G]^T [H]^{-1}[G]\{q\} = \{\bar{f}\} \tag{8}$$

where the load vector is denoted as

$$\{\bar{f}\} = \int_{\Gamma_{t_e}} \{\bar{t}\}^T \{u\} d\Gamma$$

The stiffness matrix $[K]$ will be rank deficient if it's rank is less than n-l where n is the number of degrees of freedom and l is the number of rigid body modes. The necessary condition for $[K]$ to have sufficient rank is $m \geq n$-l, where m is the number of independent β-stress coefficients. Evaluation of the $[K]$ involves inversion of the $(m$ x $m)$ $[H]$ matrix, and it is advantageous to use a minimum number of β- parameters for improved efficiency. There are also indications that increased number of β-parameters lead to stiff elements. Various choices for $[P]$ and $[L]$ matrices and the effectiveness of this formulation are shown in Ghosh and Mukhopadhyay (1993).

3.2 Elasto-Plasticity

Only small deformation elasto-plasticity, governed by J_2 flow theory with isotropic hardening, has been considered in this paper. It is evident that the elastic compliance matrix $[S]$ in equation (4) should be changed to the elastic-plastic tangent compliance matrix $[\mathbf{S^T}]$ to incorporate the effects of state and internal variables in the model. A rate form of the constitutive relations for isotropic hardening can be expressed as:

$$\dot{\epsilon} - \dot{\epsilon}^p = \mathbf{S} : \dot{\sigma} \qquad (9)$$

and,

$$\dot{\epsilon}^p = \frac{9}{4} \frac{\sigma' \otimes \sigma'}{H\bar{\sigma}^2} : \dot{\sigma} \qquad (10)$$

where ϵ^p is the plastic strain rate, $\mathbf{S}$ is the fourth order elastic compliance tensor, σ' is the deviatoric Cauchy stress tensor tensor, H is the plastic hardening modulus and $\bar{\sigma}$ is the current effective stress. In an incremental finite element method, the increments of the strain tensor $\Delta\epsilon$ for a given stress increment $\Delta\sigma$ can be obtained by integrating equation(10). Using a stable backward Euler numerical integration method, increment of the plastic strain tensor in the $n-$th incremental step becomes :

$$\Delta\epsilon^p = (\frac{9}{4} \frac{\sigma' \otimes \sigma'}{H\bar{\sigma}^2})^{(n+1)} : \Delta\sigma \qquad (11)$$

Both plane strain and plane stress formulations are considered in this paper.

Variational and hybrid methods for application to nonlinear problems have been extensively studied by Atluri and Murakawa (1977). In the present work, their formulations have been appropriately modified for Voronoi cell finite elements. The complementary energy functional for an element in equation (3) is now modified to yield :

$$\Pi_e^{ep}(\sigma, \mathbf{u}) = -\int_{\Omega_e} \frac{1}{2}\epsilon(\sigma) : \sigma d\Omega + \int_{\partial\Omega_e} \sigma \cdot \mathbf{n} \cdot \mathbf{u} d\partial\Omega - \int_{\partial\Omega_\Gamma} \bar{\mathbf{t}} \cdot \mathbf{u} d\Gamma_t \qquad (12)$$

In an incremental formulation, stationarity is sought for the incremented complimentary energy functional $\Pi^{ep}(\sigma + \Delta\sigma, \mathbf{u} + \Delta\mathbf{u})(= \sum_e \Pi_e^{ep}(\sigma + \Delta\sigma, \mathbf{u} + \Delta\mathbf{u})$ under the assumptions of equilibriated stress field $\sigma + \Delta\sigma$ in Ω_e and compatible displacement fields $\mathbf{u} + \Delta\mathbf{u}$ on $\partial\Omega_e$. As in section 3.1, the incremental stresses and boundary displacements are expressed in the polynomial forms as

$$\{\Delta\sigma\} = [\mathbf{P}]\{\Delta\beta\} \ in \ \Omega_e \qquad (13)$$

$$\{\Delta\mathbf{u}\} = [\mathbf{L}]\{\Delta\mathbf{q}\} \ on \ \partial\Omega_e \qquad (14)$$

Substituting equations (13) and (14) in equation (12), the stationarity condition with respect to $\Delta\sigma$ at the end of an increment, yields

$$\int_{\Omega_e} [\mathbf{P}]^T[\mathbf{S^T}]\{\delta\Delta\sigma\}d\partial\Omega = \int_{\partial\Omega_e}[\mathbf{P}]^T\{\mathbf{n}\}\{\mathbf{L}\}d\Omega\{\mathbf{q} + \Delta\mathbf{q}\} - \int_{\Omega_e}[\mathbf{P}]^T\{\epsilon(\sigma)\}d\Omega$$

$$- \int_{\Omega_e}[\mathbf{P}]^T\{\Delta\epsilon(\Delta\sigma)\}d\Omega = \{\mathbf{f_1}\} \ in \ \Omega_e \qquad (15)$$

where $\{\delta\Delta\sigma\}$ in an iterative process is given as:

$$\{\delta\Delta\sigma\} = [\mathbf{P}]\{\delta\Delta\beta\} = (\{\Delta\sigma\})^{i+1} - (\{\Delta\sigma\})^i \qquad (16)$$

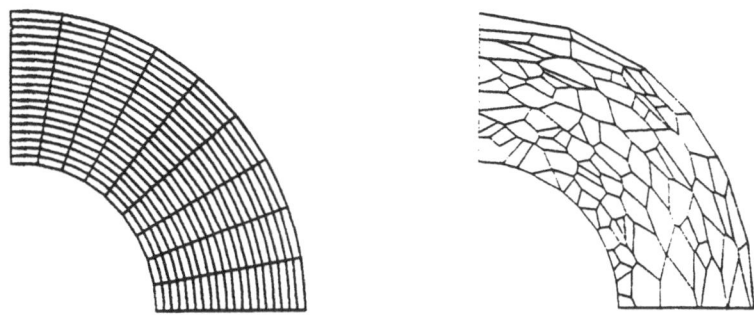

Figure 3: (a) Conventional and (b) Voronoi meshes for the pressure vessel problem

$\{\delta\Delta\beta\}$ is obtained by substituting equation (16) in equation (15). Global displacements are obtained by substituting equation (14) in the global complimentary energy functional $\Pi^{ep}(\sigma + \Delta\sigma, \mathbf{u} + \Delta\mathbf{u})$ and setting its first variation with respect to $\Delta\mathbf{q}$ to zero. This yields

$$\int_{\partial\Omega_e}[\mathbf{L}]^T\{\mathbf{n}\}\{\Delta\sigma\}d\Omega = \int_{\partial\Omega_\Gamma}[\mathbf{L}]^T\{\bar{\mathbf{t}}\}d\Omega - \int_{\partial\Omega_e}[\mathbf{L}]^T\{\mathbf{n}\}\{\sigma\}d\Omega \qquad (17)$$

from which the displacement increment is solved iteratively as

$$(\{\Delta\mathbf{q}\})^{i+1} = (\{\Delta\mathbf{q}\})^i - \{\delta\Delta\mathbf{q}\} \qquad (18)$$

3.3 A homogeneous elastic-plastic problem

A thick cylinder under the conditions of plane strain elasto-plasticity is subjected to an internal presure that varies from zero to a maximum of 18 dN/mm². The internal and external radii of the pressure vessel are 100 mm and 200 mm respectively. Material properties are Young's Modulus (E) = 21000 dN/mm², Poisson Ratio (ν)=0.3, Yield stress (σ_Y)=24 dN/mm² and Plastic Modulus (H)=0. This problem has been solved by Owen and Hinton (1980). Figure 3 shows a conventioal mesh and a Voronoi mesh (generated from a set of arbitrary points) for the problem. Figure 4(a) shows the evolution of inner radius with the applied pressure and figure 4 (b) plots the hoop stress as a function of the radial distance at the final increment. Excellent agreement is obtained between the results of Vornoi cell finite element and the displacement FEM by the authors and also by Owen and Hinton (1980).

4 Composite Voronoi Elements with Second Phase

A composite Voronoi element formulation for incorporating the effect of the second phase materials in the matrix is developed. Two alternative approaches are discussed in this paper. The first is based on the introduction of a transformation strain in the

280

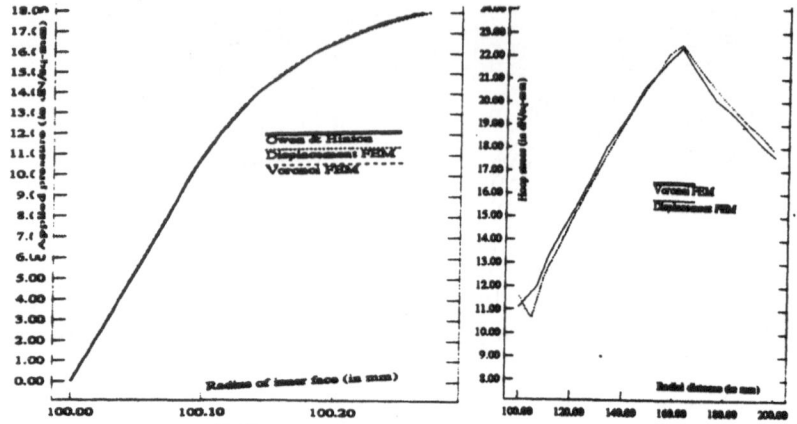

Figure 4: (a) Evolution of inner radius and (b) Hoop stress vs radial location

regions of material discontinuity. The second approach is a direct implementation
of the heterogeneity through introduction of traction discontinuity constraint at the
interface.

4.1 Transformation Strain Method
Subject to a prescribed stress field σ in a heterogeneous domain. the second phase
occupying a domain $\Omega_c \subset \Omega_e$ (figure 2), exhibits an additional non-stress causing
eigenstrain ϵ^*, given by,

$$\{\epsilon^*\} = [\mathbf{S}(\mathbf{x})] : \{\sigma\} - [\mathbf{S}^0] : \{\sigma\} = [\mathbf{\Delta S}] : \{\sigma\} \tag{19}$$

where $[\mathbf{S}(\mathbf{x})]$ is the location dependent elastic compliance of the inhomogenuous
composite material and $[\mathbf{S}^0]$ is the elastic compliance of the homogenuous matrix
material. For the composite element, the complimentary energy functional in equation (3) is modified as

$$\Pi_e^{c1} = -\int_{\Omega_e} B(\sigma)d\Omega + \int_{\partial\Omega_e} \sigma.n.ud\partial\Omega + \int_{\Gamma_{t_e}} \bar{t}.ud\Gamma - \int_{\Omega_c} \sigma\cdot\epsilon^* d\Omega + \int_{\Omega_c} \frac{1}{2}\epsilon^*\cdot\Delta\mathbf{S}^{-1}\cdot\epsilon^* \tag{20}$$

ϵ^* is defined only in the composite domain Ω_c. In addition to the approximating
functions for the stress and displacement variables defined in section 3.1, the *eigen
strain* is interpolated as,

$$\{\epsilon^*\} = [\mathbf{R}]\{\lambda\} \tag{21}$$

Setting the first variation of Π_e^{c1} to zero yields the following two equations,

$$[\mathbf{H}]\{\beta\} = [\mathbf{G}]\{q\} - [\mathbf{Q}]\{\lambda\} \tag{22}$$

$$[\mathbf{Q}]^T\{\beta\} = [\mathbf{T}]\{\lambda\} \tag{23}$$

where,

$$[\mathbf{Q}] = \int_{\Omega_c} [\mathbf{P}]^T[\mathbf{R}]d\Omega$$

$$[\mathbf{T}] = \int_{\Omega_c} [\mathbf{R}]^{\mathrm{T}} [\mathbf{\Delta S}]^{-1} [\mathbf{R}] d\Omega$$

A coupled set of equations result, which is solved for displacements, eigen strains and consequently the stresses. Only elasticity problems have been solved with this method.

4.2 Direct implementation of interface constraint

Along a bonded composite interface $\partial\Omega_c$, the stress and strain fields are discontinuous, while the displacement and traction fields are continuous. In the transformation strain formulation, strain fields are discontinuous while stress fields are continuous. For enhanced accuracy in the solution, a discontinuous stress field is now introduced into the modified complimentary energy. Traction continuity along the composite boundary is ensured by introducing this constraint into the modified complimentary energy Π_e through the use of Lagrange multipliers. Consider a discontinuous stress field $\boldsymbol{\sigma}$ and a displacement field $\mathbf{u}'$ on $\partial\Omega_c$ satisfying *apriori*, equilibrium conditions of $\boldsymbol{\sigma}$ in Ω_c and $\Omega_e - \Omega_c$ but not necessarily in $\partial\Omega_c$. The complimentary energy functional in equation (3) is modified to accommodate the constraint as,

$$\Pi_e^{c2} = \Pi_e - \int_{\partial\Omega_c} (\boldsymbol{\sigma}^m - \boldsymbol{\sigma}^c) \cdot \mathbf{n} \cdot \mathbf{u}' d\Omega \tag{24}$$

$\boldsymbol{\sigma}^m$ and $\boldsymbol{\sigma}^c$ designate stresses in the matrix and second phase respectively and $\mathbf{n}$ is the unit normal on $\partial\Omega_c$ out of the composite domain. The discontinuous stress field in an element is expressed as,

$$\{\boldsymbol{\sigma}\} = [\mathbf{P}]\{\boldsymbol{\beta} + L(\mathbf{x})\boldsymbol{\beta}'\} \text{ on } \Omega_{\mathbf{e}} \tag{25}$$

where $L(\mathbf{x})$ defines the discontinuity as,

$$
\begin{aligned}
L(\mathbf{x}) &= 0 \ for \ \mathbf{x} \subset \Omega_e - \Omega_c \\
&= 1 \ for \ \mathbf{x} \subset \Omega_c
\end{aligned}
$$

The displacement field $\mathbf{u}'$ written as,

$$\{\mathbf{u}'\} = [\mathbf{L}]\{\mathbf{q}'\} \text{ in } \partial\Omega_c \tag{26}$$

First variation of Π_e^{m2} yields,

$$[\mathbf{H_e}]\{\boldsymbol{\beta}\} + [\mathbf{H_c}]\{\boldsymbol{\beta}'\} = [\mathbf{G_e}]\{\mathbf{q}\}, \tag{27}$$

$$[\mathbf{H_c}]\{\boldsymbol{\beta} + \boldsymbol{\beta}'\} = [\mathbf{G_c}]\{\mathbf{q}'\}, \tag{28}$$

$$[\mathbf{G_c}]^{\mathrm{T}}\{\boldsymbol{\beta}'\} = \{\mathbf{0}\} \tag{29}$$

Subcripts on $\mathbf{H}$ and $\mathbf{G}$ indicate the domain of integration. These equations are solved for displacements and stresses.

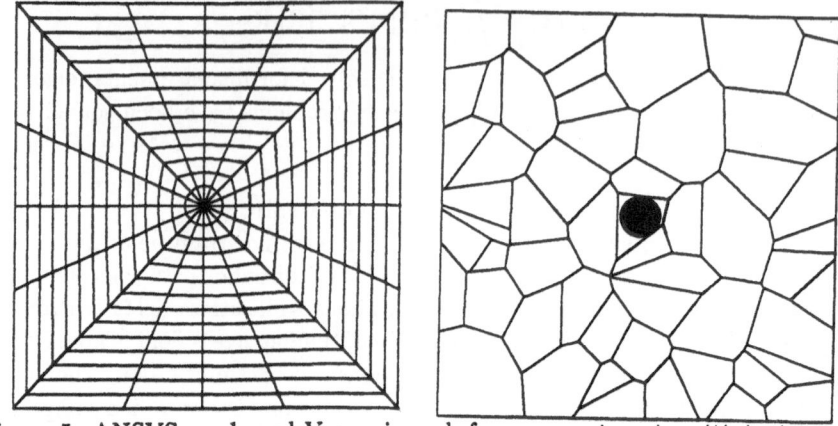

Figure 5: ANSYS mesh and Voronoi mesh for square domain with inclusion

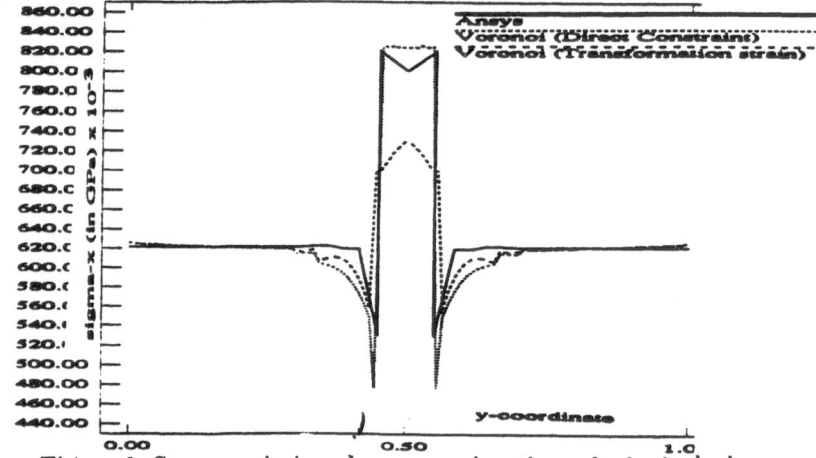

Figure 6: Stress variation along a section through the inclusion

4.4 Elasticity example

A square domain (1m x 1m) with a circular domain inclusion is subjected to uniform stretching of 0.005 m per increment for 10 increments. Plane stress calculations are performed. The matrix material has properties (E_m=69 GPa, ν_m=0.33) and the inclusion has properties (E_c=133 GPa, ν_c=0.285). Results with the two approaches in the composite Voronoi cell FEM formulation have been compared with displacement FEM results, generated using ANSYS Software package. Figure 5 shows the ANSYS mesh and a Voronoi cell mesh. Figure 6 shows the variation of longitudinal stress along a section through the middle of the inclusion. The direct constraint method is seen to improve results significantly when compared with the eigen strain method.

4.5 Elasto-plasticity example

In this example a square edge packing composite, subjected to periodic boundary conditions is analyzed. The material parameters for the matrix are (E_m=69 GPa, ν_m=0.33, σ_Y =43 MPa) and for the inclusion are (E_c=410 GPa, ν_c=0.3). The post-

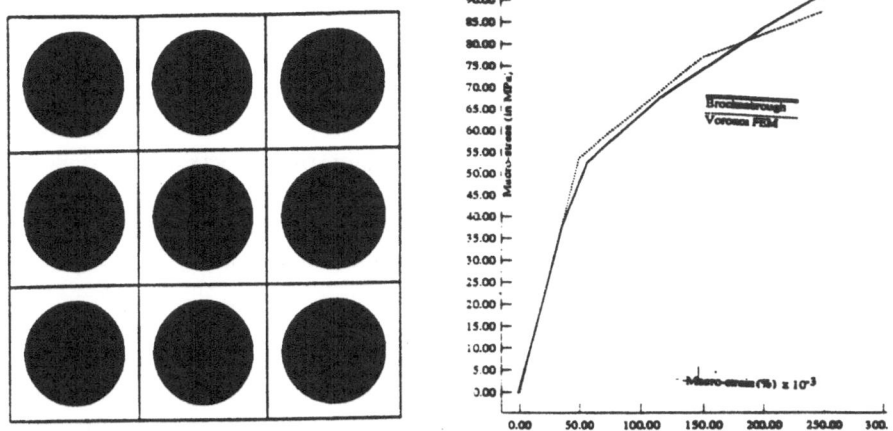

Figure 7: (a) Voronoi mesh and (b) Macroscopic stress-strain behavior for square edge packing composite

yield stress-strain law for the matrix material is

$$\bar{\sigma} = 0.266\bar{\epsilon}^{0.3}$$

Results are compared with those of Brockenbrough, Suresh and Wienecke (1991). The direct constraint method is used to incorporate the effect of the inclusions. Figure 7 (a) shows the Voronoi mesh for the problem and 7(b) shows a comparison of the corresponding effective stress-strain behavior.

5 Conclusion

A Voronoi cell finite element method is developed for two-dimensional heterogeneous materials with arbitrary microstructures. Formulations are developed for elasticity and small deformation elasto-plasticity. The effect of the second phase in each element is incorporated through two alternative formulations viz. a tranformation strain method and and direct constraint method. The latter is observed to perform better in terms of accuracy when compared with results from displacement finite element method. Overall the Voronoi cell FEM offers an excellent repository for analysis, that evolves naturally from a heterogeneous microstructure.

6 Acknowledgements Sponsorship of this work by the US Army Research Office through grant DAAL03-91-G-0168 (Program Director : Dr. K.R. Iyer) and by the US National Science Foundation through grant MSS-9196137, is gratefully acknowledged.

7 References

Atluri, S. and Murakawa, H, (1977) On hybrid finite element models in

284

nonlinear solid mechanics, in **Finite Elements in Nonlinear Mechanics** (eds P.G. Bergan et.al.), Tapir, vol. 1, pp. 3-41.

Budiansky, B. (1965) On the elastic modulii of some heterogeneous materials, **J. Mech. Phys. Solids**, 13, 223-227.

Brockenbrough, J.R., Suresh, S. and Wienecke, H.A. (1991) Deformation of metal-matrix composites with continuous fibers: geometrical effects of fiber distribution and shape **Acta Metall. Mater.**, 39, 5, 735-752.

Chen, H.S. and Acrivos, A. (1978) The effective elastic moduli of composite materials containing spherical inclusions at non-dilute concentrations **Int. J. Solids and Structures**, Vol. 14, 349-364.

Christensen, R.M. and Lo, K.H. (1979) Solutions for effective shear properties in three phase sphere and cylinder models. **J. Mech. Phys. Solids**, 27, 315-330.

Ghosh, S. and Mukhopadhyay, S. N. (1993) A material based finite element analysis of heterogeneous media involving Dirichlet Tessellations. **Computer Meth. in Applied Mechanics and Engineering**, 104, 211-247.

Ghosh, S. and Mukhopadhyay, S.N. (1991) A two dimensional automatic mesh generator for finite element analysis of random composites. **Computers and Structures**, 41, 2, 245-256.

Green, P.J. and Sibson, R., (1978) Computing dirichlet tessellations in the plane. **The Computer Journal**, 21, 168–173.

Hashin, Z. and Strikman, S. (1963) A variational approach to the theory of the elastic behavior of multiphase materials. **J. Mech Phys. Solids**, 11, 127-140.

Hill, R. (1965) A self consistent mechanics of composite materials. **J. Mech Phys. Solids**, 13, 213-222.

Owen, D.R.J. and Hinton, E., (1980) Finite elements in plasticity, Theory and Practice, Pineridge Press Ltd., Swansea, U.K.

Pian, T. H. H. (1964) Derivation of element stiffness matrices by assumed stress distribution. **AIAA Journal**, 2, 1333-1336.

Spitzig, W.A., Kelly, J.F. and Richmond, O. (1985) Quantitative characterization of second phase populations. **Metallography**, 18, 235-261.

Tandon, G.P. and Weng, G.J. (1988) A theory of particle reinforced plasticity. **J. Appl. Mech.**, 55, 126-135.

Teply, J.L. and Dvorak, G.J. (1988) Bounds on overall instantaneous properties of elastic-plastic composites. **J. Mech. Phys. Solids**, 36, 29-58.

Tvergaard, V. (1990) Analysis of tensile properties for a whisker-reinforced metal matrix composite. **Acta Metall. et Mater.**, 38, 185-194.

RANDOM STRESS FIELDS WITHIN GRANULAR MEDIA

G. AUVINET
Instituto de Ingenieria, UNAM, Mexico
and Laboratoire de Géomécanique
Ecole Nationale Supérieure de Géologie, Nancy, France

Abstract
The concept of stress within a granular material is reviewed. Local stresses at the grain level are introduced and expressed in terms of contact and intragranular forces. It is shown that these local stresses form a family of random fields which converge towards the stress field of continuum mechanics as the size of the area in which they are computed increases. The rate of this convergence, also called mechanical scale effect, is evaluated. Taking into account the results of numerical simulations and experimental data obtained with the Schneebeli material, it is also shown that the variance of local stresses mainly reflects the random variations of porosity within the medium (geometric scale effect). Thus a simple dependence exists for granular media between geometric and mechanical scale effects. Some practical implications are discussed.
Keywords: Micromechanics, Stress, Granular Media, Random Field, Contact Forces, Scale Effect.

1 Introduction

Limitations of continuum mechanics when applied to particulate media are obvious. It is generally accepted that, for these materials, stress can only be defined when a "sufficient" number of grains are considered. The meaning of "sufficient" in this context has however always been rather vague. It was therefore considered worthwhile to try to clarify this problem within a probability theory framework, using numerical simulations as well as experimental results, and to define in probabilistic terms the conditions in which, continuum mechanics can be used to describe the internal actions within these materials.

2 Forces and local stresses within granular media

Marsal (1973) introduced the concept of "intragranular" force F_i defined as the vector sum of the contact forces P_{ij} on the cap of a particle i detached by an intersection plane Θ (Fig 1). Considering an area A on plane Θ, local stresses can then be defined as:

$$\bar{\sigma}_z = \frac{1}{A} \sum_{i=1}^{n_A} F_{zi}$$

285

D. Breysse (ed.), Probabilities and Materials, 285–291.

$$\bar{\tau}_{zx} = \frac{1}{A} \sum_{i=1}^{n_A} F_{xi} \tag{1}$$

$$\bar{\tau}_{zy} = \frac{1}{A} \sum_{i=1}^{n_A} F_{yi}$$

where $\bar{\sigma}_z$, $\bar{\tau}_{zx}$ and $\bar{\tau}_{zy}$ are respectively the normal and shear local stresses; n_A is the number of grains intersected within area A and F_{xi}, F_{yi}, and F_{zi} are the components of F_i in the reference system x,y,z (Fig 1).

In statistically homogeneous media and homogeneous stress fields, local stresses are random variables that are expected to converge towards stresses of continuum mechanics σ_z, τ_{zx} and τ_{zy} as A increases (mechanical scale effect). They form a family of stationary spatial stochastic processes or random fields (one for each value of A). The variance of these fields depends on the specific value of A. As a matter of fact, this variance tends asymptotically to be inversely proportional to A. This is so because a local stress associated to an area A can be considered as the average of local stresses associated to N smaller contiguous areas a. When a reaches dimensions such that local stresses associated to those smaller areas can be considered as practically independent, the variance of this average stress is inversely proportional to N and therefore to A. A similar argument can be used, together with the central limit theorem, to conclude that local stresses tend to be normally distributed when area A increases.

Taking into account that stress transfer occurs only through solid matter, local stresses will be principally concentrated in areas of plane Θ where local porosity is low. It seems then natural to write those stresses as the sum of a "geometric" and a "mechanical" components. For normal stress:

$$\bar{\sigma}_z = \sigma_z \frac{1 - \mu_A}{1 - \mu} + \varepsilon \tag{2}$$

where μ_A is the local value of porosity $= 1 - A_s/A$; A_s is the area of solids intersected within A; μ is the average porosity of the medium and ε is a random variable with zero expectation and variance dependent on A.

The first term of eq. 2 is the expected value of $\bar{\sigma}_z$ for a uniform distribution of stresses within the intersected solids, while ε accounts for random load concentrations on some particular grains. The variance of the normal local stress can then be written:

$$\text{var}[\bar{\sigma}_z] = \frac{\sigma_z^2}{(1-\mu)^2} \, \text{var}[\mu_A] + \text{var}[\varepsilon] - \frac{2}{1-\mu} \, \text{cov}[\mu_A, \varepsilon] \tag{3}$$

3 Geometric scale effect

The variance of local porosity appears in the first term of eq.3 .The reduction of this variance when A increases, called geometric scale effect, has been evaluated previously using random processes theory and numerical simulation (Auvinet, 1989)

In each point X of the medium, a stochastic characteristic function $K_0(X)$ was defined as:

$K_0(X) = 1$ if X belongs to the pores

$K_0(X) = 1$ if X belongs to the grains

(4)

Then, if μ is the porosity of the medium,

$$E\{K_0(X)\} = \mu \tag{5}$$

$$\text{var }[K_0(X)] = \mu(1-\mu) \tag{6}$$

Local porosity in a volume V within the medium could then be defined as

$$\mu_V = \frac{1}{V}\int_V K_0(X)dX \tag{7}$$

It was shown that, for an assembly of spheres, the autocorrelation coefficient of spatial stochastic process $K_0(X)$ is isotropic and can be expressed as:

$$\rho(h) = \exp\left[\frac{-Sh}{4\mu(1-\mu)}\right] \tag{8}$$

where h is a scalar (distance between two points of the medium), μ is the expected value of porosity and S is the specific surface of the medium (average surficial area of the grains per unit volume).

Variance of the integral of eq 7 is:

$$\text{var }[\mu_V] = \frac{\mu(1-\mu)}{V^2}\int_V \rho(\tau)d\tau \tag{9}$$

Similar considerations can be made for porosity μ_A defined in an area A within a plane $\ni$ intersecting the medium:

$$\mu_A = \frac{1}{A}\int_A K_0(X)dX \tag{10}$$

$$\text{var }[\mu_A] = \frac{\mu(1-\mu)}{A^2}\int_A \rho(\tau)d\tau \tag{11}$$

As volume V and area A increase, it can be shown that these expressions tend asymptotically to:

$$\text{var }[\mu_V] = \frac{128\mu^4(1-\mu)^4}{VS^3} \tag{12}$$

$$\text{var } [\mu_A] = \frac{32\mu^3(1-\mu)^3}{AS^2} \tag{13}$$

The first term of eq 3 can then be expressed simply as a function of the specific surface of the medium. This term can also be expressed in terms of simple stereological parameters of the medium (Auvinet, 1992)

4 Tests on the Schneebeli material

Using a technique developed by Faugeras and Gourves, Mezghani (1987) performed tests to measure local stresses within the Schneebeli material, i.e. a 2-D piling of cylindrical rods. Rods having diameters of 2, 3 and 4 cm were mixed in the same proportion (by weight). Contact forces acting on a "particle" were evaluated measuring the force required to overcome lateral friction and to induce displacement of the rod along its longitudinal axis. Forces on groups of particles were measured by pressing the edge of a metallic plate on the material. Measurements were made with plates 1, 2, 5, 7 and 10 cm long. Repeating these measurements it was possible to obtain the coefficient of variation of local normal stress (Fig 2).

These experimental results can be interpreted on the basis of the considerations set forth in part 3. For a 2-D material, the expressions equivalent to eqs. (2), (3) and (13) are (Auvinet, 1986):

$$\bar{\sigma}_y = \sigma_y \frac{1-\mu_L}{1-\mu} + \varepsilon \tag{14}$$

$$\text{var}[\bar{\sigma}_y] = \frac{\sigma_y^2}{(1-\mu)^2} \text{var}[\mu_L] + \text{var } [\varepsilon] - \frac{2}{1-\mu} \text{cov } [\mu_L, \varepsilon] \tag{15}$$

$$\text{var } [\mu_L] = \frac{2\pi\mu^2(1-\mu)^2}{LS} \tag{16}$$

where: L: length of the segment on which the stress acts (length of plate); μ_L : local value of porosity = 1 - L_s/L; L_s: length of solids intersected within L; S specific perimeter of discs, ε random variable with zero expectation.

Theoretical values given by eq. 16 were compared to measured values (Fig 2). The dotted line corresponds to the first term of eq 15. For very short plates, the influence of the second and third terms are significant, meaning that local load concentrations contribute to the stress variance. However, as L increases, the variance of local stresses tends to reflect mainly the variance of local porosity, since the variance of ε and the covariance term vanish.

5 Stochastic model of local stresses variations within a granular material

The results presented suggest that in a granular material the variance of local stresses is caused by two phenomena:

- load concentrations on some particular grains, probably associated to the existence of chains of stressed particles within the medium as shown by Dantu (1957)

- Local variations of porosity: local stresses tend to concentrate in areas with low porosity.

The first phenomenon seems to vanish very rapidly as the size of the area A in which local stress is computed increases, thus the second and third terms of eq 15 can be progressively neglected. When this condition is reached, a simple stochastic model can be defined to describe the variations of local stresses defined in an area A within the medium.

Eq 2 is reduced to :

$$\bar{\sigma}_z = \sigma_z \frac{1-\mu_A}{1-\mu} \tag{17}$$

where μ_A is the random local value of porosity, so:

$$E\{\bar{\sigma}_z\} = \sigma_z \tag{18}$$

$$var[\bar{\sigma}_z] = \frac{\sigma_z^2}{(1-\mu)^2} var[\mu_A] \tag{19}$$

and, taking into account Eq 13:

$$var[\bar{\sigma}_z] = \frac{\sigma_z^2}{(1-\mu)^2} \frac{32\mu^3(1-\mu)^3}{AS^2} = \frac{32\mu^3(1-\mu)\sigma_z^2}{AS^2} \tag{20}$$

and the coefficient of variation of $\bar{\sigma}_z$ is :

$$cv[\bar{\sigma}_z] = \sqrt{\frac{32\mu^3(1-\mu)}{AS^2}} \tag{21}$$

As mentioned before, it can be further considered that the field representing $\bar{\sigma}_z$ is approximately Gaussian. The difference between local stress and the continuum mechanics stress will then be such that:

$$P[|\bar{\sigma}_z - \sigma_z| > \varepsilon\sqrt{var[\bar{\sigma}_z]}\,] = 2[1 - \Phi(\varepsilon)] \tag{22}$$

where $\Phi(\varepsilon)$ is the cumulative standard normal distribution function. This expression allows to evaluate in probabilistic terms the validity of substituting the resultant of contact forces per unit surface by the stresses of continuum mechanics for a given area A.

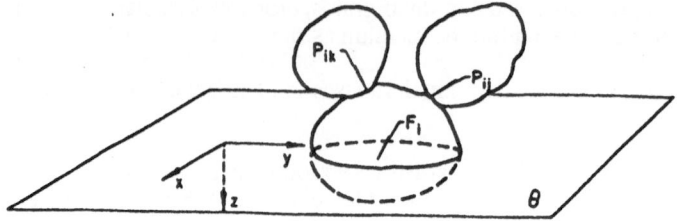

Fig 1 Contact and intragranular forces

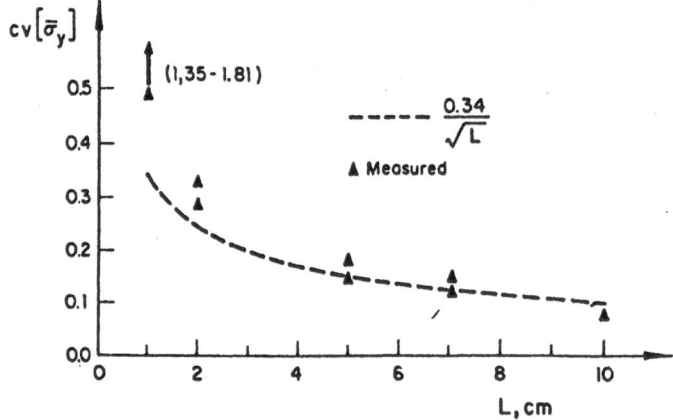

Fig 2 Theoretical and measured coefficients of variation of local normal stresses

These results have obvious practical implications. They can be used for example to estimate the error in stresses measured using load cells placed within coarse geomaterials or to define the minimum size of these cells for a given required accuracy. It should be emphasized however that the results presented here do not account for singularities of the stress field induced by the presence of the measurement devices due to arching and other similar phenomena.

6 Conclusions

Statistical relations between internal forces existing within granular materials and the stresses of continuum mechanics have been established. Local stresses undestood as the components of the resultant of contact forces per unit area were introduced. They were shown to form stochastic fields with a variance dependent on the area in which they are computed. When this area increases, the variance of these local stresses decreases reflecting mainly the local variations of porosity. Local stresses then tend to be Gaussian and converge towards the continuum mechanics stresses.

7 References

Auvinet, G. (1986) "Estructura de los medios granulares", Doctoral Thesis, UNAM, Mexico D.F.

Auvinet, G. (1989) "Geometric scale effect in granular media", **Powder and Grains**, Proc. Int. Conf. on Micromech. of Gran. Media, Clermont Ferrand, France, 29-34

Auvinet, G. (1992) "Grain concentrations and stress transfer within particulate media", **Marsal Volume,** Sociedad Mexicana de Mecanica de Suelos, Mexico D.F., 155-160

Dantu, P. (1957) "Contributions à l'étude mécanique et géométrique des milieux pulvérulents", Proc. 4th Int. Conf. Soil. Mech. and Found. Eng., London, U.K.

Faugeras, J.C., and Gourvés, R. (1980) "Mesure des contraintes au sein d'un massif analogique de Schneebeli", **Rev. Fra. de Géotechnique**, No 11, France

Marsal, R.J. (1973) "Mechanical properties of rockfill", Embankment-Dam Engineering, **Casagrande Volume,** John Wiley and Sons, New York, N.Y., USA

Mezghani, F. (1987) "Milieux granulaires, analyse statistique de l'état de contrainte macroscopique au sein d'un milieu granulaire", Doctoral Thesis, Université de Clermont II, Aubière, France.

STATISTICAL MODELING FOR THE FLOW
OF SHORT FIBERS COMPOSITES

A. POITOU*, F. CHINESTA**, F. OLMOS***
* Laboratoire de Mécanique et Technologie, E.N.S.Cachan, CNRS, université Paris VI
61 avenue du président Wilson, 94235 Cachan. FRANCE
** Universidad Politécnica de Valencia, Camino de Vera S/N 46071 Valencia SPAIN
*** Universitat de Valencia, Artes Graficas, 46071 Valencia SPAIN

Abstract

Numerical results are given for the flow of fiber composites modelled as suspensions of
non spherical particles. In this framework, because the many particles rotate, their state
of orientation is described with a statistical approach. We used these methods to
compute coupled solutions in which the orientation of the particles is affected by the
flow and the flow itself depends on the orientation of the particles. The computation
methods involve an augmented lagrangian approach and a streamline upwind petrov
galerkin formulation to solve the convective orientation equation.

1. Introduction

The modeling of flow induced orientation in short fibers reinforced thermoplastics is of
major interest in composite processing. It permits both to foresee the orientation state
of the fibers in injected molded pieces and to define the mechanical behavior of the
composite which is namely known to be strongly coupled with the fibers orientation .
this behavior is evaluated in the framework of dilute or semi-dilute suspensions of non
spherical particles in a newtonian fluid in which the orientation of the particules is
described with a probability density fonction. Resulting equations involve the coupling
of an elliptic boundary value problem with a convection type equation. The elliptic
problem is associated with the equilibrium equations whereas the convection equation
describes the time evolution of the anisotropic viscosity tensor.

The aim of this paper is to give numerical results concerning the orientation of the
particules in a Poiseuille flow for which the coupling between the particules orientation
and the behavior of the composite have been taken into account.

Section 2 summarizes the general equations of anisotropic suspensions (viz. (i) a
standard volume average homogeneisation procedure and (ii) a statistical description of

D. Breysse (ed.), Probabilities and Materials, 293–303.
© 1994 Kluwer Academic Publishers.

the orientation). We do not exhibit new results in this section but we give a derivation of the equations for dilute suspensions which uses the results given by Eshelby for the calculation of the stress in an elastic ellipsoidal inclusion in a linear elastic matrix subjected to a homogeneous displacement field at infinity. This very classical method in solid mechanics seems to have been very seldom applied in fluid mechanics. Section 3 gives the numerical methods which have been applied to compute the solution. An augmented lagrangian method coupled with the Usawa algorithm have been applied to account for the incompressibility of the material. A Streamline Upwind Petrov Galerkin (SUPG) method has been applied to solve the orientation equation which is of convective nature.

Notations:

Vectors are written in bold characters : e.g. $\mathbf{u}$ (velocity field)
Matrices are written in shadow characters : e.g. σ (extra stress tensor)
Tr denotes the trace operator : e.g. $\mathrm{Tr}(\mathbf{A}) = A_{ii}$
^{T} denotes the transpose operator : e.g. $\mathbf{A}^T = A_{ji}$
 : e.g. $\mathbf{u}^T$ denotes the line vector whereas $\mathbf{u}$ denotes the column vector
Scalar product is written : $\mathbf{u}^T \mathbf{v} = u_i\, v_i$
Diadic product is written : $\mathbf{u}\, \mathbf{v}^T = u_i\, v_j$
Strain rate tensor : $\mathbf{d}(\mathbf{v}) = \{\mathbf{grad}\ \mathbf{v} + (\mathbf{grad}\ \mathbf{v})^T\}/2$
Vorticity tensor : $\Omega(\mathbf{v}) = \{\mathbf{grad}\ \mathbf{v} - (\mathbf{grad}\ \mathbf{v})^T\}/2$

Every other particular notations are introduced throughout the text.

2. Constitutive law for anisotropic suspensions

Molten thermoplastics composites are commonly modelled as suspensions of axi-symetric particles. A complete derivation of the equations can be found in Giesekus (1962), Hinch and Leal (1973), Batchelor (1970,1971), Dinh and Armstrong (1984). Two main steps permit to derive macroscopic constituve equations for anisotropic suspensions: (i) a now standard homogeneisation procedure which leads to volume average quantities if the orientation of the particles is perfectly defined and (ii) a statistic description of the orientation.

Volume average

Basic equations are obtained with a volume average procedure. Let $\mathcal{V}$ be a representative volume of our macroscopic scale, which contains many particles located in $\mathcal{V}_i$. Let τ, $\mathbf{u}$ and $\mathbf{d}(\mathbf{u})$ denote the microscopic stress tensor, velocity vector and strain rate tensor and $\mathbf{n}$ be an outwards normal vector to the particles' boundary $\partial\mathcal{V}_i$. For a given orientation of every fiber in the suspension and if inertia terms can be neglected, the corresponding macroscopic variables $\mathbb{T}$, $\mathbf{v}$, $\mathbf{d}(\mathbf{v})$, are defined by :

$$\mathbf{v} = \frac{1}{\mathcal{V}} \int_{\mathcal{V}} \mathbf{u} \, dv \tag{2-1}$$

$$\mathbf{d}(\mathbf{v}) = \frac{1}{\mathcal{V}} \int_{\mathcal{V}} \mathbf{d}(\mathbf{u}) \, dv = \frac{1}{\mathcal{V}} \int_{\mathcal{V} - \sum_i \mathcal{V}_i} \mathbf{d}(\mathbf{u}) \, dv + \frac{1}{\mathcal{V}} \sum_i \int_{\mathcal{V}_i} \mathbf{d}(\mathbf{u}) \, dv$$

$$= \frac{1}{\mathcal{V}} \int_{\mathcal{V} - \sum_i \mathcal{V}_i} \mathbf{d}(\mathbf{u}) \, dv \tag{2-2}$$

$$\mathbb{T} = \frac{1}{\mathcal{V}} \int_{\mathcal{V}} \tau \, dv = \frac{1}{\mathcal{V}} \int_{\mathcal{V} - \sum_i \mathcal{V}_i} \tau \, dv + \frac{1}{\mathcal{V}} \sum_i \int_{\mathcal{V}_i} \tau \, dv$$

$$= \frac{1}{\mathcal{V}} \int_{\mathcal{V} - \sum_i \mathcal{V}_i} \tau \, dv + \frac{1}{\mathcal{V}} \sum_i \oint_{\partial \mathcal{V}_i} \tau \, \mathbf{n} \, \mathbf{x}^T \, dv \tag{2-3}$$

If the ambient fluid is assumed to be newtonian, the microscopic constitutive law within the fluid reads:

$$\tau = -p \, \mathbb{I} + 2 \, \eta \, \mathbf{d}(\mathbf{u}) \tag{2-4}$$

This leads to the expression of the macroscopic pressure and extra stress tensor:

$$P = \frac{1}{\mathcal{V}} \int_{\mathcal{V}} p \, dv \qquad \text{and} \qquad \mathbb{T} = 2 \, \eta \, \mathbf{d}(\mathbf{v}) + \Sigma_p \tag{2-5}$$

P is the volume averaged pressure and Σ_p describes the contribution of the particles to the macroscopic stress:

$$\Sigma_p = \frac{1}{\mathcal{V}} \sum_i \int_{\partial \Omega_i} \tau \, \mathbf{n} \mathbf{x}^T \, dv \tag{2-6}$$

Determination of Σ_p requires to solve a microscopic problem. For spheroidal particles, Σ_p can be calculated by the method given by Batchelor. We give here a simpler derivation which is based on the results given by Bilby et al. (1976) for linear incompressible elasticity, at the limit when the stiffness of the inclusion tends to infinity:

$$\mathcal{S} \, \Sigma_p = 2 \, \mu \, \phi \, \mathbf{d} \, (\mathbf{v}) \tag{2-7}$$

In this expression $\mathcal{S}$ is the forth order Eshelby tensor which depends on the shape of the spheroid only. It is to be noted that $\mathcal{S}$ is singular but can be inverted on the subspace of deviatoric tensors. If $\mathbf{p}$ denotes a unit vector along the axis of the spheroid, one can show that $\mathcal{S}$ depends on 3 parameters $\lambda_1, \lambda_2, \lambda_3$ only:

$$\mathscr{E}\, \Sigma_p = \lambda_1\, \Sigma_p + \lambda_2\, \mathrm{Tr}\, (pp^T\, \Sigma_p)\, (\mathrm{Id} - 3pp^T)$$

$$+ \lambda_3\, (pp^T\, \Sigma_p + \Sigma_p\, pp^T - 2\, \mathrm{Tr}\, (pp^T\, \Sigma_p)\, pp^T\,) \tag{2-8}$$

Thus

$$\Sigma_p = \mu_1\, d\,(v) + \mu_2\, \mathrm{Tr}\, (pp^T\, d\,(v))\, (\mathrm{Id} - 3pp^T)$$

$$+ \mu_3\, (pp^T\, d\,(v) + d\,(v)\, pp^T - 2\, \mathrm{Tr}\, (pp^T\, d\,(v))\, pp^T \tag{2-9}$$

where μ_1, μ_2, μ_3 are calculated as functions of the shape of the spheroid.

Orientation distribution

The particles' orientation is described through the probability distribution of fiber orientation $\psi(p)$. If p denotes a unit vector aligned along the symmetrical axis of the particle, and if brownian motion can be neglected, the equations of evolution for p read:

$$\frac{dp}{dt} = \Omega(v)\, p + \lambda\, \{d(v)\, p - (p^T d(v)p)\, p\} \tag{2-10}$$

where $\lambda = \dfrac{r^2-1}{r^2+1}$ and $r = \dfrac{a}{b}$ is the shape factor of the spheroid

Equation (2-10) can be rewritten in a lagrangian form:

$$p = \frac{\mathbb{E}p_0}{(p_0{}^T\, \mathbb{E}^T\mathbb{E}\, p_0)^{\frac{1}{2}}} \tag{2-11}$$

where p_0 is the orientation of the particule at time 0 and $\mathbb{E}$ is a generalized deformation gradient defined by:

$$\begin{cases} \dfrac{d\mathbb{E}}{dt} = (\Omega(v) + \lambda\, d(v))\, \mathbb{E} \\ \mathbb{E}(0) = \mathrm{Id} \end{cases} \tag{2-12}$$

The equation of evolution of $\psi(p)$ is:

$$\frac{d\psi}{dt} + \frac{\partial}{\partial p}\Big(\psi\, \frac{dp}{dt}\Big) = 0 \tag{2-13}$$

The macroscopic and orientation averaged extra stress tensor σ is then defined by:

$$\sigma = \int \mathbb{T}\,\psi(\mathbf{p})\,d\mathbf{p} \tag{2-14}$$

Let $\mathbf{a}$ and $\mathbf{a_4}$ be the second order and forth order orientation tensor.

$$\mathbf{a} = \int \mathbf{p}\,\mathbf{p}^T\,\psi(\mathbf{p})\,d\mathbf{p} \qquad \text{and} \qquad \mathbf{a_4} = \int (\mathbf{p}\,\mathbf{p}^T)\otimes(\mathbf{p}\,\mathbf{p}^T)\,\psi(\mathbf{p})\,d\mathbf{p} \tag{2-15}$$

From equation (2-11), a straightforward calculation yields:

$$\mathbb{B}\left\{\frac{d}{dt}(\mathbb{B}^{-1}\mathbf{a}\,\mathbb{B}^{-T})\right\}\mathbb{B}^T = \frac{d\mathbf{a}}{dt} - (\mathbf{\Omega}(v) + \lambda\,\mathbf{d}(v))\,\mathbf{a} + \mathbf{a}\,(\mathbf{\Omega}(v) + \lambda\,\mathbf{d}(v))$$
$$= -2\,\lambda\,\mathbf{a_4}\,\mathbf{d}(v) \tag{2-16}$$

Thus a general expression for σ can be written which is the constitutive relation of the suspension:

$$\sigma = 2\eta_I\left\{\mathbf{d}(v) + N_p\,\mathbf{a_4}\,\mathbf{d}(v) + N_s[\,\mathbf{d}(v)\,\mathbf{a} + \mathbf{a}\,\mathbf{d}(v)]\right\} \tag{2-17}$$

To close the set of equations (2-16) and (2-17), one assumes a closure approximation. For example, if the fibers are almost aligned, the quadratic closure approximation is valid:

$$\mathbf{a_4} = \mathbf{a}\otimes\mathbf{a} \tag{2-18}$$

This constitutive equation has the same basic structure as that of anisotropic fluids introduced by Hand [12]. Thus, though the microscopic model is generally not valid for thermoplastics composites because the fiber concentration is too large for the composite to be considered as a dilute or even a semi-dilute suspension, equations (2-13), (2-14) remain valid as a first approximation, provided that the rheological parameters η_I, N_p and N_s are determined experimentally. In this paper, we will consider the case of short fibers composites for which N_s is negligible with respect to N_p and for which the aspect ratio of the particule is quasi-infinite, leading to $\lambda = 1$. With these asumptions, equations (2-16), (2-17), (2-18) write:

$$\sigma = 2\eta_I\left\{\mathbf{d}(v) + N_p\,\mathrm{Tr}[\,\mathbf{a}\,\mathbf{d}(v)]\,\mathbf{a}\right\} \tag{2-19}$$

$$\frac{\delta\mathbf{a}}{\delta t} = \frac{d\mathbf{a}}{dt} - \mathbf{grad}(v)\,\mathbf{a} - \mathbf{a}\,\mathbf{grad}(v)^T = -2\,\mathrm{Tr}[\,\mathbf{a}\,\mathbf{d}(v)]\,\mathbf{a} \tag{2-20}$$

3 - NUMERICAL MODELING

The problem to be solved is shown on figure 1. $\partial\Omega_1$ denotes the surface on which a given stress vector $\mathbf{F_g}$ is assumed to be prescribed. $\partial\Omega_2$ denotes the edge on which the velocity field v_g is prescribed. $\partial\Omega_2$ is divided into $\partial\Omega_{21}$ and $\partial\Omega_{22}$. On $\partial\Omega_{21}$ the fluid enters the domain (ie $v_g.\mathbf{n} < 0$) and an orientation tensor $\mathbf{a_g}$ is prescribed, whereas on

$\partial\Omega_{22}$ the fluid leaves the domain (ie $v_g.n > 0$) and no condition is prescribed on **a**. An additional strong hypothesis is made, that is usually easy to satisfy for extrusion like processes: no fluid enters the domain on $\partial\Omega_1$ (ie on $\partial\Omega_1$ the a priori unknown velocity field satisfies the condition $v_g.n > 0$). Furthermore, no mass forces have to be accounted for, because the high viscosity of molten polymers allows us to neglect them as well as the inertia terms in the balance of momentum equation.

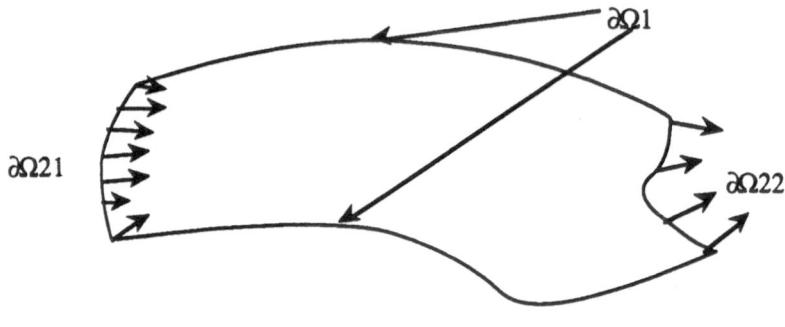

figure 1

If the orientation **a** is assumed to be known, the problem to be solved can be written as a minimisation problem under the constraint that the velocity field is incompressible:

Pbl : Find **v** in $\mathcal{V}$, minimizing J (**v**)

with $$J (v) = \int_\Omega \eta_I \{ \mathrm{Tr}[d(v)d(v)] + N_p \, \mathrm{Tr}[a\, d(v)]^2 \} - \int_{\partial\Omega_1} F_g\, v$$

$$\mathcal{V} = \{v, \mathrm{div}\,(v) = 0, v = v_g \text{ on } \partial\Omega_2\}$$

If the velocity field is assumed to be known, the problem to be solved is an evolution equation. The difficulty of this problem is the one of every history dependant fluid mechanics problem. In one hand, equation (2-20) is easy to solve from equation (2-16) in a lagrangian form and with a method of caracteristics because it is then a local equation in space. But, as far as we are interested here in steady solutions, and because the steadiness condition is not local in space in a lagrangian form, the lagrangian coordinate system is not really easy to use. In the other hand, the eulerian coordinate system is well adapted to the prescription of the steadiness of the solution, but in this case equation (2-20) is not anymore local in space. We chose here the eulerian approach so that the second problem can be written:

Pb2: find **a** such that:

$$(v^T \, grad) \, a - grad(v) \, a - a \, grad(v)^T = - 2 \, Tr \, [\, a \, d(v)] \, a \qquad \text{in } \Omega$$

$$a = a_g \qquad \text{on } \partial\Omega_{21}$$

The global solution is reached with a fixed point method in solving successively *Pb1* and *Pb2*. The initialisation of the algorithm is achieved in solving the newtonian corresponding problem (ie. in taking $a = Id/3$).

Algorithm to solve *Pb1*

Pb1 is solved with an augmented lagrangian method (Fortin, Glowinski 1982). It can be written:

$$div \, (2 \, \eta_I \, \{ d(v) + N_p \, Tr[a \, d(v)] \, a\} - grad \, p = 0$$
$$div \, (v) = 0$$

or in a more formal way:

$$A(a) \, v + B^T \, p = 0$$
$$B \, v = 0$$

Let A_r be defined as

$$A_r(a) = A(a) + r \, B^T \, B$$

Pb1 is equivalent to *Pb1'*:

Find (v,p) such that $\Im(v,p)$ is extremal with:

$$\Im(v,p) = \frac{1}{2} <A_r(a) \, v,v> + <v,B^T \, p>$$

$$v = v_g \quad on.\partial\Omega_2$$

Pb1' is solved with the finite element method. To ensure the Babushka Brezi condition a quadratic triangulation has been chosen to approximate the velocity field and the pressure is chosen to be constant per element. The minimization is achieved with a conjugate gradient algorithm as follows:

Algorithm: (if v^n and a^n are assumed to be known from a previous step)

- $p^{n+1,0}$ is taken arbitrarily
- $v^{n+1,0}$ is computed from:

$$A_r(a^n)\ v^{n+1,0} + B^T\ p^{n+1,0} = 0$$

$$v^{n+1,0} = v_g \qquad \text{on } \partial\Omega_2$$

- at the sth iteration, the descent direction is defined as:

$$w_s = B\ v^{n+1,s} \qquad \text{if } s = 0$$

$$w_s = B\ v^{n+1,s} + \lambda_s\ w_{s-1} \qquad \text{if } s \neq 0$$

$$\lambda_s = \frac{\|B\ v^{n+1,s}\|^2}{\|B\ v^{n+1,s-1}\|^2}$$

- computation of z^s:

$$A_r(a^n)\ z^s + B^T\ w_s = 0$$

$$z^s = 0 \qquad \text{on } \partial\Omega_2$$

- computation of ρ_s:

$$\rho_s = -\frac{\|B\ v^{n+1,s}\|^2}{<B\ v^{n+1,s},\ Bz^s>}$$

- actualisation:

$$p^{n+1,s+1} = p^{n+1,s} - \rho_s\ w_s$$

$$v^{n+1,s+1} = v^{n+1,s} + \rho_s\ z_s$$

The convergence test is achieved on the incompressibility condition.

Algorithm to solve $Pb2$

The main problems that arise in solving Pb2 come from its non elliptic nature and from its non linearity. It is thus impossible to use a standard Galerkin formulation which is known to be unefficient for convetion diffusion equations (Canto, Husseini 1987, Heinrich 1990, Olmos 1991).

The non linearity has been solved with a fixed point method in order to decouple each component of the orientation tensor:

Algorithm: (v^n is assumed to be known from a previous step)
  - $a^{n,0} = a^n$
  - for $s > 0$ and for each component of $a^{n,s+1}$

$$v^{nT}grad\ a^{n,s+1}_{ij} + 2\ Tr[a^{n,s}\ d(v^n)]a^{n,s+1}_{ij} = [\ grad(v^n)a^{n,s} + a^{n,s}grad(v)^T\ ]_{ij}$$

The spatial discretisation is made with an upwind scheme: if Nj are the shape functions, the weight functions are chosen to be:

$$W_j = N_j + \frac{\bar{h}}{2\ \|v^n\|}(\ v^n_x\ \frac{\partial N_j}{\partial x} + v^n_y\ \frac{\partial N_j}{\partial y}\ )$$

Where $\bar{h}$ is the average size of the element in the flow direction.

Results and discussion:

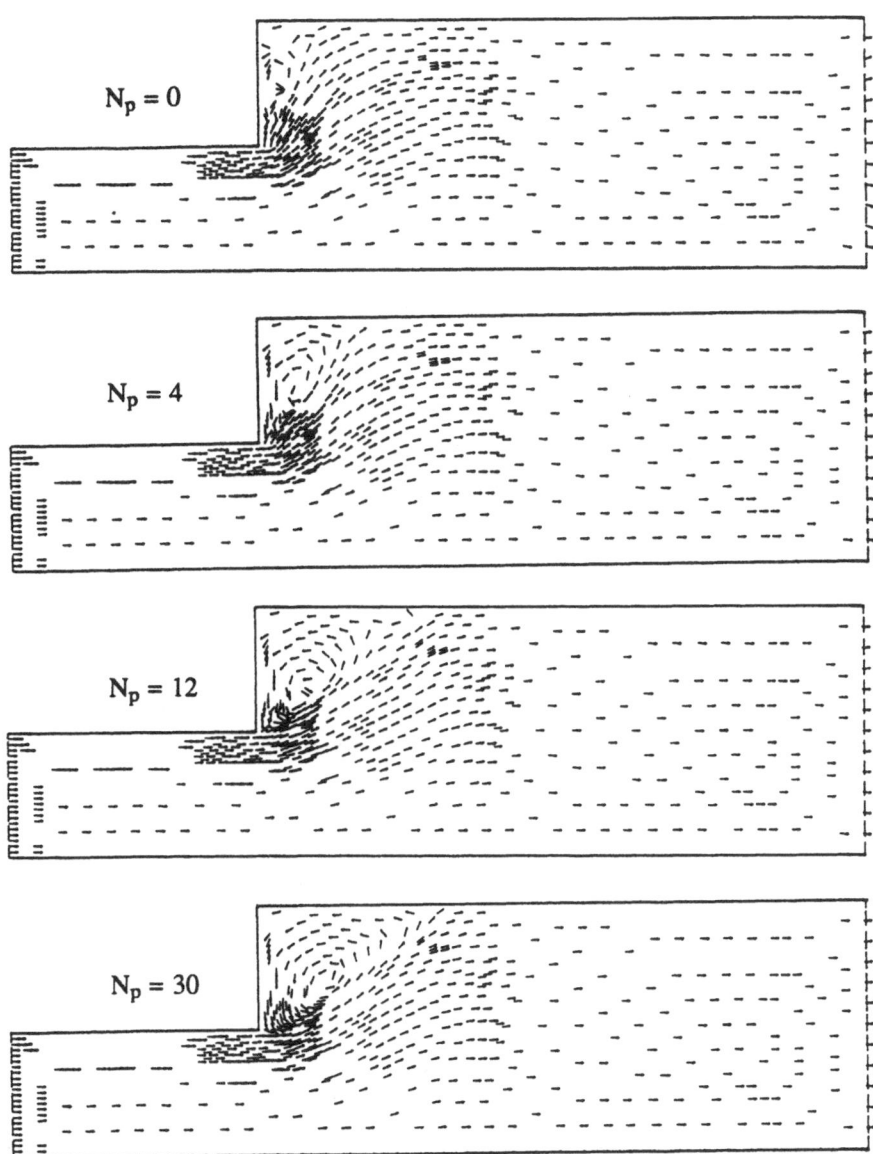

figure 2 - Flow in a contraction using the locally aligned aproximation
- influence of Np on the velocity field

The first results concern the solution of Pb1 which has been tested for a contraction in using the locally aligned aproximation:

$$\mathbf{a} = \frac{v \ v^T}{v^T \ v}$$

A fully developped velocity profile has been prescribed at the entrance of the die and we studied the influence of N_p. We obtained similar results as those given by Lipscomb et al. (1988) except that we did not find any critical value of N_p above which the convergence cannot be reached. The flow field is strongly affected by the presence of fibers, even for small Np values (figure 2)

Second resultas concern a coupled solution in a Poiseuille flow. At the entrance a parabolic velocity profile is prescribed as well as a totally disordered orientation ($\mathbf{a}$ = Id/2). The flow field is almost unaffected by the presence of fibers but the orientation varies along the flow. Figure 3 shows with ellipses the direction of eigenvectors and the eigenvalues of $\mathbf{a}$, which means that along the longest axes of the ellipse, the orientation is found with a probability equal to the aspect-ratio of the ellipse. The aligned orientation is reached exponentially except on the symetrical axis of the flow. This is a consequence of the absence of diffusion in the orientation equation perpendicular to the flow.

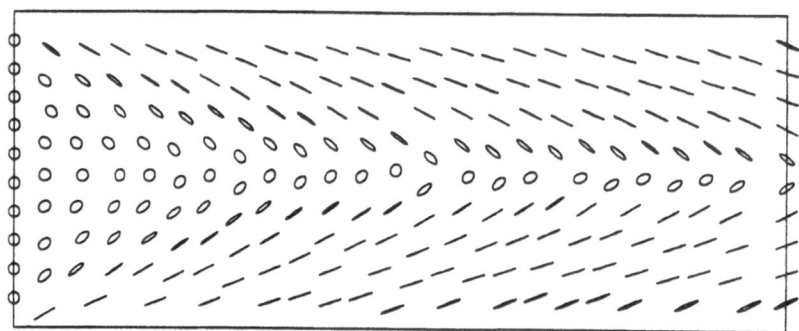

figure 3 - Orientation of fibers in a Poiseuille flow (coupled calculation)

The algorithm is general and could be applied to any other geometry but could give rise to convergence difficulties in presence of vertices inside the flow.

4 . CONCLUSION

A statistical description of the orientation is a way to handel with flow induced orientation in forming processes. The theoretical results concerning the behaviour of materials are not new but begin to be applied in numerical modellings. Further improvements should concern (i) a direct identification of the rheological parameters (ii) a derivation of a self consistant models which could provide a better prediction of the behavior for high concentrated composites (iii) a derivation of simplified but coupled models for tridimensionnal predictions in injection molding processes (Tucker 1991, Poitou et al. 1992). It should be noted that this kind of approach is not restricted to fibers composites and is applied in a very similar way for the prediction of damage in structural mechanics. In this latter case the orientation is then the normal vector to a crack's plane.

References

G.K. Batchelor, (1970), J. Fluid Mech. **41** 545-570
G.K. Batchelor, (1971), J. Fluid Mech. **46** 813-829
M.A. Bibbo, S.M. Dinh, R.C. Armstrong, (1985), J. Rheol, **29** 905-929
B.A. Bilby, J.D. Eshelby, M.L. Kolbuszewski, A.K. Kundu, (1976) Tecnophysics **28**, 265
K. Chiba, K. Nakamura, (1990), J. Non Newtonian Fluid Mech, **35** , 1-14
S.M. Dinh and R.C. Armstrong, (1984), J. Rheol. **28** 207-227
M. Fortin, R. Glowinski, (1982) **Augmented lagrangian methods : Applications to the numerical solutions of boundary value problems** - North Holland
H. Giesekus, (1962) Rheol. Acta **2** 50-62
G.L. Hand, (1961), Arch. Rat. Mech. Anal. **7** 81-86
E.J. Hinch and L.G. Leal, (1973), J. Fluid Mech. **57** 753-767
E.J. Hinch and L.G. Leal, (1976), J. Fluid Mech. **76** 187-208
G.B. Jeffery, (1922), Proc. Roy. Soc. **A-102** 161-179
G.G. Lipscomb, M.M. Denn, (1988), J. Non Newtonian Fluid Mech., **26** 297-325
A. Poitou, F. Chinesta, R. Torres, (1992), J. Materials Processing Tech., **32** 429-438
C.L. Tucker, (1991), J. non newt. fluid mech, **39** 239-268

MODELING GRANULAR FABRIC BY TENSORS AND THEIR STATISTICAL TEST

J. L. Chameau* and B. Muhunthan**
*   School of Civil Engineering, Georgia Institute of Technology, Atlanta, GA., USA
**   Department of Civil Engineering, Washington State University, Pullman, WA., USA

Abstract
It is well known that fabric is one of the most important factors in the mechanics of granular materials. This paper presents a theoretical development directed at quantifying the fabric anisotropy of granular media. An averaging technique is used in developing tensor measures that are useful for this purpose. The technique allows tensors based on solid and void phase to be treated in a unified manner. Several definitions of the fabric tensor have been proposed in the past. A discussion of these tensors is provided. It is shown that for certain fabric tensors, it is necessary to give an appropriate expression for the distribution of fabric descriptors. For other types of tensors, this is not necessary. The void phase tensors are useful from an experimental point of view. In its development, the void phase distribution function is limited to a second order tensor term. If the fluctuations of the distribution are high, higher order terms may be necessary for a complete description. A statistical test is proposed to determine the adequacy of the choice of the rank of tensors in describing a particular distribution.
Key words: Granular, Statistics, Fabric, voids, soils, stereology

1 Introduction

It is now widely accepted that fabric is one of the most important concepts in the mechanics of granular materials. It has been discussed and identified in terms of several quantities such as contact normal, branch vector (connecting the centroids of particles in contact), orientation of non-spherical particles and so on (Figure 1). All of these quantities are related to the solid phase of the material because it forms the skeleton of the material. Efforts have been directed to characterize the fabric anisotropy in terms of a tensor representing these quantities and many important effects of fabric have been quantified (e.g., Rothenburg and Bathurst, 1990; Chang and Misra, 1990). Yet, most of these analyses were confined to numerical simulations of simplified geometries, especially discs (2-D) or spheres (3-D). A further drawback of these studies is the lack of experimental means to measure any of the parameters.

Most stress-strain models in soil mechanics use parameters based on the void phase as constitutive variables in representing the fabric of soils. The critical state soil mechanics framework (CSSM) (Schofield and Wroth, 1968) uses void ratio as a constitutive variable. The use of a scalar parameter such as the void ratio makes these models inherently incapable of modeling the effect of the fabric anisotropy of soils. A second order tensor based on void phase is proposed in this study to describe fabric anisotropy. This void tensor is useful to incorporate the effects of fabric in soil constitutive models. An additional feature of the development is that

305

D. Breysse (ed.), Probabilities and Materials, 305–319.
© 1994 *Kluwer Academic Publishers.*

the parameters of the fabric description can be obtained from stereological observations. Detailed development of the theory and its applications have been presented elsewhere (Muhunthan, 1991; 1993 a & b; Muhunthan and Alwail, 1992).

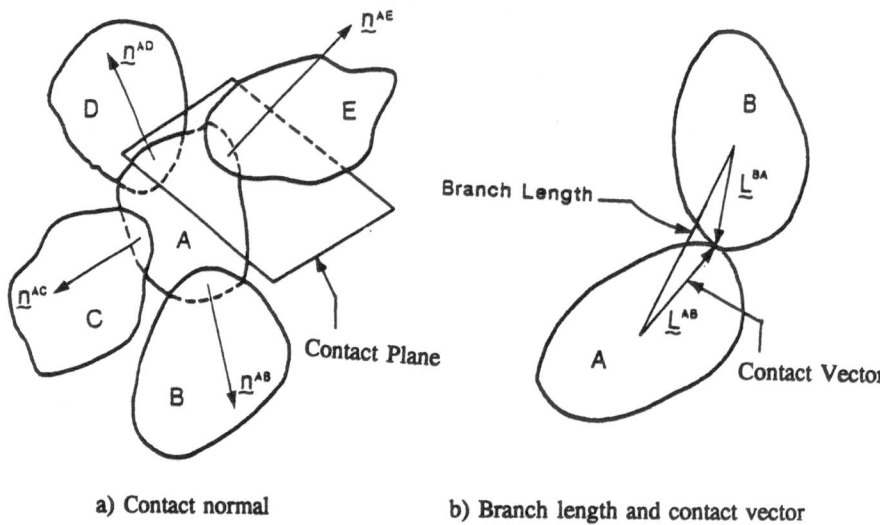

a) Contact normal   b) Branch length and contact vector

Fig. 1. Solid phase fabric descriptors

It is not intended to describe the distributions accurately but rather to characterize them by tensors, which could then be related to macroscopical quantities. The current analysis assumes that the distribution of void ratio in space can be approximated by terms involving up to a second order tensor. If the form of the distribution itself is the aim, then other forms and measures must be employed at the cost of simplicity. Although the present choice has been found to be sufficient for practical purposes, its applicability may be limited for cases involving high fluctuations of the distribution function. A statistical test to determine the adequacy of the choice of tensor terms up to a certain order is also proposed.

2 Fabric tensor

Let us consider a micro volume V, usually called the representative element volume, REV, (Note: for a two dimensional case REV becomes a representative element area REA) which is assumed homogeneous and has a large number (say N) of particles sufficient for statistical treatment. The REV may be of any shape. i.e. sphere, cube, etc. In this study a unit sphere is used as an REV.

There are two contacts at each contact point, as indicated by the two unit normal vectors in Fig. 2. It is assumed that the surface of V is chosen such that it does not pass through any contact point. If a particle cut by this surface has p contacts lying within V and a total of q contacts, its contribution to the number N of particles within V will be the fraction p/q (i.e., N may not be an integer). Let us assume that the total number of contacts in V is C.

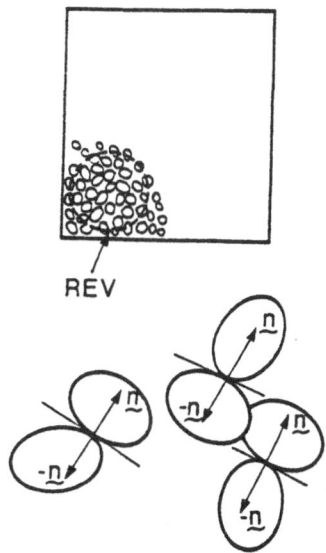

REV

Fig. 2. Contact normal and REV for an assembly of grains

The direction of a unit normal at a contact can be described by spherical coordinates 1, θ, ϕ with respect to some fixed rectangular cartesian system (Fig. 3). Let $C_{\theta\phi}$ be the total number of contacts in V whose unit normals lie in the elemental solid angle $d\Omega = \sin\theta d\theta d\phi$ within the limits θ to $\theta+d\theta$, ϕ to $\phi+d\phi$. The symmetry condition:

$$C_{\theta\phi} = C_{(-\theta)(-\phi)} \tag{1}$$

must apply, since at any contact point two contact normals exist. It is assumed that for a sufficiently large volume (and hence sufficiently large N) $C_{\theta\phi}$ is given by a smooth function of θ,ϕ:

$$C_{\theta\phi} = C \ E(\theta,\phi)\sin\theta d\theta d\phi \tag{2}$$

or as:

$$C_{\theta\phi} = C \ E(\Omega)d\Omega \tag{3}$$

where $E(\Omega)$ is a function of θ,ϕ. The summation of $C_{\theta\phi}$ over all the contacts in the domain should equal C and thus:

$$\int_0^{2\pi} \int_0^{\pi} E(\theta,\phi)\sin\theta d\theta d\phi = 1 \tag{4}$$

or

$$\int E(\Omega)d\Omega = 1 \tag{5}$$

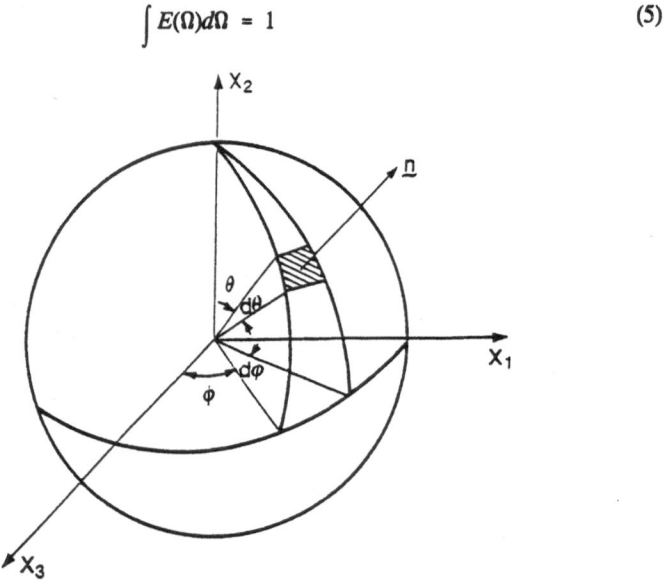

Fig. 3. Spherical REV enclosing particles

Hence $E(\Omega)$ can be considered as a probability density function for the distribution of contacts. To conform with the notation used later a weighting function $f(\Omega)$ is introduced as:

$$f(\Omega) = CE(\Omega) \tag{6}$$

thus:

$$C = \int f(\Omega)d\Omega \tag{7}$$

or explicitly:

$$C = \int_0^{2\pi} \int_0^{\pi} f(\theta,\phi)Sin\theta d\theta d\phi \tag{7a}$$

If the frequency of contacts is uniform at all angles (as would be expected in an isotropic assembly), $E(\Omega)$ has the constant value $1/4\pi$, and $f(\Omega)$ becomes:

$$f(\Omega) = \frac{C}{4\pi} \tag{8}$$

When sand is deposited under gravitational forces, particles are more likely to form vertically oriented contacts than horizontal contacts. Preferential orientations can also evolve as a result of shearing deformations, as these induce relative inter-particle movements as a function of direction. In these situations, granular materials become anisotropic. The probability density function given by Eq. 8 is only applicable to an isotropic case and a general description is required for the anisotropic case. Many approaches have been adopted to achieve this including the concept of fabric tensor (Kanatani, 1984; Satake, 1982; Rothenbug, 1980).

Satake(1982) defined a fabric tensor by:

$$\omega_{ij} = \frac{1}{C}\sum_R n_i n_j \tag{9}$$

where $n_i(i=1,2,3)$ are the cartesian components of a unit vector n in the REV, and C is the total number of contact points in the REV. Since the REV is assumed to contain a sufficiently large number of contact normals, this equation can be written as:

$$\omega_{ij} = \int E(\Omega)n_i n_j d\Omega \tag{10}$$

where $E(\Omega)$ is the probability density function (Eq. 5) of contact normals. ω_{ij} may be regarded as an average of $n_i n_j$ in REV. These and other types of fabric tensors can be obtained from a general formulation based on the mathematical concept of averaging (Marle, 1981; Muhunthan, 1991).

2.1 Averaging concept and fabric tensors

As shown above, the concept of "fabric tensor" evolved as a means to describe the discrete granular medium in a continuum sense. In general it could be defined as an average of any physical quantity in a domain under consideration. The concept of mathematical averaging will now be used to develop a general scheme for possible descriptors of the microstructure of granular materials. It is also shown that the existing formulations of fabric tensors are a subset of this general framework. Engineering measurements are usually made at a macroscopic scale. The general approach of defining a macroscopic quantity at a point x at time t, is to take the mean value of the corresponding quantity at the micro level over some elementary volume around point x. Different approaches are available to achieve this. The treatment presented herein, follows the mathematical formalism presented by Marle (1981).

The essential element of the mathematical description is a weighting function $m(\Omega)$, or a probability distribution function similar to the one defined earlier. It is assumed integrable locally and over the whole space such that:

$$\int m(\Omega)d\Omega = 1 \qquad (11)$$

The granular soil may contain several phases. Let $a_\alpha(\Omega)$ be a microscopic physical property associated with a phase α. It is defined at each point of the spatial domain occupied by a granular medium, but with non zero values only in the subdomain occupied by the phase α. This quantity can be of any tensorial order. By definition, the macroscopic field variable $<a_\alpha\, m(\Omega)>$ associated with a microscopic property $a_\alpha(\Omega)$ is expressed via a weighting function $m(\Omega)$ as:

$$<a_\alpha\, m(\Omega)> = \int a_\alpha(\Omega)m(\Omega)d\Omega(n) \qquad (12)$$

This general equation, to obtain a macroscopic quantity from a corresponding microscopic one, can be used to develop different tensors based on both solid and void phase with an appropriate choice of α. Most fabric tensors proposed in the literature (e.g., Rothenburg 1980; Satake 1982; Nemat-Nasser and Mehrabadi 1983) are based on the solid phase of granular media.

It often becomes a difficult task to obtain an appropriate expression for the weighting function or the probability density function in the analysis of random packing of granular materials. Instead, an alternate approach to specify the function in terms of tensors has been proposed by researchers (e.g. Kanatani, 1984)

2.2 General formulation for weighting function

Any single valued function $f(\Omega)$, square integrable on the unit sphere, can be expanded in a Fourier series, convergent in the mean, in spherical harmonics (Abramowitz and Stegun, 1965; Muhunthan, 1991). This expression for the scalar valued function $f(\Omega)$ can also be written in terms of cartesian coordinates using tensor notation (Kanatani 1984; Leckie and Onat 1981):

$$f(n) = \frac{C}{4\pi} [1 + \phi_{ij} \, n_i n_j + \phi_{ijkl} \, n_i n_j n_k n_l + \ldots] \qquad (13)$$

where the components of the unit vector n are given by $n_1 = \sin\theta\cos\phi$, $n_2 = \sin\theta\sin\phi$, $n_3 = \cos\theta$ and ϕ_{ij}, ϕ_{ijkl} are symmetric traceless even order tensors. This form is especially useful for constitutive modeling purposes.

The use of higher order tensors in the above expression allows the fluctuations of the density function to be taken into account and yields an accurate representation of the distribution. However, the use of these terms is not mathematically convenient. For normal applications it is sufficient to keep terms up to a second order tensor. For this case Eq. (13) reduces to:

$$f(\Omega) = \frac{C}{4\pi} [1 + \phi_{ij} \, n_i n_j] \qquad (14)$$

which can be rewritten as:

$$f(n) = \frac{C}{4\pi} [(\delta_{ij} + \phi_{ij}) \, n_i n_j] \qquad (15)$$

or as:

$$f(n) = \frac{C}{4\pi} [F_{ij} \, n_i n_j] \qquad (16)$$

Some studies (Kanatani 1978; Mullenger 1978) use F_{ij} as the fabric tensor. It must however be noted that this tensor formulation is the result of limiting the terms in the description of the function $f(\Omega)$ to two. If fluctuations are high, such a choice may not be sufficient. Substitution of $E(\Omega)$ in terms of $f(\Omega)$ in Eq. 10 results in:

$$\omega_{ij} = \frac{1}{C} \int f(n) n_i n_j d\Omega(n) \qquad (17)$$

This formulation can also be obtained from Eq. 12 by substituting $n_i n_j$ for a_α, with α being the solid phase, and $f(\Omega)$ for $m(\Omega)$. Substitution of Eq. (16) into Eq. (17) gives the relation

$$F_{ij} = \frac{3}{2} [5\omega_{ij} - \delta_{ij}] \qquad (18)$$

and further substitution of Eq. (18) back into Eq. (16) results in:

$$f(\Omega) = \frac{15}{2}[\; \omega_{ij}\; n_i n_j \;-\; \frac{1}{5}] \qquad\qquad (19)$$

A complete derivation of these relations appears elsewhere (Muhunthan 1991; Satake, 1982).

It is seen that $f(\Omega)$ is determined from ω_{ij} uniquely and further the form of Eq. (18) may not be appropriate for $\omega_i < 1/5$, where ω_i denotes the principal values of ω_{ij}. This however, does not happen for most practical problems (Kanatani, 1984).

The freedom to choose a weighting function $f(\Omega)$ with an associated C leads to different interpretations of the fabric tensors. The fabric tensor discussed thus far had the number of contact normals in the denominator. Instead, it could have been defined in terms of the area or the volume of solid particles. Correspondingly, the C term in the weighting function (Eq. 14) will have different interpretations as a function of the definitions used. Hence, C is useful in unifying all the different tensor formulations. The ultimate choice of a particular description has to be made based on physical as well as experimental considerations.

2.3 Higher measures using void phase

As mentioned previously, void ratio is used as a measure of fabric in granular materials. It was also pointed out that this measure is inadequate in describing several phenomena associated with their deformation. To improve upon this description, second order measures based on the solid phase(i.e, contact normal, contact vector, etc) have been used as discussed above. However, it will now be suggested that the extension to higher measures can be attempted on void phase. The void ratio is used as a constitutive variable in the framework of critical state soil mechanics (CSSM) (Schofield and Wroth, 1968) and thus the approach taken here will extend the CSSM concepts naturally (Muhunthan and Chameau, 1992). An added incentive to the use of a measure based on void phase is that it would provide a means to define some form of a unified second order measure for all particulate media including clays and silts. It is difficult to define a contact normal or a contact vector for clayey type materials. It is however, convenient to measure the void space distribution.

The first step towards developing a tensor in terms of void phase is the assignment of a directional vector to a single void. This is not as straightforward as that for a contact normal or a contact vector. It is possible to define a representative vector for a void in many ways. The representation proposed by Konishi and Naruse (1987) and used by Muhunthan and Chameau (1992) is adopted herein. Let us imagine a "unit void" surrounded by several contacting particles. In general, the void space in a granular material can be a continuous space with a complex configuration and it may have to be divided into a number of "unit voids". The boundary of the void consists of n curved segments (arcs) linked to each other at contacting points of particles. To quantify the size, shape and orientation of the voids, a void with n curved segments is replaced by a polygon with n vectors along the chords (Fig. 4). The vectors of the replaced polygon pass through the associated solid particles. A unit vector h, representing the average of the n chord vectors (Fig.4), is chosen as the void vector. As with the case of vectors based on the solid phase, this vector is anisotropically distributed. Accordingly, a void distribution function, V (Ω), can be defined as:

$$V(\Omega) = \frac{C}{4\pi} \left[1 + \phi_{ij} \, h_i h_j + \phi_{ijkl} h_i h_j h_k h_l + \ldots \ldots \right] \tag{20}$$

where ϕ_{ij}, ϕ_{ijkl} are symmetric traceless tensors based on void phase similar to those defined earlier. They are termed void tensors of the second, fourth and higher order (Muhunthan, 1991) respectively. C is obtained by substituting $V(\Omega)$ for $f(\Omega)$ in Eq. 7. The higher rank tensors ϕ_{ijkl}, ϕ_{ijkl}..., describe the higher order fluctuations of the void space distribution. As in the case of solid phase descriptors, for practical applications a second order approximation is preferred. For a second order description $V(\Omega)$ becomes:

$$V(\Omega) = \frac{C}{4\pi} \left[1 + \phi_{ij} \, h_i h_j \right] \tag{21}$$

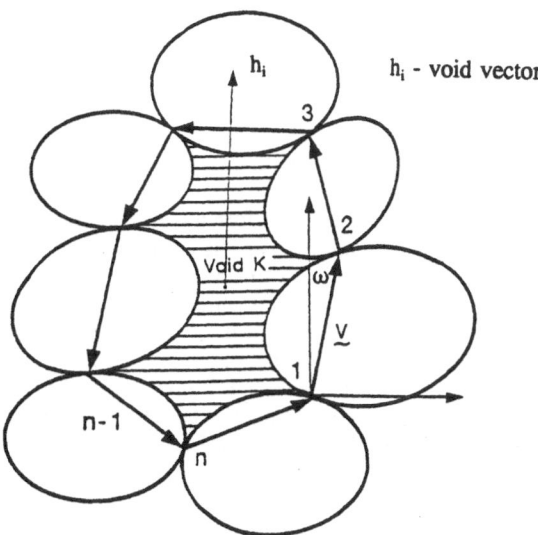

Fig. 4. Illustration of a void vector

The last expression for void space distribution (Eq. 21) can be specifically written in terms of porosity (n) or void ratio (e) distribution used commonly in soil mechanics (Muhunthan and Chameau, 1992):

$$n(\Omega) = n_0 \left[1 + \phi_{ij} h_i h_j \right] \tag{22}$$

or as:

$$e(\Omega) = e_0 \left[1 + \phi_{ij} h_i h_j \right] \tag{23}$$

where n_0 and e_0 are the mean porosity and void ratio of the soil mass respectively. The directional distribution of the two parameters is given by the fabric tensor associated with it. Notice that $C = 4\pi n_0$ and $4\pi e_0$ respectively for these choices. For an isotropic distribution of grains, the components of the fabric tensor are zero and Eqs. (2) and (3) reduce to n_0 and e_0 respectively as assumed in conventional soil models. The development shows that granular media can be characterized by (n_0, ϕ_{ij}) or (e_0, ϕ_{ij}). An experimental procedure based on stereological observations to determine the parameters of description has been presented elsewhere (Muhunthan and Alwail, 1992; Muhunthan, 1991, 1993b).

3 Statistical Test for the Fitness of the Distribution

In reality, the distribution function of contact normals or void is discontinuous. As emphasized in this paper, the aim is to characterize $f(\Omega)$ or $V(\Omega)$ by smooth functions with certain physical meaning. Therefore, only a small number of terms need be retained in Eq. 20, but how many of them are sufficient? In order to answer this question, it is best to resort to a statistical test. There are many such tests available (Lehmann, 1986). The statistical test proposed by Kanatani (1984) is used in the present study. The discussion that follows is directed at the general function $f(\Omega)$.

The simplest problem is the "test of uniformity". Suppose the computed ϕ_{ij} is very small and the distribution is almost "uniform" or "isotropic", i.e. $f(\Omega) = C/4\pi$ in the three dimensional case or $f(\Omega) = c/2\pi$ in the two dimensional case. Then how small should ϕ_{ij} be in order to conclude that the true population is uniform and that the computed non-zero is a statistical fluctuation due to the finite size of the data? This problem is solved by calculating the "likelihood ratio", i.e. the probability that the observed data are generated by the distribution calculated up to the term of ϕ_{ij}. If the ratio is too small, it cannot be concluded that the true population is uniform. Then, the second term must be retained. The same process also applies to higher terms. For example, in order to test whether the term of ϕ_{ijkl} can be neglected or not, the likelihood ratio with respect to the distribution up to the term of ϕ_{ij} vs the distribution up to the term of ϕ_{ijkl} is computed, and so on.

Let λ be the likelihood ratic. It is known that $-2\log\lambda$ behaves according to the χ^2-distribution (Lehmann, 1986). Its degree of freedom is the number of independent parameters whose nullity is to be tested, if the number N of independently observed data is sufficiently large. It is also known that $-2\log\lambda$ can be expressed as a quadratic form of the parameters to be tested. In this case the statistic to be tested becomes N times the "Fisher information matrix", if N is sufficiently large (Lehmann, 1986).

First, let us consider the test of uniformity. Let $F(n)$ be the distribution calculated up to the second term from independently observed N data:

$$F(n) = \frac{C}{4\pi}\left[1 + \phi_{ij}n_i n_j \right] \tag{24}$$

The Fisher information matrix in this case is a tensor defined by

$$I_{ijkl} = \int \frac{1}{F}\frac{\partial F}{\partial \phi_{ij}}\frac{\partial F}{\partial D_{kl}}dn \mid D_{mn}=0 \tag{25}$$

Substitution of Eq. (24) yields

$$I_{ijkl} = \frac{C}{5}\delta_{(ij}\delta_{kl)} \tag{26}$$

where the parenthesis () denotes the symmetrization of the indices between them (Kanatani, 1984; Muhunthan, 1991). Hence, the statistic to be tested is $NI_{ijkl}\phi_{ij}\phi_{kl}$, or

$$\frac{2NC}{15}\phi_{ij}\phi_{ij} \tag{27}$$

Let $\chi^2_c(p,\alpha)$ be the value of χ^2 whose upper probability is α, i.e. Prob $\{ \chi^2 > \chi^2_c (p,\alpha)\} = \alpha$, χ^2 obeying the χ^2-distribution of degree of freedom p. If the value of Eq. (27) is larger than $\chi^2_c(5,\alpha)$, the observation is "significant", i.e. the distribution cannot be regarded as uniform, under "significance level" α. A test for ϕ_{ijkl} is obtained similarly.

4 Illustration of void distribution characterization

The soil investigated in this study is a Champlain sea sensitive clay obtained by block sampling from St. Marcel, Quebec. The soil was cut into small cylindrical specimens with a thin wire, quick frozen in Freon 22 cooled by liquid nitrogen, freeze fractured, and sublimated.

Four standard consolidation tests were performed on the clay with a 1.5 load increment ratio every 24 hours to 23, 124, 421, and 1452 kPa. The stresses were then released in several stages and each sample left under a load of 4 kPa. Specimens were cut from the samples and prepared for microscopy. Observations were made on a JEOL 25 scanning electron microscope on intact sample as well as at stresses of 124, 421 and 1452 kpa. Graphic representations of porosity (void fraction) made on horizontal and vertical sections at different stress levels are shown in Figs. 5 and 6, respectively.

A typical section (for 124 kPa) is shown in Fig. 7 with the chosen REV. The stereological technique proposed by Muhunthan and Chameau 1992b) was applied to this section to obtain the fabric tensor. A set of probe lines were drawn at 10^0 intervals. The total length of the interceptions of each test ray with the pores was measured. The average length of the interceptions per unit length of the test line was evaluated and resulted in the following tensor.

$$(\phi_{ij})_{124} = \begin{bmatrix} 0.509 & 0.13 & 0.295 \\ 0.13 & 0.509 & 0.295 \\ 0.295 & 0.295 & -1.020 \end{bmatrix} \qquad (28)$$

Detailed derivation of the above tensor is presented by Muhunthan (1991) and Muhunthan and Alwail (1992).

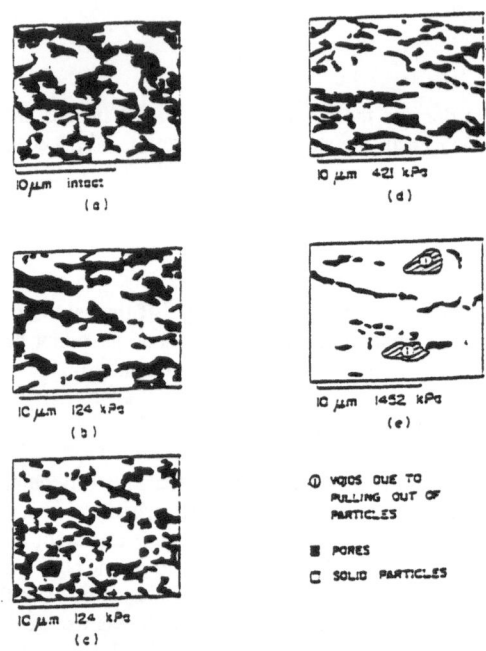

Fig. 5. Graphic representation of porosity on horizontal section (Delage and Lefebvre 1984)

Further, as part of the calculations the value of C (Eq. 21) can also be obtained. This in turn can be used to obtain the value of n_0. For the above section, the value of C for was obtained as 2.927 with a corresponding porosity n_0 of 0.233. This value agrees well with the experimental value of the porosity reported by Delage and Lefebvre. The presence of non zero terms in ϕ_{ij} indicates that porosity of the soil is anisotropically distributed.

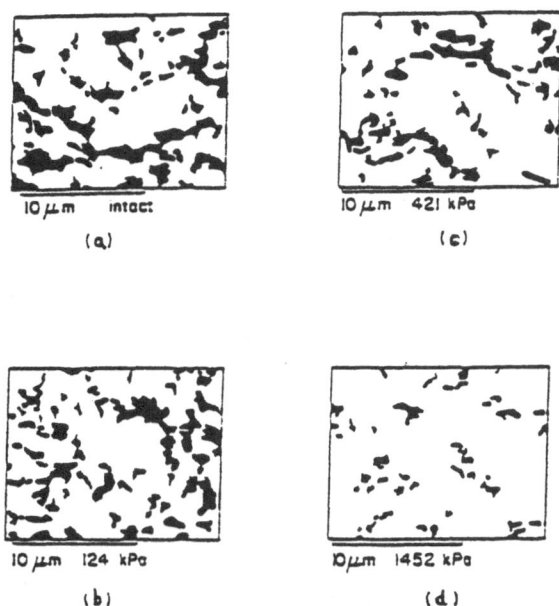

Fig. 6. Graphic representation of porosity on vertical section (Delage and Lefebvre, 1984)

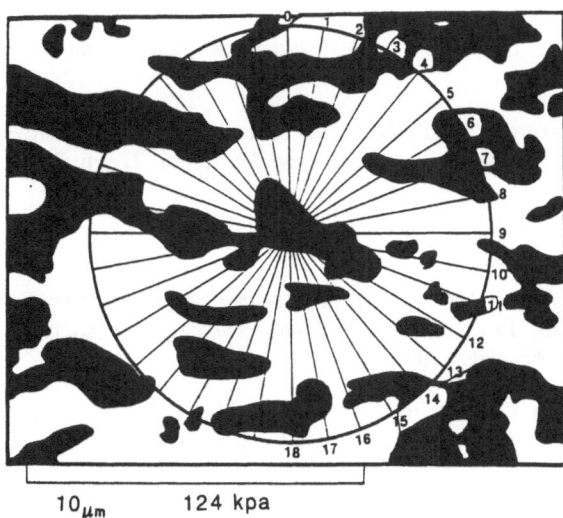

Fig. 7. Stereological analysis on section (Champlain clay, 124 kPa)

318

5 Conclusions

A unified framework for the modeling the fabric of granular media is presented. The presentation suggests that granular media can be characterized via two measures: a scalar measure (C) giving the mean value of a fabric descriptor and an associated tensor measure (ϕ_{ij}) giving its orientational distribution. The fabric descriptors can be either based on the solid phase (contact normal, contact vector, etc) or void phase (void vector). The choice of these is dependent on the physics of the deformation process.

In most current studies emphasis is placed on the solid phase parameters in the development of fabric tensors to describe the orientation. This form is not useful for experimental determination of the parameters of description.

In soil mechanics the void ratio is used as a constitutive variable. The development presented herein suggests that this description could be improved with the choice of an associated tensor describing the distribution of void ratio. This type of formulation is doubly advantageous in that the parameters of description can be obtained from microscopic observations. In the development of the void tensor the expression for the distribution function has been limited to a term involving a second order term. For cases involving high fluctuations in the data this may not be satisfactory. A statistical test is proposed to determine the adequacy of the choice of limiting terms up to a certain order.

The proposed fabric description and its experimental determination are useful for extending behavior models like CSSM. The approach allows a microscopic model to be linked to a continuum based model in a natural manner. With the advances in hardware and software of image analyzers the technique will prove to be efficient.

6 References

Abramowitz, M. and Stegun, I. A. (1965). *Handbook of Mathematical Functions*, Dover, New York.

Delage, P. and Lefebvre, G. (1984). "Study of the structure of a sensitive champlain clay and its evolution during consolidation." *Can. Geot. Jour.*, Vol. 21, pp. 21-35.

Kanatani, K. (1984). "Distribution of directional data and fabric tensors." *Int. Jour. of Eng. Sci.*, Vol. 22, No. 2, pp. 149-164.

Konishi, J. and Naruse, F. (1987). "A note on fabric in terms of voids." *Micromechanics of Granular Materials, Proc. of the U.S/Japan Seminar on the Micromechanics of Granular Materials*. Sendai-Zao, Japan, pp. 39-46.

Leckie, F. Å., and Onat, E. T. (1981). "Tensorial nature of damage measuring internal variables.", *Proc. IUTAM Symp. Physical Nonlinearities in Structures*, Senlis, Springer, pp. 140-155.

Lehmann, E. L. (1986). *Testing Statistical Hypotheses*, Wiley, NY.

Marle, C. M., (1981). "From the pore scale to the macroscopic scale; Equations governing multiphase fluid flow through porous media.", *Proc. Euromech.* 143, Delft, Verruijt, A., and Barends, F. B. J., pp. 57-61.

Muhunthan, B. (1993a). "Micromechanics of granular media," *Proc. 3rd Pan American Congress of Applied Mechanics* (PACAM III), Sao Paulo, Brazil, Jan 4-8.

Muhunthan, B. (1993b). "A new three-dimensional modeling technique for studying porous media," *Proc. Digital Image Processing: Techniques and applications in Civil Engineering*, ASCE, Hawaii, 228-235.

Muhunthan, B. (1991). "Micromechanics of Steady State, Collapse and Stress-Strain Modeling of Soils," Ph.D. dissertation, Purdue University, IN, 222 pp.

Muhunthan, B., and Alwail, T. (1992). "Use of image analysis in mathematical characterization of orientation data." *Scanning*, 14, 291-297.

Muhunthan, B., and Chameau, J. L. (1992). "Mathematical characterization of fabric and its use in mechanics of geomaterials." *Proc. 9th Eng. Mech. Conf.*, ASCE, College station, TX, 725-728.

Mullenger, G., (1978), "A condition for a continuum model of granular structure." *Proc. US-Japan Seminar on Mechanics of Granular Materials*, Tokyo, S.C. Cowin, Satake, M., (Eds.), pp. 282-290.

Rothenburg, L. (1980). "Micromechanics of idealized granular systems," Ph.D. Thesis, Carleton University, Ottawa, Canada, 332 pp.

Rothenburg, L. and Bathurst, R.J. (1990). "Observations on stress-force-fabric relationships in idealized granular materials." *Mech. Materials*, Vol. 9., pp. 65-80.

Satake, M., (1982). "Fabric tensor in granular materials." *IUTAM Symp. on Deformation and Failure of Granular Materials*, Delft, pp. 63-68.

Schofield, A. N., and Wroth, C.P. (1968). *Critical State Soil Mechanics*, Mc Graw- Hill, London.

STATISTICAL FRACTURE MECHANICS - BASIC CONCEPTS AND NUMERICAL REALIZATION

A. CHUDNOVSKY AND M. GORELIK
University of Illinois at Chicago, P.O. Box 4348, Chicago, IL 60680, USA

Abstract
The apparent randomness of brittle fracture is closely associated with the distribution of defects on various scales within a solid. The presence of microdefects is modeled by a random field of specific fracture energy γ following the framework of Statistical Fracture Mechanics (SFM). A brief summary of SFM is presented. SFM is the only model which explicitly incorporates the fractographic information, e.g. fractal characterization of fracture surfaces in the probabilistic description of brittle fracture. At the same time, the model has limitations in engineering applications, mainly due to its mathematical complexity. In this paper the Monte Carlo technique is employed to overcome these limitations. It allows one to combine the physical insight and modeling of the fracture mechanisms in SFM with the flexibility of the Monte Carlo method. Probability distributions of the fracture parameters such as a critical load, critical crack length, and fracture toughness are simulated and compared with experimental observations. Dependency of the conventional measure of fracture toughness on roughness of crack profiles, specimen and grain size, as well as load level is discussed. The ambiguity of the concept of fracture toughness in a probabilistic setting is addressed.
Keywords: Fracture Mechanics, Statistics, Monte Carlo Method, Simulation, Crack Propagation, Brittle, Scatter.

1 Introduction

Development of Fracture Mechanics in the last few decades provides much better understanding of micromechanisms leading to failure.Crack initiation mechanisms map, Chen et al. (1993), gives a simple classification of fracture scenarios. Two extreme cases at which the defects play quite different role could be mentioned.

The first case is a so-called cooperative fracture, where the intensity of damage formed as a response to the stress concentration at the tip of a propagating crack is much greater than the intensity of the pre-existing defect population. The crack propagation is then inseparable from the evolution of the damage accompanying the crack. Although the statistics of defect locations, sizes and orientations with respect to crack tip controls the crack growth, the behavior of the "crowd" of defects is highly reproducible. This strongly cooperative phenomenon is modeled, for example, by the Crack Layer Theory, Chudnovsky (1984), which operates with integral characterization of the damage array. The propagation of a crack coupled with an evolution of the damage zone is described by a set of deterministic equations (see, for example, Kadota and Chudnovsky (1992)).

Another extreme mode of failure is a perfectly brittle fracture, when a crack propagation is controlled by a pre-existing field of defects and does not cause noticeable changes to

D. Breysse (ed.), Probabilities and Materials, 321–338.

this field. In this case the random location and orientation of the individual microdefects result in an irregular, stochastic crack trajectory, scatter of the main fracture parameters and a scale effect.

There are difficulties in the direct application of the Continuum Mechanics to the modeling of brittle fracture since the crack path is usually quite irregular, in many cases "non-differentiable".

An approach to bridge the linear Fracture Mechanics with the statistical strength theories has been proposed by Chudnovsky (1973). It has gradually evolved into the formulation of Statistical Fracture Mechanics (SFM), Chudnovsky and Kunin (1987, 1992). The mathematical complexity of SFM in its present form significantly limits the engineering application of the theory. An employment of the Monte Carlo simulation technique presented in this paper opens the door for the application of SFM to a wide variety of practical problems. The new method is examined by comparison of computer simulation of the statistics of crack instability with the experimental results reported by Mull et al. (1987).

2 On the source of irregularities of crack trajectories

It is commonly believed that there is a certain similarity between the turbulence in a fluid flow and the fracture in the solid state since the characteristic features of critical phenomenon such as hierarchy of interacting defects, stochastization and a distinct scale effect are present in both processes.

An irregular stochastic appearance of the fracture surfaces is well documented and widely discussed phenomenon. One may suggest two potential sources of such appearance:

1) intrinsic instability of crack growth process, and/or
2) an impact of a randomly distributed array of the pre-existing material inhomogeneities (defects).

Here we examine the likelihood of the above sources.

Stability of a crack trajectory with respect to smooth perturbations for a two-dimensional problem has been analyzed by Cotterell and Rice (1978). However, in many cases irregularities of the fracture surfaces could be visualized as a sequence of the sharp kinks. According to our numerical analysis, crack trajectories are also stable with respect to a kink-type perturbation, Gorelik (1993).

Thus, within the assumptions of linear Fracture Mechanics, erraticism of brittle fracture surfaces can not be attributed to the instability of the crack trajectory . Therefore it can be concluded that the main source of stochasticity in brittle fracture is a random distribution of the material inhomogeneities (microdefects).

The presence of the microdefects within the Continuum Mechanics context can be modeled by a random field of the specific fracture energy (γ-field). It does not effect much the thermo-elastic and other volume average properties of material, but may strongly influence the strength and toughness.

3 Statistical Fracture Mechanics

SFM is based on a few natural assumption regarding the process of brittle fracture Chudnovsky and Kunin (1992):

(a) the crack path is random, i.e. the crack randomly selects a path from

a set of virtual paths;
(b) crack advance along a particular path consists of a sequence of local failures in front of the crack tip, controlled by the Griffith criterion

$G_1 = 2\gamma$,

(c) the local failures are random events due to the random field $\gamma(x)$.

It is important to note that the ERR $G_1[x_k, \omega]$ determined from elastostatic solution should be employed in the condition (b). Indeed, it is known that the ERR for dynamic crack propagation $G_1^{dyn} \approx G_1^{stat}\left(1 - \dfrac{v}{c_R}\right)$ with $v/c_R < 1$, where v and c_R stand for the crack growth rate and Rayleigh wave speed, respectively. Thus the condition $G_1^{stat} \geq 2\gamma$ suffices the crack extension, at least quasi-statically.

One of the main building blocks of SFM is a Crack Propagator (CP) $P(\underline{x}, \underline{X})$. Let us consider a loaded solid containing a crack ω with its tip at $\underline{x}$ (Fig. 1).

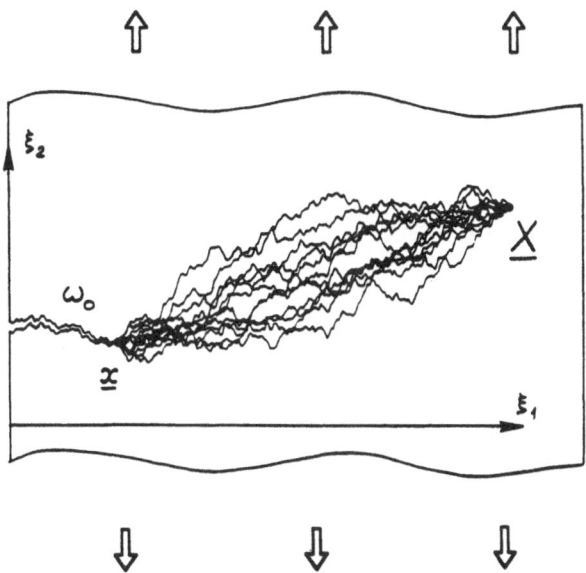

Fig. 1. A sample set of virtual crack paths leading from point $\underline{x}$ to point $\underline{X}$.

First we define a conditional CP $P(\underline{x}, \underline{X}| \omega)$ as the probability that the crack extends from $\underline{x}$ through another point $\underline{X}$ along a path ω, i.e. the condition ERR $\geq 2\gamma$ is met at every point of the crack trajectory ω. Thus we are interested in the probability of the conjunction of the events:

$$P(\underline{x}, \underline{X} \mid \omega_o) = \text{Prob}\left\{\bigcap_{k=1}^{n}(G_1[\underline{x}_k \mid \omega_o] > 2\gamma[\underline{x}_k \mid \omega_o])\right\} \qquad (1)$$

where $\underline{x}_k$ is a point from ω_o, and $G_1[\underline{x}_k \mid \omega_o]$ is the ERR of the crack with its tip at $\underline{x}_k$, and n is a number of subintervals between $\underline{x}$ and $\underline{X}$. According to Chudnovsky (1973), the expression (1) can be evaluated as:

$$P(\underline{x}, \underline{X} \mid \omega_o) = \exp\left\{-\int_{x_1}^{X_1} \text{Prob}\{2\gamma(\xi_1) \geq G_1(\xi_1 \mid \omega_o)\}\frac{d\xi_1}{r}\right\} \qquad (2)$$

It is assumed that γ is a statistically homogeneous random field with a correlation distance r much smaller than the crack size.

Assuming that a crack "searches" for the easiest direction at each point of its trajectory, one concludes that the distribution of γ along the crack trajectory is sampled from the minimal values of the γ-field (see paper by Kunin in this volume for details). Thus we employ Weibull distribution (one of the most popular distributions of extremes) for the values of γ at every point of the crack trajectory:

$$F(\gamma) = \begin{cases} 1 - \exp\left[-\left(\Gamma\left(1+\frac{1}{\alpha}\right)\frac{\gamma - \gamma_{min}}{\langle\gamma\rangle - \gamma_{min}}\right)^{\alpha}\right], & \gamma > \gamma_{min} \\[2mm] 0, & \gamma < \gamma_{min} \end{cases} \qquad (3)$$

where $\alpha > 0$, $\langle\gamma\rangle$ and γ_{min} are the shape factor, mean and minimal values of the γ-field along the crack path, respectively. Our parametrization of the Weibull distribution slightly departs from conventional one.

If one assumes that the crack trajectories are mutually exclusive (single path fracture), then the CP $P_\omega(\underline{x}, \underline{X})$ can be written using formula of total probability as:

$$P_\omega(\underline{x}, \underline{X}) = \sum_{\Omega} P(\underline{x}, \underline{X} \mid \omega_o) \, \text{Prob}(\omega \mid \omega_o) \qquad (4)$$

where $\text{Prob}\{\omega \mid \omega_o\}$ stands for the probability that the crack "chooses" a path ω among all virtual paths extending from $\underline{x}$ to $\underline{X}$, and Ω is a set of all virtual paths ω.

In a continuum based model, the set Ω is uncountable, and so the sum in Eq. (4) becomes an integral (Chudnovsky (1973), Chudnovsky and Kunin (1987))

$$P_\omega(\underline{x}, \underline{X}) = \int_{\Omega_{\underline{x}, \underline{X}}} P(\underline{x}, \underline{X} \mid \omega) \, \mu(d\omega) \qquad (5a)$$

Similar consideration has been applied for multiple crack fracture, i.e. for the case when crack trajectories are formed simultaneously. The assumption of the mutually independent crack paths then results in the following formulation of CP

$$P_\omega(\underline{x}, \underline{X}) = \exp\left\{ -\int_{\Omega_{\underline{x}, \underline{X}}} P(\underline{x}, \underline{X} \mid \omega) \, \mu(d\omega) \right\} \qquad (5b)$$

All fracture mechanics parameters can be expressed in terms of CP.
Representation (5a) of the CP leads to three major tasks of SFM:

1) selection of a realistic set Ω of virtual crack trajectories together with

a probabilistic measure $d\mu(\omega)$ on it,

2) determination of the conditional crack propagator $P(\underline{x},\underline{X}\mid \omega)$, and
3) evaluation of the functional integral in (5a) and (5b).

Then the probabilities of crack initiation, crack arrest, distributions of critical load, toughness parameters, time to failure etc. can all be expressed in terms of the CP.
The main difficulty in the evaluation of the crack propagator, besides a proper choice of Ω, is a procedure of functional integration, Chudnovsky (1973). There are analytical methods which allow one to reduce functional integration to solving a partial differential equation of diffusion type or even ordinary differential equation for some cases, Gelfand and Jaglom (1956), Chudnovsky and Kunin (1987). However, this simplification is possible only for a certain type of virtual crack trajectories Ω. Namely, those should belong to the class of Brownian trajectories (Wiener processes). Such trajectories do not always adequately model the fracture paths. For example, Fig. 2 shows a superposition of 27 fatigue crack trajectories from 27 identical single edge notch specimens made of short fiber reinforced polyester, Mull et al. (1987).
Each specimen has been fatigued up to failure under identical conditions. Thus the observed trajectories can be considered as a sample set from Ω. The trajectories of Fig. 2 are apparently smoother than the Brownian paths shown in Fig. 3 for comparison.

In a case when the set Ω can not be approximated by the Wiener process (Brownian trajectories), an analytical evaluation of functional integral becomes a very difficult problem. Furthermore, even if the above approximation is acceptable, the solution has to be performed for every particular physical problem, since there is no general solution which could be "tailored" for the prescribed condition.

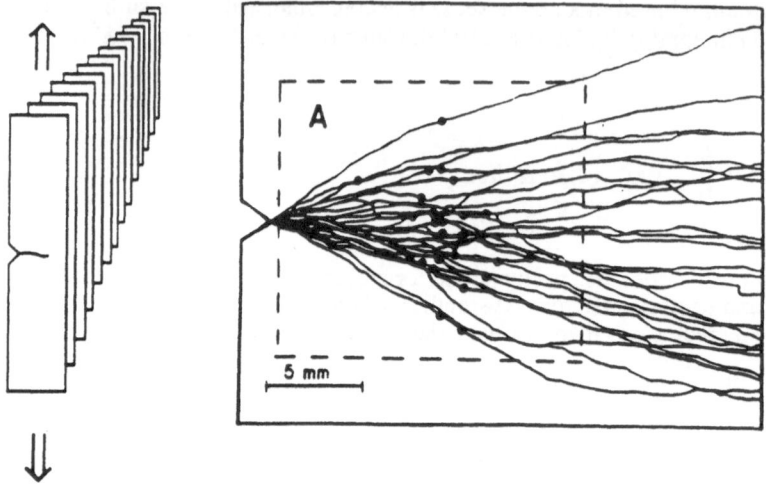

Fig. 2. An ensemble of macroscopically identical specimens cracked under identical fatigue load and the superposition of the profilograms of the crack traced from each individual specimen (only one trajectory takes place in each specimen).

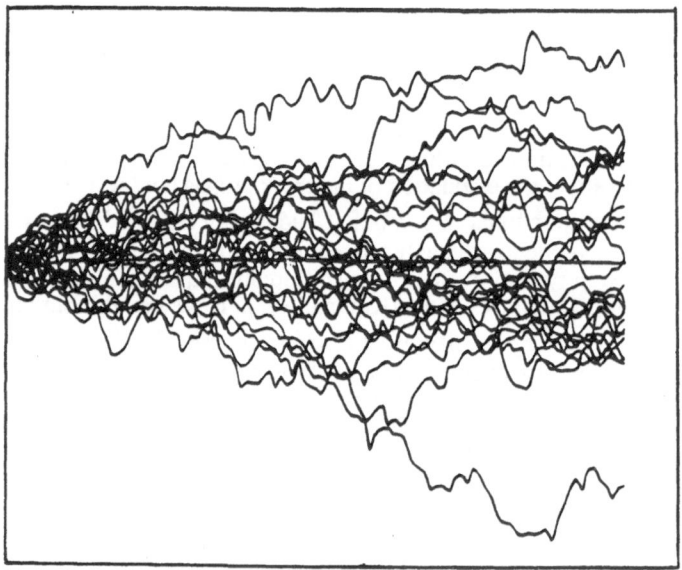

Fig. 3. A sample set of the realizations of Wiener process

In SFM statistical description of the γ-field includes: 1) the statistical parametrization of the virtual crack trajectories, and 2) an evaluation of the parameters α, $<\gamma>$ of pointwise Weibull distribution and the correlation distance r along the virtual crack trajectories (minimal value of γ is taken to be zero $\gamma_{min}=0$).

The problem of adequate modeling of actual crack trajectories has been addressed by Kunin and Gorelik (1991). Because of the apparent discrepancy between the observed fracture paths and Brownian trajectories, it was suggested to use partially smoothed Wiener paths. Partial smoothing can be achieved by means of fractional integration of Wiener process. Let w (x) be a conventional Wiener process. The fractional integral of w (x) to the power λ is defined as (Lavoie et al (1976))

$$w_{\lambda}(x) = \frac{1}{\Gamma(\lambda)} \int_{o}^{x} w\ (\xi)\ (x - \xi)^{\lambda-1}\ d\xi \tag{6}$$

with $0\leq \lambda \leq 1$, where $\lambda=1$ corresponds to the conventional Cauchy integral, and $\lambda<1$ determines different degrees of smoothing. An example of realizations of $w_{\lambda}(x)$ for various λ is shown in Fig. 4. Parameter λ is simply related to the fractal dimension d of $w_{\lambda}(x)$: d = 1.5 - λ (for $\lambda \leq 0.5$).

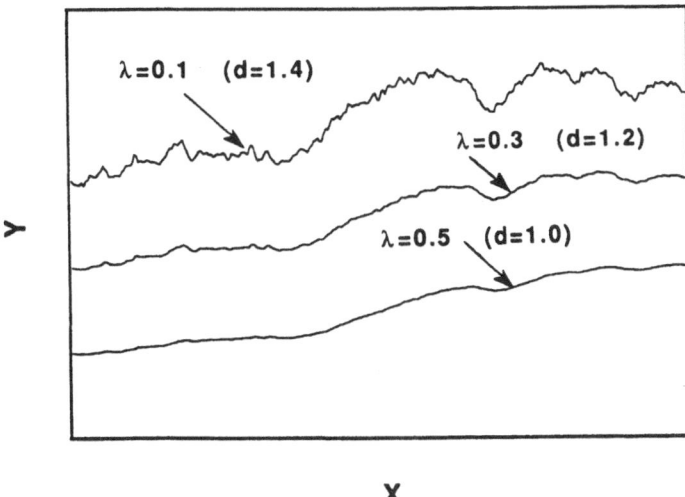

Fig. 4. Three realization of the fractional Wiener process with λ = 0.1, 0.3, and 0.5. Corresponding fractal dimension is indicated in paranthesis.

It can be shown that the variance of $w_\lambda(x)$ is given by

$$\text{var}[w_\lambda(x)] = D_\lambda \, x^{2\lambda} \tag{7}$$

where

$$D_\lambda = \frac{D}{(2\lambda+1)\, \Gamma^2(\lambda+1)} \tag{8}$$

and D is a diffusion coefficient of the "underlying" Wiener process $w(x)$ ($\text{var}[w(x)]=Dx$).

The relationships (7) and (8) allow one to reconstruct the parameter l corresponding to the actual fracture process by plotting the variance of the experimental trajectories vs. x in the log-log scale. For the set of trajectories depicted in Fig. 2 the value of λ was found to be 0.4. A computer simulation of 27 realizations of $w_\lambda(x)$ with $\lambda=0.4$ (d=1.1) is shown in Fig. 5. The simulated sample set of $w_\lambda(x)$ closely resembles the set of experimental trajectories of Fig. 2.

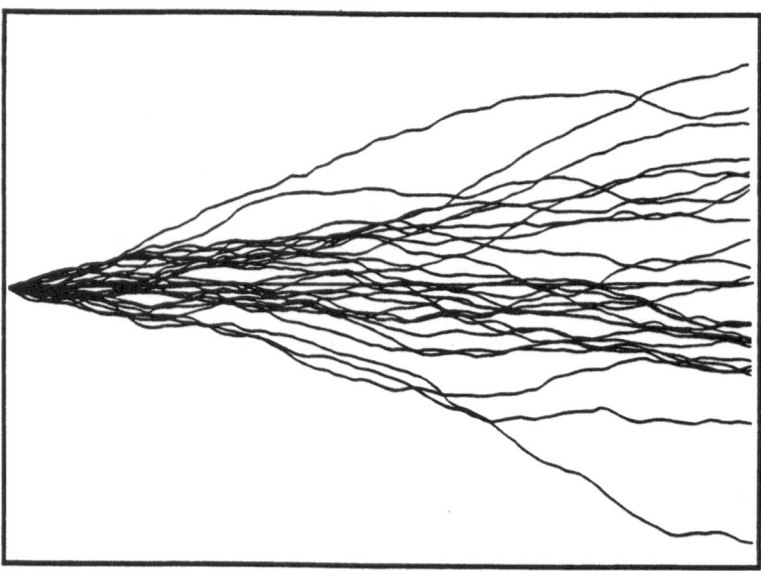

Fig. 5. A sample set of the realizations $w_\lambda(x)$ with $\lambda=0.4$, resulting from a fractional integration of Wiener process.

4   Monte Carlo technique

In this paper we propose to employ the Monte Carlo technique to overcome the difficulties the SFM has faced. This allows one to combine the physical insight and understanding of the fracture mechanisms reflected by the SFM with the flexibility of the Monte Carlo method. Following the spirit of the Monte Carlo technique functional integration as a mean of the averaging over Ω is substituted by a computer simulation of a set of N virtual crack paths with the following estimation of the frequency of events when crack extends from point $\underline{x}$ to $\underline{X}$ along these trajectories. As number N of numerical "experiments" increases, "experimental" frequency in principle converges to the value of the propagator $P(\underline{x}, \underline{X})$.

Three major tasks of SFM then can be reformulated for the Monte Carlo method:

1) computer simulations of the adequate crack trajectories and generation of the realizations of random γ-field values along them;

2) evaluation of elastic ERR at every point of the simulated crack trajectory and its comparison with the γ-field realization along the trajectory;

3) replacement of functional integration in (5) by statistical averaging of the output parameters, e.g. critical crack length, fracture toughness, etc.

To illustrate the new approach we have revisited the statistical analysis of the experiment reported by Mull et al. (1987).

The objective of this effort is to simulate the probability density of the crack tip position at the instance of crack instability. Description given below shows how three main tasks of the method were approached (see Gorelik (1993) for details).

Task 1

Realizations of $w_\lambda(x)$ (partially smoothened Brownian trajectory (Wiener process)) (see Fig. 5) are used to simulate the crack trajectories. Local values of γ-field along trajectories are obtained by means of the Weibull random numbers generator. Parameters $<\gamma>$ and α of the γ-field are found by an optimization of the output statistics. The correlation distance r is evaluated from the fractographic analysis of the fracture surface.

Task 2

Evaluation of the ERR for an irregular crack trajectory is a difficult problem. In general, it can be solved only numerically, e.g. using finite element method or boundary collocation method. However, solution of the elastic problem for each crack trajectory and variable crack tip position along the trajectory requires considerable amount of computer time. This approach becomes impractical for a large number (hundreds or, even thousands) of crack trajectories. Therefore, a reasonable approximation to the exact solution for the ERR should be adopted.

To simplify the problem we approximate the ERR at every point of an irregular crack trajectory by that of a straight cut from the notch tip to the current point with the smooth kink directed along the tangent to the crack trajectory at the same point. This approximation is well justified for the trajectories under consideration (Fig. 5) based on the solutions reported by Cotterell and Rice (1978), Murakami (1987), Rubinstein (1989). An example of the ERR evaluated for a simulated trajectory (solid line), and the realization of $2\gamma(x)$ along the same trajectory are shown in Fig. 6.

330

Dashed curve represents the ERR along the same trajectory with no short wave perturbations. One can observe that the fluctuations of the ERR due to the short wave perturbations have a much smaller effect on the critical crack tip position than the scatter

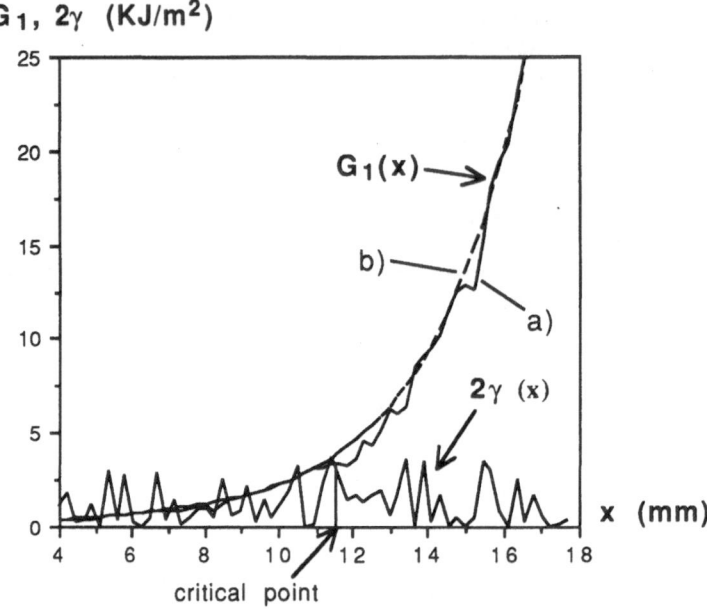

Fig. 6. ERR vs. crack depth along a simulated crack trajectory a) with and b) without the random kinks, superimposed with the realization of the γ-field along the same trajectory.

of the γ-field (Fig. 6). Let x^c be the abscissa of the first point on the crack trajectory after which $G_1(x) > 2\gamma$ for all $x > x^c$. The distance l^c from the notch to the crack tip at $x = x^c$ is defined as the critical crack length, with x^c being the critical depth.

Task 3

For each of N simulated crack trajectories the value of the critical crack length l^c is found according to the criterion stated above. Set of the crack instability points (x^c_i, y^c_i), $i=1\div N$ represents a statistical output of the model in this case. Both one-dimensional (for x^c_i and y^c_i separately) and two-dimensional probability densities can be readily obtained.

Based on comparison of the first two moments of the experimental and simulated critical crack depths x^c_i, $i=1\div27$ the optimal parameters of Weibull distribution for γ-values were found $\alpha= 0.65$, $<\gamma>= 1.0$ KJ/m^2. Based on these parameters the simulation was performed for N=3000 trajectories, and the output (x^c_i, y^c_i), $i=1\div3000$ was used to build a two-dimensional distribution of critical crack tip locations. The contour lines of the probability density of critical crack tip positions are shown in Fig. 7.

A sample set of 27 critical ERR values G_{1c} has been generated using the optimal parameters $\alpha=0.65$ and $<\gamma>=1.0$ KJ/m^2. The mean value $<G_{1c}>=6.8$ KJ/m^2 and the standard deviation $\sigma(G_{1c})=3.5$ KJ/m^2 are in a good agreement with the estimate reported by Mull et al. (1987) ($<G_{1c}>^{exp}=6.9$ KJ/m^2, $\sigma(G_{1c})^{exp}=3.9$ KJ/m^2).

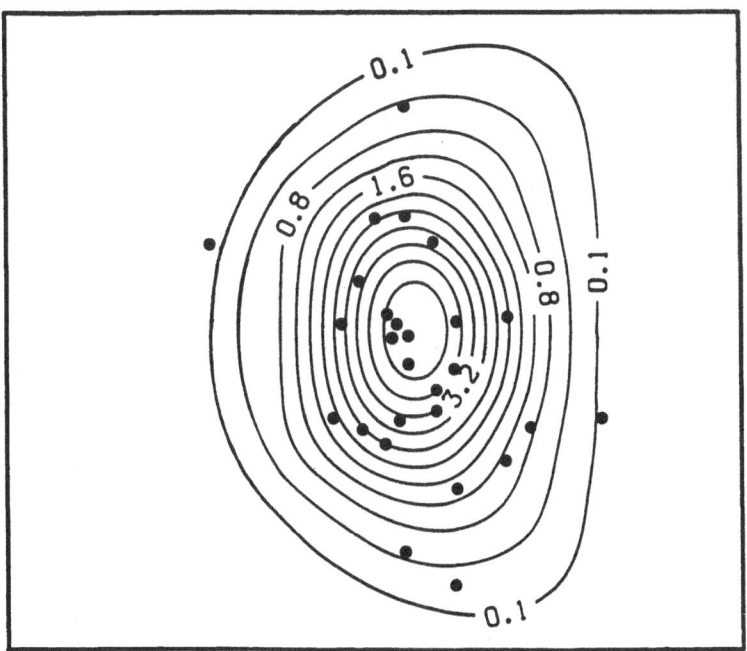

Fig. 7. Domain A from Fig. 3. Contour lines represent the equal levels of probability density for the crack tip position at the instance of crack instability (Monte Carlo simulation). Dots indicate the experimental points.

5 Illustrative examples

The output of the simulation procedure, described above, is a sample set. Thus, probability distribution of any critical fracture parameter can be readily reconstructed. Therefore, the model may be utilized to study the features of brittle fracture associated with chance. Among those are the scale effect, the significant scatter of fracture toughness, critical ERR dependency on the loading conditions, difference between the ERR values at the crack arrest and crack initiation. Modeling of some of these phenomena is presented in the examples below.

5.1 Fracture toughness dependency on the roughness of crack trajectories

The roughness of the crack trajectory (short-wave perturbations) can be correlated with its fractal dimension d. To investigate the fracture toughness dependency on the fractal dimension statistical trials were performed for the sets of the virtual crack trajectories having various roughness, d=1.5, 1.4, 1.3, 1.2, 1.1, and 1.0 (150 crack trajectories have been simulated in each set). These values of d correspond to the crack paths ranging from the Brownian trajectories (d=1.5) to completely smooth (d=1.0) profiles. Some examples of simulated trajectories corresponding to various values of d are shown above: d=1.5 in Fig. 3, d=1.0, 1.2, and 1.4 in Fig. 4, d=1.1 in Fig. 5.

With increase of fractal dimension similar trends are observed for the mean values of the critical ERR G_{1c} and the critical crack depth x_c (Figs. 8, 9). Both increase non-linearly with d. Value of $<G_{1c}>$ varies by as much as 70%. In the same time, normalized standard deviations of G_{1c} and x_c are practically not effected by the change of d.

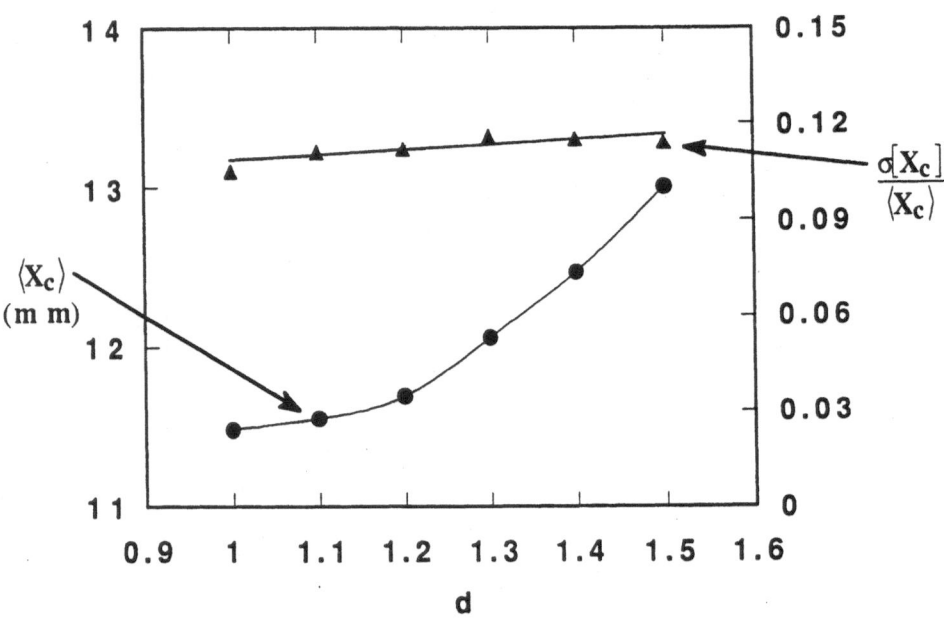

Fig. 8. Effect of the fractal dimension d on the statistics of the critical crack depth.

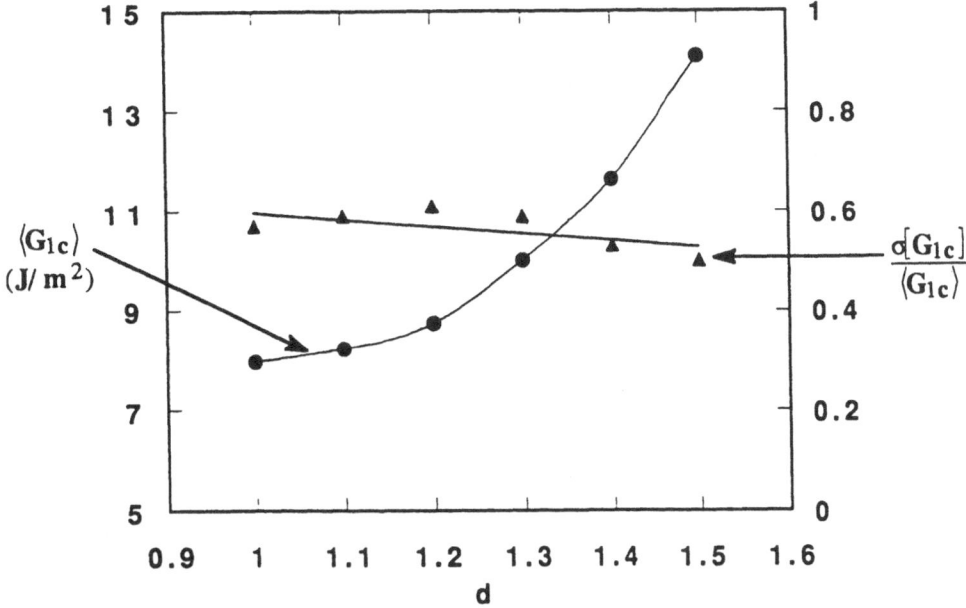

Fig. 9. Effect of the fractal dimension d on the statistics of the fracture toughness G_{1c}.

5.2 Effect of the load level on the critical ERR G_{1c} and crack depth x_c values

The computational algorithm outlined in the Section 4 was also used to study the load level effect on the critical ERR. A sample set of G_{1c} values was generated in 100 numerical experiments. The variations of the mean value $<G_{1c}>$ and standard deviation $\sigma[G_{1c}]$ are shown in Fig. 10. Both increase with the increase of applied load σ^∞. Constant level of the mean value $2<\gamma>$ of the specific fracture energy, indicated by a dashed line, corresponds to the Griffith model.

Effect of σ^∞ on the probability distribution of G_{1c} is even more pictorial. The probability densities for G_{1c} resulting from computer simulation for 4 MPa and 15 MPa load levels are shown in Fig. 11. Probability density of the surface energy γ, which is a statistical input of the model, is shown on the same figure for comparison.

Besides the mean value and the variance, the shape of the probability distribution of G_{1c} significantly changes with the variation of applied load. While the probability density of G_{1c} is nearly symmetric for σ^∞=4 MPa, it has a well pronounced positive skewness for σ^∞=15 MPa (Fig. 11).

334

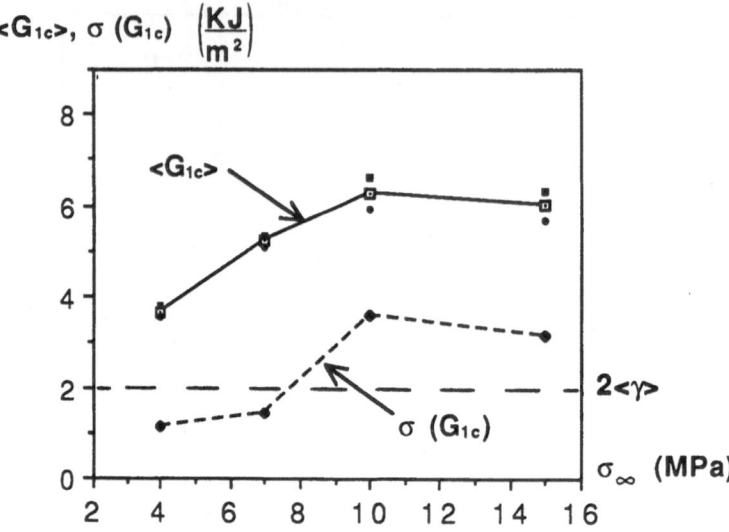

Fig. 10. The effect of the remote load on the mean value $<G_{1c}>$ and the standard deviation $\sigma(G_{1c})$.

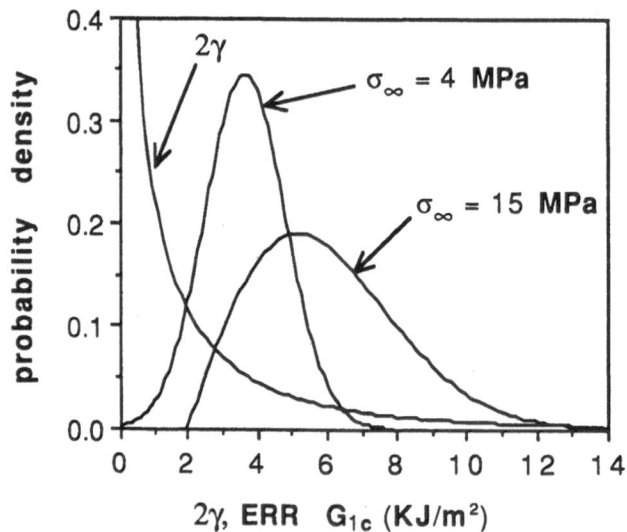

Fig. 11. Probability density of G_{1c} values based on 100 statistical trials of crack propagation up to the point of instability for two load levels σ^{∞}.

5.3   Scale effect

One of the typical features of brittle fracture, associated with intrinsic stochasticity of the process, is a scale effect. It is manifested in dependency of fracture toughness parameters (K_{1c}, G_{1c}) on the macroscopic size and shape of the specimen.

Linear elasticity is commonly accepted for modeling of brittle fracture. Conventional elastic models do not have an internal scale, and, therefore, can not directly account for material heterogeneities on the microscale. In SFM those are accounted for by the random field of specific fracture energy γ with the correlation distance r_o. Thus, in the frame of SFM, a change in the correlation distance r_o of the γ-field in the specimen with fixed dimensions is equivalent to the corresponding change in size of the specimen under fixed r_o. Specifically, n-times increase in r_o is equivalent to n-times decrease in specimen's dimensions. It should be noted that change of r_o may be interpreted as a replacement of the specimen's material by another one with the same elastic properties, but different microstructure, while the specimen size and shape are fixed.

The scale effect in our model was studied by changing the relative correlation distance r_o/W. 100 statistical trials were performed for $r_o/W = 0.003$ and $r_o/W = 0.009$. For each value of r_o/W 100 crack trajectories were simulated, and local values of the surface energy γ and the ERR G_1 were generated for every trajectory. Comparing G_1 with 2γ pointwise along a particular path, a set of critical crack depths and corresponding critical ERR values were obtained. Statistical output in the form of probability densities (approximated by Weibull distribution) of critical crack depth x_c and critical ERR G_{1c} is shown in Fig. 12 and 13, respectively. Distribution of x_c for smaller r_o is shifted to the left and has a larger relative scatter. Distribution of G_{1c} for smaller r_o is also shifted to the left, but difference in relative scatter is less pronounced.

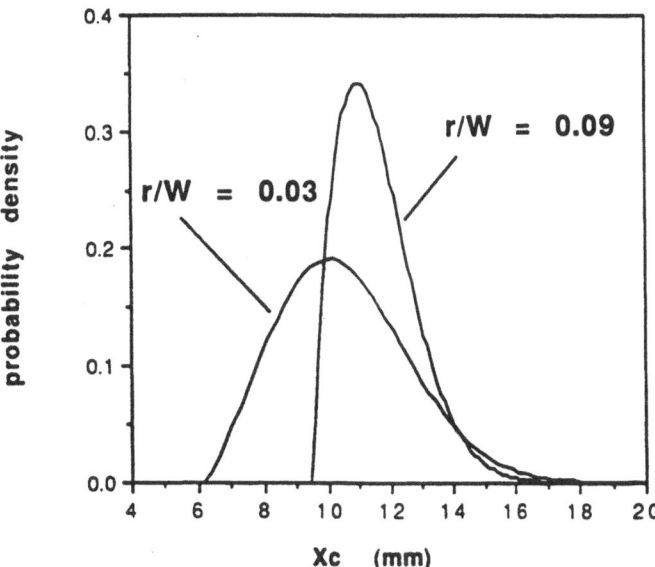

Fig. 12. Probability density of the critical crack depth x_c for two correlation distances r.

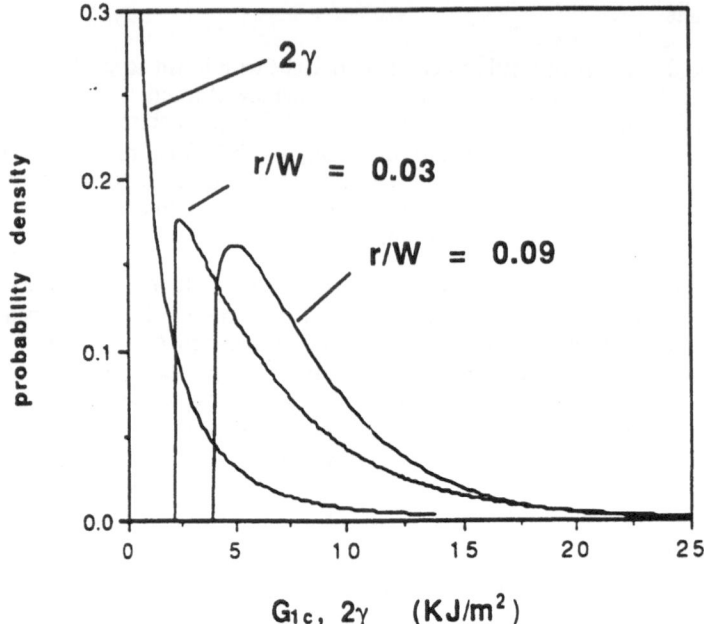

Fig. 13. The critical ERR G_{1c} distributions for two r. Probability density of specific fracture energy 2γ is shown for comparison.

5.4 An alternative definition of the critical ERR

An alternative definition of the critical ERR may be proposed based on the experimental procedure for evaluation of the critical load. In this section critical ERR G_{1c}^{*} is defined as an ERR corresponding to crack initiation after the last arrest in a load-controlled ramp test. Its probability density obtained numerically is shown in Fig. 14a.

In the frame of deterministic description of brittle fracture, i.e. constant value of γ, the definition of critical ERR G_{1c}, used in the section 4, and the present one for G_{1c}^{*} are identical. However, the situation is different when the specific fracture energy is a random field. This is illustrated in Fig. 14, where simulated probability densities of G_{1c} (a) and G_{1c}^{*} (b) are compared. Both the mean value $<G_{1c}^{*}> = 10.8$ KJ/m^2 and the standard deviation $\sigma[G_{1c}^{*}] = 7.2$ KJ/m^2 of G_{1c}^{*} are greater than the corresponding values for G_{1c} ($<G_{1c}> = 6.3$ KJ/m^2, $\sigma[G_{1c}] = 3.8$ KJ/m^2).

Thus, the service conditions of the structural element should be included in the formulation of a measure of toughness.

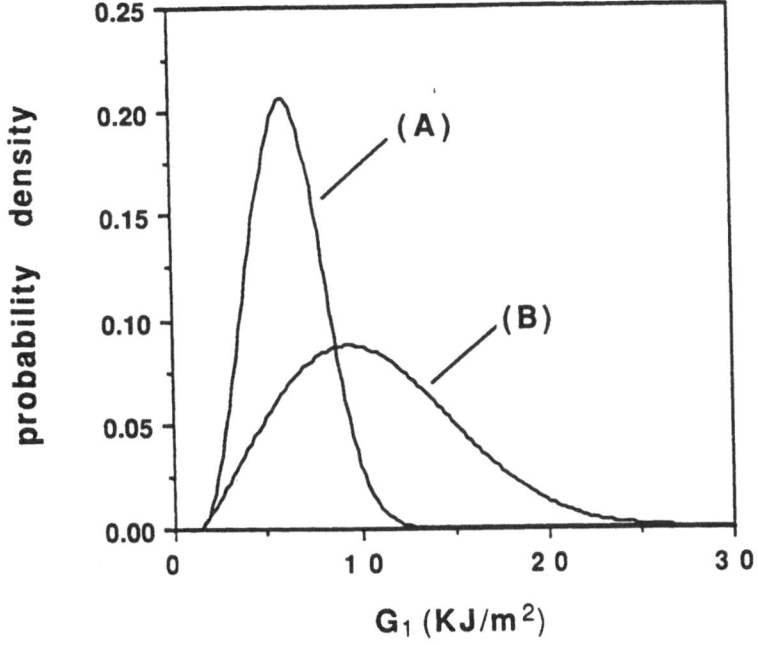

Fig. 14. Probability densities (simulation) of the critical ERR corresponding to: (A) last crack arrest in fatigue test, and (B) last crack arrest in ramp test.

6   Conclusions

• Brittle fracture is one of the best illustrations for the role of a chance in physical phenomena. The main source of stochasticity in fracture is the population of the microdefects resulting from particular technological process.

• Statistical Fracture Mechanics bridges the conventional Fracture Mechanics with the weakest link theories. The elements of Fracture Mechanics in SFM reflect the fracture localization phenomena. The concepts of the Weakest Link are employed in SFM to account for the mechanism controlled by chance.

• The Monte Carlo method significantly broadens the application of SFM. It provides a new tool for the realistic evaluation of material toughness, structural reliability, critical load, etc. The method, can be applied to model a wide variety of Fracture problems.

• Technique of statistical modeling, outlined in this paper, allows to model efficiently the process of brittle fracture. Variety of fracture parameters can be simulated and their statistics evaluated. Computer-simulated experiment allows to study different effects typical for the brittle fracture, such as scale effect, load level effect, etc. The method can be applied to a broad range of loading conditions, specimen geometries, and types of fracture surfaces.

• Two observations can be made based on the results of statistical modeling:

1) although the Griffith's criterion has been employed point-wise in the model, conventional fracture toughness parameter G_{1c} can not be identified with the mean

338

value of the specific fracture energy $2<\gamma>$. It is related to the fact that the fracture events are controlled by the peaks of the γ-field.

2) the dependency of G_{1c} on the applied load and the specimen shape and size suggests that the critical value of ERR G_{1c} is not a material property. The true material parameters are the mean value of γ and its standard deviation.

7 References

Chen,T.J., Bosnyak, C.P. and Chudnovsky, A. (1993) A fatigue crack initiation mechanisms map for polycarbonate. **J. of Appl. Pol. Sc.**, 49, pp. 1909-1919.

Chudnovsky, A. (1984) Crack Layer Theory. NASA Report 174634.

Chudnovsky, A. (1973) On the fracture of solids, in **Studies on Elasticity and Plasticity** (ed. L. Kachanov), Leningrad Univ. Press, Leningrad, USSR, Vol. 9, pp. 3-41.

Chudnovsky, A. and Kunin, B. (1987) A probabilistic model of brittle crack formation. **J. of Appl. Phys.**, Vol. 62, 10, p. 4124.

Chudnovsky, A. and Kunin, B. (1992) Statistical Fracture Mechanics. in **Microscopic Simulation of Complex Hydrodynamic Phenomena** (eds. M. Mareschal and B.L. Holian), Plenum Press, New York, Series B: Physics Vol. 292, pp. 345-360.

Cotterell, B. and Rice, J.R. (1978) Slightly curved or kinked cracks, Report MRL E-110, Brown University.

Gelfand, I.M. and Jaglom, A.M. (1956) Integration in functional spaces and its applications in quantum physics. **Russ. Math. Surv.**, Vol. 11, 48, pp. 48-69.

Gorelik, M. (1993) **Novel Application of the Monte Carlo Method in Statistical Fracture Mechanics,** Ph.D. Thesis, University of Illinois at Chicago.

Kadota, K. and Chudnovsky, A. (1992) Constitutive equations of crack layer growth. **J. of Pol. Eng. Sc.**, 32, pp. 1097-1104.

Kunin, B. and Gorelik, M. (1991) On representation of fracture profiles by fractional integrals of a Wiener process. **J. of Appl. Phys.**, 70 (12), pp. 7651-53.

Lavoie, J.L., Osler, T.J. and Tremblay, R. (1976) Fractional derivatives and special functions. SIAM Review, 18, pp. 240-268.

Mull, M.A., Chudnovsky, A. and Moet, A. (1987) A probabilistic approach to the fracture toughness of composites. **Phil. Mag. A**, Vol. 56, 3, pp. 419-43.

Rubinstein, A. A. (1991) Mechanics of the crack path formation. **Int. J. of Fract.**, Vol. 47, pp. 291-305.

Rubinstein A.A. (1989) Crack path effect on material toughness. **ASME Applied Mechanics Division,** Paper No. 89-WA/APM-43.

Stress Intensity Factors Handbook (ed. Y. Murakami), Pergamon Press (1987), Vol. 1, pp. 118-127.

A STOCHASTIC MODEL FOR CONTROLLED DISCONTINUOUS CRACK GROWTH IN BRITTLE MATERIALS

BORIS I. KUNIN
University of Alabama in Huntsville, Huntsville, USA

Abstract
Statistical Fracture Mechanics aims at describing fracture of brittle solids with complex microstructure and pronounced combination of scatter and 'scale effect' for conventional fracture parameters such as fracture toughness, critical energy release rate, etc. It achieved significant progress in prediction of distributions of critical loads, crack lengths, displacements, etc., for a stressed structural element under a variety of conditions. The present paper addresses another important question: predicting life-time scatter for a stressed brittle structural element. We model an observed mode of slow crack growth in brittle materials, namely a Markovian stochastic pattern of a microscopic random jump, followed by a random waiting time, followed by a random jump, and so on. The waiting times are related, on physical grounds, to random energy barriers at the arrest points, whereas random magnitudes of the jumps are treated within the existing framework of Crack Diffusion Theory. The transition probability density for the resulting random process of crack growth is shown to satisfy a differential equation which admits a solution in quadratures. A simple illustrative example is considered.
Keywords: Brittle Fracture, Stochastic Crack Growth, Life Time Scatter

1 Introduction

Statistical Fracture Mechanics (SFM) aims at describing fracture of brittle solids with complex microstructure and a pronounced combination of scatter and 'scale effect' for such fracture parameters as fracture toughness K_{IC}, critical energy release rate G_{IC}, etc. The approach originated in Chudnovsky (1973) emphasizes the microheterogeneity of the material was brought up as the source of two macroscopically observable phenomena: 1) scatter of critical (causing failure) loads, crack lengths, etc., in identical experiments and 2)randomness and erraticism of crack trajectories[1]. Fluctuating specific fracture energy has been chosen as a manifestation of the microstructure.

---

[1]The significance of crack trajectory erraticism is evident, for example, from the strong effect of kinks on the energy release rate, Wu (1978).

D. Breysse (ed.), Probabilities and Materials, 339–360.

It would be tempting to introduce specific fracture energy as a random function $\gamma(\underline{x},\underline{n})$ of a point $\underline{x}$ and the direction $\underline{n}$ normal to a cut. Together with a crack growth criterion, $\gamma(\underline{x},\underline{n})$ would determine the stochastic geometry of crack paths as well as the probability of crack formation along a particular path. Probabilities of various fracture events, e.g. that of ultimate failure, can be then expressed as averages over all paths the crack may possibly follow. A serious weakness of this program stems from the need to determine experimentally parameters of the random fracture energy $\gamma(\underline{x},\underline{n})$, which is not directly measurable. Indeed, for each point $\underline{x}$, $\gamma(\underline{x},\underline{n})$ is an even function on a unit sphere $|\underline{n}| = 1$ whose crudest non-constant approximation is quadratic: $\gamma(\underline{x},\underline{n}) = \underline{n} \cdot A \cdot \underline{n}$, where the symmetric random matrix $A = A(\underline{x})$ contains three different random entrees in the two-dimensional case and six in the 3D case. Even for a macroscopically homogeneous solid (A independent of $\underline{x}$), the first two moments and covariances for these random entrees amount to 12 empirical constants in the 2D case (42 in the 3D case) - an unattractive situation.

In Chudnovsky (1973), this difficulty has been overcome through an entirely different characterization of $\gamma(\underline{x},\underline{n})$. Firstly, instead of treating the shape of an actual crack path as a consequence of the function $\gamma(\underline{x},\underline{n})$, it was proposed to use crack trajectories (experimentally observable!) as part of the characterization of $\gamma(\underline{x},\underline{n})$. Secondly, it was proposed that only minimal values of $\gamma(\underline{x},\underline{n})$ with respect to $\underline{n}$ play a role, since the crack advances locally in a direction of least resistance (see Chudnovsky and Kunin (1987) for the discussion in terms of the energy barrier). Thus, the random function $\gamma(\underline{x},\underline{n})$ can be characterized by a random field $\gamma(\underline{x})$ whose pointwise distribution belongs to one of the three known families of distributions of minimal values, namely, the Weibull distribution (*ibid.*).[2] Further along the road, a simplifying assumption that the random field $\gamma(\underline{x})$ has a small correlation distance allows evaluation of crack formation probabilities in a closed form.

In Chudnovsky and Kunin (1987, 1989), Kunin and Gorelik (1991), and Kunin (1992), the ideas put forth in Chudnovsky (1973) were carried out in some detail in the form of the Crack Diffusion Theory (CDT). The concept of Crack Propagator (the probability of crack formation between two arbitrary points) emerged as a convenient building block entering formulations of all of the probability distributions considered in CDT.

Interesting work on effective modeling of the random field of specific fracture energy has been done recently, Jeulin (1993).

So far, SFM achieved significant progress in prediction of distributions of critical loads, crack lengths, displacements, etc., for a stressed structural element under a variety of conditions. The present paper addresses another important question: predicting life-time scatter for a stressed brittle structural element; it builds, in part, on concepts put forth in CDT. Alternative approaches to stochastic crack growth may be found, for example, in Lin and Yang (1983, 1985) and references therein.

---

[2]For a macroscopically homogeneous material, the parameters of the Weibull distribution do not depend on the point $\underline{x}$.

In the paper, we attempt to model an observed mode of slow crack growth in brittle materials (see, for example, Calomino and Brewer (1992) and Calomino, et al. (1993)), namely, a Markovian stochastic pattern of a microscopic random jump, followed by a random waiting time, followed by a random jump, and so on. We relate the waiting times, on physical grounds, to random energy barriers at the arrest points (Sec 3), and treat the random magnitudes of the jumps within the existing framework of CDT (Sec 2). This leads, in Sec 4, to a description of crack growth as a random process whose transition probability density satisfies a hyperbolic PDE (in contrast to the parabolic Kolmogoroff equations, which govern the transition probabilities in Lin and Yang (1983, 1985)). The relation to the probabilistic prediction of life-time is standard and is discussed in Sec 5. Section 6 contains a simple illustrative example of the specimen-·loading geometry, for which explicit solutions are produced and analyzed. In Sec 7, we offer some constructive criticism of the model.

2 Crack Propagator

The concept of Crack Propagator (CP) has been introduced in Chudnovsky and Kunin (1989) as the probability of a spontaneous crack formation between two arbitrary points of a loaded two-dimensional solid. In this paper, we are only concerned with crack formation along a straight line. A particular form of CP will be introduced here and employed in later sections.

Let us consider a loaded solid with a straight crack (see Fig. 1). Below we will refer to the coordinate of the crack tip as 'crack tip position/location', 'crack depth', or 'crack length' interchangeably. Assume that the loading is either symmetrical or antisymmetrical relative to the crack line to justify rectilinear crack growth along the x-axis. Suppose that the crack tip is presently at x. Denote by $<X|x>$ the probability that the crack would jump from x to X *if its growth is initiated*. We interpret the jump as a sequence of local failures immediately ahead of the current crack tip and adopt a Griffith-type criterion for an infinitesimal crack advance: at a current crack tip position x', $x \le x' \le X$, the potential energy release rate (ERR) $G_l(x')$ should exceed the specific fracture energy $2\gamma(x')$ for an advance to be possible. Inhomogeneity of the material on the microscale is reflected in the assumption that γ fluctuates from point to point, i.e. that $\gamma(x')$ is a random field with a small correlation distance r. For a macroscopically homogeneous material, the statistics of the γ-values should be independent of a point in the solid, i.e. the distribution function $F_{\gamma(x)}(\gamma)$ should be independent of x, $F_{\gamma(x)}(\gamma) = F(\gamma)$. Then the conditional probability $<X|x>dX$ of the crack excursion from x to a new crack tip position between X and $X+dX$, *provided* that the crack growth got initiated, equals the probability of the crack growth criterion being satisfied everywhere between x and X and violated somewhere between X and $X+dX$ (see Chudnovsky and Kunin (1987, 1989) and Kunin (1992) for details):

342

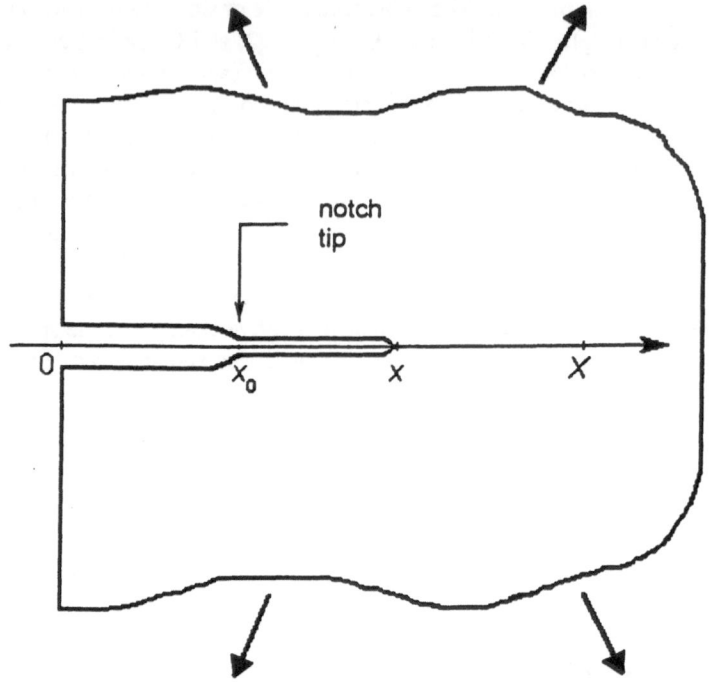

Fig. 1 Specimen-loading configuration (the solid may be either bounded or unbounded). Initial crack (notch) had length x_0 at time $t = 0$.

$$<X|x>dX$$
$$= \text{Prob} \left[G_1(x') \geq 2\gamma(x') \,,\, \text{all } x', \, x < x' \leq X \right]$$
$$\times \text{Prob} \left[G_1(x') < 2\gamma(x') \,,\, \text{some } x', \, X < x' \leq X + dX \right]$$

$$= \begin{cases} \exp\left(-\int_{x}^{X} U(x') \frac{dx'}{r}\right) U(X) \frac{dX}{r} \,,\, & x \leq X \\ \\ 0 \,,\, & x > X \end{cases} \tag{1}$$

where $U(x')$ is the probability of arrest at a point x',

$$U(x') = \text{Prob}\left(G_1(x') < 2\gamma(x')\right)$$

$$= 1 - F\left(G_1(x')/2\right) \tag{2}$$

The first equality in Eq (1) reflects independence of the γ-values on intervals $x \leq x' \leq X$ and $X < x' \leq X + dX$ (though 'infinitesimal', dX is a continuum entity, hence an implicit assumption $dX \gg r$). For $x > X$, the probability $<X|x>dX$ of jumping backward is zero, as we model materials in which cracks do not cure.

The following useful property of CP can be directly verified using Eq (1):

$$\langle X|x'\rangle\langle x'|x\rangle = \tfrac{1}{r} U(x') \langle X|x\rangle \tag{3}$$

for any $x \leq x' \leq X$.

One can justify (see Chudnovsky and Kunin (1987)) assigning a Weibull distribution to the values of γ at every point:

$$F(\gamma) = 1 - \exp\left(-\left(\Gamma\!\left(1+\tfrac{1}{\alpha}\right)\frac{\gamma - \gamma_{min}}{\gamma^* - \gamma_{min}}\right)^{\alpha}\right) \tag{4}$$

if $\gamma \geq \gamma_{min}$ and $F(\gamma) = 0$, if $\gamma < \gamma_{min}$. Here γ^* and γ_{min} are the average and minimal values of γ, respectively, and α is known as a 'shape parameter'; they characterize the scatter of material's strength on a microscale; . $\Gamma(\cdot)$ is the Γ-function. Thus, from Eq (2),

$$U(x') = \exp\left(-\left(\Gamma\!\left(1 + \tfrac{1}{\alpha}\right)\frac{G_1(x')/2 - \gamma_{min}}{\gamma^* - \gamma_{min}}\right)^{\alpha}\right) \tag{5}$$

if $G_1(x') \geq 2\gamma_{min}$ and $U(x') = 1$ if $G_1(x') < 2\gamma_{min}$. One obtains an explicit expression for CP by substituting Eq (5) into Eq (1):

$$\langle X|x\rangle = \exp\left\{-\int_{x}^{X}\exp\left(-\left(\Gamma\!\left(1 + \tfrac{1}{\alpha}\right)\frac{G_1(x')/2 - \gamma_{min}}{\gamma^* - \gamma_{min}}\right)^{\alpha}\right)\frac{dx'}{r}\right\}$$

$$\times \tfrac{1}{r}\exp\left(-\left(\Gamma\!\left(1 + \tfrac{1}{\alpha}\right)\frac{G_1(X)/2 - \gamma_{min}}{\gamma^* - \gamma_{min}}\right)^{\alpha}\right) \tag{6}$$

if $x \leq X$. If $x > X$, then $\langle X|x\rangle = 0$.

Note that a limiting transition to the case of a homogeneous material with no microstructure (constant γ) may be understood physically in more than one way. One is to suppose that $r \to \infty$. This cannot be achieved within the assumptions of the model. Indeed, by construction, the model applies to scales large relative to r. A second interpretation involves letting both r and the variance of γ tend to zero simultaneously. This one is consistent with the model.

Remark. For examples of experimental evaluation of the parameters γ^*, γ_{min} and α of the γ-field, see Kunin (1992).

Later we will need the first two moments of CP, i.e. those of the distribution $\langle X|x\rangle$ as a function of X. Before their evaluation, however, we have to settle the issue of whether the zeroth moment of CP equals 1 as should be expected from a probability distribution.

2.1 Zeroth moment of CP; stable vs. unstable configurations
One may notice that [see Eq (1)]

$$
\langle X|x\rangle = \left\{ \begin{array}{ll} -\dfrac{d}{dX}\exp\left(-\displaystyle\int_{x}^{X}U(x')\dfrac{dx'}{\Gamma}\right), & x \leq X \\[4mm] 0\,, & x > X \end{array} \right.
\tag{7}
$$

whence

$$
\begin{aligned}
m_0 &= \int_{x}^{\infty}\langle X|x\rangle dX \\[2mm]
&= -\exp\left(-\int_{x}^{X}U(x')\dfrac{dx'}{\Gamma}\right)\Bigg|_{X=x}^{\infty} \\[2mm]
&= 1 - \exp\left(-\int_{x}^{\infty}U(x')\dfrac{dx'}{\Gamma}\right)
\end{aligned}
\tag{8}
$$

The latter expression does not necessarily equal 1, though it is certainly positive, since U is a positive function.

To resolve this seeming contradiction, recall that $\langle X|x\rangle$ represents the probability density of the crack being arrested at X once it has initiated from x. However, there is a possibility that the crack will not be arrested at all, as would be typical for such specimen-loading configurations as uniform tension or pure bending. For a finite width specimen, 'no arrest' means the crack reaching the right edge; for a half-plane, which extends to the right, 'no arrest' means the crack extending to $+\infty$ in a single excursion (see Fig 1). In terms of our fracture criterion, the above possibility means that the ERR exceeds twice the values of the γ field at all points of the solid to the right of x. Thus, for certain configurations, a non-zero probability (depending on x) should be assigned to $X = +\infty$; this probability is equal to the second term in Eq (8).

This observation suggests the following characterization of stable and unstable load specimen configurations (terminology proposed in Chudnovsky and Kunin (1987,1989)): a configuration is *stable* if $m_0(x)\equiv1$, otherwise it is *unstable*. In this paper, we restrict ourselves to stable configurations. Commonly these occur when a solid is loaded through prescribed displacements, though an unbounded solid with a crack loaded by localized forces may also produce a stable configuration.

2.2 First moment of CP
The mean crack arrest depth for cracks initiated at x is

$$m_1(x) = \int_x^\infty X <X|x> dX$$

$$= x + \int_x^\infty \exp\left(- \int_x^X U(x') \frac{dx'}{\Gamma}\right) dX \tag{9}$$

The second equality results from using Eq (7) and integrating by parts.

Notice that $m_1(x) > x$, which agrees with common sense since all possible arrest locations are to the right of x.

2.3 Second moment of CP
For the mean square of the crack arrest depth one has

$$m_2(x) = \int_x^\infty X^2 <X|x> dX$$

$$= x^2 + 2\int_x^\infty X \exp\left(- \int_x^X U(x') \frac{dx'}{\Gamma}\right) dX \tag{10}$$

The second inequality results from using Eq (7) and integrating by parts.

3 Duration of crack arrest

Now we will consider what happens to a crack after it is arrested.

In accordance with the fracture criterion adopted above, arrest at a point x occurs, if ERR $G_1(x)$ is smaller than the random value 2γ at x. Let us make a physical assumption that, while the crack stays arrested, some submicroscopic fracture processes are at work in the immediate vicinity (linear dimension ~r) of the crack tip. The effect of these processes is to decrease the value of γ at x with time until, at some random instance, the energy barrier $2\gamma - G_1(x)$ drops to zero and crack initiation occurs. Let us assume that the probability of crack initiation at x (provided it has been arrested there) during an infinitesimal time interval dt is proportional to dt and that the coefficient of proportionality is a function of both the 'virgin' value of γ at the moment of arrest and of $G_1(x)$:

$$\text{Prob [initiation from x during dt} \mid \gamma(x) = \gamma] = \lambda\left(\gamma, G_1(x)\right) \frac{dt}{\tau} \tag{11}$$

Here τ is a characteristic time scale for the hypothetical submicroscopic processes and is introduced explicitly to render λ dimensionless. In the absence of other relevant dimensional constants, λ would have to be a function of $\gamma/G_1(x)$. One may also expect

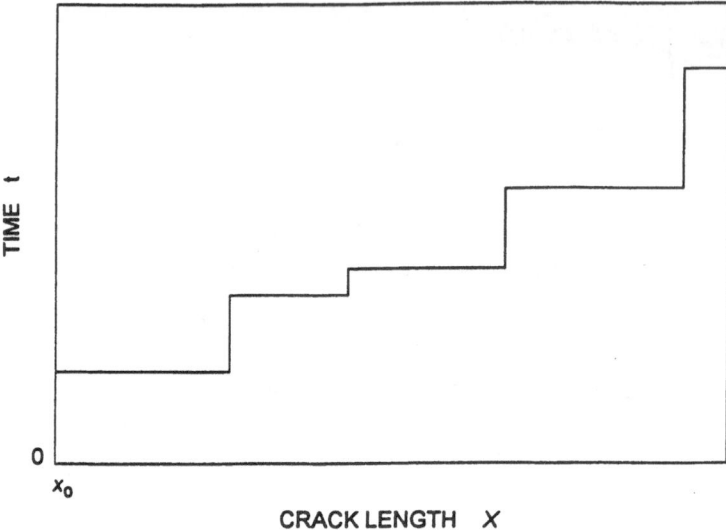

Fig. 2 Schematics of the dependence of crack length on time to be reproduced by the model.

the value of λ at x to depend on the relative energy barrier $[2\gamma - G_1(x)]/G_1(x)$. Perhaps, $\lambda = G_1(x)/[2\gamma - G_1(x)]$ is the simplest dependence that renders large λ (short arrest) for a negligible barrier and *vice versa* (long arrest). It is not our intention to examine the structure of the function λ in this paper. In fact, only the average value of λ over all possible values of γ at x will play a role in the ultimate equations below. We denote this average by $\Lambda(x)$,

$$\Lambda(x) = \int_{\gamma_{min}}^{\infty} \lambda\left(\gamma, G_1(x)\right) dF(\gamma) \tag{12}$$

4 Crack length distribution as a function of time

In this section, the central character of this paper is introduced, and equations for it are derived.

Let us consider a loaded solid described in Sec 2 (see Fig 1). Denote by $\xi(t)$ the random position of the crack tip at a time t. The preceding discussion implies that the random process $\xi(t)$ has monotonically growing piecewise constant realizations (see Fig 2). Furthermore, one can recognize in $\xi(t)$ a combination of two Poisson processes: duration of each arrest (vertical segments in Fig 2) is governed by a Poisson process, whose intensity $\lambda(\gamma, G_1(x))/\tau$ depends on x, whereas the magnitude of each jump (horizontal segments in Fig 2) is governed by a Poisson process with the crack depth X in place of a time-like parameter and the variable intensity $U(X)/r$.

Denote by $p(X,t)$ the probability density of finding the crack tip at a depth X at a time t, assuming that at $t = 0$ the tip was at x_o:

$$p(X,t)dX = \text{Prob}\,[X \le \xi(t) \le X + dX \mid \xi(0) = x_0] \tag{13}$$

To relate $p(X,t+dt)$ to the distribution $p(\cdot,t)$ of crack locations at t, notice first that the crack tip ends up at the depth X at the time $t+dt$ in one of the following mutually exclusive ways: during dt, it either jumped to X from one of the depths x between the notch tip and X, $x_o \le x < X$, or it was at X at the time t and remained there. By the formula of total probability,

$$p(X,t + dt) = \int_{x_0}^{X} \text{Prob}\,[\xi(t + dt) = X \text{ and } x \le \xi(t) \le x + dx]$$

$$+ \text{Prob}\,[\xi(t + dt) = X \text{ and } \xi(t) = X] \tag{14}$$

Notice that the 'jumping probabilities', which are summed up by the integral, are infinitesimal, whereas the probability of 'remaining where one is' during a short time is almost one. Specifically, writing each of the probabilities as a product of appropriate conditional probabilities, one arrives at (see Appendix 1 for the details)

$$p(X,t + dt)$$

$$= \int_{x_0}^{X} \left(<X|x> \frac{\Lambda(x)}{\tau}\, dt\right) p(x,t)\, dx + \left(1 - \Lambda(X) \frac{dt}{\tau}\right) p(X,t) \tag{15}$$

where $\Lambda(x)dt/\tau$ is the average probability of the crack taking off from x during dt [see Eq (12)].

Carrying $p(X,t)$ in Eq (15) to the left and dividing by dt, one obtains

$$\frac{\partial p(X,t)}{\partial t} = \int_{x_0}^{X} <X|x> \frac{\Lambda(x)}{\tau}\, p(x,t)\, dt - \frac{\Lambda(X)}{\tau}\, p(X,t) \tag{16}$$

By differentiating Eq (16) and using properties of CP, one arrives at (see Appendix 2 for the details)

$$\frac{\partial^2 p(X,t)}{\partial X \partial t} + \left(\frac{U(X)}{\tau} - \frac{U'(X)}{U(X)}\right) \frac{\partial p(X,t)}{\partial t} + \frac{\Lambda(X)}{\tau} \frac{\partial p(X,t)}{\partial X}$$

$$+ \frac{1}{\tau}\left(\Lambda'(X) - \frac{U'(X)}{U(X)} \Lambda(X)\right) p(X,t) = 0 \tag{17}$$

We are interested in the solution of Eq (17) that satisfies

$$p(X,0) = \delta(X - x_0) \tag{18a}$$
$$p(X,t) = 0 , \quad X < x_0 , \quad t \geq 0 \tag{18b}$$

where $\delta(X)$ is the Dirac delta-function. Eq (18a) expresses the assumption that we deal with an ensemble of identically precracked or notched solids and that the notch tip has the depth x_0 (see Fig 1). Eq (18b) states that the crack never becomes shorter than the original notch.

Application of the Laplace Transform in the time variable t to Eq (17) results in a first order linear ODE in the space variable X for the function $P(X,s) = \mathcal{L}_t[p(X,t)]$ (see the end of Appendix 2). Thus $P(X,s)$ can be written as an explicit expression in quadratures and $p(X,t)$ as the inverse Laplace transform of that.

In Sec 5 we solve Eqs (17), (18) under very special circumstances to illustrate the model's output. However, in any applications, the model would be only a crude approximation of reality. Thus, at any time t, only integral characteristics of $p(X,t)$ may represent reliable information. With that in mind, we finish this section with some discussion of the first spatial moments of $p(X,t)$. If, as may be expected for a stable configuration, the crack length distribution at every instant is bell-shaped, then its first moment would give us the motion of the bell's center of gravity, and the central second moment would describe the evolution of the bell's width in time.

A discussion of spatial moments of $p(X,t)$ must actually begin from the zeroth moment. Indeed, with the possibility of cracks reaching the edge of the solid (or $+\infty$) in a finite time, one cannot be sure *a priori* that the zeroth moment equals 1 at all times (c.f. the discussion of the zeroth moment of CP at the end of Sec 2).

4.1 Zeroth moment of $p(X,t)$
To evaluate the zeroth moment

$$M_0(t) = \int_{-\infty}^{\infty} p(X,t) \, dX = \int_{x_0}^{\infty} p(X,t) \, dX \tag{19}$$

integrate Eq (16) in X from x_0 to ∞. Reversing the order of integration in the right hand side and using the first line in Eq (8), one gets

$$\frac{dM_o(t)}{dt} = \int_{x_o}^{\infty} [m_o(x) - 1] \frac{\Lambda(x)}{\tau} p(x,t) \, dx \qquad (20)$$

We restricted ourselves to stable configurations, hence $m_o(x) = 1$ [see the discussion after Eq (8)]. Therefore the right hand side of Eq (20) equals zero, i.e. $M_o = $ const. But

$$M_o(0) = \int_{-\infty}^{\infty} p(X,0) \, dX$$

$$= \int_{-\infty}^{\infty} \delta(X - x_o) \, dX = 1 \qquad (21)$$

[the second equality is due to Eq (18a)], whence

$$M_o(t) \equiv 1 \qquad (22)$$

We reiterate that Eq (22) is valid for stable configurations only.

4.2 Evolution of the mean crack length

For a stable configuration, the average crack tip position at a time t is given by

$$M_1(t) = \int_{-\infty}^{\infty} X \, p(X,t) \, dX = \int_{x_o}^{\infty} X \, p(X,t) \, dX \qquad (23)$$

For an unstable configuration, $p(X,t)$ would represent only part of the probability distribution!) Multiplying Eq (16) by X and integrating it in X, then changing the order of integration in the right hand side and using Eq (9), one obtains

$$\frac{dM_1(t)}{dt} = \int_{x_o}^{\infty} [m_1(x) - x] \frac{\Lambda(x)}{\tau} p(x,t) \, dx$$

$$= \int_{x_o}^{\infty} \left(\int_{x}^{\infty} \exp\left(-\int_{x}^{X} U(x') \frac{dx'}{\tau} \right) dX \right) \frac{\Lambda(x)}{\tau} p(x,t) \, dx \qquad (24)$$

Though not closed with respect to $M_1(t)$, Eq (24) assures us at least that $dM_1/dt > 0$, i.e. that the average crack tip position moves to the right with time, as should have been expected. For the special case considered in Sec 6, Eq (24) becomes a closed equation for $M_1(t)$.

4.3 Second moment of $p(X,t)$

For a stable configuration, the second spatial moment of $p(X,t)$ is

350

$$M_2(t) = \int_{-\infty}^{\infty} X^2 \, p(X,t) \, dX = \int_{X_o}^{\infty} X^2 \, p(X,t) \, dX \qquad (25)$$

Multiplying Eq (16) by X^2 and integrating it in X, then changing the order of integration in the right hand side and using Eq (10), one obtains

$$\frac{dM_2(t)}{dt} = \int_{X_o}^{\infty} [m_2(x) - x^2] \frac{\Lambda(x)}{T} \, p(x,t) \, dx$$

$$= 2 \int_{X_o}^{\infty} \left(\int_{x}^{\infty} X \exp \left(- \int_{x}^{X} U(x') \frac{dx'}{T} \right) dX \right) \frac{\Lambda(x)}{T} \, p(x,t) \, dx \qquad (26)$$

For the special case considered in Sec 6, Eq (26) becomes a closed equation for $M_2(t)$.

5 Life-time scatter

Consider a loaded solid with a crack, such as shown in Fig 1. For a stable configuration, we do not expect a catastrophic failure, so let us consider a situation in which the presence of too deep a crack would result in the solid's (read: "structural element's") rejection.

Let us define *solid's failure* in terms of our formalism as the crack reaching certain depth L. Let T be the random moment of failure, i.e. the first instant when $\xi(t)$ equals L. Actually, for all later times T, $T>T$, the crack length will be no less than L, since we assumed that the crack cannot shrink: $\xi(t) \geq L$ for all $T > T$. Thus we can write the distribution function for the moment of failure:

$$F_T(T) = \text{Prob } [T \leq T]$$
$$= \text{Prob } [\xi(T) > L]$$

$$= 1 - \int_{X_o}^{L} p(X,T) \, dX \qquad (27)$$

6 Illustrative example: constant energy release rate

Suppose that we deal with a specimen-loading configuration, for which ERR is independent of the crack length, $G_I(x) = $ const. An example of such a configuration is shown in Fig 3.

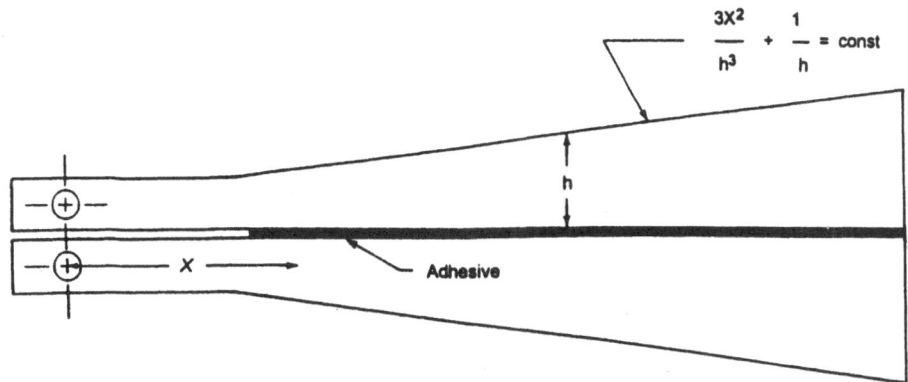

Fig. 3 An example of a specimen-loading configuration with ERR independent of the crack length, ASTM (1988). Loading is displacement-controlled. In a properly conducted test the crack grows strictly within the adhesive strip.

The functions $U(x)$ and $\Lambda(x)$, which enter Eq (17), depend on x through the dependence on x of ERR $G_I(x)$ only [see Eqs (5), (12)]. Therefore both of them must be constant: $U(x) = U_o$, $\Lambda(x) = \Lambda_0$. If we put $r_o = r/U_o$ and $\tau_0 = \tau/\Lambda_0$, then Eq (17) becomes

$$\frac{\partial^2 p}{\partial X \partial t} + \frac{1}{r_o}\frac{\partial p}{\partial t} + \frac{1}{\tau_0}\frac{\partial p}{\partial X} = 0 \tag{28}$$

The physical meaning of r_o, τ_0 can be deduced from the bi-Poisson nature of the random process $x(t)$ (see the second paragraph of Sec 4): r_o is the average crack jump size, and τ_0 is the average duration of an arrest. This, in turn, implies that the average crack tip position should be moving to the right with the speed r_o/τ_0, which is confirmed below [see Eq (33)].

One solves Eq (28) subject to the conditions Eq (18) by applying the Laplace transform in both t and X. The solution is

$$p(X,t) = \left(\delta(X-x_0) + \frac{1}{r_0} H(X-x_0) \sqrt{\frac{r_0}{\tau_0}\frac{t}{X-x_0}} I_1\left(2\sqrt{\frac{X-x_0}{r_0}\frac{t}{\tau_0}}\right) e^{-\frac{X-x_0}{r_0}} \right) e^{-\frac{t}{\tau_0}} \tag{29}$$

where $H(X)$ is the Heaviside unit step function,

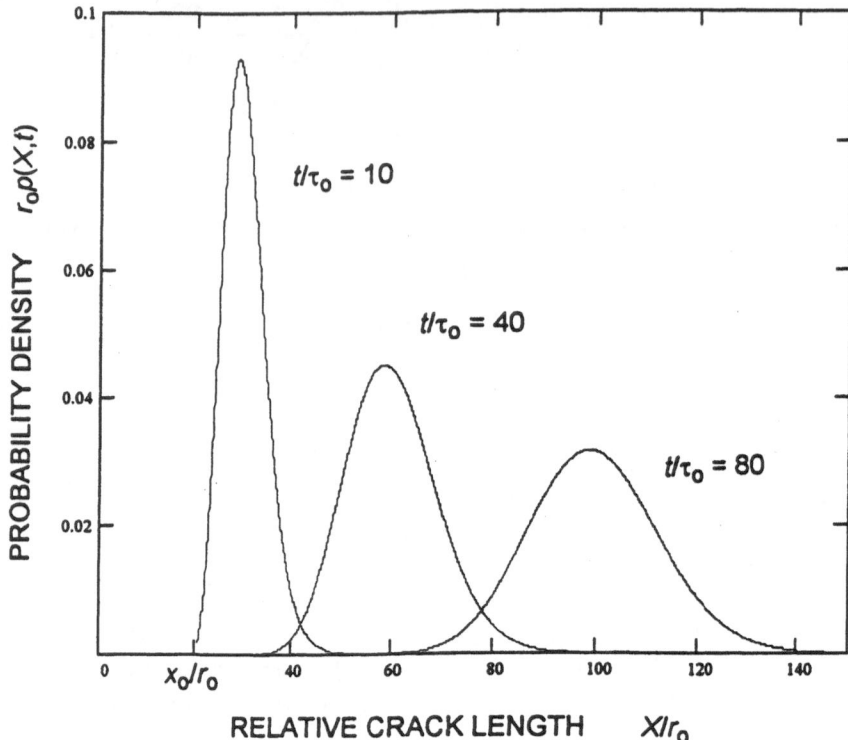

Fig. 4 Probability densities of crack length at various times (constant ERR). The δ-functional part of $p(X,t)$ [c.f. Eq (29)] corresponds to the non-zero probability that the crack length remains equal to x_0. It is concentrated at x_0 and decreases exponentially with time. Here these probabilities are small ($5 \cdot 10^{-5}$, 10^{-17}, 10^{-35} for $t/\tau_0 = 10, 40, 80$, respectively) and are not shown. The value of x_0/r_0 (chosen to be 20) is responsible for the simultaneous shift of the graphs to the right by 20 and plays no other role.

$$H(x) = \begin{cases} 1, & x \geq 0 \\ 0, & x < 0 \end{cases} \tag{30}$$

and $I_1(\cdot)$ is a Bessel function. Notice that the δ-functional part of $p(X,t)$ in Eq (29) corresponds to the finite probability of the crack tip remaining at x_0 (this probability decreases exponentially with time).

Figure 4 illustrates the evolution of the crack length distribution with time as predicted by Eq (29).

One can use the known asymptotic behavior of $I_1(z)$ for large z,

$$I_1(z) = (2\pi z)^{-\frac{1}{2}} e^z \left(1 + O(\tfrac{1}{z})\right) \tag{31}$$

to find the behavior of the solution, Eq (29), for either times much larger than τ_0 or crack lengths much larger than r_o. Namely,

$$p(X,t) = \frac{1}{2} \pi^{-\frac{1}{2}} \frac{1}{r_0} \left(\frac{X - x_0}{r_0}\right)^{-\frac{3}{4}} \left(\frac{t}{\tau_0}\right)^{\frac{1}{4}} \exp\left(-\left(\sqrt{\frac{X - x_0}{r_0}} - \sqrt{\frac{t}{\tau_0}}\right)^2\right)$$

$$\times \left(1 + O\left(\sqrt{\frac{r_0}{X - x_0}} \frac{\tau_0}{t}\right)\right) \tag{32}$$

6.1 Evolution of the mean, standard deviation, and skewness of the crack length distribution

Mean crack length

The first moment of CP, in the case $U(x) = U_o$, equals [see Eq (9)]

$$m_1(x) = x + \int_x^\infty \exp\left(-\frac{U_0}{\Gamma}(X - x)\right) dX$$

$$= x + r_0 \tag{33}$$

Now, the first line of Eq (24) becomes, as $\Lambda(x) = \Lambda_o$,

$$\frac{dM_1(t)}{dt} = \int_{x_0}^\infty \frac{r_0}{\tau_0} p(x,t) \, dx$$

$$= \frac{r_0}{\tau_0} M_0(t)$$

$$= \frac{r_0}{\tau_0} \tag{34}$$

The mean crack length at $t = 0$ equals the notch length x_o, $M_1(0) = x_o$, therefore

$$M_1(t) = x_0 + \frac{r_0}{\tau_0} t \tag{35}$$

Standard deviation of the crack length

The second moment of CP, in the case $U(x) = U_o$ equals [see Eq (10)]

$$m_2(x) = x^2 + 2 \int_X^\infty X \exp\left(-\frac{U_0}{\Gamma}(X - x)\right) dX$$

$$= x^2 + 2r_0x + 2r_0^2 \tag{36}$$

(by parts). Notice also that, from Eqs (33),(36), the standard deviation of CP (i.e. that of a random jump) is $[m_{2c}(x)]^{1/2} = (m_2(x) - [m_1(x)]^2)^{1/2} = r_0$.

Now the first part of Eq (26) becomes, as $\Lambda(x) = \Lambda_0$,

$$\frac{dM_2(t)}{dt} = \int_{x_0}^\infty (2r_0x + 2r_0^2)\, \frac{1}{\tau_0}\, p(x,t)\, dx$$

$$= 2\frac{r_0}{\tau_0} M_1(t) + 2\frac{r_0^2}{\tau_0} M_0(t)$$

$$= 2\frac{r_0^2}{\tau_0^2} t + 2\frac{r_0}{\tau_0}(x_0 + r_0) \tag{37}$$

Since $M_2(0) = x_0^2$, we have the following solution to Eq (37):

$$M_2(t) = r_0^2 \left(\frac{t^2}{\tau_0^2} + 2\left(\frac{x_0}{r_0} + 1\right)\frac{t}{\tau_0} + \frac{x_0^2}{r_0^2} \right) \qquad \bullet \tag{38}$$

Finally, the standard deviation of the crack length scatter at a time t is

$$\sigma(t) = \left(M_2 - M_1^2 \right)^{\frac{1}{2}}$$

$$= r_0 \sqrt{\frac{2t}{\tau_0}} \tag{39}$$

Thus, for times $t \gg \tau_0$, the standard deviation of the crack length grows as the square root of the mean increase, $M_1(t) - x_0$, of the crack length itself.

Skewness of the crack length distribution

Just as for $m_1(x)$ and $m_2(x)$, one easily obtains the following expression for the third moment of CP, in the case $U(x) = U_0$:

$$m_3(x) = \int_x^\infty X^3 <X|x> \, dX$$

$$= x^3 + 3r_o x^2 + 6r_o^2 x + 6r_o^3 \tag{40}$$

Incidentally, the central third moment of CP equals zero (in the case $U(x) = U_o$): $m_{3c} = m_3 - 3m_2 m_1 + 2m_1^3 = 0$, as follows from Eqs (40), (36), (33).

Similarly to the above treatment of $M_1(t)$ and $M_2(t)$ one obtains

$$M_3(t) = r_o^3 \left(\frac{t^3}{\tau_o^3} + 3 \left(\frac{x_o}{r_o} + 2 \right) \frac{t^2}{\tau_o^2} + 3 \left(\frac{x_o^2}{r_o^2} + 2 \frac{x_o}{r_o} + 2 \right) \frac{t}{\tau_o} + \frac{x_o^3}{r_o^3} \right) \tag{41}$$

Thus, the central third moment of crack length distribution is

$$M_{3c}(t) = M_3 - 3M_1 \sigma^2 - M_1^3$$

$$= 6r_o^3 \frac{t}{\tau_o} \tag{42}$$

to the left, and the dimensionless skewness coefficient equals [using Eqs (42), (39)]

$$S(t) = \frac{M_{3c}^{\frac{1}{3}}}{\sigma} = \left(\frac{9}{2} \frac{\tau_o}{t} \right)^{\frac{1}{6}} \tag{43}$$

i.e. relative skewness decreases with time.

6.2 Life-time scatter

In the case of constant ERR, one obtains an explicit expression for the distribution function $F_T(T)$ for the moment of failure T introduced in Sec 5. Namely, Eq (27), which relates $F_T(T)$ to $p(X,T)$, together with the explicit expression for $p(X,t)$ [Eq (29)] gives

$$F_T(T) = 1 - e^{-\frac{T}{\tau_o}} \left(1 + \int_{z(x_o)}^{z(L)} I_1(z) \, e^{-\frac{\tau_o}{4T} z^2} \, dz \right) \tag{44}$$

after the substitution $z = z(X) = 2[XT/(r_o \tau_o)]^{1/2}$.

356

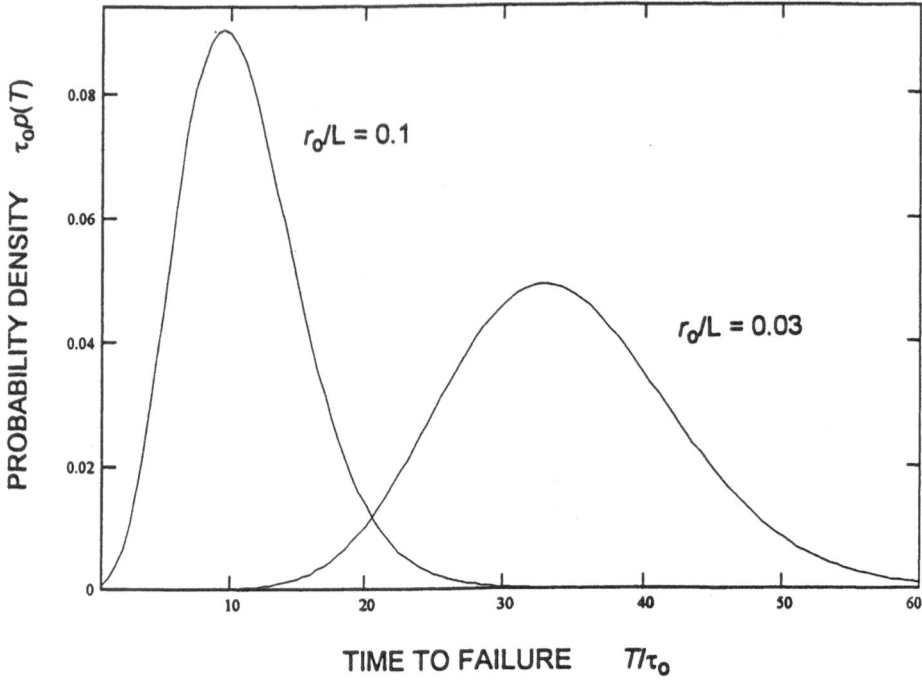

Fig. 5 Probability density of the time to failure $p(T) = dF_T(T)/dT$ for two values of r_0 in the case of constant ERR.

Figure 5 shows the dependence of the life-time scatter on r_0 predicted by the model. This dependence is of interest for the following reason. Under identical loading conditions, r_0 is proportional to the correlation distance r of the γ-field, $r_0 = r/U_0$, and r may be interpreted as an indirect manifestation of the material's architecture. Thus, one may view Fig 5 as an illustration of the effect of the material's microstructure on the specimen's life-time.

7 Discussion

The proposed model is only a first step in the direction that the author would like to explore. It has several limitations that have to be addressed.

Firstly, the model describes crack growth under static loading. Assuming quasistatic load variation, e.g. fatigue, would result in only minor complications in deriving an analog of Eq (17). However, one may expect, on physical grounds, that the characteristic time scale τ would then differ from the static one and possibly depend on such parameters as the loading rate. Further steps in this direction require reconsideration of the Poisson-type assumptions put forth in Sec 3.

Secondly, the brittleness of a material is expressed in the assumptions that there is either no 'process zone' around the crack tip (during crack excursions) or the zone has negligible size (~r, during crack arrest). This is certainly an extreme type (labeled 'solo crack fracture' in Chudnovsky and Kunin (1989)) among crack propagation processes. Another extreme would be the case when the intensity of damage formed as a response to the stress concentration at the crack tip is much greater than the intensity of pre-existing damage ('cooperative fracture,' *ibid.*). Even for brittle materials the truth may be somewhere in between. In particular, the presence of a 'process zone' may result in stages of truly quasistatic crack growth (i.e. no discernible jumps on a scale >> r during a finite time interval). Thus, incorporation of effects of a 'process zone' is another challenging task.

Finally, heterogeneity of the material on a microscale is manifested not only in fluctuations of local strength, but also in the tortuosity of crack trajectories. In fact, cracks may deviate significantly, in a random fashion, from a rectilinear path even under pure mode loadings. An attempt to quantify the ability of a material to 'confuse' crack paths has been made in Crack Diffusion Theory. Incorporation of the latter into the present model would certainly produce a better approximation to reality.

Acknowledgement

The author is thankful to Dr. N. Berger for advice concerning Eq (28). Financial support from the UAH Research Institute Research Mini-Grant is gratefully acknowledged.

Appendix 1. Derivation of Eq (15)

Let us consider, in Eq (14), the probability under the integral sign first. This probability of jumping during dt to X from one of the depths x', $x \leq x' \leq x+dx$, equals to the product of (a) the probability $<X|x'>$ of the crack jumping during *dt* from some x', $x \leq x' \leq x+dx$, to X, *provided* that the crack has been sitting at x' for some time prior to t and then initiated again during dt, (b) the probability of crack initiation from such an x' during dt, *provided* that the crack has been sitting at x' at the time *t*, and (c) the probability $p(x,t)dx$ of the crack being between x and $x+dx$ at the time *t*:

$$\text{Prob}\,[\xi(t + dt) = X \text{ and } x \leq \xi(t) \leq x + dx]$$
$$= <X|x> \text{Prob}\,[\text{initiation from } x' \text{ during dt} \mid \xi(t) = x]\, p(x,t)\, dx \qquad (A1.1)$$

The probability in the middle is the average of the conditional probabilities of initiation from x' during *dt* over all possible values that $\gamma(x')$ may assume (the formula of total probability),

Prob [initiation from x' during dt | $\xi(t) = x$]

$$= \int_{\gamma_{min}}^{\infty} \text{Prob [initiation from } x' \text{ during dt | } \xi(t) = x' \text{ and } \gamma(x') = \gamma] \, dF(\gamma)$$

$$= \int_{\gamma_{min}}^{\infty} \left\{ \lambda(\gamma, G_1(x')) \frac{dt}{\tau} \right\} dF(\gamma)$$

$$= \Lambda(x) \frac{dt}{\tau} \tag{A1.2}$$

The last equality uses the definition of $\Lambda(x)$, Eq (12) and the fact that x' is infinitesimally close to x. Eqs (A1.2), (A1.1) yield the desired form of the integrand in Eq (15).

Consider now the second term in Eq (14):

$$\text{Prob } [\xi(t + dt) = X \text{ and } \xi(t) = X]$$
$$= 1 - \text{Prob [initiation from } X \text{ during dt | } \xi(t) = X] \, p(X,t)$$
$$= \left(1 - \Lambda(X) \frac{dt}{\tau}\right) p(X,t) \tag{A1.3}$$

which is its form in Eq (15) [the last equality uses Eq (A1.2) for $x' = X$].

Appendix 2. Derivation of Eq (17)

From Eq (3) written for the triple $x_o \leq x \leq X$, one obtains

$$<X|x> = \frac{1}{\tau} U(x) \frac{<X|x_o>}{<x|x_o>} \tag{A2.1}$$

Substitute the above expression into Eq (16) to obtain

$$\frac{\partial p(X,t)}{\partial t} = <X|x_o> \int_{x_o}^{X} \frac{U(x)}{\tau} \frac{\Lambda(x)}{\tau} \frac{p(x,t)}{<x|x_o>} \, dx - \frac{\Lambda(x)}{\tau} p(X,t) \tag{A2.2}$$

Divide Eq (A2.2) by $<X|x_o>$ and denote

$$q(X,t) = \frac{p(X,t)}{<X|x_o>} \tag{A2.3}$$

Then Eq (A2.2) reads

$$\frac{\partial q(X,t)}{\partial t} = \int_{x_0}^{X} \frac{U(x)}{\tau} \frac{\Lambda(x)}{\tau} q(x,t) \, dx - \frac{\Lambda(X)}{\tau} q(X,t) \qquad (A2.4)$$

Differentiating Eq (A2.4) in X one obtains the following equation for q:

$$\frac{\partial^2 q(X,t)}{\partial X \partial t} = \left(\frac{U(X)}{\tau} \frac{\Lambda(X)}{\tau} - \frac{\Lambda'(X)}{\tau} \right) q(X,t) - \frac{\Lambda(X)}{\tau} \frac{\partial q(X,t)}{\partial X} \qquad (A2.5)$$

Substituting into Eq (25) the expressions for q, $\partial q/\partial X$ and $\partial^2 q/\partial X \partial t$ as implied by Eq (A2.3) and using Eq (1) one arrives at Eq (17).

Applying the Laplace transform in t to Eq (17) and using the initial condition Eq (18a), one obtains the following first order linear ODE in X for $P(X,s) = \mathcal{L}_t[p(X,t)]$:

$$a(X,s) \frac{\partial P(X,s)}{\partial X} + b(X,s) \, P(X,s) = c(X) \qquad (A2.6)$$

where

$$a(X,s) = s + \frac{\Lambda(X)}{\tau} \qquad (A2.7)$$

$$b(X,s) = \left(\frac{U(X)}{\tau} - \frac{U'(X)}{U(X)} \right) s + \frac{1}{\tau} \left(\Lambda'(X) - \frac{U'(X)}{U(X)} \Lambda(X) \right) \qquad (A2.8)$$

$$c(X) = \left(\frac{U(x_0)}{\tau} - \frac{U'(x_0)}{U(x_0)} \right) \delta(X - x_0) + \delta'(X - x_0) \qquad (A2.9)$$

References

ASTM Standard 3433 (1988) **Annual Book of ASTM Standards**, v 15.06, 227-33.

Calomino A, and Brewer D (1992) Controlled crack growth specimen for brittle systems, **J of American Ceramic Soc.**, v. 75, no. 1, 206-208.

Calomino A, Brewer D, and Ghosn L (1993) Fracture behavior of ceramics under displacement controlled loading, **NASA Technical Memorandum 105565**.

Chudnovsky A (1973) On fracture of solids, in **Studies on Elasticity and Plasticity** (ed LM Kachanov), Leningrad University Press, Leningrad, USSR, 3-41 (in Russian).

Chudnovsky A and Kunin B (1987) A probabilistic model of brittle crack formation, **J Appl Phys**, 62(10), 4124-29.

Chudnovsky A and Kunin B (1989) On applications of probability in fracture mechanics, in **Computational Mechanics of Probabilistic and Reliability Analysis**, (eds WK Liu and T Belytschko), Elmepress International, Washington, 395-415.

Jeulin D (1993) Fracture statistics models and crack propagation in random media, **Proc. ASCE/ASME/SES Meeting, June, 1993, Charlottesville, VA**.

Kunin B (1992) A probabilistic model for predicting scatter in brittle fracture, **PhD Thesis**, Univ of Illinois at Chicago, Chicago, IL.

Kunin B and Gorelik M (1991) On representation of fracture profiles by fractional integrals of a Wiener process, **J Appl Phys**, 70(12), 7651-53.

Lin YK and Yang JN (1983) On statistical moments of fatigue crack propagation, **Eng Fracture Mech**, v. 18, no. 2, 246-56.

Lin YK and Yang JN (1985) A stochastic theory of crack propagation, **AIAA Journal**, v. 23, no. 1, 117-24.

Ostoja-Starzewski, M. (1992) **Stud. Appl. Mech.**, 31, 71-80.

Wu CH (1978) Maximum-energy-release-rate criterion applied to a tension-compression specimen with crack, **J Elasticity**, 8, 235.

CRACK PATTERN RELATED UNIVERSAL CONSTANTS[1]

G. Frantziskonis
University of Arizona, Department of Civil Engineering and Engineering Mechanics,
Tucson, Arizona 85721, USA   fran@ccit.arizona.edu

Abstract
The paper examines the implications of material heterogeneity on the crack pattern
formed during progressive failure of materials. The displacement gradients of the micro-
medium are considered to be random stationary fields or fractal fields. Both formulations
yield interesting and, surprisingly, analogous results. Two variables are found to affect
the crack pattern that will form under loading. The first one is the fluctuations, i.e.
coefficient of variation of the displacement gradient fields. The second one is
correlations, i.e. correlation length of the random fields as compared to the size of the
structure/specimen, or lower, upper cutoffs and fractal dimension of the fractal fields.
The (macro) crack pattern that will develop is governed by the interplay of fluctuations
and correlations, independent of the elastic constants (for deterministic local Poisson
ratio) and fairly independent of the crack formation criterion. Universal constants are
found for the coefficients of variation. For values above the universal ones the material
will develop a highly irregular crack pattern with multiple crack intersections and/or self-
intersecting cracks. This is typically observed in testing of high strength concretes, high
strength rocks, metals of ultra high hardness and certain ceramics, composites. Under
dynamic load and/or under load control the highly irregular crack pattern is accompanied
by an "explosive" burst of chunks of material and high energy release. For values below
the universal ones, a single crack (may be accompanied by branches) will develop. This
is typical of low strength concretes, soft rocks, low hardness metals, certain ceramics,
composites. Energy release is relatively low - characteristic of low strength materials.

Keywords: fracture / heterogeneity / random fields / scaling / universality

1. Introduction and Review

In general, every material is inherently heterogeneous due to the presence of
microstructure. Micro- and meso-scopically, if interest is in mechanical properties at

---

[1]Also includes a brief overview of parts of recent work by the author.

D. Breysse (ed.), Probabilities and Materials, 361–375.
© 1994 *Kluwer Academic Publishers.*

362

scales much larger than the atomic, heterogeneity for materials like ceramics, rocks, concretes, composites means several things: size and properties of grains, aggregates, pores, microcracks, interfaces, interactions with discontinuities and surfaces. These influence the analysis of such materials significantly, both experimentally (in the choice of scales of observation) and theoretically (limitations of homogenization methods, for example). The fracture behavior of such materials is of substantial technological importance, so it continues to be the subject of intensive research. Microcracking, crack bridging, crack arrest are some of the many mechanisms that absorb energy during the fracture process. Physical reasoning suggests that these mechanisms are affected by the heterogeneity of the material before (macro) fracture activation. Thus in a general sense heterogeneity contributes to the energy absorption mechanisms which then contribute to the tendency of the (macro) crack network (defined subsequently) to follow a tortuous path.

The first problem examined in this paper can be stated as follows. We consider a brittle material loaded at a low loading rate, either by controlling the external load or the external displacement. At some point a macrocrack network will form. Does this network form in a dynamic fashion or quasi-statically?

Let us consider the process of crack formation and propagation in a specimen of brittle rock (for example) loaded externally at a low load rate. The overall force displacement response for such a material typically shows an ascending part which is more or less linear until a peak load is reached. The post peak response is typically non linear. If the sample is loaded by controlling the external force a tortuous macro (visible by the naked eye) crack network develops at or very near the peak. High strength brittle materials disintegrate or fragment at or near the peak load. The macro-crack formation and propagation is very abrupt in these cases of load controlled testing. Thus although the external load may be applied at low rate (statically) macro-crack propagation may be dynamic in this case. If the specimen is loaded by controlling the external displacement, due to the brittleness of the material it is difficult to achieve slow macro-crack propagation, especially for high strength brittle materials, i.e. ultra high strength concretes. This is also due to the fact that loading frames control the displacement by continuously adjusting the load, and this results in a sawtooth type of loading.

The above discussion addresses the case of low rate external load application. For a specimen loaded at a high loading rates, macro-crack propagation is dynamic. Thus it may be that, although the present analysis contradicts it for the low rate external load application case, a macro-crack network in a brittle material develops (always, i.e. under both low and high external load rate application) in a fast rate dynamic fashion. If this is the case, there may not be enough time for the stress/strain fields to redistribute to an equilibrium state until the crack propagates further. Of course, this is a heuristic argument that is difficult to verify experimentally. In dynamic fracture mechanics literature, i.e. Kanninen (1985), it is reported that, in general, the stress intensity decreases with increasing crack velocity and becomes zero when it reaches the Rayleigh velocity. Thus, initially we examine the case of dynamic crack propagation where there is no stress redistribution and subsequently the case where crack propagation is "slow" so that stress redistribution in front of crack and/or notch tip(s) takes place. It may be argued that the "truth" is somewhere in between although some other "truths" may be equally competent. Then, for fast propagation case under slow external load, if we neglect influence of body forces/inertia effects, it may be argued that the macro-crack pattern is determined by the stress/strain field prevailing before macro-crack initiation.

Of course, near boundaries the behavior is expected to be different. Boundary effects are important. The present analysis does not consider such effects thus the results presented herein are valid far from boundaries. Numerical solutions for boundary effects will be presented elsewhere.

If the specimen/structure is loaded dynamically, the macrocrack network develops dynamically and several factors influence its characteristics - the waves propagating in the medium influence the crack pattern significantly. Especially near surfaces strong surface waves usually result in disintegration of the material from the surfaces inward. Such problems are difficult to treat analytically.

Localization of deformation into narrow zones (within the context of continuum mechanics) has received intensive theoretical and experimental attention. Theoretically, the onset and post-bifurcation solution of the governing homogeneous equations is sought - such solution(s) allow localization of deformation and/or damage into narrow bands. When "traditional" constitutive equations are used the wavelength of the standing wave solution remains undetermined and the solution in the post-localization regime is ill-posed and mesh sensitive when numerical solutions are sought.

The need for a length scale or scales present in the constitutive equations for solid materials has been realized long ago. The so-called Cosserat theory appeared at the beginning of this century (1909), Scheefer (1962), where higher order deformation gradients (through couple stresses) and relevant length scale(s) are introduced. Since the 1950's the subject of "couple stress" and gradient theories has received much attention. A great part of the early work was published in the journal Archives of Rational Mechanics and Analysis. Despite the significance of the gradient theories i.e. in surface related phenomena, surface energy, Mindlin (1964), wave propagation, certain difficulties in implementation discouraged wide use of the theories. However, the central role of gradients in localization and fracture phenomena has been realized, i.e. Aifantis (1984, 1992 and references cited there). The role of the so-called internal length entering in gradient theories is important with respect to localization phenomena, Muhlhaus and Vardoulakis (1987), Muhlhaus and Aifantis (1991), Vardoulakis and Frantziskonis (1992). The role of the internal length is also paramount with respect to surface related phenomena, Frantziskonis and Vardoulakis (1992). In another approach, length is introduced in the constitutive equations through non-local concepts, i.e. Bazant and Pijaudier-Cabot (1988 and references cited there). Also, introduction of viscosity introduces a length into the formulation, i.e. Needleman (1988).

In a recent study, Frantziskonis (1993a, b), it was shown that under certain conditions a statistical approach to material heterogeneity yields results analogous to the gradient theories. These conditions call for small heterogeneity fluctuations allowing a spatial Taylor series expansion for the micro-strain field Ψ_{ij} that yields constitutive equations in the form

$$\dot{\sigma}_{ij} = C_{ijkl}\dot{\varepsilon}_{kl} - C_{ijkl}\bar{\nabla}^2 \dot{\varepsilon}_{kl}$$

(1)

where

$$\bar{\nabla}^2 = l_k^2 \frac{\partial^2}{\partial x_k^2}$$

(2)

and $\dot{\sigma}_{ij}, \dot{\varepsilon}_{ij}$ are the time rates of stress, macro-strain, C_{ijkl} is the constitutive tensor for the incrementally linear (elastic or elastic plastic) case, and $l_k, k=1,2,3$ are the spatial correlation lengths of Ψ_{ij} for the general three-dimensional case (repeated indices imply summation). Derivation of (1), (2) is given in Frantziskonis (1993b). Since they address the continuum limit case of the present study, the implications of (1), (2) are discussed briefly. For statistical isotropy, $l_1 = l_2 = l_3 = l$, equation (1) reduces to the gradient constitutive equations, i.e. equation (58) in Aifantis (1992) or equation (4.18) in Vardoulakis and Frantziskonis (1992) for the plasticity case, where l is termed as the "internal material length." In contrast to gradient theories, the statistical formulation, Frantziskonis (1993a,b) introduces a natural evolution law for l. The physical interpretation of such evolution is change in the initial spatial correlation due to microcrack growth/initiation during loading. Also, the case that heterogeneity is a power law scaling field (the only mathematically admissible form of scaling), for small fluctuations, was examined in Frantziskonis (1993a,b) yielding results similar to equation (1) where the lower and upper cutoffs and the fractal dimension of the field enter the formulation.

Constitutive equations (1) and (2) imply homogenization - the real heterogeneous material is replaced by a homogeneous one. Due to the nature of the formulation for localization when homogeneous equations, i.e. (1), (2), are used the only information on the geometry of the localized zone is its thickness and its orientation. No insight into the tortuous geometrical features or into the structure of the localized zones at a range of scales is provided. A natural question is how important is the geometry of the localization and/or fracture process. Assuming it is important (this is discussed in the sequence) it may be concluded that homogeneous equations cannot provide adequate information on fracture formation. Then the implications of spatial heterogeneity should be addressed in experimental and theoretical studies of fracture formation and propagation.

Is the geometry of fracture important? In concrete for example its brittle (ultra high strength concretes) or quasi-brittle (normal concretes) behavior is associated with tortuous fracture surfaces dominated mainly by the size and distribution of aggregates. Often, high strength brittle materials fracture in an "explosive" manner, i.e. they fragment or disintegrate into pieces at a range of scales from "dust" to pieces of the size of the tested specimen. Relevant fracture surface or fracture network parameters are magnification dependent since such surfaces have information at a wide range of scales.

Then the total fracture surface area is larger than the nominal one. This has significant implications for mechanics based models for fracture and localization. Not only should there be a multiplicative factor for energy involved in the fracture or localization process, but recognition of the tortuosity of the relevant path leads to a better understanding of the involved mechanisms. Several experimental studies address the problem of roughness and "texture" of fracture surfaces/networks.

For many materials, fracture surfaces show fractal character although the subject is at the "embryonic" stage (this may be the reason for some controversy in the literature as discussed below) and deserves further research, Chermant and Coster (1978), Mandelbrot et al (1984), Allen and Scholfield (1985), Underwood and Banerji (1986), Allen et al (1987), Pande et al (1987a,b), Blinc et al (1988), Lung and Mu (1988), Mecholsky et al (1988), Skjeltorp and Meakin (1988), Wang et al (1988), Davidson (1989), Alexander (1990), Castano et al (1990), Tsai and Mecholsky (1991), Lange et al (1992). The materials examined in these references are various metals, microsphere monolayers, cement gels, mortar, rocks, ceramics and ceramic composites. There is some controversy with respect to whether fractal properties of fracture surfaces relate to fracture parameters. For example, Tsai and Mecholsky (1991), Lung and Mu (1988), Meckolsky et al (1988), Lange et al (1992) support the idea, while Alexander (1990), Davidson (1989), Pande et al (1987) do not. It should be noted that the relevant experimental results are sensitive to several factors, especially to the observation scale(s).

Since the spatial characteristics of fracture surfaces depend on deformational spatial characteristics of the material before (macro)fracture it is interesting to examine the implications of material state (heterogeneity) before fracture. This has been realized and a branch of statistical physics has recently focused on the subject. For review on the subject we refer to the edited books by Herrmann and Roux (1990) and Charmet et al (1990). Here, in general, the domain of interest is discretized into a lattice, heterogeneity is introduced statistically and the elasticity equations are solved incrementally. Various conditions for crack propagation have been examined. A lattice discretization is more suitable than a finite element one since finite elements are more appropriate and efficient when material response can be homogenized and deformation gradients are not prominent. The fracture patterns that emerge are often fractal in character. At first it may appear surprising that fractal fracture patterns emerge even though the underlying heterogeneous fields are not considered to have scaling properties. However, the experience from (not so) different evolution problems, i.e. DLA (diffusion-limited aggregation), dynamic growth of surfaces, provide some insight. Several issues yet remain poorly understood.

2. Formulation

We examine a fracture network in two dimensions. The possibly scaling properties of the network are considered to hold within the range limited by a lower and by an upper cutoff. The lower cutoff is governed by a length scale characteristic of the material, i.e. a fraction of the grain size in a granular medium. The upper cutoff is in general much larger than the lower one and could be equal to or a fraction of the size of a material specimen or structure. We consider that the network limited by the cutoffs, i.e. as observed in an experiment, is part of the "mathematical" fractal where there are no cutoffs. Thus, it may be considered as a pre-fractal in the terminology of Mandelbrot. We consider the forms of irregular brittle fracture shown in figure 1. The "final" crack

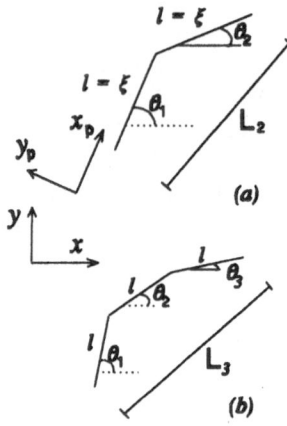

Figure 1: Forms of Irregular Brittle Fracture

consists of repetitions (statistically, spatially and also dilatationally for the present fractal case) of the basic forms shown in figure 1. It is noted that the properties of a unique tortuous crack are addressed (it may be accompanied by branches). If several, i.e. an ensemble which is often the case, such cracks develop then the analysis holds for each of those individual cracks. It can be shown, Frantziskonis (1993c), that case (a), of phenomenological lenght L_2 requires less energy than case (b) (of length L_3) and higher order ones, thus it is preferable. Then, for branches of equal length, in the deterministic case the fractal dimension D would be, Mendelbrot (1983)

$$D = \frac{Log[N]}{Log[\frac{1}{r}]} = \frac{Log[2]}{Log[2Cos(\frac{\theta_2-\theta_1}{2})]}$$

(3)₁

where, figure 1a, $r = l/L_2$, $N = 2l/l = 2$. For the statistical fractal case, the "generalized" fractal dimension is defined as (figure 1, Frantziskonis 1993c)

$$D = \frac{Log[2]}{Log[<2Cos(\frac{\theta_2 - \theta_1}{2})>]}$$

$$(3)_2$$

where $\theta_2 - \theta_1$ is the angle change along two successive "bends" of the crack network and <.> denotes ensemble average. It is noted that $(3)_2$ is a necessary but not sufficient condition for the problem at hand. However, the value of D depends strongly on the spatial variation of the fracture angle θ, i.e. θ_1, θ_2. It is the experiment that provides information on such variation. For the three loading cases examined herein (tension, pure shear, dilatation) we can consider constant mean value of θ, (boundary effects are neglected) and we obtain, after expanding $Cos[\partial\theta/\partial x_p]$ around its mean to a Taylor series and keeping up to second order terms, details are given in Frantziskonis (1993c)

$$D = \frac{Log[2]}{Log[2(1 - \frac{1}{8}l^2V)]}$$

$$(4)_1$$

where

$$V = <(\frac{\partial Sin\theta}{\partial x} - \frac{\partial Cos\theta}{\partial y})^2>$$

$$(4)_2$$

The fracture angle θ is decided from the fracture criterion. In linear isotropic elasticity the principal directions of stress and strain coincide. Thus a criterion based on principal directions of strain would yield identical results (statistically) if the same criterion was based on principal directions of stress. Furthermore, extreme shear strains occur at planes inclined 45° from the planes of extreme normal strains. Then, the relative angle $\theta_2 - \theta_1$ remains unchanged, and a fracture criterion based on extreme normal strains would yield statistically identical results with the same criterion based on extremes values of shear strain. Thus any linear combination of fracture criterion that leaves the relative angle $\theta_2 - \theta_1$ unchanged will yield identical results (statistically). Then, we consider the simplest possible and physically reasonable criterion based on the orientation of the extreme normal strain, thus

$$Tan2\theta = \frac{2\psi_{xy}}{\psi}$$

$$(5)$$

where

$$\psi = \psi_{xx} - \psi_{yy}.$$

In order for the analysis to be complete, the statistical properties of the relevant kinematic quantities should be known. For specific materials the details of the micro-structure may provide information on such properties. However, mainly due to lack of relevant experimental information, and difficulties in describing the geometry of micro-structure mathematically, the relevant statistical properties are assumed herein, as Gaussian (covariance) or scaling fields, and the validity of the assumptions are examined through physical reasoning. In passing, it is noted that Gaussian spatial processes approximate many physical problems adequately.

3. Crack propagation

The following cases are examined:

(a) dynamic crack propagation in an incompressible material - here introduction of a stream function in two dimensions is mathematically convenient and it is considered to be a random field with Gaussian covariance structure for the fluctuations.

(b) dynamic crack propagation in a compressible material - here the micro-displacements are treated in a way similar to the statistical treatment of velocities in the theory of turbulence.

(c) dynamic crack propagation in a compressible material - the micro-displacements are considered to have scaling properties, thus no characteristic length is present except for a lower and an upper cutoff.

(d) slow crack propagation - the properties of the kinematic fields in the vicinity of a crack tip or notch are important in this case.

From equations (4) and (5) for the four cases mentioned above, expressions for D result. Some of the details of the rather lengthy algebra are given in Frantziskonis (1993c) and are not repeated herein since the purpose of the present paper is to examine the implications and possible existence of "universal" values.

Since a definite similarity between cases (a), (b) and (c) has been identified, we only present here the results for cases (a) and (d), and the load cases shown in figure 2.

3.1 Dynamic Crack Propagation - Incompressible Material

For uniaxial tension, and for case (a) of material incompressibility the following equation results

$$D = \frac{Log[2]}{Log[2(1 - \frac{1}{4}\frac{l^2}{L^2_{\psi_{xx}}}\frac{Var[\psi_{xx}]}{<\psi_{xx}>^2})]} \tag{6}$$

where l denotes the lower cutoff of material heterogeneity and $L_{\psi_{xx}}$ is the correlation

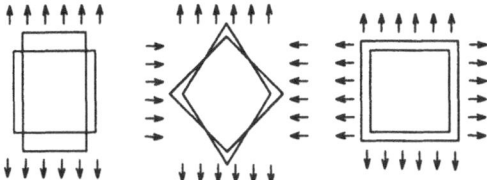

Figure 2: (a) Uniaxial Load, (b) Pure Shear, (c) Dilatation

length for the microstrain Ψ_{xx}. As can be seen, D is influenced by the interplay of correlations $l/L_{\Psi_{xx}}$ and fluctuations $\sqrt{Var[\psi_{xx}]}/<\psi_{xx}>$ which is the coefficient of variation of the micro-strain field Ψ_{xx}. This interplay is discussed in detail subsequently.

For pure shear,

$$D = \frac{Log[2]}{Log[2(1 - \frac{1}{32}(\frac{3}{2})^2 \frac{\pi}{16} \frac{l^2}{L_{\Psi_{xy}}^2} \frac{Var[\psi_{xy}]}{<\psi_{xy}>^2})]} \tag{7}$$

and for the dilatation case

$$D = \frac{Log[2]}{Log[2(1 - \frac{9}{16} \frac{l^2}{L^2})]} \tag{8}$$

where

$$L = \sqrt{2} \, L_{\Psi_{xx}}$$

As can be seen from Eqn (8) for the dilatation case only the correlations affect D. This is due to the fact that in the dilatation case, isotropic external load is imposed and since the fluctuations are also isotropic and homogeneous they do not appear in the expression for D. Thus no matter how small the fluctuations the same (statistically) fracture pattern will evolve. Also, the covariance structure influences the scaling properties of the fracture network (the factor 9/16 comes out in this case but slightly different ones for other types of covariance structure for the fluctuations).

The above results were based on the assumption that material is locally incompressible and that the relevant stream function ϕ has a Gaussian covariance structure and a corresponding correlation length. Differentiation of ϕ introduces covariance structures for the displacement fields with a "hole" with a vanishing integral scale but non vanishing correlation, Lumley (1970), pp 70. The involved assumptions were made for the sake of mathematical convenience only and may not be realistic. It is noted that the compressible case and the scaling case avoid such "problems." However, they yield results similar to the above. Note that no length scales appear in the scaling case, but the scaling exponents do.

In the analysis presented above, the expression of D was shown to be of the form $Log[2]/Log[2(1-\zeta)]$. For $1 \leq D \leq 2$ it follows that $0 \leq \zeta \leq (2-\sqrt{2})/2 \approx 0.29$. The value $\zeta = 0$ implies $D = 1$ or a "straight" crack pattern. Any value such that $\zeta \geq 0.29$ implies multiple self intersecting cracks and may imply fragmentation or disintegration of the material. The best way to evaluate the analysis and the corresponding assumptions would be direct comparison with experimental data. The following can be identified. First, due to the statistical approach to the problem a large number of experiments have to be performed, for each of the loading cases examined herein (and possibly additional ones). Further, except for the dilatation case (where only the spatial correlation structure is needed), information on the fluctuation statistics of the micro-strain fields should be available. It is possible to obtain information on the strain field via, for example, interferometric techniques, Castro-Montero et al (1992), or diffraction techniques, Budhu (1992). No such complete series of tests is available, to our knowledge, at this time. However, relevant information can be extracted from the analytical expression of D, the loading cases examined, available experimental information and numerical results reported in the literature. From the dilatation case, for the spatially uncorrelated case where $l=L$, we have for the incompressible material case $\zeta = 9/16 = 0.56 > 0.29$. This implies that such a material would develop multiple crack intersections when loaded in dilatation mode. There is no physical evidence to support such an implication. One may wonder if such a formulation is valid for high strength fragmenting brittle materials, i.e. high strength concrete. But, such materials are known to be compressible. Then we may conclude (unless experimental information supports the opposite) that the incompressible material formulation is physically unreasonable for many materials. Then, how about the compressible material formulation. In this case for the spatially uncorrelated case we have, Frantziskonis (1993c), $\zeta = 1/8 = 0.125 < 0.29$ and this yields for dilatation $D \approx 1.24$. Thus this case suggests that the maximum possible value of D in dilatation mode is 1.24. If this case can be supported by experimental data, then the 1.24 would be a "universal" number describing the fracture network in dilatation mode (for spatially uncorrelated disordered materials). Although dilatation experiments are not (traditionally) performed often, it seems that many high strength brittle materials will fragment in dilatation, and multiple crack intersections will develop under low rate external load application. If this is the case then the compressible material formulation may be unreasonable (for many materials). Similar arguments can be employed for the case of scaling kinematic fields,

Frantziskonis (1993c). Thus next we present the slow crack propagation case.

3.2 Slow Crack Propagation

Considering a sharp notch in an elastic body, the displacement field in the immediate vicinity of the notch varies as, Atkinson et al (1988)

$$u_x \sim r^\lambda, \quad u_y \sim r^\lambda \tag{9}$$

and the strain fields vary as

$$\psi_{xx}, \psi_{yy}, \psi_{xy} \sim r^{\lambda-1} \tag{10}$$

The above expressions hold for a very small distance in the vicinity of the notch or crack tip. Let the distance where (9) and (10) hold be denoted as ξ_0. Statistically, we can consider that (9), (10) hold in the root mean square sense for displacements, strains. Then the covariance of the displacements, for distances from the notch/crack tip up to ξ_0 has the form $C_{u_x}, C_{u_y} \sim r^{-2\lambda}$. Following the procedure of the previous section we obtained equations (11), (12) and (13) for uniaxial tension, pure shear and dilatation respectively

$$D = \cfrac{Log[2]}{Log[2(1 - \cfrac{1}{8(1+\nu)^2} \cfrac{(1+\lambda)^2(3+2\lambda)}{1+2\lambda} (\cfrac{\xi_0}{\xi})^2 \cfrac{Var[\psi_{xx}]}{\langle\psi_{xx}\rangle^2})]} \tag{11}$$

$$D = \cfrac{Log[2]}{Log[2(1 - \cfrac{1}{32} \cfrac{(1+\lambda)^2(3+2\lambda)}{1+2\lambda} (\cfrac{\xi_0}{\xi})^2 \cfrac{Var[\psi_{xx}]}{\langle\psi_{xx}\rangle^2})]} \tag{12}$$

$$D = \cfrac{Log[2]}{Log[2(1 - \cfrac{1}{16} \cfrac{(1+\lambda)^2(3+2\lambda)}{1+2\lambda} (\cfrac{\xi_0}{\xi})^2]} \tag{13}$$

where $l = \xi$, figure 1, and ν denotes Poisson ratio considered constant. A more general formulation calls for ν being variable (spatially), although even in the present formulation the macroscopic (integrated over some finite volume) Poisson ratio is not constant. Note that for the incompressible material case ν does not appear in the final expressions for D.

In deriving the above equations the following anzatz have been made: (a) equations (9), (10) hold in the root mean square sense, (b) as discussed in the following, the coefficient of variation $\sqrt{Var[\psi_{xx}]}/<\psi_{xx}>$ with respect to a fixed coordinate system is governed primarily by the macroscopic strain field which in turn is governed by the imposed boundary conditions, (c) transverse and longitudinal displacement correlations are the same. These assumptions have a statistical rather than fracture mechanics basis. Especially (b) assumes that heterogeneity is "strong" enough that the statistical properties of the relevant kinematic fields "penetrate" into the immediate vicinity of the crack or notch tip. This does not, however, imply that the fluctuations are large. The linearity of the problem makes this assumption reasonable, since the same linear constitutive equations are considered to hold far from and in the vicinity of crack/notch tips. If crack tips dominate crack propagation, then $\lambda = 0.5$. If crack branching is important then λ may have different values on different regions which may lead to multiscaling.

Here it may be appropriate to mention the work of Mosolov (1991) where starting with an assumption about the stress and strain singularities in the vicinity of a crack tip, the relevant singularity exponents are related to the fractal dimension of the crack. It is noted that nonlinear material behavior effects have similar implications on the singularity exponents, i.e. Broek (1986), pp 246. Then, the approach followed herein is different since the singularities are considered to hold statistically for a heterogeneous solid.

For finite domains it is not clear what the implications are near the boundaries since both the statistical properties of the kinematic fields and the crack tip properties are influenced strongly by the boundaries. Such questions are very important and difficult to examine analytically. Numerically such problems are currently being studied. Another problem is that the value of λ is governed either by notch effects, Atkinson et al [1988] or by plasticity effects near the crack tip.

4. Conclusions: On "Universality"

The above results indicate (cf. following discussion) that under low rate external load the "slow" crack propagation case - the stresses/strains redistribute before further crack propagation - seems to be the consistent one. For example, in dilatation mode, equation (13), all values of D between one and two are possible depending on the value of λ and on the ratio ξ_0/ξ. It is known that stress/strain distribution at the tip of a crack/notch propagating at high speed is different from the static case. According to the dynamic fracture literature, i.e. Broek (1986), the stress intensity will go to zero at the Rayleigh velocity. However, in general, the stress/strain distribution is not largely affected by crack velocity up to a large percent (more than 50) of the Rayleigh velocity. Then, even for load control tests in brittle materials it may be that the crack propagation is not fast enough so that stress redistribution does not take place. For fragmenting materials note that the wave velocities are high (due to high elastic modulus) thus even in this case crack propagation may not be fast enough to avoid stress redistribution during propagation.

In short, the analysis implies that when specimens are fractured either by controlling the external load or the external displacement (at a low loading rate) the following may be occurring: (a) crack propagation is slow, (b) crack propagation is dynamic but such that the dynamic effects on stress/strain intensity are small.

Let us assume that one is capable to vary precisely the heterogeneity properties in a material. For example, this could be done by varying the size, distribution, density, mechanical/chemical properties of aggregates in concrete, as well as the size and distribution of pores, initial microcracks etc. Then certain combinations would yield what is known today as ultra high strength concrete, which under loading disintegrates or fragments as discussed in the previous sections.

We now examine the possible existence of "universal" constants characterizing the transition from a material combination that will develop a single crack (may be accompanied by branches) to the one that will develop multiple crack intersections at a range of scales. To our knowledge the term "universal" has different interpretations in various scientific or engineering fields. Trying to avoid such an issue, it is hoped that our definition of universality used in this paper is clear.

As can be seen from equations (11)-(13) if universal values exist they are specific to the external loading conditions. For dilatation (13) in the limit of multiple crack intersections two of the possible solutions are: solution with $\xi_0/\xi = 1$ and $\lambda = 0.548..$ and solution with $\xi_0/\xi = 1.02..$ and $\lambda = 0.5$. Note that ξ_0/ξ can have values greater than one (this is not the case for the ratio l/L appearing in (6)-(8)). A value of $\lambda = 0.548..$ could imply a nonlinear power law constitutive response for the micro-medium with an exponent $n = 1.216...$ Although this may be possible we strongly believe and/or consider that the constitutive response of the micro-medium for brittle materials is linear. Then the value $\xi_0/\xi = 1.02..$ marks the transition to multiple crack intersections. This value indicates a highly disordered material where the heterogeneity in the vicinity of a crack tip is dominating rather than the singularity itself. This is irrespectively of the size of the crack, thus in order to achieve or exceed such a limit heterogeneity must be present at all scales between the lower and upper cutoffs. Here we note that high strength concretes have a well-graded distribution of aggregates and this agrees well with our results. With this in mind for the dilatation case it is straight forward to show from (11) and (12) that the "universal" value for the coefficient of variation for the tension case is $0.88..$ and for pure shear $\sqrt{2}$, which are relatively high values for disorder. Through post-mortem information - by studying the properties of crack networks - one can identify how far a certain material stands from the limit (universal) values, and this provides a well-grounded quantitative measure of material microstructure quality. Perhaps such concepts pave the way to new frontiers of research in material characterization. For example, the measured D for mortars and normal strength concretes being in the range 1.1 - 1.2 agree with our analysis - such materials stand far from developing multiple crack intersections during fracturing.

Acknowledgments

The research reported herein was supported by Grant No. MSS-9157237 (PYI) from the

National Science Foundation, Washington, D.C., and by the Hughes Aircraft Company. Many of the symbolic computations were performed using the program *Mathematica*.

References

Aifantis E.C., 1984, "On the Microstructural Origin of Certain Inelastic Materials," *J. Engr. Mat. Tech.*, 106, 326-330.

Aifantis E.C., 1992, "On the Role of Gradients in the Localization of Deformation and Fracture," *Int. J. Eng. Sci.*, 30, 1279-1299.

Alexander D.J., 1990, in Quantitative Methods in Fractography, STP 1085, B.M Strauss and S.K. Patantunda editors, 39-51.

Allen A.J. and Scholfield P., 1985, in Scaling Phenomena in Disordered Solids, R. Pynn and A. Skejeltorp editors, 77-85.

Allen A.J., Oberthur R.C., Pearson D., Schofield P. and Wilding C.R., 1987, "Development of Fine Porosity and Gel Structure of Hydrating Cement Systems," *Phil. Mag. B*, 56, 263-288.

Atkinson C., Bastero J.M., and Martinez-Esnaola J.M., 1988, "Stress Analysis in Sharp Angular Notches Using Auxiliary Fields, *Engr. Fracture Mechs.*, 31, 637-646.

Bazant Z.P. and Pijaudier-Cabot G., 1988, "Nonlocal Continuum Damage, Localization Instability and Convergence," *J. Appl. Mech.*, ASME, 55, 287-293.

Budhu M., 1992, private communication

Blinc R.G., Lahajnar G. and Zumer S., 1988, "NMR Study of the Time Evolution of the Fractal Geometry of Cement Gels," *Phys. Rev. B*, 38, 2873-2875.

Broek D., 1986, *Elementary Engineering Fracture Mechanics*, Nijhoff Publ., 4th ed.

Castano V.M., Martinez G., Aleman J.L and Jimenez A., 1990, "Fractal Structure of the Pore Structure of Hydrated Portland Cement Pastes," *J. Matls. Sci. Lett.*, 9, 1115-1116.

Castro-Montero A., Jia Z. and Shah S.P., 1992, "Evolution of Damage in Brazilian Test Using Holographic Interferometry," in ASCE 1992 Engr. Mechs Conf., edited by L.D. Lutes and J.M Niedzwecki, pp 612-615.

Charmet J.C., Roux S., and Guyon E., 1990, editors, *Disorder and Fracture*, Plenum.

Chermant J.L. and Coster M., 1978, "Fractal Objects in Image Analysis, Proc. Int. Symp. on Quantitative Metallography, 125-137, Florence.

Davidson D.L., 1989, "Fracture Surface toughness as a Gauge of Fracture Toughness: Aluminum-particulate SiC Composites," *Mat. Sci.*, 24, 681-687.

Frantziskonis G. and Vardoulakis I., 1992, "On the Micro-structure of Surface Effects and Related Instabilities," *Europ. J. of Mechanics*, A-Solids, 11, 21-34.

Frantziskonis G., 1993a, in "Damage in Composite Materials," G. Voyiadgis editor, Series Studies in Applied Mechanics, Elsevier, pp 137-160.

Frantziskonis G., 1993b, "Heterogeneity and Implicated Surface Effects - Statistical, Fractal Approach and Relevant Analytical Solution," *Acta Mechanica*, in press.

Frantziskonis G., 1993c, "On Scaling Phenomena in Fracture of Heterogeneous Solids," *Europ. J. of Mechanics*, in press.

Herrmann H.J. and Roux S. Editors, 1990, Statistical Models for the Fracture of Disordered Media, North-Holland.

Kanninen M.F., 1985, *Advanced Fracture Mechanics*, Elsevier.

Lange D.A., Jennings H.M. and Shah S.P., 1992, "Relationship Between Fracture Surface Roughness, Fracture Behavior of Cement Paste, Mortar," *J. Amer. Cer. Soc.*, in press.

Louis E. and Guinea F., 1987, "The Fractal Nature of Fracture," *Europh. Lett*, 3, 871-877.

Lumley J.L., 1970, *Stochastic Tools in Turbulence*, Academic Press.

Lung C.W. and Mu Z.Q., 1988, "Fractal Dimension Measured with Perimeter-area Relation and Toughness of Materials, *Phys. Rev. B*, 38, 11781-11784.

Mandelbrot B.B., 1983, *The Fractal Geometry of Nature*, W.H. Freeman and Co., New York.

Mandelbrot B.B., Passoja D.E and Paullay A.J., 1984, "Fractal Character of Fracture Surfaces of Metals, *Nature*, 308, 721-722.

Mecholsky J.J., Mackin T.J. and Passoja D.E., 1988, "Self-similar Crack propagation in Brittle Materials,"in Advances in Ceramics, Fractography of Glasses and Ceramics, 22, 127-134.

Mindlin R.D., 1964, "Micro-Structure in Linear Elasticity," *Arch. Rational Mech. Anal.*, 4, 50-78.

Mosolov, A.B., 1991, "Fractal Griffiths Cracks," *Sov. Phys. Tech. Phys.*, translated in the Americ. Instit. Phys., (1992) pp 754-755.

Muhlhaus H.B. and Vardoulakis I., 1987, "The Thickness of the Shear Band in Granular Materials," *Geotechnique*, 37, 271-283.

Muhlhaus H.B. and Aifantis E.C., 1991, "The Influence of Micro-structure Induced Gradients on the Localization of Deformation in Viscoplastic Materials," *Acta Mechanica.*, 89, 217-231.

Needleman A., 1988, "Material Rate Dependence and Mesh Sensitivity in Localization Problems," *Comp. Meth. Appl. Mech. Engr.*, 67, 69-86.

Pande C.S., Richards L.E., Louat N., Dempsey B.D. and Schwoeble A.J., 1987a, "Fractal Characterization of Fractured Surfaces," *Acta Metall.*, 35, 1633-1637.

Pande C.S., Richards L.E., and Smith S., 1987b, "Fractal Characteristics of Fractured Surfaces, " *J. Mater. Sci. Lett.*, 6, 295-297.

Sayles R.S. and Thomas T.R., 1978, "Surface Topography as a Nonstationary Random Process, *Nature*, 271, 431-434.

Scheefer H., 1962, "Das Cosserat-kuntinuum," *Zeit. Argew. Math. Mech.*, 47, 485-498.

Skjeltorp A.T. and Meakin P., 1988, "Fracture in Microsphere Monolayers Studied by Experiment and Computer Simulation," *Nature*, 335, 424-426.

Takayasu H., 1986, "Pattern Formation of Dendritic Fractals in Fracture and Electric Breakdown," in Fractals in Physics, L. Pietronero and E. Tosatti editors, Elsevier.

Tsai Y.L and Mecholsky J.J., 1991, "Fractal Fracture of Single Crystal Silicon," *J. Mater. Res.*, 6, 1248-1263.

Underwood E.E. and Banerji K., 1986, "Fractals in Fractography," *Matls. Scien., Engr.*, 80, 1-14.

Vardoulakis I. and Frantziskonis G., 1992, "Micro-structure in Kinematic-hardening Plasticity, *Europ. J. of Mechanics*, A-Solids, 11, 467-486.

Wang Z.G., Chen D.L., Jiang X.X., Ai S.H. and Shih C.H., 1988, Relationship Between Fractal Dimension and Fatigue Threshold Value in Dual-phase Steels," Scripta Metall., 22, 827-832.

CRACK GROWTH SIMULATIONS IN CONCRETE AND ROCK

JAN G.M. VAN MIER, ADRI VERVUURT and ERIK SCHLANGEN
Delft University of Technology, Department of Civil Engineering,
Stevin Laboratory, Delft, The Netherlands

Abstract
In the paper the lattice model developed in the Stevin Laboratory is outlined. In the model, the material is discretised as a network of brittle breaking beam elements. F(
simulating fracture in highly disordered materials like concrete and rock, the materia
structure is mimiced in great detail. The generated (two-dimensional) structure of th
material is projected on top of a regular triangular lattice, or on top of a lattice with
beams of random length. The amount of detail included in the material structure
determines the size of the beam elements used in the lattice. Obviously the compute
time will increase with the size of the lattice, and the available computer capacity
mainly determines the size of the lattices that can be analysed. Fracture is simulated
by removing in each load step the beam with the highest stress over strength ratio.
This implies that fracture is completely brittle, and the results obtained so far indica
a close relationship between the amount of detail included in the projected material
structure and the computed softening behaviour of concrete in tension. In the
laboratory the model is used for assessing the response of fracture tests, for example
for determining the correct response parameter in a displacement controlled
experiment. Some examples of analyses are included in the paper.
Keywords: Lattice model, disorder, particle distributions, fracture, concrete, rock,
uniaxial tension, boundary conditions.

1 Introduction

In 1941 Hrennikov developed a simple model in which the continuum was discretized into a network of bar elements. The various bars could appear in different
configurations as can be seen in the original paper, Hrennikov (1941). The great
advantage of the method was that for many insolvable problems in elasticity
suddenly an approximate solution could be obtained. However, due to a lack of
sufficient computer power at that time, the method remained largely unused.
Recently, with the development of more powerful computers, in combination with

D. Breysse (ed.), Probabilities and Materials, 377–388.
© *1994 Kluwer Academic Publishers.*

new modelling and solution techniques (e.g. Meakin (1991), Herrmann (1991)), the method has become an ubiquitous tool in materials science. In 1989 a summerschool was organised in Corsica (Charmet, Roux & Guyon (1990)), where theoretical physisists tried to convince engineers that with these models fracture could be handeled in a more simple and straightforward manner. Inspired by this event, we decided to build a model as tool in experimental fracture studies of concrete and rock. The model has proven to be quite powerful in analysing fracture mechanisms under various combinations of tensile and shear loading, especially in conjunction with experimental observations Schlangen (1993), Schlangen & Van Mier (1992a), Schlangen & Van Mier (1992b), Van Mier (1991a), Van Mier (1991b), Van Mier, Vervuurt & Schlangen (1993). In this paper, the model developed in the Stevin Laboratory will be outlined. More specifically attention will be given to modelling disorder, and some results of simulations will be shown.

2 Lattice Model

2.1 Particle structure

Basic to the model is the generation of a particle structure which resembles the structure of the materials that are studied as closely as possible. Obviously for concrete and rocks different structures have to be generated. When a planar section through a concrete block is made, it seems that the aggregate particles are swimming in a cement matrix. However, it must be realised that in principle a three-dimensional particle stacking exists as shown (in two dimensions) in Figure 1. The role of the

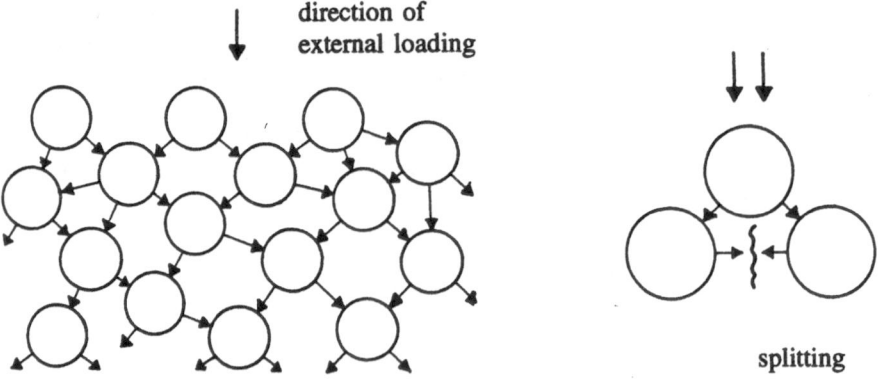

Fig. 1. Two-dimensional particle stacking and splitting forces when external compression is applied, after Reinhardt (1977).

cement is to bind the individual aggregate particles together, but usually the cement layers between the particles will be rather thin. This is caused by the fact that in general a well balanced aggregate grading is selected, where smaller particles fill the space between the larger aggregates. A commonly used particle grading for concrete is the Fuller distribution,

$$p = 100 \ (D/D_{max})^{1/2} \qquad \qquad ...(1),$$

where p is the percentage by weight passing a sieve with aperture diameter D and D_{max} is the largest aggregate size.

The Fuller distribution contains relatively many small particles, which implies that more cement is needed to glue the aggregates together. Based on the Fuller distribution, Walraven (1980) derived the distribution of intersections of aggregate particles (which were assumed to be spherical) in a planar section. The function represents the probability P_c that an arbitrary point in the concrete sample, lying in an intersection plane, is located in an intersection circle with diameter $D < D_0$,

$$P_c(D,D_0) = P_k * \{1.455*D_0^{0.5}*D_{max}^{-0.5} - 0.50*D_0^2*D_{max}^{-2} + 0.036*D_0^4*D_{max}^{-4} +$$

$$0.006*D_0^6*D_{max}^{-6} + 0.002*D_0^8*D_{max}^{-8} + 0.001*D_0^{10}*D_{max}^{-10}\}$$

$$...(2).$$

In Figure 2 a generated particle distribution for a concrete containing 75 volume % of aggregate particles with a maximum size $D_{max} = 16$ mm is shown. The circles are randomly positioned in the area, starting with the largest circles. The minimum distance between the centers of two neighbouring circles A and B is taken $1.1(D_A + D_B)/2$, as suggested earlier by Hsu (1963).

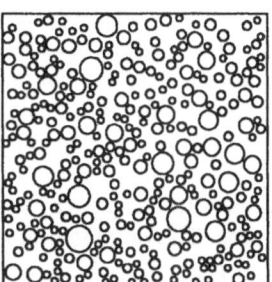

Fig.2. Generated particle structure.

For sandstone, a sedimentary rock, a similar situation arrises. Sand particles are bonded together with a cement consisting mainly of quartz and calciumcarbonate. The process where the particles are bonded together is called diagenesis or consolidation, e.g. see Heijnen (1992). In the case of sedimentary rock, however, the distribution of the sand particles seems more uniform.

Equation 2 is a simple manner to generate a particle structure in a planar section. Recently De Schutter & Taerwe (1993) published a method based on Delaunay triangulation which gave a good approximation of the distributions of intersections of aggregate particles in a plane. However one should realize that such a two-dimensional representation of the material is essentially not correct. Especially for external compressive loading, neglecting the third dimension seems not allowed. In that case the splitting forces are the main source of crack growth in the composite (see Fig. 1), and this effect is completely neglected in the two-dimensional representation of aggregate particles swimming in a matrix. However for simulating tensile fracture the latter approach seems to give a good approximation of 'reality'. The fracturing of the concrete is then determined by the bond between aggregate and matrix, and interactions between aggregate particles seem of less importance.

2.2 Regular lattice overlay

In the lattice model developed at the Stevin Laboratory (Schlangen (1993), Schlanger & Van Mier (1992a,b), Van Mier, Vervuurt & Schlangen (1993)), the continuum is discretized in a network of brittle breaking beam elements. The method is derived from work by Herrmann (1991). However, for our application a regular triangular lattice was chosen rather than the regular square lattice proposed by Herrmann. The advantage of a triangular lattice is that a better prediction of the Poisson's ratio of the material is obtained. When fixed nodes are used, i.e. when bending moments can be transferred in the nodes, a Poisson's ratio of 0.16 is calculated. In contrast, when the original model by Hrennikov is used where the bars are connected through hinges, the Poisson's ratio is equal to 1/3.

The triangular lattice is overlaid over the generated material structure as shown in Figure 3. Subsequently different properties are assigned to the beam elements depending on their location in the projected material structure. For example, a beam element projected in an aggregate circle will get the tensile strength of that aggregate as well as the corresponding Young's modulus and Poisson's ratio. The size of the beams in the lattice is determined by comparing the stiffness of a lattice with that of a continuum element that it should represent. In general the thickness of the beams is taken equal to the thickness of the continuum element and the height of the lattice beams is adjusted. In most analyses the height of the beam was found to be equal to $0.68*\ell$. The length ℓ of the beams depends on the size of the smallest aggregate particle included in the material structure. In general the length is one third of the size of the smallest aggregate. In the present examples aggregates between 3 and 8 mm are included in the mesh. The complete tuning procedure for the stiffness is given in Schlangen (1993), and partly in Schlangen & Van Mier (1992b).

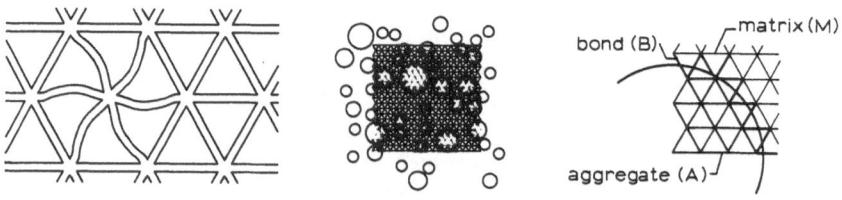

Fig.3. Regular triangular lattice and overlay on generated material structure.

2.3 Random lattice overlay

Instead of the regular lattice also a random lattice can be used. The regular lattice has the apparent disadvantage that once a crack starts to develop, it will tend to follow the mesh lines of the lattice. Deviations occur only when stiff regions are encountered during crack growth. By using a random lattice this problem can be circumvented. The random lattice that was used in our analyses was developed at KFA Jülich by Moukarzel & Herrmann (1992), Vervuurt, Van Mier & Schlangen (1993). Starting point for constructing the random lattice is a square grid of size s*s mm^2. In each cell a point is selected at random as shown in Figure 4. Subsequently the random lattice is defined by connecting always the three points that are closest to each other using the Voronoi construction (Fig. 4a.). The randomness can be varied by decreasing the area in a square cell where a point is chosen at random (see Fig. 4b.). The random lattice can be used in two different ways. The first application is to project it on top of the generated material structure as described in the previous section for the regular lattice. Another option is to use the lattice directly. The heterogeneity in the lattice is then defined by the random beam length. In that case all the beams get the same failure strength and the same Young's modulus.

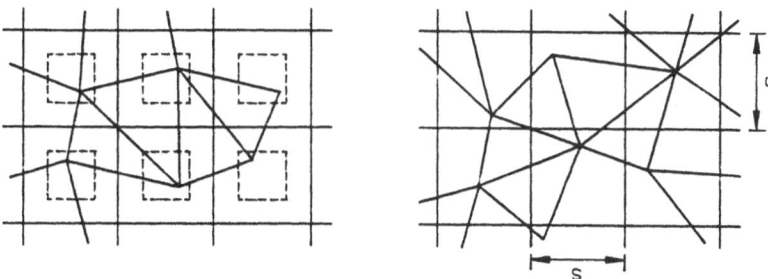

Fig.4. Construction of a random lattice (a) and variation of randomness (b).

2.4 Simulating fracture

After the lattice has been constructed, the simulation can be carried out. Because of the lengthy analysis, only the part of the specimen where cracks are expected to grow is modelled with the lattice. The remainder of the specimen is modelled with simple plane stress elements that are available in the finite element package DIANA that was used in the analysis. The analysis is fully linear elastic and deterministic. After a test load has been applied, stresses in the beam elements are determined. In each beam element a combination of normal and flexural stress is calculated following

$$\sigma_t = F/A + \alpha*(|M_i|, |M_j|)_{max}/W \qquad \qquad ...(3),$$

where A is the cross-sectional area of the beams and W is the sectional modulus. The factor α is included in order to adjust the amount of bending that is taken into account. In a parameter study it was found that α determines the length of the tail of a computed softening diagram, see Schlangen & Van Mier (1992b). In the present

382

analyses α = constant = 0.05. However also the cross-section of the beams is important, but these effects have not been studied in detail at the moment this paper was written. Fracture is simulated by removing the beam element with the highest stress σ_t relative to its tensile strength f_t. Subsequently a new test load is applied to the new lattice and the next beam is removed. This process continues until complete fracture occurs. The external load that can be carried by the lattice can be calculated in each load step by multiplying the test load P with a factor $\beta * f_t/\sigma_t$. The parameter β is a scaling factor for the global maximum stress, and is tuned through an analysis of a simple uniaxial tension test, Schlangen & Van Mier (1992b). Finally the rather jagged load-displacement diagram is smoothened. The exact procedure is described in Schlangen (1993). It should be mentioned that beam theory is applied in the analysis of σ_t in spite of the fact that the beams in the lattice are rather high beams (h = 0.68*ℓ). At present the effect of the slenderness of the beams is analysed.

3 Numerical Simulations

The lattice model has proven to give correct simulations of crack patterns in concrete and rock specimens subjected to uniaxial tension or combined tension and shear. From the analyses global 'ductile' response is obtained, even though a perfectly brittle local fracture law is used. The 'ductility' can be explained from crack face bridging, which seems to be a direct consequence of the heterogeneity of the material. Crack face bridging was recently demonstrated in tensile experiments, Van Mier (1991a,b), Van Mier, Vervuurt & Schlangen (1993). The computed load-displacement diagrams are, however, still too brittle. This brittleness can be explained, at least partially, from the so-called small particle effect, see Van Mier, Vervuurt & Schlangen (1993). The small particle effect means that the additional ductility caused by bridging around smaller sized aggregate particles, i.e. below the lower cut-off, is neglected in the analyses. However, deviations probably appear as

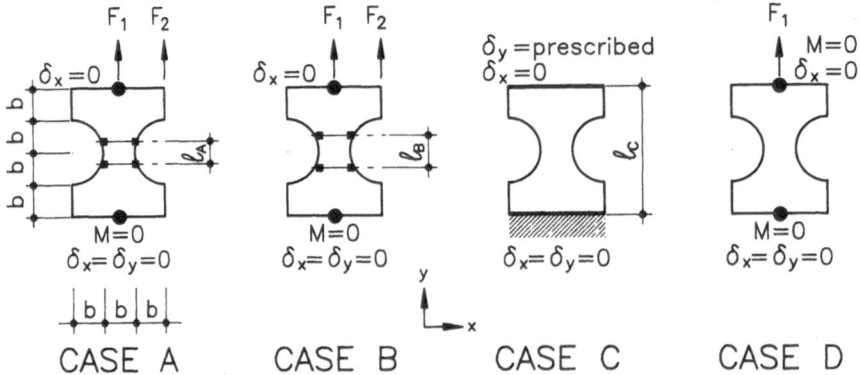

Fig.5. Specimen geometry and boundary conditions analysed: (a) control length l = 35 mm; (b) control length l = 50 mm; (c) parallel end-platen displacement; and (d) freely rotating loading platens; b = 50 mm.

well because the 3D fracture process is modelled in two dimensions. In this paper the effect of particle distribution in the generated material structure is shown, and the influence of loading platen rotations in tensile tests on concrete is demonstrated. The specimen geometry is derived from recent experiments carried out by Carpinteri & Ferro (1992) and is shown in Figure 5. Carpinteri and Ferro tested specimens of different size. The specimen thickness was always t = 100 mm. In the present paper only simulations of the smallest specimen size are included, namely b = 50 mm. The tests in Torino were controlled in a rather complicated manner. The specimens were loaded by three hydraulic jacks in order to maintain a uniform deformation distribution in a 50 mm wide zone as indicated in Figure 5b. F_2 indicates the location of the second hydraulic jack. Carpinteri and Ferro claimed that through this test procedure directly strain-softening curves for concrete could be measured. Through this remark, they implicitly stated that structural effects had no significant effect on crack growth in their tests. Crack growth was not monitored in their experiments.

3.1 Effect of particle distribution
For boundary condition C, parallel end platen displacement, three analyses were carried out on meshes with different particle distributions. Three different distributions were generated using eq. (2). The maximum particle size was 8 mm, the minimum aggregate size included in the analysis was 3 mm. A regular triangular lattice was projected on top of each of the material structures. The following material parameters were used: f_t = 1.25 MPa, 5 MPa and 10 MPa for the bond zone beams, the matrix beams and the aggregate beams respectively; E = 25 GPa, 25 GPa and 70 GPa for these three phases; the beam thickness is equal to the specimen thickness t = 100 mm, the beam height h = 0.68*ℓ, and ℓ = 5/3 mm.

The load-displacement diagrams of the three different analyses are shown in Figure 6. The crack patterns at two stages of loading are shown in Figure 7 for each of the three simulations. Depending on the distribution of aggregate particles, the simulated crack patterns develop in different manners. Yet, the global stress-crack opening diagrams show large similarities. In fact the result is comparable to the scatter found in tests on laboratory specimens.

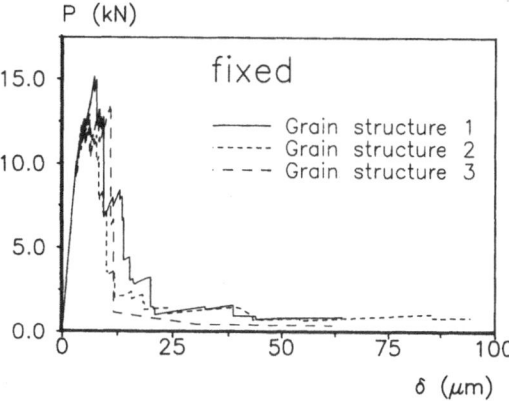

Fig.6. Load-displacement diagrams for three simulations of boundary condition C.

3.2 Effect of boundary rotations in uniaxial tension

The effect of allowable rotations outside the crack zone is shown in Figure 8. Because the results of boundary condition A and B were almost identical, the result of boundary condition B is omitted. Cracking is shown at two stages: first for 200 beams removed, and secondly at the end of the analysis when full separation occurs.

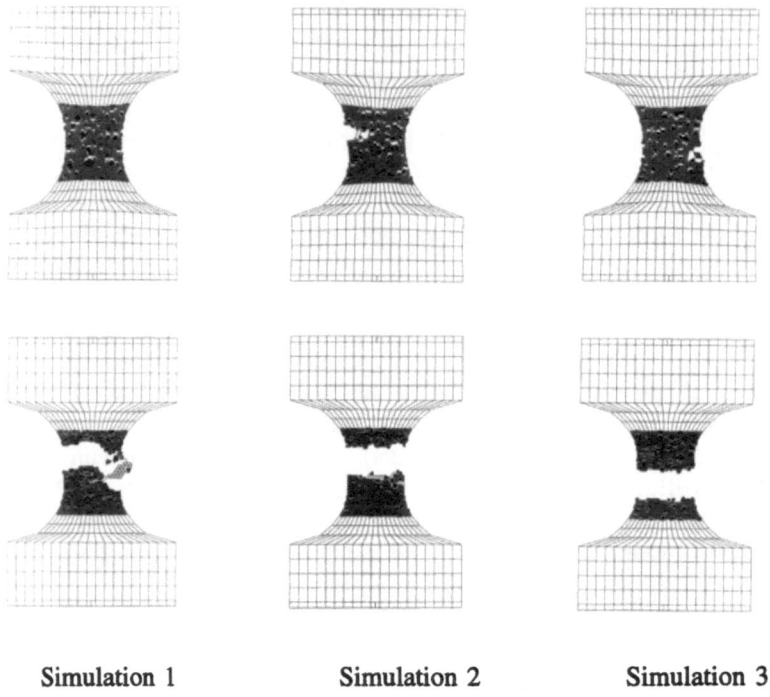

Simulation 1   Simulation 2   Simulation 3

Fig.7. Two stages of cracking for each of the three simulations. The upper figure is always at the stage where 200 beams are removed, the lower figures show the crack patterns at final separation.

The results of the analyses A and C, where uniform deformations are kept over a part of the specimen length (A) or over the entire specimen length (C), show similar behaviour. Note that in analysis A only uniform deformations are kept in a central zone of the specimen. This implies that still rotations may occur at the specimen/ loading platen interface. Although the details of the crack patterns are different, in general large stress-redistributions may occur under these boundary conditions and highly distributed crack patterns are observed. In contrast, when the loading platens are allowed to rotate freely, the specimen seems to fracture at the weakest spot, and only a limited amount of side cracking is found (Figure 8c). The computed fracture energies are also different. In table I, the computed fracture energy up to an average axial crack opening of 100 μm are shown for the four different boundary conditions of Figure 5. It may be obvious that the fracture energy decreases when the amount of side cracking decreases. This is most clearly visible when the simulations of boundary condition D (free rotating end-platens) are compared with the other three boundary conditions.

This result may indicate that the behaviour of a tensile test cannot be transfered directly to other situations. In fact this observation is in agreement with conclusions drawn before by the first author, Van Mier (1986). Recently experiments in Lulea showed the effect of boundary rotations, Daerga (1992). The conclusion of Carpinteri and Ferro seems therefore not permitted, and indeed the application of continuum softening models seems rather debatable. The structural effects are very prominent,

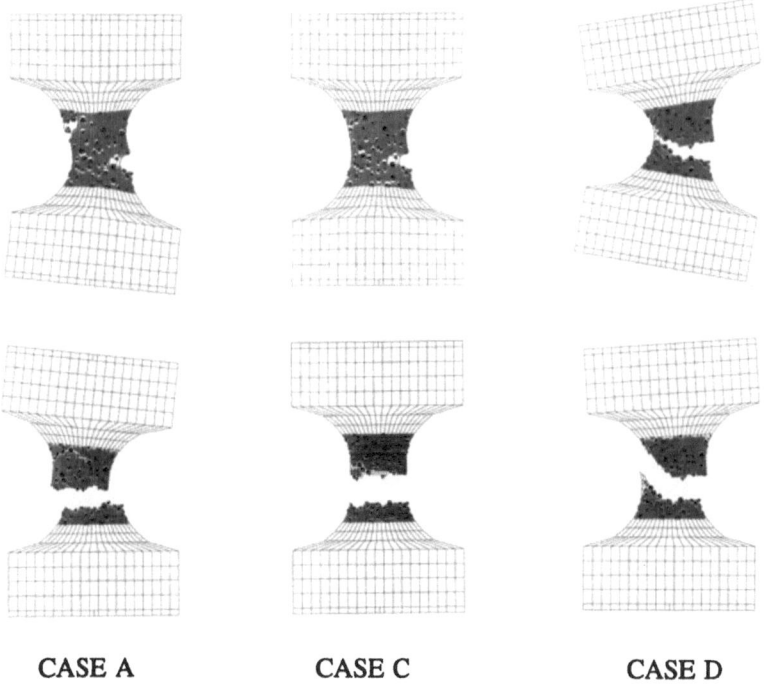

CASE A   CASE C   CASE D

Fig.8. Cracking for three simulations with particle structure 2 for boundary condition A, C and D. For each analysis cracking at two stages is shown: at 200 beams removed and at complete separation.

and cannot be neglected, not even in a 'simple' test like a uniaxial tension test. Recently it was shown that with a Cosserat continuum modelling tensile fracture localization is not possible, De Borst & Mühlhaus (1991). The fracture patterns probably suggest what might be the reason for the fact that the higher order terms in the Cosserat continuum are not activated in tension. The bridging effects that appear in the macroscopic cracks, as can for example be seen from the two analyses with boundary condition A and C in Figure 8, are responsible for flexural stresses in the crack region. These details are missed when the crack is modelled as a line crack, as is done in a Cosserat continuum. It seems very likely that the higher order terms are activated when the Cosserat continuum is combined with a particle structure, which implies however that the continuum principle is left. As in the lattice model simulations, the heterogeneity will then deviate the crack from a plane, and it will be obvious that the higher order flexural terms wil become activated. This also explains why the factor α in our fracture law (eq. 3) has the largest effect in the tail of the softening diagram. The experiments by Van Mier (1991a,b) clearly indicated that this is the region where bridging is most prominent.

Table I. Computed "fracture energies". The deformations are based on a measuring length of 35 mm in all cases. The fracture energy has been calculated as the area under the stress-crack opening diagram up to 100 μm. The cross-sectional area is taken equal to 50*100 = 5000 mm².

Simulation	grain-structure	G_f (N/m)
CASE A (ℓ_{meas} = 35 mm)	1	40.0
	2	44.0
CASE B (ℓ_{meas} = 50 mm)	1	40.6
	2	43.8
CASE C (fixed)	1	35.3
	2	32.5
CASE D (rotating)	1	29.8
	2	11.6

4 Conclusion

In this paper a simple lattice model for simulating fracture in concrete and sandstone is outlined. The heterogeneity of the materials is introduced by superimposing a regular triangular or random lattice on top of a generated two-dimensional particle structure of the materials under consideration. With the model crack face bridging in tension can be simulated. Early experiments Van Mier (1991a,b) have shown that crack face bridging is the main toughening mechanism in this class of quasi-brittle materials. With the model the response of laboratory scale experiments can be better understood. An example is given of the effect of loading platen rotations of the cracking in uniaxial tensile tests. The simulations indicate that when the rotational stiffness of the loading platens decreases, a large decrease in fracture energy is observed. In other words, testing between rotating end-platens yields a result where the most critical cross-section of a specimen will fail. In general the amount of side cracking will be less and the peak load will be reduced under such conditions. The results are in agreement with experimental observations. The main conclusion is that macroscopic fracture properties of heterogeneous composites cannot be measured directly in experiments. Moreover, it seems rather debatable whether the micromechanics model can be used to derive a macroscopic fracture law for these materials in view of the specimen geometry and boundary condition dependency of the problem. In that case, the analyses of fracture behaviour of, for example, large concrete structures, should be carried out with a micromechanics model. With the current state of computational possibilities this is of course impossible. However following this route, the engineer will be certain that a more reliable result is obtained.

5 References

Carpinteri, A. and Ferro, G. (1992) Apparent Tensile Strength and Fictitious Fracture Energy of Concrete: a Fractal Geometry Approach to Related Size Effects, In Fracture and Damage of Concrete and Rock (FDCR-2), H.P. Rossmanith (ed.), Chapman & Hall/E&FN Spon, 1993, pp. 86-94.

Charmet, J.C., Roux, S. and Guyon, E. (Ed.) (1990) Disorder and Fracture, Plenum Press, New York.

Daerga, P.A. (1992) Some Experimental Fracture Mechanics Studies in Mode I of Concrete and Wood, Licentiate Thesis, Lulea University of Technology.

De Borst, R. and Mühlhaus, H.-B. (1991) Continuum Models for Discontinuous Media, In Fracture Processes in Concrete, Rock and Ceramics, J.G.M. Van Mier, J.G. Rots and A. Bakker (eds.), Chapman & Hall/E&FN Spon, London/New York, pp. 601-618.

De Schutter, G. and Taerwe, L. (1993) Random Particle Model for Concrete Based on Delaunay Triangulation. Materials & Structures (RILEM), 26(156), 67-73.

Heijnen, W.M.M. (1992). Natural Mineral Aggregates for Concrete. In New developments in Concrete PATO, pp. 1-10.

Herrmann, H.J. (1991) Patterns and Scaling in Fracture, In Fracture Processes in Concrete, Rock and Ceramics, J.G.M. Van Mier, J.G. Rots and A. Bakker (eds.), Chapman & Hall/E&FN Spon, London/New York, pp. 195-211.

Hrennikov, A. (1941) Solution of Problems of Elasticity by the Framework Method. J. Appl. Mech., A169-A175.

Hsu, T.T.C. (1963) Mathematical Analysis of Shrinkage Stresses in a Model of Hardened Concrete. Journal of the American Concrete Institute, 3, 371-390.

Meakin, P. (1991) Simple Models for Material Failure and Deformation, In Fracture Processes in Concrete, Rock and Ceramics, J.G.M. Van Mier, J.G. Rots and A. Bakker (eds.), Chapman & Hall/E&FN Spon, London/New York, pp. 213-229.

Moukarzel, C. and Herrmann, H.J. (1992), A Vectorizable Random Lattice, Preprint No. HLRZ 1/92.

Reinhardt, H.W. (1977) Anspruche des Konstrukteurs an den Beton, hinsichtlich Festigkeit und Verformung. Beton, 5, 195-199.

Schlangen, E. (1993) Experimental and Numerical Analysis of Fracture Processes in Concrete, PhD thesis, Delft University of Technology.

Schlangen, E. and Van Mier, J.G.M. (1992a) Experimental and Numerical Analysis of Fracture of Cement-Based Composites. Cement & Concrete Composites, 14(2), 105-118.

Schlangen, E. and Van Mier, J.G.M. (1992b) Micromechanical Analysis of Fracture of Concrete. Int. J. Damage Mechanics, 1(4), 435-454.

Van Mier, J.G.M. (1986) Fracture of Concrete under Complex Stress. HERON, 31(3), 1-90.

Van Mier, J.G.M. (1991a) Crack Face Bridging in Normal, High Strength and Lytag Concrete, In Fracture Processes in Concrete, Rock and Ceramics, J.G.M. van Mier, J.G. Rots and A. Bakker (eds.), Chapman & Hall/E&FN Spon, London/New York, pp. 27-40.

Van Mier, J.G.M. (1991b) Mode I Fracture of Concrete: Discontinuous Crack Growth and Crack Interface Grain Bridging. Cement & Concrete Research, 21(1), 1-15.

Van Mier, J.G.M., Vervuurt, A. and Schlangen, E. (1993) Analysis of Fracture Mechanisms in Particle Composites, In Micromechanics of Concrete and Cementitious Composites, C. Huet (ed.), Presses Polytechnniques et Universitaires Romandes, Lausanne, pp. 159-170.

Vervuurt, A., Van Mier, J.G.M. and Schlangen, E. (1993) Analysis of anchor pull-out in concrete. Materials & Structures (RILEM), accepted for publication.

Walraven, J.C. (1980) Aggregate Interlock: A Theoretical and Experimental Analysis, PhD thesis, Delft University of Technology.

REPORT ON THE SECOND ROUND TABLE SESSION: MODELS FOR FRACTURE

A. CHUDNOVSKY* and F. HILD**
*Department of Civil Engineering, Mechanics and Metallurgy, Chicago, U.S.A.
**Laboratoire de Mécanique et Technologie, Cachan, France.

Abstract

This paper is the report on the round table discussion about Models for Fracture: should one use deterministic or stochastic concepts, at what scale? The authors tried to be as close as possible to the actual discussion that took place.

Keywords: Fracture, Scale, Micromechanics, Predictions, Stochastic formulation, Deterministic formulation.

1 Introduction by Prof. A. Chudnovsky

Let me start the round table discussion on Models and Fracture. It is the second round table discussion since we started yesterday and today it would be a natural continuation of that previous one. I believe the purpose of a round table discussion is to address some challenging questions, maybe even provocative questions, deliberately provocative in order to focus on the critical issues we may agree or disagree about.

To start the discussion I would like to refer to Max Bohr regarding the status of the concept of the chance in classical Physics: classical Physics attempted without success to get rid of the chance; today, the hierarchy of values changed: the probability and uncertainty appear to be more fundamental than deterministic concepts. That was said with respect to the development of Quantum Mechanics, where probabilistic ideas indeed are more fundamental than deterministic ones. However, an opposite opinion exists in the field of Applied Mechanics. Let me quote one famous Russian scientist and engineer V. I. Feadosiev: '... All the efforts (in terms of time and money) invested into the development of mathematical techniques for a probabilistic evaluation of failure or buckling may be wasted since the probabilistic information on the initial imperfections is

389

D. Breysse (ed.), Probabilities and Materials, 389–401.

highly uncertain.' His point is that if one tries to account for the uncertainties which exist in real systems, the probabilistic results become so uncertain that may be it is not worth the efforts to get such a vague and fuzzy result.

To summarize this opinion regarding the status of probability in Applied Mechanics, I would like to rephrase a famous statement of Michelangelo Buonaroti: '... *A sculptor is not a person who <u>can</u> create a sculpture, it is a person who <u>cannot</u> live without doing sculptures.*' In our field it would sound as follows: '*The probabilistic approach should be employed not in the problems it can be applied but for the problems that cannot be solved otherwise.*' The two opposite opinions regarding the place of probabilistic concepts in Mechanics are open for discussion. However, in this form, the subject is too broad. I want to focus our discussion on Fracture Phenomena.

Large scatter of the observable quantities and well pronounced scale effects are characteristic features of Fracture. Irregular (fractal) fracture surfaces are also commonly observed in most engineering materials. Thus it appears that Fracture Phenomena is an appropriate field for the application of probabilistic techniques. The chance in Fracture is conventionally accounted for by various modifications of the 'Weakest Link' arguments. The deterministic elements of brittle fracture are well depicted by Fracture Mechanics. What are the limitations of the above mainstreams of Fracture research? What aspects of Fracture Phenomena a) have not been addressed, or b) have been addressed but not answered?

There is an additional philosophical question I want to offer for discussion. One of the critical elements of any physical theory is the possibility of its experimental verification or disapproval. Only negative results of multiple attempts to disprove a theory give a confidence in its correctness. Probabilistic models of Fracture, however, avoid such a scrutiny. There are always enough adjustable parameters to reach an agreement with the observations (the same can be said about most of the deterministic models as well). Thus modeling of Fracture is rather descriptive (i.e. phenomenological) and lacks the predictive power. Is it an inherent limitation resulting from the extreme complexity of the phenomena we are dealing with? Should we accept the status quo or is there a hope to elevate it to the status of a physical theory? If there is a hope, in what direction(s) do we need to launch our research?

2 Discussion

Prof. Soulie

I am just a geotechnician, but I accepted the invitation to participate [to this discussion] because we are faced with the same problems as you, we have highly variable materials, and for many reasons we have followed different routes to tackle the problem of fracture. In the past we were not interested in this kind of problem in Soil Mechanics: sands have no cracks unless you have some cementitious sand. However, the problem of fracture is interesting even in Soil Mechanics for some specific cases, for example the fracture of soils at sub–zero temperature. I remember some years ago we were looking at the mechanical behavior of clay in soils at a temperature of $-160°C$ (study of the possibility to put liquefied gas directly in underground cavities). I can tell you that at this temperature clay is very brittle, and we got a lot of problems with cracks. I still do not have the solution to the problem. I have also a question to ask: you usually study the propagation of a crack driven by external forces applied on specimens or samples. However, the stresses can be induced by different causes (e.g. the thermal stresses). Do you believe that the characterization of these cracks can be similar to that for the cracks induced by external forces? I am not sure. It seems to me the physics [of the problem] is different. Are the characterization of crack propagation, fractal geometry of fracture surfaces independent of the causes of the cracks?

Prof. C. Huet

I see no reason for which it should be different from what happens in brittle materials at room temperature, and the same kind of stresses. I understand you have thermal stresses and we do know how to calculate thermal stresses, for instance the energy release rate G in a field of thermal stresses. It should not be different at low temperatures because it is only a problem of the position of the glass transition temperature.

Prof. M. Soulie

The near crack tip fields are really temperature dependent. So, at the end, I will not be surprised if the geometry of the cracks will be also temperature–dependent.

Prof. C. Huet

Yes, the geometry of the crack of course, especially if you have Γ or G_c represented by a function in space. But there are other situations for which these parameters are functions of space, and we can deal with them very easily. The only question is to know if the basic criterion (i.e. Griffith criterion) is still valid. In my opinion it is valid.

Prof. A. Chudnovsky

I would take Prof. Huet's response with some caution. In the crack driving force we have active part and resistive parts. Energy release rate, resulting from the thermo–elastic stress analysis is practically the same (if we know how to perform the stress analysis). However, the resistive part which depends on the particular fracture mechanism may change with temperature. This is I believe the point of Prof. Soulie's question.

Prof. C. Huet

What you call the resistive part is the dissipated energy. Once you are able to recognize and to calculate how the energy is dissipated and what is the amount of dissipated energy by each small increment of crack progression you can do the same. It is an easy matter from the thermodynamics point of view to write the equations in such a way to show that, whatever the dissipative mechanism, the basis of the Griffith criterion still works because it is standing upon the universal laws. Just like my assimilation conditions are standing upon the universal balance equations. It is the same for the Griffith law. The Griffith law conveniently extended is standing upon the universal balance equations, whatever the behavior of the material may be. So if you have a good mastering of the Griffith law, you can do everything.

Prof. A. Chudnovsky

We are in agreement regarding the applicability of the Griffith criterion. The open question still would be how the specific fracture energy G_c depends on particular fracture mechanisms. For example in thermal fracture or fracture due to swelling when water is absorbed or due to compaction, drying and so on. So it is an open field. It seems to me that it is a good suggestion that we should look more carefully on the fracture mechanisms, and identify the specific fracture energy for various mechanisms.

Prof. H. Mihashi

Yesterday [in Session 2], Prof. Chudnovsky emphasized that simple models based upon a physical background are most important. I do agree with his opinion. Now I try to give my opinion on the two questions proposed by Prof. Chudnovsky. First, fracture of materials usually depends on the microstructure. It has been already discussed during Session 2. We can think about three different levels, [viz.] macro, meso and microscopic levels. If one thinks about that structural hierarchy, it is natural to introduce the probabilistic models to discuss the Fracture Phenomena. Assume that one establishes a comprehensive probabilistic model that accounts for various physical mechanisms, and then finds that the distribution of strength has a quite small scatter. In such a case it is not always necessary to use probabilistic models. We can use deterministic models for practical purposes. Maybe that is one of the reasons why practitioners use deterministic models: probabilistic models are more complex. In principle, probabilistic aspects in Fracture are due to the following two reasons: first of all, there is a dynamic randomness that leads to a scatter in the fracture time. This randomness was not addressed in this workshop so far. Secondly, there is a geometrical randomness of material defects (e.g. size, shape, distance of some meso and/or micro defects). The Weakest Link Model has been used quite often for such kind of materials. However, I think it is not so easy to use directly classical WLM for highly heterogeneous materials such as concrete. Crack initiation from some micro defects in such materials does not mean global failure. We have to think about some crack propagation process. What kind of model should be used for that fracture behavior? I think we have to be concerned not so much with the mathematical elegance but rather with more realistic models even if they become more complex.

The problem is that it is not so easy to determine the material parameters by some standardized tests for complex models. That is the reason why it is more difficult to apply probabilistic models to practical problems. One can get some ideas about the sensitivity of the model to certain parameters linked with some physical meaning by a numerical simulation and a qualitative analysis of the model. It may provide some answers for practitioners regarding the morphological features underlying the observed scatter. Predictions based upon probabilistic models are not always necessary to give some quantitative values directly. Probabilistic models can give some ideas about the framework of that phenomenon.

Prof. D. Breysse

In the synthetic paper of Session 4, I tried to separate micro and macro models, trying to show what we are choosing when we build a model. When we write a model it is just a scale where we can put some data. Many speakers have pointed out that the most important point is to know what are the available data. First, these data can result from a morphological analysis, from a structural study, from many other things like mechanism analyses. This is our starting point to build the model. Then we have to begin to build something which takes this information at a micro level and transports it to a macro level where we want the information (i.e. the response). What seems the most important point to me is knowing that the information can be taken at many different scales. At what scale do we have to introduce the phenomenology to have a given response? For failure, we have seen that we have many different alternatives. We can just say for instance that time to failure is a random variable, and we fit the experimental distribution of time to failure of the global response and we have a probabilistic macro model. If this is good, and if this can be a predictive model, I think we do not need micromechanics. But in many cases, this is not sufficient, and we have to go and try to get more information.

I will go back to the presentation of [Prof.] van Mier just as a basis for discussion. It seems to me that it is a very good example of what we can do with the models and what are the basic points concerning identification and prediction with models. He presented his model such that we could understand what he was doing at all levels. The parameters can be separated between geometry, elastic properties, material strength and also some numerical parameters due to the beam model. For each parameter he explained that he is able to identify them by three different methods, by mapping for the geometry (it is completely correct, I think), by fitting for the elastic properties (fitting with the Continuum) or reverse engineering. A small problem was about the characteristics of the various phases, but this is not important in this kind of applications because he has shown that this model was not built to predict all the results of behavior in any case, but for specimens used fro the identification. The model is then used to predict similar behaviors of specimens with different sizes but made of the same material. This is an example, once this has been told, of a predictive model in a very specific domain. We lack for each model this kind of methodology which makes us able to say: 'this model can predict that feature with the information taken at a given level.' For instance, [Prof.] Kunin has shown that the correlation length can be related to scatter in time to failure. This is a key element for modeling.

Prof. G.I. Schuëller
Prof. Soulie is not a materials scientist and myself am also not a materials scientist. I am a structural engineer and in that respect user of the product. The user of the product has some influence on the product, what Material Science is delivering to us. The structural

engineers have to make a decision, I presume like the soil engineers. Therefore quantitative information is much better than a qualitative one, no matter how large the scatter is. The key issue is to use as much information as we can. If we turn around the argument, we may say that without the probabilistic analysis one neglects some information which is available. Who nowadays wants to neglect information? I do not see the contradiction between deterministic models and statistical models, and that particularly refers to Fracture Mechanics; the modeling being done at a macro scale and there I understand more than on a meso or micro level. We do have physical models, and we try to develop the physical models as good as we can. As I mentioned in yesterday's discussion we are not always successful, but I think every scientist or engineer will agree that the physical models we are developing are not perfect. There are always approximations coming with it. As we develop the field, we can drop some of these approximations which have been introduced at first. On the other hand we do have a problem if we talk about failure probabilities and then I come back to make decisions on how to design a particular component. We also have the statistical uncertainty or the uncertainties in the probabilistic models. They are of two types, [viz.] there is a modeling uncertainty related to the physics and there is a modeling uncertainty associated with the probability (statistical inference). We have to predict failure probabilities or reliabilities to say it more positively, and based on all these uncertainties to estimate the parameters on which we base our decisions. Apparently, it is also associated with the uncertainty. We have to know about these uncertainties.

What is the future research? The information on initial defects is the final information the user is concerned with. A lot of research in the so–called crossroads of materials and structural engineering is spent on identifying initial defects. Experimental methods have to be developed for this and they have been developed. Some of them have contradictory results. Probabilistic models have to be developed for describing the initial defects. We have heard today about a number of models and I would not like to evaluate those, but we have heard a number of times that we have to make the assumption on this or that, e.g. on the probabilistic model, the Poisson distribution assumption, etc... We have to ask ourselves the question how accurate are these assumptions with respect to the physical reality. This is one of the key issues in my opinion.

396

Prof. A. Chudnovsky

There is one remark I want to make as a continuation of yesterday's discussion. I agree with Prof. Shueller, if we do not use probabilistic approaches we are missing some information which is available. However, when we do use probabilistic approaches we can introduce misinformation which we are not aware of and that is another danger. Thus we need to find a compromise.

Prof. M. Ostoja-Starzewski

I just want to say one thing very shortly on the philosophical side but also on the foundation side in Mechanics. I do not know if we can separate the probabilistic Physics from the classical one. The simple cases I want to quarrel around are computer experiments rather than physical ones. In a computer we have to use some mechanisms which are non–random. We just cannot generate random numbers; strictly speaking they are pseudo–random. So, basically, I do not disagree with the panel but we have a very deep problem here. We are simulating randomness with non–random numbers strictly speaking, coming from a strange attractor, something like that. Some pseudo-random sequence is all very nice, so why do we do it for a real material? We do it because there is some intrinsic randomness (e.g. Γ–field, elastic constants). I think the challenge which is still waiting for us is to find the rules of generation of randomness, which really are of some deterministic nature. On the other hand, for applications one might start with some adequate approximation of a random field.

Prof. A. Chudnovsky

I am not sure that I follow your last statement. Do you mean that on the very basis, on the very fundamental level of the probabilistic laws you are going to find some deterministic law?

Prof. M. Ostoja-Starzewski

Maybe some analog of a strange attractor in the spatial temporal sense. I do not know. Why do we use randomness, for example Poisson random field of defects? This is just a mathematical model, but if we simulate a random process in time we might just do it using random numbers which come from a random number generator which is non–random.

Prof. A. Chudnovsky

It seems that we are coming back to the very beginning of the discussion or disagreements between Bohr and Einstein. What is more fundamental: determinism or randomness? What is the basis?

Prof. M. Ostoja-Starzewski

Yes. I think this observation points out that maybe we cannot ever separate them.

Prof. A. Chudnovsky

So, maybe it is the same.

Prof. D. Jeulin

I think we cannot decide whether something is random or deterministic. It cannot be decided, you have no experiments that will decide something is random or deterministic. What is important is to choose the technique or model which can solve your problem. And when you look at scatter in results, it is probably better to use non–deterministic approaches to handle with heterogeneities. But you cannot decide something is random or not [...].

I want to come back after the choice of a method to the construction of the model. There are different steps, in the construction. The model itself is deterministic in general. First you have to exhibit the construction of the model, for instance the microstructure. You must say: I have a polycrystal or a multi–phase material, so you have to give the key points of the construction. You must also introduce the mechanisms, the basic Physics I mean, for instance you will [choose] a given fracture criterion, or various criteria. Then you have to use parameters in the model. You have to do this explicitly, not only implicitly. You have to know exactly the construction, the parameters, the criteria you are using. If it is not done explicitly, you may forget something, the model will escape from you and you cannot control the results. This is the first part, the construction of the model that has to be explicit. The second part is that you have to estimate them. You have to design some experiments. The problem is that when you deal with random models, you cannot make an identification of the model using a single experiment. It is not possible because what you are expecting is a heterogeneous result, some scatter, so you have to repeat the experiments. If you look at several realizations you will say 'OK I can now try and see if there is a random process behind this [experiment].' For one single realization you cannot. So when you use probabilistic models, you need to repeat experiments. This is a difference with deterministic approaches: the cost is much higher because you have to make a lot of experiments. And once you have made the proper tests, you can hope to estimate the parameters. The last point for the model is to check its validity. To check the validity of the model, you need to predict some other properties that were not used in the phase of estimation and to compare the predictions with reality. This has to be done, otherwise you have been using a multiparameter model, you could estimate the parameters from the data, if you add more parameters, you will be more satisfied with the model but you cannot predict anything. So the three steps for the construction of a model construction–estimation–validation have to be satisfied. The key point for probabilistic approaches is the cost. Of course, the model may be more complex, but I am not sure it is so complex. If you are familiar with statistics, you can handle a lot of things. It is the cost for the estimation of the parameters and the test because you need to have repetitive experiments. And maybe one challenge for the future is not only on probabilistic characteristics, but to have the facilities to get more data on a small scale. I do agree with [Prof. Schueller] concerning the information. If you can design experimental devices where you can get a fracture strength or any kind of physical or mechanical properties you need on a very small scale. I do not speak about one cubic micron, but you have to define this scale of course. To know for instance the fracture energy on the small scale, and to be able to repeat this kind of experiments on the same material just as [Prof.] Soulie showed yesterday concerning other physical properties like water content on very large samples with a regular grid. If we can get such information in the future, we can get more data to design new models, probabilistic models, on a small scale, and to test the models. If you just keep information on a macroscopic scale I think you forget a lot of information.

Prof. D. Breysse
Coming back to the problem of cost, which has been quoted by Profs. Schueller and [Jeulin], I think that the cost of simulation is not the main problem. The main cost is the cost to reach new information, and the real question is: is it required to get them? For instance, if we are working on nuclear power plants, we will try to reach the curve of defect distribution of size or position because we know that the probabilistic approach or probabilistic modeling of life duration of pipes is important. And that we know that the cause of the life duration is mainly at this scale. On the other hand, if we consider that it is a problem of crack propagation, is it interesting to use probabilistic models? We can

also be satisfied by a macro random model which describes this propagation. What are the reasons for choosing one model or another?

Prof. D. Jeulin

I think it is interesting [to use micromechanics] if you want to have new data to validate the model or to estimate the parameters of the model. I am not sure the cost is so low. Because you have to design special devices, special experiments. It is not very easy to design experiments to measure accurately on a small scale. To validate a model on a small scale, it is important to have data on a small scale. If you just consider macroscopic data you have a kind of inverse problem to solve and you have no [unique] solution.

Prof. D. Breysse

I fear that it is a problem with no end. I fear that if we go at this lower scale the random evolution will also be explained by some random data at [even] a lower scale (e.g. positions of 'micro–micro' cracks). We will never end. So we have to stop and say 'at this scale it is random.' The cause may be related to a given mechanism.

Prof. D. Jeulin

You have to define the scale of your observations.

Prof. C. Huet

I want to try and give an answer to the last question by [Prof.] Breysse. Because it depends on what is your goal or what is your situation. With the example of nuclear power plants, you are in the conditions in which you have an ordinary problem of quality control. It is for that kind of problem that, at the beginning, the statistical tools have been developed. In this case, you have almost constant boundary conditions. The situation around your material is almost the same all the time because you have a machinery working under the same conditions, and if you apply purely statistical tools, you will get something. But if you try to apply these same statistical tools that you have obtained in the observation during your work in the nuclear power plants to what will happen to the same material for instance in a dam or in an airplane, this will not work. Therefore you need more predictive models. And it is a reason for which you have to go inside the material and do some micromechanics of the phenomena that are [taking place] in the material before going on to probabilistic tools. And I think that we need a multidisciplinary approach, multidisciplinary teams and people well acquainted with probabilistic methods, well acquainted with micromechanics, very often they are not the same persons, and than you can do something valid. If a mathematician is doing things without any idea of what is going on at a micro level, there will be 'postdictions' not predictions. But if you can go inside at some level that is relevant to your problem, nothing prevents you from applying the tools of probability at a micro level. Generally you will have fewer parameters.

Prof. M. Soulie
I would like to go back to the choice of a method. As [Prof. Huet] said, we have to get a multidisciplinary team. We have no choice between probabilistic and deterministic approaches. We have to combine both. From my personal experience, 10–15 years ago, I was involved with some people on the design of an earth dam in the US with the US Army Corps. They asked the team to look at the probabilistic approach to design an earth dam, a very complex earth dam on very bad foundations. The reason was that most of the people were coming from NASA. At NASA they had a very good practice of this kind of probabilistic approaches. We started to look at different problems (e.g. internal erosion, hydraulic fracturing, contact between compact soil and concrete). The problems were very complex, and in fact we were not able to solve most of the problems in a probabilistic way. We worked for two years on this project. After two years, the result was that around 10% was solved by using probabilistic approaches and 90% were solved by conventional approaches. However, every one agreed that after these two years even though most of the problems were solved by conventional approaches, we were analyzing them in a probabilistic manner. Just reasoning in a probabilistic manner and trying to understand what is happening was sufficient to use, in a better way, a conventional approach. From this experience we concluded that statistics is a tool but we have to combine these with a conventional approach based upon a lot of experience. If some people come to me with some probabilistic approach which contradicts conventional approaches, I will be very hesitant.

Prof. G.I. Schuëller
I think it is a very valid point you made. Randomness should be validated in any case.

Prof. A. Bolle
But there is a lack of information ...

Prof. G.I. Schuëller

I fully agree with you. Lack of information is no information. For this reason *Bayesian* statistics has been developed and *Bayesian* statistics adheres to the Kolmogorov theorems: it is a mathematical tool. Yet, the prior distribution which we are putting in, if we do not have any information we use uniform *a priori*. As we get more information, we have an update. At the ultimate it is identical to classical statistics, if you have sufficient information.

3 Concluding Remarks by Prof. A. Chudnovsky

It is my impression that we all basically agreed that it does not look promising to consider fracture models as just black boxes describing fracture with application of probabilistic approaches. Where one utilizes some physical models, micromechanics and some kind of understanding of the fracture mechanisms, one finds that there is no universality in fracture behaviors. Thus we do not have a universal theory yet. It is still a combination of art and science. Art in a sense of creating a model that combines the physical understanding and the available experimental results with the imagination to compensate for the missing information. It seems that we have enough challenges and time ahead of us before statistics of Fracture becomes a true physical theory.

PROBABILISTIC MODELS AND IMAGE ANALYSIS: TOOLS TO DESCRIBE LIQUID PHASE SINTERED MATERIALS

M. COSTER, J-L. QUENEC'H, CHERMANT and D. JEULIN†
LERMAT, URA CNRS n°1317, ISMRa, Caen, France
† Centre de Géostatistique, ENSMP, Fontainebleau, France

Abstract
Liquid phase sintering is a method often used to elaborate materials in powder metallurgy. As the properties of the materials strongly depend on the microstructure of the grains, the evolution of the grain growth has been studied and modelled for many years. Unfortunately, the experimental conditions do not fit correctly with the hypothesis of this model.
  We have choosen to follow the growth of materials using probabilistic multiphased models which do not depend on the experimental parameters. Several liquid phase sintered materials were studied by automatic image analysis. The parameters of 3 models (Boolean model, Mosaic model and Dead Leaves model) were estimated and the coarsening kinetics of WC-Co may be described by a Boolean scheme with poissonian grains. These models do nevertheless not apply to materials with spherical grains.
Keywords: Sintering, WC-Co, Carbides, Image analysis, Probabilistic models, Boolean model, Mosaic model, Dead Leaves model.

1 Introduction

The liquid phase sintering is a process often used to obtain materials by powder metallurgy. This sintering process consists of heat treatment of a mixing of two powders during some time at a temperature lightly above the lower melting point of the powders. This process can be roughly divided into two steps : densification by pore elimination and coarsening of refractory phase. This last step has been studied many times by a theoretical way using "physical models" and experimentally. But in this paper, a different approach is proposed in order to study the liquid phase sintering using "geometric models" to describe the microstructure and follow its evolution during the coarsening.

2 The physical models of sintering and stereological analysis

In the physical models of liquid phase sintering derived from those proposed by Lifshitz and Slyozov (1961), Wagner (1961) and Hanitzsch and Kahlweit (1969), the hypothesis are the following : small and constant volumic fraction of dispersed phase noted $V_v(X)$ and spherical

D. Breysse (ed.), Probabilities and Materials, 403–414.
© 1994 Kluwer Academic Publishers.

shape of this phase. But in real case, the initial hypothesis are not verified (shape, state of dispersion and $V_v(X)$ value).

All these models give the evolution of the size of dispersed phase X in terms of mean size and size distribution. The mean size, as a function of time t at constant temperature T, changes according to the following power law :

$$\bar{d}^n - \bar{d}_0^n = kt = k_0 \exp(-E/RT) \tag{1}$$

where n is an exponent function of the growing mechanism and k is the rate constant which follows an Arrhenius law.

In order to perform granulometric measurements by automatic image analysis, it is necessary to obtain a segmented image from thresholded image. Solutions exist for materials with spherical or rounded grains, for example by watershed algorithm on image distance of phase X (Beucher, 1990). But the problem is more difficult for polyhedral grains like WC (Gauthier et al, 1993).

3 The probabilistic models

3.1 Generalities

The probabilistic models were first introduced by Matheron (1967) and were afterwards developed and used by many other scientists, mainly those of the Ecole des Mines of Fontainebleau (Jeulin, 1979, 1981, 1991). From a mathematical point of view, these models are random closed sets (RACS) (Serra, 1982) and follow the classical laws of the probabilistic theory. The RACS are so characterized by a function called the Choquet's capacity T(K) which represents the probability that a given compact K hits the RACS X :

$$T(K) = 1 - Q(K) = \mathcal{P}(K \cap X \neq \emptyset) = 1 - \mathcal{P}(K \subset X^c) \tag{2}$$

where X^c is the complementary set of X.

The analytical expressions of T(K) is known for some models as a function of the parameters of the models. An interesting and useful property of T(K) is that two RACS having the same T(K) for all K are theorically and experimentally undistinguishable.

All the models used in this paper are defined in the Euclidean space. They have few parameters and have good stereological properties : a model defined in $\mathbb{R}^3$ induce the same kind of model in a sub-space, $\mathbb{R}^2$ for example. We may so acceed to the spatial laws from polished sections.

The classical methods of image analysis (Serra, 1982; Coster and Chermant, 1985) allow to estimate T(K) from polished sections : we can show that :

$$\mathcal{P}(K \subset X^c) = A_A(X^c \ominus K) \tag{3}$$

where $\ominus$ represents the erosion operation in image analysis and A_A is the surface area.

Therefore we may determine experimentally the Choquet's capacities for several different K and for few models (Quenec'h et al, 1993). So we acceed to the parameters of the tested model. The comparison between the experimental and theorical Choquet's capacities allows to estimate the good quality of the modelling of a given material for a chosen model.

3.2 The models used

Because all the tested materials were biphased, we used only probabilistic multiphase models. The three models tested are the following : a Boolean scheme with convex primary grain, a bicolor Mosaic and a two-phased Dead Leaves model.

3.2.1 The Boolean scheme

The Boolean model is a very well known model and is often used (Jeulin, 1979, 1991; Bretheau and Jeulin, 1989; Stoyan et al, 1987).

 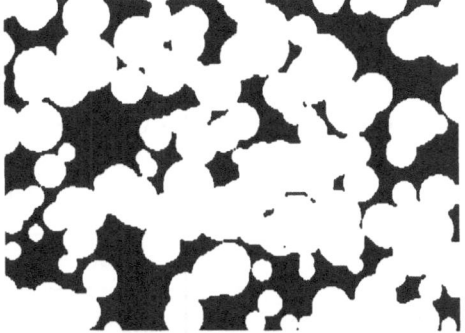

Fig.1. Simulation of a Boolean model in $\mathbb{R}^2$ with poissonian grains. Fig.2. Simulation of a Boolean model in $\mathbb{R}^2$ with circular grains.

To build a Boolean scheme, a Poisson point process of intensity θ is needed. In each point a primary grain is implanted : we assume that the primary grains are convex and we used two different shapes of grain, poissonian or spherical, according to the microstructure of the material. In the case of spherical grains, several grain size distributions are tested: single size, uniform, exponential or linear distributions of parameter a (Jeulin, 1979), in which a is the maximum diameter of the grain. This model depends on two parameters : θ, the parameter of the Poisson process and a, the parameter of the primary grain. Figures 1 and 2 represent simulations of Boolean scheme with poissonian and circular grains.

The primary grain X' is defined by its average volume, $\overline{V}(X')$, its surface area, $\overline{S}(X')$ and its integral of mean curvature, $\overline{M}(X')$. For spheres, these mean values are obtained from the moments of orders 1, 2 and 3 of the distributions which are shown in Table 1.

From the moments, we have $\overline{V}(X')$, $\overline{S}(X')$ and $\overline{M}(X')$:

$$\overline{V}(X') = \pi/6\ E(X^3),\quad \overline{S}(X') = \pi\ E(X^2),\quad \overline{M}(X') = 2\pi\ E(X) \qquad (4,\ 5,\ 6)$$

In the case of Poisson primary grains of parameter λ, we have :

$$\overline{V}(X') = 6/(\pi^4\lambda^3),\quad \overline{S}(X') = 24/(\pi^3\lambda^2),\quad \overline{M}(X') = 3/\lambda \qquad (4',\ 5',\ 6')$$

The use of different compacts K allows to estimate the parameters of the Boolean scheme, when the analytical expressions of T(K) are known :

-if K is a point P :

$\quad Q(P) = q = \exp(-\theta\ \overline{V}(X'))$ \hfill (7)

$\quad$ where q is the surface area of the phase X^c.

-if K is a segment ℓ of size ℓ :

$$Q(\ell) = q \exp(-\theta\ell \, \frac{\overline{S}(X')}{4}) \tag{8}$$

-if K is a pair of point h separated by a distance h, the mean geometric covariogram K(h) is defined by (Jeulin, 1979) :

$$K(h) = \frac{\pi}{6} \left[\int_h^\infty a^3 f(a)da - \frac{3}{2}h\int_h^\infty a^2 f(a)da + \frac{1}{2}h^3\int_h^\infty f(a)da \right] \tag{9}$$

in which f(a) is the distribution function of the grains.
For poissonnian grains :

$$K(h) = K(0) \exp(-\pi \lambda h) \tag{9'}$$

K(h) is obtained experimentally from the covariance Q(h) :

$$K(h)/K(0) = (2 - Ln\ Q(h)/Ln\ q) \tag{10}$$

-if K is a hexagon of size r, H(r):

$$Q(H(r)) = q \exp(-\theta(3r\,\frac{S(X')}{4} + \frac{3\sqrt{3}}{4\pi}\,r^2\,\overline{M}(X'))) \tag{11}$$

Table 1. Moments of order 1, 2 and 3 for several distribution laws.

Distribution law	$E(X)$	$E(X^2)$	$E(X^3)$
Single size	a	a^2	a^3
Uniform	$\dfrac{a}{2}$	$\dfrac{a^2}{3}$	$\dfrac{a^3}{4}$
Exponential	a	$2a^2$	$6a^3$
Linear ($0 \le k \le 1$)	$\dfrac{a\ (1+k)}{3}$	$\dfrac{a^2}{6}\dfrac{(1-k^3)}{(1-k)}$	$\dfrac{a^3}{10}\dfrac{(1-k^4)}{(1-k)}$

The expressions of the specific number of connectivity in $\mathbb{R}^3$, $N_V(X)$ and in $\mathbb{R}^2$, $N_A(X)$ are known for the Boolean scheme (Miles, 1976) :

$$N_V(X) = q\theta(1 - \frac{\theta\overline{M}(X')\overline{S}(X')}{4\pi} + \frac{\pi\theta^2}{6}\,(\frac{\overline{S}(X')}{4})^3) \tag{12}$$

$$N_A(X) = q\theta(\frac{\overline{M}(X')}{2\pi} - \theta\,\frac{\pi}{4}\,\frac{\overline{S}(X')^2}{16}) \tag{13}$$

In the case of images digitalized on an hexagonal grid, $N_A(X)$ is given by :

$$N_A(X) = q\theta(\frac{\overline{M}(X')}{2\pi} - \theta\,\frac{\sqrt{3}}{2}\,\frac{\overline{S}(X')^2}{16}) \tag{14}$$

In order to test the Boolean scheme, the linearity of the function Ln $Q(\ell)$ must be verified : $\theta\overline{V}(X')$ and $\theta\overline{S}(X')$ may be then estimated from relations (7) and (8). In order to obtain more information, it is necessary to introduce assumptions about the shape of the primary grain, poissonian or spherical and in this last case about the size distribution. The values of $\overline{V}(X')$, $\overline{S}(X')$ and $\overline{M}(X')$ are then calculated from Table 1 and relations (4), (5) and (6) and allow estimations of a and θ by several ways.

3.2.2 The Poisson Mosaic model
The Poisson Mosaic model (Jeulin, 1987, 1991) is built from a Poisson

plane tesselation of the $\mathbb{R}^3$ space.

Let $D(\omega)$ be a straight line with orientation ω, let $\lambda d\omega$ be the density of the isotropic Poisson point process defined on this line; a tesselation of space is build by the planes $\Pi(\omega)$, defined at each point of the Poisson process, and perpendicular to the lines $D(\omega)$. Each polyhedron is afterwards allocated randomly and independently to one or another phase according to their specific areas (Figure 3). The two parameters of the model are then λ, the parameter of the polyhedra and q, the volumic fraction of one phase. It is obvious that the Mosaic model may be applied only to the materials which have polyedrical grains.

Fig. 3. Simulation of a Poisson Mosaic model in $\mathbb{R}^3$ with poissonian grains.

The analytical expressions of T(K) is known for the three compacts that we used to test the Mosaic model :
For $K = \ell$:

$$Q(\ell) = q \exp(-\pi\lambda p\ell) \tag{15}$$

with $p = 1 - q$
For $K = h$, the centered covariance $\bar{C}(h)$ is given by :

$$\bar{C}(h) = C(h) - p^2 = Q(h) - 1 + 2p - q^2 \tag{16}$$

$$\bar{C}(h) = \sigma^2 \exp(-\pi\lambda h) \tag{17}$$

where σ is the variance pq.
For K=T, the triplet of points corresponding to the vertices of an equilateral triangle of size h, the third order centered moment is given by :

$$\bar{W}_3(h_1, h_2) = p(1 - 3p + 2p^2) \exp(-\frac{3}{2} \pi\lambda h) \tag{18}$$

These three compacts allow to estimate three different values of λ, q being known by the only measure of A_A. Moreover, the Mosaic is a symmetric model : λ is the same for both phases. This allows to test the good fit of the Poisson Mosaic for a given material.
The expression of the number of connectivity in $\mathbb{R}^2$, $N_A(X)$ is known but $N_V(X)$ is not reached from the parameters of the poissonian Mosaic.

$$N_A(X) = \frac{\sqrt{3}}{2} \pi^2\lambda^2 q(1 - q)(1 - 2q) \tag{19}$$

3.2.3 The Dead Leaves model (DLM)

We used the homogenehous Dead Leaves Tesselation model which was introduced by Matheron (1968) and developed by Serra (1982) and Jeulin (1979, 1991).

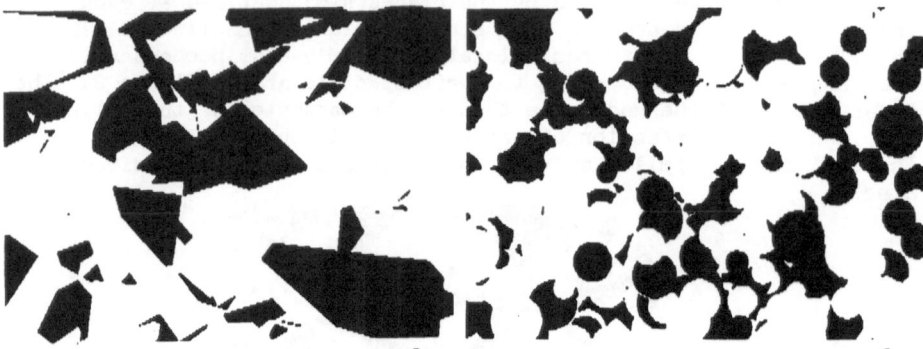

Fig. 4. Simulation of a DLM in $\mathbb{R}^2$ with poissonian grains.

Fig. 5. Simulation of a DLM in $\mathbb{R}^2$ with circular grains.

This tesselation is built as follow : for both phases, primary grains of parameters a1 and a2 are implanted randomly with respective density θ1dt and θ2dt. The grains appeared between t and t + dt may cover a part of the former grains. If the implantation continues to stability, a random tesselation of space is obtained. a1, a2, θ1 and θ2 are the four parameters of the model. In order to simplify the problem, we have considered that the two densities of implantation θ1 and θ2 were similar. We used consequently a two-phased Dead Leaves model with three parameters: a1, a2 and θ. Figures 4 and 5 show simulations of Dead Leaves models with poissonian and circular grains.

In the case of this model, only the expressions of the covariance of both phases, $C_1(h)$ and $C_2(h)$, and the crossed covariance, $\gamma(h)$, are helpful to test the model.

$$C_1(h) = \frac{p(1-2p)\ K_1(h)/K_1(0) + 2p^2}{2 - p\ K_1(h)/K_1(0) - q\ K_2(h)/K_2(0)} \tag{20}$$

$$\gamma(h) = p - C_1(h) = q - C_2(h) \tag{21}$$

if we note $r_1(h) = K_1(h)/K_1(0)$ and $r_2(h) = K_2(h)/K_2(0)$, we have :

$$\Phi(h) = \frac{\gamma(h)}{p(1 - p)} = \frac{2 - (r_1(h) + r_2(h))}{2 - p(r_1(h)) - q(r_2(h))} \tag{22}$$

$\Phi(h)$ may be experimentally determined using the relations (21), (22) and the experimental covariance.

In order to test the model, we can suppose that a1 = a2 = a. Then we can say that $r_1(h) = r_2(h) = r(h)$:

$$\Phi(h) = \frac{2(1 - r(h))}{2 - r(h)} \tag{23}$$

The study of relation (23) allows to test the good fit of the Dead Leaves model.

In the case of a DLM with Poisson grains, we have :

$$r(h) = \exp(-\pi a h) = \frac{2(1 - \Phi(h))}{2 - \Phi(h)} \tag{24}$$

The test of Ln r(h) confirmes or invalidates this first hypothesis. But the expressions of r1(h) and r2(h) allow also to test other hypothesis in the case of Poisson grains :

In the case where a1 ≠ a2, then we will suppose in a first time that a1 » a2 (resp. a2 » a1); so r1(h) « r2(h) (resp. r2(h) « r1(h)). Therefore r2(h) and a2 (resp. r1(h) and a1) are estimated using Φ(h) and relation (22) for high values of h. Moreover, the sum a1 + a2 may be evaluated from Φ(h) as it is known that :

$$\lim_{h \to 0} \frac{d\ \Phi(h)}{d(h)} = \pi(a1 + a2) \tag{25}$$

Knowing the sum a1 + a2 and one of those grain parameters, it is easy to calculate the second one and to compare the functions Φ(h), estimated by relation (22) and determined experimentally.

When a1 cannot be neglected with respect to a2 (resp. a2 w.r. to a1), the only way to test the two-phased Dead Leaves model is to proceed with iterations on a1 and a2, knowing of a1 + a2, by means of a least square method. The good fit of the model is tested as above.

4 Application to liquid phase sintered materials

4.1 Coarsening of WC-Co
WC-Co is an heavy metal elaborated by liquid phase sintering and very often used in industry. It is consequently of great interest to follow the evolution of the microstructure during the coarsening.

Fig.6. Micrograph of WC 15 wt% Co sample sintered at 1450°C.

As shown in Figure 6, the carbide grains are polyedral and the comparison between the simulations of Figures 1, 3 and 4 shows that it is not useless to ask the question whether WC-Co is modelled by one of the three probabilistic models studied.

Our samples were sintered at 1450°C in a graphite resistance furnace under a pressure of approximatively 10^{-4} Pa. The sintering time varied from 1 to 100 hours and the standard composition was 15% in weight of Co. The samples were polished and oberved with a Scanning Electron Microscope (JEOL JSM T 330A). 20 micrographs, obtained by backscattering mode, were afterwards analysed by means of an image processor NS-1500, which works with an hexagonal grid.

4.1.1 WC-Co tested by a Dead Leaves model
The function r(h) is estimated from the experimental covariance C(h) and relation (24), and Ln r(h) is plotted as a function of h. The

curves are not linear, so a Dead Leaves model with the same primary grain for both phases do not fit with WC-Co cermets. Then the hypothesis of two different grains for WC and Co was examined.

The derivation of $\Phi(h)$ at h = 0 gives a1+a2 for the samples. While neglecting a1 w.r. to a2 (or a2 w.r. to a1), we obtain a simplified expression of (22) with only one parameter. The calculation of this parameter gives always a result higher than the sum a1+a2. A negative value of the parameters a1 or a2 being physically impossible, we can therefore not neglect one parameter w.r. to the other.

At last, we iterated a1 and a2 as explained in 2.2.3, but all the curves obtained (Figure 7) were very far from the experimental one.

So we concluded that the Dead Leaves model was not adapted to describe the structure of WC-Co. It is not surprising; the model implies by construction possibility of inclusions of one phase into another, which is not compatible with the physical properties of sintering.

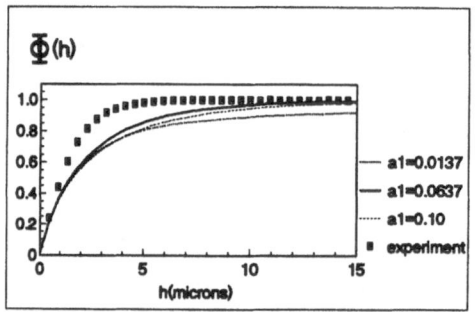

Fig.7. Comparison between $\Phi(h)$ experimentally determined and calculated by iterations for a WC 15 wt% Co sample sintered 100 h at 1450°C (a1+a2 = 0.1637).

4.1.2 WC-Co tested by a Mosaic model
The samples were eroded by three different compacts ℓ, h and T. The mean of the three values of λ, obtained from relations (15), (17) and (18) for each sample, is presented in Table 2 with its coefficient of variation CV, which is the ratio of the standard deviation to the mean.

Table 2. Mean value, $\bar{\lambda}$, and coefficient of variation, CV, of the parameter λ for the Mosaic model, estimated for WC 15 wt% Co sintered at 1450°C.

	sintering time (h)	1	2.5	3.5	7	14	30	100
WC	$\bar{\lambda}$ (μm^{-1})	0.59	0.60	0.60	0.52	0.39	0.26	0.21
	CV (%)	19	13	19	19	12	10	09
Co	$\bar{\lambda}$ (μm^{-1})	0.56	0.55	0.53	0.46	0.31	0.22	0.17
	CV (%)	25	18	26	24	30	24	27

For a given sintering time, the parameter λ is approximately the same for both phases but the coefficients of variation CV are different and seem better when the phase WC is tested.

4.1.3 WC-Co tested by a Boolean model

The linearity of the functions Ln $Q(\ell)$ and Ln $K(h)/K(0)$ was examined for both phases WC and Co. The erosions by ℓ and h give very good results when the phase Co is tested. So we have tested a Boolean poissonian scheme with the particle of WC as primary grains. The parameter a is estimated by different methods (relations (8), (9') and (11)). The mean values of λ, $\bar{\lambda}$, are summarized in the Table 3. The parameter θ is estimated from relation (7) using $\bar{\lambda}$.

Table 3. Mean values, $\bar{\lambda}$, and coefficient of variation, CV, of the para-meter λ for the Boolean model measured for WC 15 wt% Co sintered at 1450°C.

sintering time (h)	1	2.5	3.5	7	14	30	100
$\bar{\lambda}$ (μm^{-1})	0.37	0.39	0.38	0.33	0.26	0.17	0.14
CV (%)	21	33	34	42	31	17	22

4.1.4 Conclusion

We saw above that the Dead Leaves model is not convenient to describe the sintering of WC-Co, but a choice has to be made between the Mosaic and the Boolean models. First we noticed that the coefficient CV for the Mosaic are not the same for both phases. Furthermore, the simula-tion (Figure 4) shows that the grain boundaries are aligned, which is not verified on the real structure. At last, when the numbers of connectivity $N_A(X)$ are estimated from relations (14) and (19), and compared to the experimental values (Figure 8), the approximation by a Boolean model seems to give better results. So we conclude that WC-Co may be described by the Boolean scheme with Poisson primary grains.

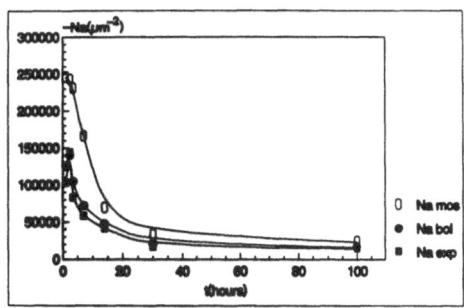

Fig.8. Connectivity numbers in R^2 measured, Na exp, and estimated for the Boolean model, Na bol, and the Mosaic model, Na mos.

The parameters of the model allow to estimate the sintering

parameters : kinetics exponant, n, and activation energy E. The values obtained are in good agreement with previous studies (Exner & Fischmeister, 1966, and Coster, 1969): values of n = 3 and E = 300 kJ are found; it can be concluded that the coarsening kinetics of WC-Co was controlled by the reactions at the interfaces (Quenec'h et al, 1992).

4.2 Materials with spherical grains

Figures 9 and 10 show micrographs of carbides sintered with a liquid phase : TiC-Co and TiC-Ni. The shape of the TiC grains is very different from WC grains in WC-Co. The particles are not so angular and are almost spherical. The symmetry of both phases in WC-Co is no more observed. It is obvious that the Mosaic model can not describe the structure of any of these three materials since this model gives only polyedral grains. So, only the Boolean scheme and the Dead Leaves model with spherical primary grains were tested.

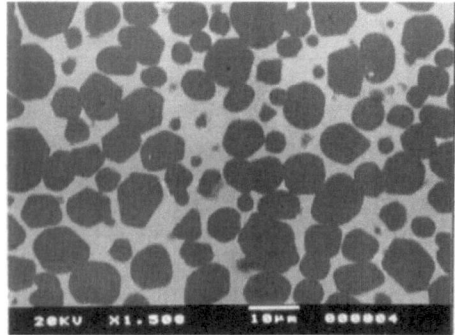 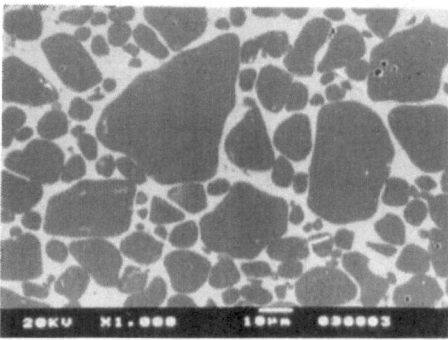

Fig. 9. Micrograph of TiC-Co sample sintered 4 hours at 1450°C.

Fig. 10. Micrograph of TiC-Ni sample sintered 8 hours at 1400°C.

4.2.1 Modelling by a Dead Leaves model

In the case of materials with spherical grains, we have tested only the hypothesis in which the two phases have the same distribution. But the comparison between the experimental and theoretical curves of the function Φ shows that the Dead Leaves model did not apply to these materials. This is coherent with the physical properties of sintering which imply no possibility of inclusion of the liquid phase into the grains.

4.2.2 Modelling by a Boolean model

The functions $Q(\ell)$ were experimentally measured; we observed that they were exponential in all cases. In order to conclude more surely, we compared the experimental $K(h)$ with covariograms obtained from theoretical distributions (Jeulin, 1979) using for the value of the parameter a of the Boolean model the experimental range of $K(h)$.

TiC-Ni

The theoretical mean geometric covariograms show that the curve which describes at best the distribution of Ni particles is the covariogram of poissonian primary grains; so, the Ni phase would be described by a distribution of polyhedral grains. But the values of λ, the parameter of the Poisson Boolean model, estimated from relations (8), (9') and

(11) are scattered and moreover $N_A(X)$ calculated from relation (14) is far from the experimental values. With regard to TiC particles, no theoretical K(h) fits with the experimental one. Consequently, both phases in TiC-Ni cermets cannot be modelled by a Boolean scheme.

TiC-Co

The same observations for the Co phase were done as for the Ni phase in TiC-Ni and it was concluded that the Co phase in TiC-Co cermets could not be described by a Boolean scheme.

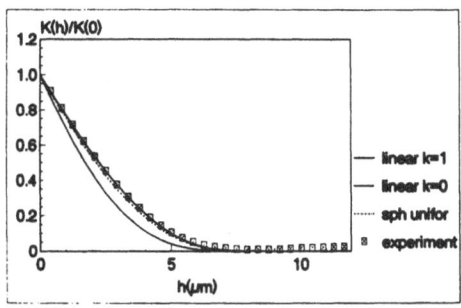

Fig. 11. Variations of experimental and theoretical K(h) functions for TiC-Co sample sintered 4 hours at 1450°C.

On the other hand, the description of the TiC phase by a Boolean model is possible in a limited domain of sintering time : figure 11 shows experimental points with theoretical K(h)/K(0) functions for a sample sintered 4 hours at 1450°C. A very good fit is found between the points and a linear distribution of the spherical primary grains of the Boolean model with k=1. For the sample sintered 8 hours at the same temperature, there is a good agreement with a linear distribution with k=0; in the case of longer sintering times, the experimental K(h) differs totally from the theoretical laws. The parameters a, estimated from relations (8) and (11), are close to the experimental range of K(h), but the numbers of connectivity in $\mathbb{R}^2$ calculated from relation (14) deviate from the experimental one. This last result can be minimized because $N_A(X)$ is very sensitive in this range of values to a small variation of structure.

Then, some samples give relatively good results but all TiC-Co samples modelled by a Boolean scheme are not described by the same distribution law. This fact can be explained by a modification of the size distribution by coalescence, and the structure would become "non Boolean" for long coarsening times.

5 Conclusion

In this paper, a systematic investigation of sintered carbides by probabilistic models was presented. The description of WC-Co structures by a Boolean poissonian model is very good whereas only the TiC-Co structures can be approximatively described by spherical Boolean scheme, with change in distribution of primary grains. This difference could be explained by the nature of boundaries in carbide phases :

"perfect" joints in WC and ordinary joints in TiC. The problem is more difficult for TiC-Ni samples because a change in shape of grains occurs during coarsening and any model can take this fact into account.

6 References

Beucher, S. (1990) **Thèse de Doctorat**, Ecole des Mines de Paris.
Bretheau, T. & Jeulin, D. (1989) Caractéristiques morphologiques des constituants et comportement à la limite élastique d'un matériau biphasé Fe/Ag. **Rev. Phys. Appl.**, 24, 861-869.
Coster, M. (1969) **Thèse de Docteur Ingénieur.**, Université de Caen.
Coster, M. & Chermant, J.L (1985) **Précis d'analyse d'images.** Les Presses du CNRS.
Exner, H.E. & Fischmeister, H. (1966) Gefügeausbildung von gesinterten WC-Co Hartelegierungen. **Arch. Eisenhüttenwes.**, 37, 417.
Gauthier, G., Quenec'h, J-L., Coster, M. & Chermant, J-L. segmentation of grain boundaries in WC-Co cermets. To be published in **Acta Stereol.**, 1993.
Hanitzsch, E. & Kahlweit, M. (1969) Ageing of precipitates, in **Symposium on Industrial Crystallisation**, p130-141.
Jeulin, D. (1979) **Thèse de Docteur Ingénieur**, Ecole des Mines de Paris.
Jeulin, D. (1981) Mathematical morphology and multiphase materials. **Stereol. Iugosl.**, 3/Suppl 1, 265-286.
Jeulin, D. (1987) Anisotropic rough surface modelling by random morphological functions. **Acta Stereol.**, 6, 183-189.
Jeulin, D. (1991) **Thèse de Docteur ès Sciences Physiques**, Université de Caen.
Lifshitz, I.M. & Slyosov, V.V. (1961) The kinetics of precipitation from supersaturated solid solutions. **J. Phys. Chem. Solids**, 15, 35-50.
Matheron, G. (1967) **Eléments pour une théorie des milieux poreux**, Masson et Cie, Paris.
Matheron, G. (1968) **Schéma Booléen séquentiel de partition aléatoire.** Rapport interne N-83. C.M.M., Ecole des Mines de Paris.
Miles, R.E. (1976) Estimating aggregate and overall characteristics from thick sections by transmission microscopy. **J. Microsc.**, 107, 227-233.
Quenec'h, J-L., Coster, M., Chermant, J-L. & Jeulin, D. (1992) Study of the liquid phase sintering process by probabilistic models: application to the coarsening of WC-Co cermets. **J. Microsc.**, 68, 3-14.
Quenec'h, J-L., Coster, M., Chermant, J-L. & Jeulin, D. (1993) Etude de la cinétique de frittage à l'aide de modèles d'ensembles aléatoires. **Rev. Met.- Sci. Génie Mat.**, 90, 611-621.
Serra, J. (1982) **Image analysis and mathematical morphology**, Academic Press, London.
Stoyan, D., Kendall, W.S. & Mecke, J. (1987) **Stochastic geometry and its application.** J. Wiley and Sons, Chichester.
Wagner, C. (1961) Theory of precipitate change by redissolution. **Z. Elektrochem.**, 65, 581-591.

PROBABILISTIC FAILURE PREDICTIONS IN BRITTLE MATERIAL UNDER MULTIAXIAL LOADING

J. LAMON
Laboratoire des Composites Thermostructuraux, UMR 47 (CNRS-SEP-UB1), Pessac, France.

Abstract
Statistical - probabilistic approaches to brittle failure under multiaxial loading conditions are discussed. Focus is placed more particularly on the Multiaxial Elemental Strength Model which considers the microstructural flaws as physical entities, as opposed to the well-known Weibull statistics. A computerized version of this model allows determination of failure probability for complex loading geometries. Implications for the prediction of failure in ceramics and fiber reinforced ceramic matrices, as well as the contribution of the preponderant factors are examined. Application to lifetime predictions is also addressed.
Keywords :Probability, Statistics, Failure, Brittle Fracture, Multiaxial Stress, Heterogeneities.

1 Introduction

Prediction of failure is a statistical - probabilistic problem for essentially brittle materials including monolithic ceramics and ceramic composites as a result of the presence of randomly distributed inherent microstructural defects. Under predominantly tensile loading conditions these defects always cause fracture. As a consequence, these materials exhibit an erratic fracture resistance, as well as important scale effects, thus requiring failure probability computations for reliability analysis and structural design.

For isotropic materials several statistical - probabilistic approaches to brittle failure have been proposed. They can be grouped into two main families : (i) purely statistical theories essentially aimed at providing a description of scatter in strength data (Weibull's theory), and (ii) more fundamental approaches which consider the flaws as physical entities. In recent years, the Weibull model has been the most widely used for statistical description of failure strength data in preponderantly uniaxial stress states, for ceramics and fibers. The main fundamental

D. Breysse (ed.), Probabilities and Materials, 415–426.
© 1994 *Kluwer Academic Publishers.*

approaches include the flaw size theories and the elemental strength approaches. In these fundamental approaches, the individual role of defects is described locally at the microscopic scale by fracture mechanics relations involving the defect type, the local stresses and a failure criterion. Since position of the critical flaw is unknown, and cannot be predicted, a probability of occurrence is associated to individual flaws, according to the characteristics of the flaw populations.

The flaw size theories are based upon statistics of crack sizes and location. Failure probability is derived from the flaw size distribution by applying fracture mechanics relations (Evans and Langdon, 1976). These flaw size theories are still in their infancy.

The elemental strength approaches use the elemental strength concept for the description of fracture - inducing flaw populations. Assuming that the flaws can be characterized by a flaw extension stress S, distribution of the fracture - inducing ones is described by a flaw density function g(S). Several treatments have been proposed to obtain strength distribution functions from the distribution in S (Argon and McClintock (1966), Batdorf (1977), Coleman (1958)).

In most elemental strength treatments, the elemental strength is defined as the strength of a volume element containing the flaw (Evans and Langdon (1976) and references therein). In his model, Batdorf considered the remote uniaxial normal fracture stress of a given crack (Batdorf and Crose (1974)).

Recently, a Multiaxial Elemental Strength Model was derived by Lamon and Evans (1983). The Multiaxial Elemental Strength Model is an approach to the determination of the flaw density function, for the most general stress state. The elemental strength was defined as a combination of the normal and shear stress components operating upon the flaws, through recent concepts of non-coplanar crack extension.

A finite element statistical post processor (CERAM) was developed for failure prediction and reliability analysis based upon this Multiaxial Elemental Strength Model (Lamon et al. (1989)). The statistical Weibull model (Barnett-Freudenthal approximation) for multiaxial loading was also incorporated.

The primary intent of the present paper is therefore to discuss probabilistic - statistical predictions of brittle failure under multiaxial loading and more particularly using the Multiaxial Elemental Strength Model.

2 Statistical - probabilistic approaches to multiaxial failure

2.1 Weibull's equations

In the original Weibull treatment of multiaxial failure, failure probability is calculated by averaging the normal tensile stress in all directions :

$$P_f = 1 - \exp\left[- \int_V k \int_A \sigma_n^m \, dA \, dV \right] \tag{1}$$

where dA is an elemental area on a unit solid sphere and $k = (2m + 1)/2\Pi$ $(1/\sigma_0)^m$, V is the volume. The exponent m is referred to as the Weibull modulus,σ_0 is a normalized factor.

The only justification given by Weibull is adopting a power-law representation for the stress function was that **good agreement with experimental results was obtained**. This expression provides no description of the physical fracture situation. Contribution of the fracture inducing flaw populations is not considered.

In eq. (1), integration is performed over half the surface area of the unit sphere where the normal stress σ_n is tensile, neglecting regions where the normal stress is compressive. Eq. (1) is a shear-insensitive description of multiaxial fracture.

Barnett et al. (1967) and Freudenthal (1968) suggested an alternative simple approximation for handling multiaxial fracture statistics. In this approach, the principal stresses are assumed to act independently in each principal direction. The failure probability is calculated from the product of the individual survival probabilities, in the direction of the tensile components, leading to the following equation (volume analysis) :

$$P_V = 1 - \exp\left[- \int \left[\left(\frac{\sigma_1}{\sigma_{OWV}} \right)^{mV} + \left(\frac{\sigma_2}{\sigma_{OWV}} \right)^{mV} + \left(\frac{\sigma_3}{\sigma_{OWV}} \right)^{mV} \right] dV \right] \qquad (2)$$

The Weibull model has been found inadequate by several authors in multiaxial stress states. A number of investigators have obtained contradictory results in multiaxial failure predictions with the normal tensile stress averaging method when using material parameters obtained from uniaxial tests. Depending on the material considered, failure predictions were either conservative or optimistic.

The Barnett-Freudenthal approximation has been criticized by several authors. As it ignores interaction of principal stresses, it is considered that it should predict lower failure probabilities than the Weibull model (Batdorf (1977)).

2.2 The Multiaxial Elemental Strength Model

Failure probability is derived from the following basic equation :

$$P = 1 - \exp\left\{ - \left[\int_V dV \int_o^s g(S)\, dS \right] \right\} \qquad (3)$$

where V is the volume, S is the observed strength, and $g(S)$ is the number of flaws with a strength between S and $S + dS$.

In the Multiaxial Elemental Strength Model (Lamon and Evans, (1983), Lamon, (1985)), the flaw density function was determined for the most general multidimensional stress state, considering concepts of non-coplanar crack extension. It was considered that fracture may occur in a direction depending upon the respective magnitudes of the normal and the shear components of the local stresses operating on the flaws. The fracture criterion is based upon the maximum in the strain energy release rate G_{max} in the direction of crack propagation (Hellen and Blackburn, (1975)) :

$$G_{max} = \frac{(1 + v)(1 + x)}{4E} \ [K_I^4 + 6 K_I^2 K_{II}^2 + K_{II}^4]^{1/2} \qquad (4)$$

where $x = 3 - 4v$ under plane strain conditions, $x = (3 - v)/(1 + v)$ under plane stress conditions, E is the Young's mudulus, K_I and K_{II} are the mode I and mode II stress intensity factors.

An equivalent stress σ_E is then derived from equation (4) as the uniaxial tensile stress that would induce the same strain energy release rate G_{max} as the actual local stress field (σ_n, τ) :

$$\sigma_E = [\sigma_n^4 + 6 \tau^2 \sigma_n^2 + \tau^4]^{1/4} \qquad (5)$$

where σ_n and τ are normal and shear stress components.

Crack extension occurs when G_{max} reaches the critical value G_c. The equivalent stress attains the equivalent strength S_E.

The flaw density function of equation (3) is then expressed in terms of the distribution $g(S_E)$ of the equivalent strength S_E, considering small elements with a vol/unit sphere given by $dV = (\frac{1}{4\Pi})$ cos Ø dØ dψ (figure 1) :

$$g(S) \ dS = \frac{1}{2\Pi} \int_0^{\Pi/2} \int_0^{\Pi} g (S_E) \ dS_E \cos Ø \ dØ \ d\psi \qquad (6)$$

A functional relation for $g(S_E)$ is needed to proceed with the analysis. In a first step, a power function was assumed for $g(S_E)$ to conform with the contribution of the most severe flaws to the failures. Basically, assuming a power function for $g(S_E)$ implies that the distribution in equivalent strengths S_E follows a Weibull's description :

$$g(S_E) = m S_E^{m-1} \sigma_0^{-m} \qquad (7)$$

where m is the shape parameter and σ_0 the scale factor.

It is worth mentioning at this stage that m and σ_0 refer to the distribution in equivalent flaw strengths S_E. This is a fundamental difference with the Weibull's theory where the statistical parameters characterize the distribution of **failure stresses** considered at the **macroscopic scale** of the structure. However, other distributions in S_E may be assumed or identified. Selection of a power function for $g(S_E)$ is also the basic reason why the Multiaxial Elemental Strength may appear as an extension of the Weibull multiaxial model, although it is different in essence.

Failure probability is finally given by the following equations for surface - and volume - located flaws respectively :

$$P_S = 1 - \exp\left[-\int_A (\frac{\sigma_1}{\sigma_{OMS}})^{m_S} I_s (m_s, \frac{\sigma_2}{\sigma_1}) \, dA\right]$$

$$(8)$$

$$P_V = 1 - \exp\left[-\int_V (\frac{\sigma_1}{\sigma_{OMV}})^{m_V} I_v (m_v, \frac{\sigma_2}{\sigma_1}, \frac{\sigma_3}{\sigma_1}) \, dV\right]$$

where the subscripts S and V refer to surface and volume respectively. The functions I_S and I_V account for the orientation of the flaws respective to principal stresses and for shear sensitivity. σ_1, σ_2 and σ_3 are the principal stresses ($\sigma_1 > \sigma_2 > \sigma_3$). σ_2 and σ_3 may be compressive provided σ_1 and σ_n are tensile.

The functions I_S and I_V are needed to compute failure probability. They are expanded to :

$$I_S (m_s, \frac{\sigma_2}{\sigma_1}) = \frac{2}{\Pi} \int_0^{\Pi/2} F_s^m (\frac{\sigma_2}{\sigma_1}, \psi) \, d\psi \qquad (9)$$

$$I_V (m_v, \frac{\sigma_2}{\sigma_1}, \frac{\sigma_3}{\sigma_1}) = \frac{2}{\Pi} \int_0^{\Pi/2} \cos\emptyset \, d\emptyset \int_0^{\Pi/2} F_v^m (\frac{\sigma_2}{\sigma_1}, \frac{\sigma_3}{\sigma_1}, \emptyset, \psi) \, d\psi \qquad (10)$$

where F_V and F_S represent the functions $\frac{\sigma_E}{\sigma_1} = \frac{1}{\sigma_1} [\sigma_n^4 + 6 \sigma_n^2 \tau^2 + \tau^4]^{1/4}$

In a surface analysis, the local stresses σ_n and τ are related to the principal stresses by the following equations :

$$\sigma_n = \sigma_1 \cos^2 \psi + \sigma_2 \sin^2 \psi \qquad (11)$$

$$\tau = (\sigma_1 - \sigma_2) \cos \psi \sin \psi \qquad (12)$$

In a volume analysis, σ_n and τ are related to principal stresses by the following equations, according to figure 1 :

$$\sigma_n = \sigma_1 l_1^2 + \sigma_2 l_2^2 + \sigma_3 l_3^2 \tag{13}$$

$$\tau^2 = \sigma_1^2 l_1^2 + \sigma_2^2 l_2^2 + \sigma_3^2 l_3^2 - \sigma_n^2 \tag{14}$$

where l_1, l_2 and l_3 represent the direction cosines :

$$l_1 = \cos\varnothing \cos\psi$$

$$l_2 = \cos\varnothing \sin\psi \tag{15}$$

$$l_3 = \sin\varnothing$$

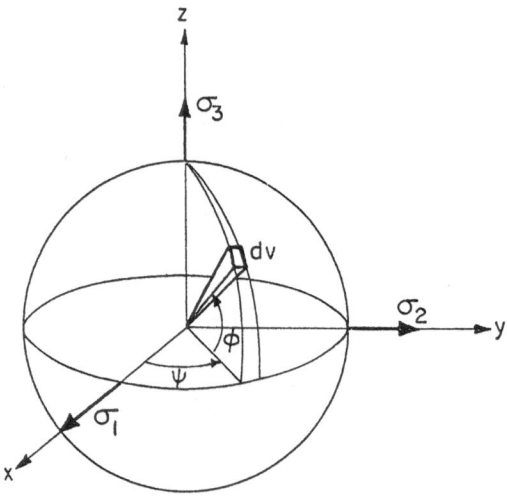

Fig. 1. Unit solid sphere element

Closed form solutions for I_S and I_V can be obtained only for specific stress-states including uniaxial, equibiaxial and equitriaxial stress states :
for uniaxial stress states :

$$I_s = \frac{2}{\Pi} \int_0^{\Pi/2} \cos^m\psi \, (1 + 4\cos^2\psi \, \sin^2\psi)^{m/4} \, d\psi \tag{16}$$

$$I_v = \frac{2}{\Pi} \int_o^{\Pi/2} \cos^m\beta \, \sin\beta \, (1 + 4\sin^2\beta)^{m/4} \, d\beta \qquad (17)$$

$$\text{with } \beta = \frac{\Pi}{2} - \varnothing$$

for equibiaxial stress-states :

$$I_s = 1$$

$$I_v = \frac{2}{\Pi} \int_o^{\Pi/2} \int_o^{\Pi/2} \cos^{m+1} \varnothing \, (1 + 4\cos^2\varnothing \, \sin^2\varnothing)^{m/4} \, d\varnothing d\psi \qquad (18)$$

for equitriaxial stress states

$$I_v = 1$$

Values for Is and Iv as defined in equations (9) and (10) were computed by gauss point integration, using specific routines incorporated into the finite element statistical post processor CERAM (table 1).

3 Determination of statistical parameters

It has been recently shown that statistical parameters are the most critical parameters for failure predictions (Lamon (1993)). The influence of failure criterion is not preponderant (Thiemeier and Brückner-Foit (1991)). The difference between various available fracture criteria is small compared to the statistical uncertainty determined by the confidence intervals of the statistical parameters. The incidence of the function used for description of the statistical distribution of equivalent flaw strengths S_E has not been examined yet, although it may be expected that the Weibull's distribution is not necessarily pertinent. Recent work performed on ceramic fibers thus evidenced the presence of extrinsic populations of flaws (Lissart and Lamon (1992)).

Sound failure predictions require appropriate statistical parameters, as a result of a strong sensitivity to the parameters due to the power form of the failure probability equations. Thus, calculations (Lamon (1993)) have shown that for shape parameters smaller than 10, as observed with most of the ceramics, a 10% uncertainty in the scale factors will lead to a discrepancy in the predicted failure probabilities comprised between 100% and 15%.

Determination of the statistical parameters requires the 3 following steps :

- strength data acquisition using mechanical tests
- fractographic examination of the specimens for fracture origin identification
- and extraction of the flaw strength parameters from the relevant strength distributions.

Table 1. Values for integrals I_S and I_V as a function of shape parameter, for various stress states

	I_S (m, $\frac{\sigma_2}{\sigma_1}$)				I_V (m, $\frac{\sigma_2}{\sigma_1}, \frac{\sigma_3}{\sigma_1}$)		
m	$\frac{\sigma_2}{\sigma_1} = 0$	0.25	0.50	0.75	uniaxial	equibiaxial	$\frac{\sigma_2}{\sigma_1} = \frac{\sigma_3}{\sigma_1} = 0.5$
1	0.7048	0.7476	0.8052	0.8881	0.5662	0.8645	0.7878
2	0.608	0.6254	0.6789	0.7964	0.431	0.7989	0.5569
3	0.5598	0.5562	0.5935	0.7209	0.3655	0.7607	0.4454
4	0.5313	0.5118	0.5332	0.6583	0.327	0.7365	0.370
5	0.5128	0.4805	0.4888	0.606	0.3017	0.7204	0.3166
6	0.5003	0.457	0.4547	0.5621	0.2841	0.7096	0.2774
7	0.4917	0.4384	0.4277	0.5248	0.2712	0.7024	0.2477
8	0.4858	0.4232	0.4057	0.493	0.2616	0.6979	0.2244
9	0.4819	0.4106	0.3873	0.4656	0.2543	0.6954	0.2056
10	0.4795	0.3998	0.3715	0.4419	0.2487	0.6945	0.1902
11	0.4783	0.3904	0.3579	0.4212	0.2443	0.6948	0.1772
12	0.4781	0.3820	0.3458	0.403	0.241	0.6963	0.1661
13	0.4786	0.375	0.3351	0.3869	0.2386	0.6987	0.1565
14	0.4799	0.3685	0.3255	0.3725	0.2368	0.7018	0.1480
15	0.4817	0.3627	0.3168	0.3597	0.2356	0.7057	0.1406
16	0.484	0.3574	0.3088	0.348	0.2348	0.7102	0.1340
17	0.4868	0.3525	0.3015	0.3375	0.2345	0.7152	0.1280
18	0.49	0.3481	0.2948	0.3279	0.2346	0.7208	0.1226
19	0.4936	0.3441	0.2885	0.3191	0.235	0.7268	0.1177
20	0.4975	0.3403	0.2827	0.311	0.2356	0.7333	0.1132

Strength data are usually measured using simple loading geometries such as 3-point or 4-point bending of bars or rods, biaxial flexure of disks, etc ... Standards have not been recommended yet for testing with respect to structural reliability analysis purposes.

Experimental distributions of strength data are established using the ranking statistics method. The measured strengths are ordered from lowest to highest. The ith result in the set of samples is assigned a cumulative failure probability P_i, calculated using an estimator.

Various estimators are available. The most appropriate is $P_i = \dfrac{i - 0.5}{N}$ which is insensitive to sample size N.

When multiple flaw populations are preexisting concurrently, strength distributions need to be separated accordingly. The censored data method proposed by Johnson (1964) can be used to determine a new rank i' for the strengths of each population by calculating a new increment Δ as soon as one or more censored strengths are encountered in the sequence of test data.

Statistical parameters are then extracted by fitting theoretical probability - strength equations to experimental strength distributions.

Since theoretical equations are restricted to simple stress-states, an alternative method based upon failure probability computations using the statistical post-processor CERAM can be applied to derive the flaw strength parameters even from failures observed with complex loading geometries. The scale factors σ_{oM} are obtained using the following equation for volume as well as for surface failure origins :

$$\sigma_{oM} = \sigma_{oA} \left[\frac{Ln\,(1 - P_{COMP})}{Ln\,(1 - P_{exp})} \right]^{1/m} \tag{20}$$

where P_{COMP} is the failure probability computed using the statistical post-processor for a dummy scale factor σ_{oA}. P_{exp} is the corresponding experimental failure probability.

Failure predictions from statistical parameters extracted using linear regression, maximum likelihood, mean strength and the CERAM-based methods showed that the best fit to experimental data was obtained with those scale factors determined using the CERAM-based method (Lamon (1993)). The lower scale factors were obtained with the maximum likelihood estimation and the mean strength based methods, thus leading to overestimation of failure probabilities, whereas the higher scale factors were obtained with the linear regression analysis, thus leading to underestimation of failure probabilities.

4 Validation

Validation of the Multiaxial Elemental Strength Model as well as the statistical post processor was performed using scaling of strength data to different specimen sizes and configurations. For this purpose, the flaw population characteristics for a given ceramic (Weibull statistical parameters m, σ_{oW} and the flaw strength parameters m, σ_{oM}) were determined from strength data measured on a given specimen size and configuration selected as reference. These flaw characteristics were subsequently incorporated into the statistical post processor for computing the failure probability-strengths of different specimen sizes and/or configurations. Comparison of the results with experimental data and theoretical calculations then allowed evaluation of the model, as exemplified on figure 2.

Various sets of experimental strength data, measured on various ceramics containing either single or multiple flaw populations and subjected to various loading conditions were used (Lamon et al. (1989 b) (1990) (1991) (1993)).

Contribution of the following factors which influence failure predictions was examined :
- the estimator for experimental failure probability,
- the uncertainty and the variability in the statistical parameters,
- the method used for determining the statistical parameters,
- the probabilistic model,
- the selection of batches,
- the finite element mesh,
- the results of stress analysis,
- the stress-to-failure (stress at fracture origin, peak stress, etc .).

The analysis showed that, with reproducible ceramics and the appropriate factors, the failure probabilities predicted with the statistical post processor were in good agreement with experimental failure data as exemplified by figure 2. In certain cases involving shearing effects, the Weibull model led to significant discrepancies between predictions and experimental results.

5 Concluding remarks

The Multiaxial Elemental Strength Model discussed in the present paper may be effectively applied if fracture occurs from direct extension of pre-existing flaws. These requirements are met with several brittle materials including multiphase materials such as continuous fiber reinforced ceramics, considering that damage and fracture are a combination of brittle events occurring in the matrix and in the fibers.

At high temperature or under aggressive environments, the flaws may grow and link up. Failure probabilities may be associated to resulting damage provided that relevant flaw strength parameters can be determined as a function of time and conditions. Failure probability measures here the chances that instantaneous failure occurs at various stages of the life of the structure, thus providing a lifetime estimate.

Failure predictions are very sensitive to flaw strength parameter estimation in approaches based upon a power form for the flaw density function.

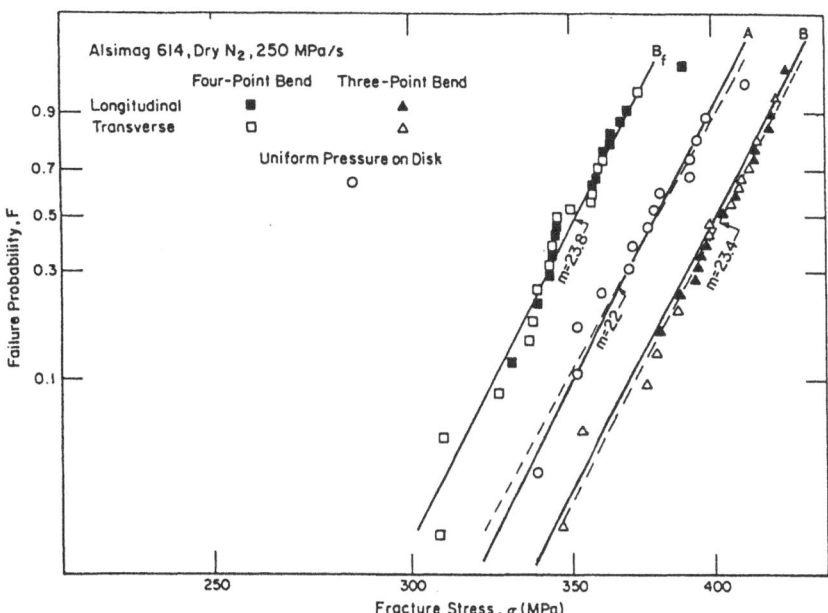

Fig. 2. Comparison of predicted strength-failure probabilities (solid lines A and B) with experimental data from flaw strength parameters extracted from the four point bending strength-data (B_f). The Multiaxial Elemental Strength Model was used for failure predictions.

References

Argon, A.S. and McClintock, F.A. (1966). **Mechanical behavior of materials,** Addison-Wesley, Reading, Mass. USA.

Barnett, R.L., Hermann, P.C., Wingfield, J.R., and Connors, C.L. (1967). **Fracture of Brittle Materials under Transient Mechanical and Thermal Loading.** Air Force Flight Dynamics Lab./TR-66-220.

Batdorf, S.B., and Crose, J.G., (1974). Statistical theory for the fracture of brittle structure subjected to nonuniform polyaxial stresses. **J. Appl. Mech.,** 41, 459-64.

Batdorf, S.B. (1977). Some approximate treatments of fracture statistics for polyaxial tension. **Int. J. Fract.,** 13, 5-11.

Coleman, B.D. (1958). On the strength of classical fibers and fiber bundles. **J. Mech. Phys. Solids,** 7, 60-70.

Evans, A.G. and Langdon, T.G. (1976). Structural ceramics. **Prog. Mater. Sci.,** 21, 171-441.

Freudenthal, A.M. (1968). Statistical approach to brittle fracture, in **Fracture,** Vol. 2 (ed. H. Leibovitz) Academic Press, New York.

426

Hellen, T.K. and Blackburn, W.S. (1975). The calculation of stress intensity factors for combined tensile and shear loading, **Int. J. Fract.**, 11, 605-617.

Johnson, L.G. (1964) **The Statistical Treatment of Fatigue Experiments,** Elsevier, New York.

Lamon, J.L. and Evans, A.G. (1983). Statistical analysis of bending strengths for brittle solids : a multiaxial fracture problem. **J. Am. Ceram. Soc.**, 66, 177-82.

Lamon, J.L. (1985) Statistical analysis of fracture of Silicon Nitride using the short span bending technique, Gas Turbine Conference and Exhibit, Houston, Texas, **Am. Soc. Mech. Eng.** paper N° 85-GT-151.

Lamon, J.L., Pherson, D., Dotta, P. (1989a) **2D and 3D Ceramic Reliability Analysis using CERAM.** Statistical Post Processor Software : Technical Documentation Final Report (Battelle Geneva).

Lamon, J.L., Melet, G. (1989 b) **CERAM Statistical Post Processor Software** : Experimental Evaluation of Reliability Analysis by CERAM 2D and CERAM 3D. Final Report (Battelle Geneva).

Lamon, J.L., (1990) Ceramics Reliability : statistical analysis of multiaxial failure using the Weibull approach and the multiaxial elemental strength model. **J. Am. Ceram. Soc.**, 73, 2204-2212.

Lamon, J.L. and Pherson, D. (1991). Thermal stress failure of ceramics under repeated rapid heatings. **J. Am. Ceram. Soc.**, 74, 1188-96.

Lamon, J.L. (1993). Probabilistic failure predictions in ceramics and ceramic matrix fiber reinforced composites, in **Life Predictions Methodologies and Data for Ceramic Materials,** ASTM STP 1201, (eds C.R. Brinkman and S.F. Duffy) Philadelphia, in press.

Lissart, N. and Lamon, J.L. (1992). Multimodal statistical analysis of fracture of Silicon Carbide fibers in Proc. JNC8 (eds O. Allix et al.) AMAC, Paris, pp. 5-16.

Thiemeier, T. and Brückner-Foit, A. (1991). Influence of the fracture criterion on the failure prediction of ceramics loaded in biaxial flexure. **J. Am. Ceram. Soc.** 74, 48-52.

VOLUME AND STRESS HETEROGENEITY EFFECTS IN CERAMICS AND FIBER-REINFORCED CERAMICS

F. HILD
Laboratoire de Mécanique et Technologie, Cachan, France.

Abstract
Failure of ceramics is due to initial flaws. These initial flaws are in general randomly distributed within the material. An expression of the cumulative failure probability can be related to the initial flaw size distribution in the framework of the weakest link theory and independent events assumption. This approach exhibits two different effects: the volume effect and the stress heterogeneity effect. These two effects can be analyzed for fiber-reinforced ceramics. Mostly, it is shown that the ultimate strength of these composites is length independent, but depends upon the stress heterogeneity.
Keywords: Ceramics, Fiber Reinforced Ceramics, Volume Effect, Stress Heterogeneity Effect, Weibull Model, Flaw Size Distribution.

1 Introduction

A large scatter in failure stress is a common feature of all brittle materials. Both the failure and the scatter are due to some initial flaws within the material. When a flaw becomes critical, the whole structure fails. In this paper, the expression of the cumulative failure probability is related to initial flaw distributions.

Under some very simple hypotheses, two- and three-parameter Weibull laws can be derived. The Weibull parameters are directly related to the statistical flaw distribution and failure criterion. The weakest link concept leads to two important properties linked with flaw distributions: the volume effect, and the stress heterogeneity effect. These effects are studied in pure tension and pure flexure.

To avoid catastrophic failure, ceramics can be reinforced by continuous ceramic fibers. Because of the presence of matrix and interface, progressive degradation of the matrix and of the fibers is possible. Here the relevant mechanisms to be considered are fiber breakage and fiber pull-out preceded by matrix cracking. An expression of the ultimate strength is analyzed in pure tension and pure flexure.

2 Reliability Analysis of Structures Made of Ceramics

The aim of this section is to analyze the reliability of structures made of ceramics and subjected to monotonic loading conditions. The study is conducted in the framework of the weakest link theory. An independent events hypothesis is made.

In this paper it is assumed that the initial flaws are randomly distributed within the structure and that the defect distribution is characterized by a probability density function

427

D. Breysse (ed.), Probabilities and Materials, 427–438.
© 1994 Kluwer Academic Publishers.

f. The function f gives the flaw size distribution. Other parameters such as orientation are not considered.

2.1 Failure of a Representative Volume Element

The cumulative failure probability, P_{F0} of a representative volume element Ω_0 of volume V_0, is the probability of finding a critical flaw within Ω_0. The probability of finding a critical flaw refers to the initial flaw size distribution f. A critical flaw is defined as a flaw whose size a is larger than the critical size a_c. In the case of monotonic loadings, the expression of P_{F0} is given by

$$P_{F0} = \int_{a_c}^{+\infty} f(a)\, da \qquad (1)$$

The critical flaw size is related to a uniaxial equivalent stress $\| \sigma \|$ by a general failure criterion

$$\| \sigma \| = g(a_c) \qquad (2)$$

where g is a positive strictly decreasing C^1 function. The function g will be specified later. For brittle materials such as ceramics the uniaxial equivalent stress $\| \sigma \|$ to be used to predict local failure may be the maximum principal stress.

2.2 Failure of a Structure

If the interaction between flaws is negligible, an independent events assumption can be made. The expression of the cumulative failure probability, P_F of a structure Ω of volume V can be derived in the framework of the weakest link theory. The expression of P_F can be related to the cumulative failure probability, P_{F0} of a link by (Freudenthal, 1968)

$$P_F = 1 - \exp\left\{ \frac{1}{V_0} \int_\Omega \ln\left(1 - P_{F0}\right) dV \right\} \qquad (3)$$

By means of Eqns. (1) and (3), a general relationship between the initial flaw size distribution and the cumulative failure probability of a structure Ω can be derived

$$P_F = 1 - \exp\left\{ \frac{1}{V_0} \int_\Omega \ln\left(1 - \int_{a_c}^{+\infty} f(a)\, da\right) dV \right\} \qquad (4)$$

A feature of Eqn. (4) is that the considered stress field is the macroscopic stress field since the interactions between flaws are neglected. This approach can therefore be used as a post-processor to any classical structural analysis (performed on the structure without defects).

2.3 Correlation with a Weibull Law

With some simple additional assumptions, Eqn. (4) leads to a Weibull law (Weibull, 1939). If the maximum flaw size is bounded ($a < a_M < +\infty$) a three-parameter Weibull law is deduced. When the maximum flaw size is not bounded, a two-parameter Weibull law is derived (Hild and Marquis, 1990, 1992).

If the maximum flaw size is bounded, there exists a threshold stress S_u below which no failure is possible

$$S_u = g(a_M) \qquad (5)$$

For a close to a_M, the function f is assumed to be equivalent to a power function $k(a_M - a)^\beta$, with $k > 0$, and $\beta > 0$. Assuming that a_c is close to a_M, the cumulative failure probability of a single link is approximately given by

$$P_{F0} \cong \left(\frac{<\| \sigma \| - S_u>}{S_0} \right)^{\beta + 1} \qquad (6)$$

where $<.>$ denotes the Macauley brackets. The cumulative failure probability of a structure can therefore be written as

$$P_F = 1 - \exp \left\{ \frac{1}{V_0} \int_\Omega \left(\frac{<\| \sigma \| - S_u>}{S_0} \right)^m dV \right\} \qquad (7)$$

Eqn. (7) corresponds to a three-parameter Weibull law where the shape parameter m is related to β by

$$m = \beta + 1 \qquad (8.1)$$

and the scale parameter S_0 is given by

$$S_0 = - g'(a_M) \left(\frac{\beta + 1}{k} \right)^{1/(\beta + 1)} \qquad (8.2)$$

The parameter β gives the trend of the flaw size distribution for large sizes. Eqns. (8) give relationships between parameters of the flaw size distribution (i.e. physical parameters) and Weibull parameters (i.e. mechanical parameters), which can be obtained through an analysis of a set of macroscopic failure tests.

If the maximum flaw size is not bounded, the previous threshold stress is equal to 0. We assume that for large values of a, the function f is equivalent to a power function Ka^{-n}, with $K > 0$, and $n > 0$. The relationship between the critical flaw size a_c and the equivalent stress $\| \sigma \|$ is supposed to be given by

430

$$g(a_c) = L(a_c)^{-(1/p)}$$

(9)

where L and p are positive constants. The cumulative failure probability of a single link is approximately given by

$$P_{F0} \cong \left(\frac{\| \sigma \|}{S_0}\right)^m$$

(10.1)

with

$$m = p(n-1)$$

(10.2)

$$S_0 = L \left(\frac{n-1}{K}\right)^{1/p(n-1)}$$

(10.3)

Here a two-parameter Weibull law is obtained. It is worth noting that the case p=2, which corresponds to a modeling using Linear Elastic Fracture Mechanics, has been studied by Jayatilaca and Trustrum (1977).

2.4 Volume and Stress Heterogeneity Effect

In the following we will use a two parameter-Weibull law. Using the results of Davies (1973), an effective volume analysis can be performed. In the framework of a Weibull model, the expression of the effective volume V_{eff} can be related to the Weibull stress heterogeneity factor H_m and to the total volume V of a structure by

$$V_{eff} = V H_m$$

(11)

where H_m is defined as (Hild et al., 1992)

$$H_m = \left\{ \frac{1}{V} \int_\Omega \| \sigma \|^m \, dV \right\} / \sigma_F^m$$

(12.1)

with

$$\sigma_F = \underset{\Omega}{\text{Max}} \| \sigma \|$$

(12.2)

The mean failure stress $\bar{\sigma}_F$ is then related to the effective volume by

$$\bar{\sigma}_F = S_0 \left(\frac{V_0}{V_{eff}}\right)^{1/m} \Gamma\left(1 + \frac{1}{m}\right)$$

(13)

where Γ corresponds to the Euler function of the second kind. Eqn. (13) describes the combined effects of volume and stress heterogeneity. Eqn. (13) accounts for the volume effect: the larger the volume, the higher the probability of finding a large flaw in a stressed region and the smaller the mean failure stress. Eqn. (13) models the stress heterogeneity effect as well: the more heterogeneous the stress field, the smaller the Weibull stress heterogeneity factor H_m, the smaller the probability of finding a critical flaw and the higher the mean failure stress. In pure flexure, the Weibull stress heterogeneity factor H_m is equal to $1/(2m+2)$. If we compare, for the same material, the mean failure stress $\bar{\sigma}_{FT}$ of a structure of volume V_1 subjected to pure tension with the mean failure stress $\bar{\sigma}_{FF}$ of a structure of volume V_2 subjected to pure flexure, we get the following relationship

$$\bar{\sigma}_{FF} / \bar{\sigma}_{FT} = \left(V_1 / V_2 \right)^{1/m} (2(m+1))^{1/m} \tag{14}$$

For monolithic ceramics, the two previous effects exist and need to be taken into account. This is particularly important when experimental data obtained on small specimens are used to design larger components.

3 Constitutive Law of Ceramic–Matrix Composites

To avoid catastrophic failure, ceramics can be reinforced by continuous ceramic fibers. Because of the presence of matrix and interface, progressive degradation of the matrix and of the fibers is possible, if the interfaces slide with low shear resistance, τ. The loading of such ceramic-matrix composites (CMCs) results in two independent damage mechanisms: matrix cracking and fiber breakage. This damage results in a reduction of the secant modulus $\bar{E}$, and the inception of permanent strains, ε_p. Furthermore, the fibers are subject to global load sharing, whereby the load transmitted from each failed fiber is shared equally among the intact fibers.

3.1 The Basic Stochastic Model
The tensile strength of a composite with a saturation density of matrix cracks is considered. The stress at which saturation occurs is denoted by S_{mc} and expressions can be found in the literature (see for instance Budiansky et al., 1986). The matrix crack spacing is L_m, within a unit cell of length L_R (Fig. 1a). The length L_R is the **recovery length** and refers to the longest fiber that can be pulled out and cause a reduction in the load carrying capacity. At reference stress, T, the recovery length is,

$$L_R = R\,T\,/\,\tau \tag{15}$$

where the reference stress T is the fiber stress in the plane of the matrix crack, R the fiber radius, and τ the sliding stress which is supposed constant over the recovery length L_R. Generally, $L_m \ll L_R$ and the stress field in intact fibers has the form illustrated in Fig. 1b. Consequently, the local stress in the fiber in the region $0 \le x \le L_m/2$ is (Cox, 1952, Kelly, 1973)

432

$$\sigma_f(T,x) = T - 2\tau x / R \tag{16}$$

If the fibers exhibit a statistical variation of strength that obeys a two-parameter Weibull law (Weibull, 1939), then the probability that a fiber would break anywhere within the recovery length L_R at or below a reference stress T is given by

$$P_F(T) = 1 - \exp\left[-\frac{1}{L_0} \int_{-L_R/2}^{L_R/2} \left\{ \frac{\sigma_f(T,x)}{S_0} \right\}^m dx \right] \tag{17}$$

where L_0 is the reference length associated with the reference volume V_0. As mentioned before, $\tau L_m / RT \ll 1$, so that Eqn. (17) can be simplified to

$$P_F(T) \cong 1 - \exp\left[-\left(\frac{T}{S_c}\right)^{m+1} \right] \tag{18}$$

where S_c is the so-called characteristic strength (Henstenburg and Phoenix, 1989) defined by

$$S_c^{m+1} = S_0^m \tau L_0 / R \tag{19}$$

The cumulative failure probability is thus independent of the total length of the composite if the total length of the composite is greater than the recovery length.

The average stress $\bar{\sigma}$ applied to the composite is related to the reference stress T by

$$\bar{\sigma} = f T \left[1 - P_F(T) \right] + f \sigma_p(T) \tag{20}$$

where $\sigma_p(T)$ denotes the component of the stress provided by failed fibers as they pull out from the matrix, and f the fiber volume fraction. For global load sharing, and neglecting the exclusion zones (Phoenix, 1992), the pull-out stress estimated by Curtin (1991) is

$$\sigma_p(T) = \frac{T}{2} P_F(T) \tag{21}$$

If we only consider the broken fibers inside the recovery length, the expression of the pull-out stress is given by

$$\sigma_p(T) = \int_0^T \frac{\sigma}{2} \frac{dP_F}{d\sigma} d\sigma = \frac{S_c}{2} \gamma\left[\frac{m+2}{m+1} ; \left(\frac{T}{S_c}\right)^{m+1} \right] \tag{22}$$

where γ [.;.] represents the incomplete gamma function. Eqn. (22) represents a **lower bound** to the expression given in Eqn. (21). The fiber breaks, which originally occurred outside the recovery length, and were then brought into the recovery length, as the recovery length increased (in proportion to T), are considered in Eqn. (21) but not in Eqn. (22). The external stress becomes

$$\bar{\sigma} = f\,T\left[1 - P_F(T)\right] + \frac{f\,S_c}{2}\,\gamma\left[\frac{m+2}{m+1}\;;\;\left(\frac{T}{S_c}\right)^{m+1}\right]$$
(23)

The ultimate tensile strength of the composite is defined by the condition

$$d\bar{\sigma}/d\bar{\varepsilon} = 0 \;\Leftrightarrow\; d\bar{\sigma}/dT = 0$$
(24)

because the reference stress T is proportional to the average strain on the composite

$$\bar{\varepsilon} = \frac{2}{L_m}\int_0^{L_m/2}\frac{\sigma_f(T,x)}{E_f}\,dx \cong \frac{T}{E_f}$$
(25)

where E_f denotes the Young's modulus of the unbroken fibers and L_m, the saturation matrix crack spacing. The ultimate tensile strength (UTS) arises when

$$\left(\frac{T}{S_c}\right)^{m+1} = \frac{2}{m+1}$$
(26)

at a recovery length

$$\left(\frac{L_R}{\delta_c}\right)^{m+1} = \frac{2}{m+1}$$
(27)

where δ_c is the so-called characteristic length defined as

$$\delta_c^{m+1} = L_0\left(S_0\,R\,/\,\tau\right)^m$$
(28)

It is worth noting that the value of $T\,/\,S_c$ at the ultimate, given by Eqn. (26), is identical to the one suggested by Phoenix (1992) based on a numerical study. The UTS becomes

$$S_{UT} = f\,S_c\,F(m)$$
(29)

with

$$F(m) = \left\{\left(\frac{2}{m+1}\right)^{1/(m+1)}\exp\left(-\frac{2}{m+1}\right) + \frac{1}{2}\,\gamma\left[\frac{m+2}{m+1}\;;\;\frac{2}{m+1}\right]\right\}$$
(30)

The UTS is thus independent of the total length of the composite and the relevant scaling stress is the characteristic stress S_c. Eqn. (29) can be obtained using Curtin's approach (Curtin, 1991) by means of Eqn. (20) and (21)

$$F(m) = \left(\frac{2}{m+2}\right)^{1/(m+1)} \frac{m+1}{m+2} \tag{31}$$

The UTS defined by Eqn. (30) and the analytic result of Curtin of Eqn. (31) are compared in Fig. 2. Although derived differently, Curtin's result and the expression of the UTS derived in this paper are very close.

3.2 Short Specimens
When the recovery length L_R becomes greater than the total length of the composite L, the previous results no longer apply, and the cumulative failure probability becomes length dependent,

$$P_F(T) \cong 1 - \exp\left[- \frac{L}{L_0} \left(\frac{T}{S_0}\right)^m \right] \tag{32}$$

Contrary to the length independent regime, a closed form solution does not exist. A typical result for the variation of strength with length is plotted on Fig. 3.

A useful approximation for very small L may be obtained by noting that, as $L \to 0$ (Hild et al., 1992)

$$S_{UT} \cong f S_0 \left(\frac{L_0}{m\,L\,e}\right)^{1/m} = S_b \tag{33}$$

where S_b denotes the classical dry bundle strength.

The solutions for short and long gauge lengths intersect in the region where $L/\delta_c \approx 1$. In practice, since δ_c is on the order of few millimeters, the UTS is length independent.

3.3 Ultimate Flexural Strength
Solutions for the ultimate strength in flexure are based on the non-linear, tensile stress/strain curve for the composite. Implicit in the use of this result is that the matrix exhibits multiple cracking on the tensile side of the specimen, but does not crack on the compressive side. Consequently, a damage variable D, reflecting the difference in Young's modulus between tension and compression, enters the solution. When crack saturation occurs in tension, D may be approximated by

$$D = 1 - f E_f / E_c \tag{34}$$

where E_c is the elastic modulus of the uncracked composite. This difference in the stress/strain law applicable to tension and compression causes a translation of the neutral plane away from the midsection of the beam.

The following calculations are based on the assumption that the global load sharing consideration takes place independently in each plane z=constant in the tensile part.

For pure four-point flexure, the strain $\bar{\varepsilon}$ between the inner loading points is related to the curvature κ by (Timoshenko and Goodier, 1970)

$$\bar{\varepsilon} = \kappa \, z \tag{35}$$

where z is the height coordinate measured from the neutral plane (Fig. 4). As a first approximation, the matrix cracking stress is assumed to be equal to 0, the stress on the tensile side of the neutral plane is thus obtained from Eqns. (20-21) as

$$\Sigma = Z - \frac{1}{2} Z^{m+2} \tag{36}$$

where $Z = \kappa \, z \, E_f / S_c$, and $\Sigma = \bar{\sigma} / f \, S_c$. Force balance dictates the position of the neutral axis, h_1 (Fig. 4), relative to the beam thickness, h. From Eqn. (36), we get

$$h_1 / h = \left[1 + \sqrt{(1-D) \frac{(m+1)(m+4)}{(m+2)(m+3)}} \right]^{-1} \tag{37}$$

When the inner span of the flexural beam L_s is sufficiently large so that $L_s > L_R$, composite failure in four-point flexure is expected to occur when the stress on the tensile surface reaches the UTS given by Eqn. (29). Imposing this failure criterion, the ultimate flexural strength (UFS) S_{UF} becomes,

$$S_{UF} = \frac{6M}{bh^2} = f \, S_c \, H(m,D) \tag{38}$$

with

$$H(m,D) = 2 \left(\frac{2}{m+2} \right)^{1/(m+1)} (h_1 / h)^2 \left[\frac{(h/h_1 - 1)^3}{1-D} + \frac{(m+1)(m+5)}{(m+2)(m+4)} \right] \tag{39}$$

where M is the flexural moment at the ultimate, b is the beam width. The flexural strength again scales with the characteristic strength, S_c, and the fiber volume fraction f. It is also explicitly dependent upon the damage variable, D, and the shape parameter, m.

Lastly, the ratio of the UFS to the UTS is given by

$$\frac{S_{UF}}{S_{UT}} = 2 \, (h_1 / h)^2 \left[\frac{(h/h_1 - 1)^3}{1-D} + \frac{(m+1)(m+5)}{(m+2)(m+4)} \right] \frac{m+2}{m+1} \tag{40}$$

436

Figures

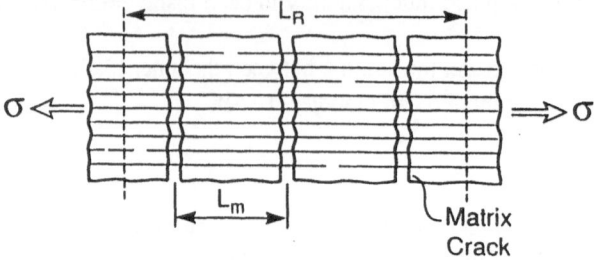

Fig. 1.a) Depiction of the recovery length L_R when the density of matrix cracks reaches saturation.

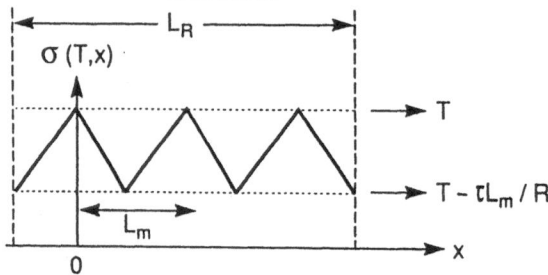

Fig. 1.b) Fiber stress field $\sigma_f(T,x)$ along a length L_R for a reference stress, T, when the fibers are intact.

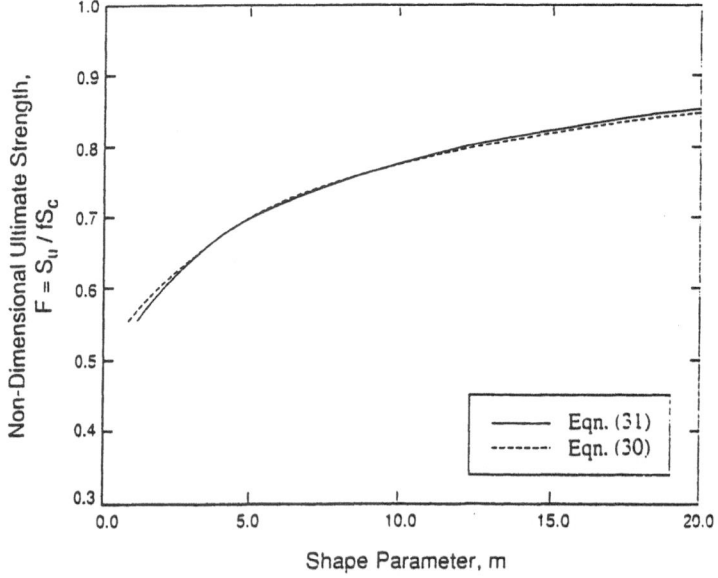

Fig. 2. Comparison between the lower bound of the ultimate tensile strength derived in this paper (Eqn. (30)) with Curtin's analytical formula (Eqn. (31)).

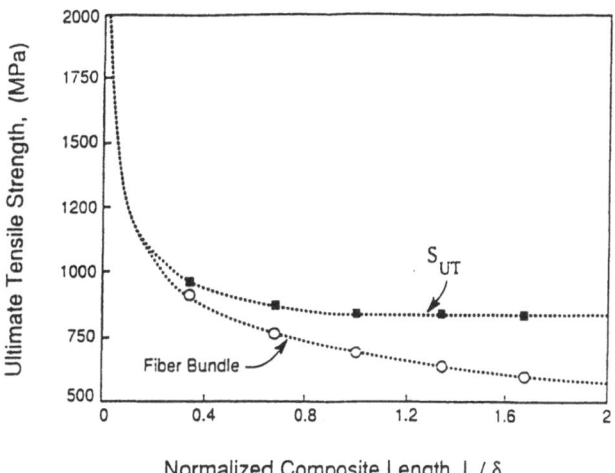

Fig. 3. Ultimate tensile strength as a function of the normalized total length of the composite. A comparison is made with a fiber bundle prediction. The calculations are based on parameters for SiC/LAS composites (Hild et al., 1992).

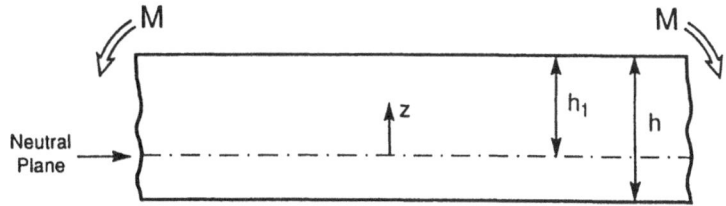

Fig. 4. Definition of the beam geometry in pure flexure.

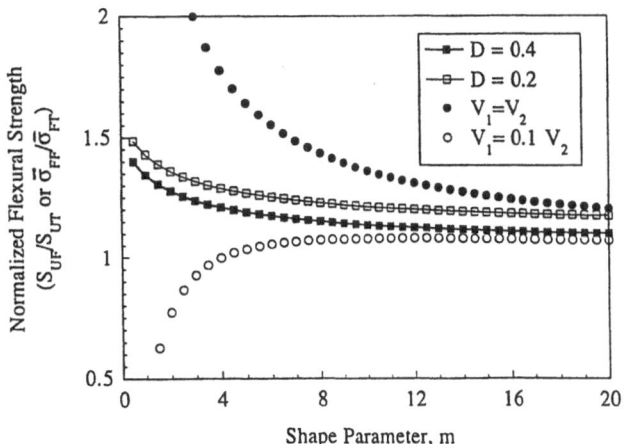

Fig. 5. Normalized flexural strength as a function of Weibull modulus m, for two different values of the damage parameter D (0.2 and 0.4). Also shown is the case of monolithic ceramics for which $V_1 = V_2$, and $V_1 = 0.1 \, V_2$.

Fig. 5 shows that CMCs exhibit a stress heterogeneity effect which is different from the stress heterogeneity effect observed in monolithic ceramics (Eqn. (14)). On the other hand, in CMCs, a volume effect is only observed for very short specimens and leads to very high strengths.

4. Acknowledgements

The author wishes to acknowledge the financial support of the DARPA through the University Research Initiative of University of California at Santa Barbara under ONR contract No. N-0-0014-86-K-0753, and of Renault through contract ENSC/24 (H5–25–511) with L.M.T. Cachan, Ecole Normale Supérieure de Cachan.

References

Budiansky, B. Hutchinson, J. W. and Evans, A. G. (1986) Matrix Fracture in Fiber-Reinforced Ceramics. **J. Mech. Phys. Solids**, 34, 167-189.

Cox, H.L. (1952) The Elasticity and the Strength of Paper and Other Fibrous Materials. **Br. J. Appl. Phys.**, 3, 72–79.

Curtin, W.A. (1991) Theory of Mechanical Properties of Ceramic Matrix Composites. **J. Am. Ceram. Soc.**, 74 [11], 2837–2845.

Davies, D. G. S. (1973) The Statistical Approach to Engineering Design in Ceramics. **Proc. Brit. Ceram. Soc.**, 22, 429-452.

Freudenthal, A. M. (1968) Statistical Approach to Brittle Fracture. **Fracture (an advanced treatrise)**, Academic Press, 2, 591-619.

Henstenburg, R.B. and Phoenix, S.L. (1989) Interfacial Shear Strength Using Single-Filament-Composite Test. Part II: A Probability Model and Monte Carlo Simulations. **Polym. Comp.**, 10 [5], 389–406.

Hild, F. Billardon, R. and Marquis, D. (1992) Stress Heterogeneity Influence on Failure of Brittle Materials. **C. R. Acad. Sci. Paris**, 315 (II), 1293-1298.

Hild, F. Domergue, J.-M. Evans, A. G. and Leckie, F. A. (1992) Tensile and Flexural Ultimate Strength of Fiber Reinforced Ceramic-Matrix Composites. **Int. J. Solids Structures**, submitted.

Hild, F., and Marquis, D. (1990) Correlation between Defect Distribution and Failure Stress for Brittle Materials. **C. R. Acad. Sci. Paris**, 311 (II), 573-578.

Hild, F. and Marquis, D. (1992) A Statistical Approach to the Rupture of Brittle Materials. **Eur. J. Mech. A/Solids**, 11 (6), 753-765.

Jayatilaca, A. D. S., and Trustrum, K. (1977) Statistical Approach to Brittle Fracture. **J. Mat. Sci.**, 12, 1426-1430.

Kelly, A. (1973) **Strong Solids,** 2nd edition, Oxford University Press, Chapter 5.

Phoenix, S.L. (1992) Statistical Issues in the Fracture of Brittle Matrix Fibrous Composites. **Comp. Sci. Techn.**, submitted.

Timoshenko, S.P. and Goodier, J.N. (1970) **Theory of Elasticity,** 3rd edition, McGraw-Hill.

Weibull, W., (1939) A Statistical Theory of the Strength of Materials. **Ingeniörsvetenskapakademiens**, Handlingar No. 153.

APPLICABILITY OF A WEIBULLIAN MODEL OF FRACTURE BY APPLICATION OF A SLOWLY GRADUAL LOAD

P. KITTL*, G. DIAZ* and V. MARTINEZ**
*Departamento de Ingeniería de los Materiales, IDIEM,
Casilla 1420, Santiago, Chile.
**Departamento de Ingeniería Mecánica,
Casilla 2777, Santiago, Chile.
Facultad de Ciencias Físicas y Matemáticas,
Universidad de Chile.

Abstract
From the independence of the cumulative probabilities of
non-fracture of a body subdivided into two arbitrary
volumes without common points, a functional equation
connecting the cumulative probability of fracture with the
material volume, for a constant stress-field is obtained.
Such expression can be generalized for the case of a
variable stress-field in the three-dimensional space. When
the solution of this equation is obtained, the so-called
specific risk of fracture function of Weibull appears,
without a presupposed analytical form. This model runs
correctly if the scale and the manufacturing process are
not modified. When the scale is varied the difficulties
appear and, in consequence, it is not possible to predict
the material behaviour. This problem is very ample and its
treatment, with empirical modifications or micro models,
can be made by means of a modification of the differential
equation -which is the basement of the Weibull statistics-
with an empirical parameter and the micro models based in
the coalescence of a definite number of cracks in some
critical volume. A model of parallel volumes is also
treated and it is shown the existence of a model
characterized by when the volume increases then the
strength increases, inversely to the commonly found.
Keywords: Probabilistic Fracture, Scale, Volume, Increase,
Decrease, Strength, Models.

1 The Principles

The cumulative probability of fracture in an isotropic
brittle body subjected to a constant stress field σ can be
determined using the following functional equation, Weibull
(1939), Kittl and Díaz (1988):

$$\tilde{F}_{12}(V=V_1+V_2, \sigma) = \tilde{F}_1(V_1, \sigma) \tilde{F}_2(V_2, \sigma) \qquad (1)$$

where V_1 and V_2 are the volumes resulting from an arbitrary

439

D. Breysse (ed.), Probabilities and Materials, 439–449.
© 1994 Kluwer Academic Publishers.

disjointed division of the body of volume V. F_{12} is the cumulative probability of nonfracture of the whole volume V, and F_1 and F_2 are the cumulative probabilities of nonfracture of volumes V_1 and V_2 respectively. The boundary condition $\tilde{F}_{12}(0,\sigma)=\tilde{F}_1(0,\sigma)=\tilde{F}_2(0,\sigma)=1$ yields easily $\tilde{F}_{12}=\tilde{F}_1=\tilde{F}_2=\tilde{F}$ and then functional equation (1) has a unique solution already indicated by Euler:

$$F(V,\sigma)=1-\tilde{F}(V,\sigma)=1-\exp\{-\frac{V}{V_0}\phi_1(\sigma)\} \tag{2}$$

where V_0 is the unit volume and $\phi_1(\sigma)$ is the specific risk of fracture function. This formula given by Weibull can be easily generalized for uniaxial variable stress fields:

$$1-F(V,\sigma)=\lim_{N\to\infty}\prod_{i=1}^{N}[1-F(V_i,\sigma_i)]$$

$$=\lim_{N\to\infty}\prod_{i=1}^{N}\exp\{-\frac{V_i}{V_0}\phi_1(\sigma_i)\}=\lim_{N\to\infty}\exp\{-\sum_{i=1}^{N}\frac{V_i}{V_0}\phi_1(\sigma_i)\} \tag{3}$$

$$\rightarrow \quad F(V,\sigma)=1-\exp\{-\frac{1}{V_0}\int_V\phi_1[\sigma(r)]\,dV\}$$

where r is the position vector, $\sigma(r)\leq\sigma$ is the stress-field, and the volume is determined by the condition $\sigma(r)\geq0$. When $F(\sigma,V)$ is experimentally known and $\sigma(r)$ is obtained through the Theory of Elasticity then integral equation (3) can be solved and, then, $\phi_1(\sigma)$ is obtained. It is worthwhile to emphasize that this Probabilistic strength of materials theory is valid in a three-dimensional space and the generalisation given by formula (3) can be written in an integral form because when σ=constant the analytical form of F is expressed by formula (2). This last allows to this theory to be the unique that can be expressed as an integral algorithm. The other models have to be numerically treated.

As regards the local probability of fracture F_L, i.e. the percentage of fractures starting in a given region of the body subjected to some stress, Oh and Finnie (1970), Kittl and Camilo (1981), since $F_L(V_1,\sigma)=n_1/N$ and $F_L(V_2,\sigma)=n_2/N$ where n_1 and n_2 are the numbers of the fractures that are started in V_1 and V_2 respectively, and N is the total number of test -with $n_1/n_2=V_1/V_2$, then the local probability of fracture is given by the following functional equation:

$$F_L(V_1,\sigma)\,V_2-F_L(V_2,\sigma)\,V_1=0 \tag{4}$$

Putting $V_1 = V$ and $V_2 = V + \Delta V$, $\Delta V \to 0$, yields:

$$F_L(V, \sigma) = F_L'(V, \sigma) \ V \tag{5}$$

whose solution is:

$$0 \leq F_L(V, \sigma) = \frac{V}{V_0} \phi_2(\sigma) \leq 1 \tag{6}$$

in which $\phi_2(\sigma)$ must be determined.

Thus the principles of Probabilistic Strength of Materials are represented by functional equations (1) and (4) -whose solution are (2) and (6)- that are independent and incompatible except in the region where $(V/V_0) \phi_1(\sigma)$ and $(V/V_0) \phi_2(\sigma)$ tend towards zero. In order that the functions become equal in that boundary then $\phi_1 = \phi_2$ and accordingly:

$$F(V, \sigma) = F_L(V, \sigma) = (V/V_0) \phi(\sigma)$$
$$\frac{V}{V_0} \phi(\sigma) \to 0 \tag{7}$$

Hence equation (7) becomes:

$$\frac{dn}{N} = \frac{\phi[\sigma(r)]}{V_0} dV \tag{8}$$

where dn is the number of fractures started in dV. The integration of equation (8) produces:

$$\frac{n}{N} = \frac{1}{V_0} \int_V \phi[\sigma(r)] \, dV; \quad r \in V[\sigma(r) \geq 0] \tag{9}$$

and the elimination of N from equation (8) and (9) yields:

$$\frac{dn}{n} = \frac{\phi[\sigma(r)] \, dV}{\int_V \phi[\sigma(r)] \, dV} \tag{10}$$
$$r \in V[\sigma(r) \geq 0]$$

From equation (10) when $\sigma(r) = \sigma = $ constant is easy to reobtain $n_1/n_2 = V_1/V_2$.

If the percentage of fractures in some volume $V_1 \leq V$ is to be ascertained, then equation (10) must be integrated in this volume, assuming that $\phi(\sigma)$ is known. If dn/ndV=1/g(r,σ) is already known then equation (10) is converted into an integral equation allowing to obtain $\phi(\sigma)$:

$$\int_{r \, \epsilon \, V[\sigma(r) \geq 0]} \phi[\sigma(r)] \, dV = g(r,\sigma) \phi[\sigma(r)] \tag{11}$$

Integral equation (11) allows to obtain $\phi(\sigma)$ excepting a constant factor. Integral equation (11) becomes trivial solely when $\sigma(r)=\sigma$, i.e. when the stress-field is uniform. In the stress space, the integral equations (3) and (11) are transformed, respectively, into:

$$\xi(\sigma) = \ln \frac{1}{1-F(\sigma)} = \frac{1}{V_0} \int_0^\sigma \phi(\Sigma) \, dV(\Sigma) \tag{12}$$

$$\int_0^\sigma \phi(\Sigma) \, dV(\Sigma) = g(\sigma) \phi(\sigma) \tag{13}$$

2 Uncertainty Principles

If $\xi(\sigma)=(V/V_0)\phi(\sigma)$ is employed as the variable in equation (2) then the distribution function is dF/dξ=exp(-ξ) and this gives:

$$\int_0^\infty e^{-\xi} d\xi = 1 ; \quad \overline{\xi} = \int_0^\infty \xi e^{-\xi} d\xi = 1 = \frac{V}{V_0} \overline{\Phi}(\sigma)$$

$$\overline{\Phi} = \frac{V_0}{V} \tag{14}$$

$$\overline{\Delta \xi^2} = \overline{(\xi-\overline{\xi})^2} = \overline{\xi^2} - \overline{\xi}^2 = \int_0^\infty \xi^2 e^{-\xi} d\xi - 1 = 2 - 1 = 1$$

This last equation supplies the following relationship:

$$\Delta \xi = (\overline{\Delta \xi^2})^{1/2} = \frac{V}{V_0} [(\phi-\overline{\phi})^2]^{1/2} = \frac{V}{V_0} \Delta \phi = 1$$

$$\Delta \phi = \frac{V_0}{V} ; \quad \frac{\Delta \phi}{\overline{\phi}} = 1 \tag{15}$$

and this expresses one of the principles of uncertainty of Probabilistic strength of materials. It is necessary to emphasize that $V\Delta\phi=V_0$ is really an uncertainty relation, because V is the uncertainty in the location of fracture and $\Delta\phi$ is a functional uncertainty. If the stress space is used:

$$\xi \leq (V/V_0)\phi; \quad 1=\overline{\xi} \leq (V/V_0)\overline{\phi}$$
$$1=\Delta\xi \rightarrow (V/V_0)\Delta\phi \tag{16}$$
$$V \rightarrow 0$$

As to the local probability of fracture, the dispersion of the number of fractures n_1 starting in volume V_1 is obtained through the following group of relationships, as it is well-known from the elementary theory of statistics:

$$n_1 = N(V_1/V)$$
$$\Delta n_1 = (n_1)^{1/2} = (N)^{1/2} (V_1/V)^{1/2} \tag{17}$$
$$\Delta n_1/n_1 = (N)^{-1/2} (V/V_1)^{1/2} = (n_1)^{-1/2}$$

This group of relationships (15) or (16) and (17) represents the principles of uncertainty of the Probabilistic Strength of Materials, none of the two principles of uncertainty depends on constants of the material. Some applications were made to Seismology, Kittl and Díaz (1987), Kittl et al. (1990), which implies the impossibility to predict earthquakes and fracture or rockburst of natural rocks such as the granite.

3 Applications and Models

A great number of works supports the mentioned theory when the volume is maintained as a constant, but when this volume varies the values of $\xi(\sigma,V)=(V/V_0)\phi(\sigma)$ does not vary with the experiences according to this theoretical functional dependence of the volume. The following functions are the ones determined with constant volume and they give a statistically acceptable χ^2:

$$\phi(\sigma) = \begin{cases} 0 & 0 \leq \sigma \leq \sigma_L \\ (\dfrac{\sigma-\sigma_L}{\sigma_0})^m & \sigma_L < \sigma < \infty \end{cases} \tag{18}$$

$$\phi(\sigma) = \begin{cases} 0 & 0 \le \sigma \le \sigma_L \\ K\left(\dfrac{\sigma - \sigma_L}{\sigma_S - \sigma}\right)^m & \sigma_L < \sigma \le \sigma_S \\ \infty & \sigma_S < \sigma \end{cases} \tag{19}$$

where the equation (18) is that one of Weibull and the equation (19) is that one of Kies-Kittl, Díaz and Morales (1988), and in these equations m and σ_0 are parameters depending on the manufacturing process of the material, σ_L is the stress under which there is no fracture, σ_s is the stress over which there is always fracture and K is the Kittl constant. By means of Weibull function, for $\sigma(r) = \sigma$, it can be calculated:

$$\bar{\sigma} = \sigma_0 (V/V_0)^{-1/m} \ \Gamma(1 + 1/m) + \sigma_L \tag{20}$$

$$\Delta\sigma = \sigma_0 (V/V_0)^{-1/m} \ [\ \Gamma(1 + 2/m) - \Gamma^2(1 + 1/m)]^{1/2} \tag{21}$$

where, in equations (20) and (21), σ and $\Delta\sigma$ as functions of V are shown.

When the stress-field is uniform, $\sigma(r) = \sigma$, the change of volume can also be studied from samples of sizes V_1 and V_2:

$$\begin{aligned} \xi(\sigma) &= (V/V_0)\phi(\sigma) \\ \ln\xi_1(\sigma) &= \ln(V_1/V_0) + \ln\phi(\sigma) \\ \ln\xi_2(\sigma) &= \ln(V_2/V_0) + \ln\phi(\sigma) \end{aligned} \tag{22}$$

and from equation (22) it is obtained:

$$\ln\xi_2(\sigma) = \ln\xi_1(\sigma) + \ln(V_2/V_1) \tag{23}$$

Hence, with the equation (23), it is possible to determine the change of the function of cumulative probability of fracture when the volume of the material changes.

In many cases the behaviour, due to a change of volume, can be described with a modification of the differential equation that solves the functional equation (1):

$$\frac{d\tilde{F}}{\tilde{F}} = -\frac{\phi(\sigma)}{V_0^{1-\lambda}}\frac{dV}{V^\lambda} \tag{24}$$

here λ is a parameter representing the empirical modification. The solution of equation (24) is:

$$\tilde{F} = \exp\left[-\frac{\phi(\sigma)}{1-\lambda}(V/V_0)^{1-\lambda}\right] \tag{25}$$

The equation (25) leads to the following distributions of probability of fracture:

$$\lambda=0:\ F(V,\sigma) = 1-\exp\{-\phi(\sigma)\frac{V}{V_0}\}\ \ \begin{array}{c}Weibull\\(1939)\end{array} \tag{26}$$

$$0\leq\lambda<1:\ F(V,\sigma) = 1-\exp\{-\frac{\phi(\sigma)}{1-\lambda}(\frac{V}{V_0})^{1-\lambda}\}\ \ \begin{array}{c}Kittl\ et\ al.\\(1987)\end{array} \tag{27}$$

$$\lambda=1:\ F(V,\sigma) = 1-\exp\{-\phi(\sigma)\ln\frac{V}{V_m}\}\ \ \begin{array}{c}Kittl\text{-}Gunther\\(1981)\end{array} \tag{28}$$
$$V_m\leq V$$

$$\lambda>1:\ F(V,\sigma) = 1-\exp\{-\phi(\sigma)(\frac{V}{V_0})^{1-\lambda}\}\ \ \begin{array}{c}Kittl\ et\ al.\\(1993)\end{array}$$
$$\frac{d\tilde{F}}{\tilde{F}} = -\frac{\phi(\sigma)}{V_0^{1-\lambda}}(1-\lambda)\frac{dV}{V^\lambda} \tag{29}$$

If the change in $\xi(\sigma,V)$ is described with the formula:
$$\ln\xi(\sigma) = \ln\phi(\sigma) + \ln f(V) \tag{30}$$

for the different above cases we have:

$$
f(V) = \begin{cases}
\dfrac{V}{V_0} & \lambda=0 & \bar{\sigma} \text{ decreases when } V \text{ increases} \\[2ex]
\dfrac{1}{1-\lambda}(\dfrac{V}{V_0})^{1-\lambda} & 0\leq\lambda\leq1 & \bar{\sigma} \text{ decreases when } V \text{ increases} \\[2ex]
\ln\dfrac{V}{V_m} & \lambda=1 & \bar{\sigma} \text{ practically without variation} \\[2ex]
(\dfrac{V}{V_0})^{1-\lambda} & \lambda>1 & \bar{\sigma} \text{ increases when } V \text{ increases}
\end{cases}
$$

$$(31)$$

In the case of the distribution of the beginning of non-linear deformation of the ASTM-516 steel, Kittl, Martínez, Díaz, Bölcich and Fernández (1993), obtained $\lambda>1$ and, therefore, the strength decreases when the volume decreases, on the contrary to a Weibull distribution. When there is no dependence with the volume $\lambda=1$ and this case is produced in cement paste, Kittl and Günther (1981), Kittl and Aldunate (1983). In the case $0\leq\lambda\leq1$ there are examples in glass and electrical porcelain, Kittl et al. (1982).

The non-dependence with the volume (Kittl and León (1987), Jayatilaka and Trustrum (1977, 1978)), which implies $\lambda=1$, can also be explained by assuming that fracture is produced when a concentration of cracks $n\Delta v=V_c$ is produced into a critical volume V_c, where n is that number of cracks and Δv is the volume assigned to each crack. Therefore:

$$
\begin{aligned}
N\Delta v &= V \\
i\Delta v &= v \\
n\Delta v &= v_c
\end{aligned}
$$

$$(32)$$

where N,i and n are integers. If the body is subjected to traction we obtain:

$$
\begin{aligned}
\sigma'(N-i)\Delta S &= \sigma N\Delta S \\
\Delta v &= L\Delta S \\
\sigma' &= \dfrac{N}{N-i}\sigma
\end{aligned}
$$

$$(33)$$

where $\Delta v=L\Delta S$ is approximately a volume whose length is L and whose section is ΔS, and σ' is the new tension after the crack propagation in the i volumes. The function $\xi(\sigma)$ for i broken volumes v is according to Weibull:

$$\xi(\sigma, V) = \frac{1}{V_0} \sum_{i=1}^{n} i \Delta v \, \phi(\frac{N}{N-i}\sigma) \tag{34}$$

when n<<N the equation (34) yields:

$$\xi(\sigma) = \frac{1}{V_0} \sum_{i=1}^{n} i \Delta v \, \phi(\sigma) = \frac{\Delta v \, n \, (n+1)}{2 V_0} \phi(\sigma) \tag{35}$$

and considering equation (32) the equation (35) becomes:

$$\xi(\sigma) = \frac{v_c^2}{2 V_0 \, \Delta v} \phi(\sigma) \tag{36}$$

in consequence:

$$F(\sigma) = 1 - \exp[-\frac{v_c^2}{2 V_0 \, \Delta v} \phi(\sigma)] \tag{37}$$

The behaviour $\lambda=1$ can also be simulated with a system of parallel zones, for example a system of two parallel bars. In this last case, the expression of the probability of collapse or fracture is given by, Kittl, Díaz and Martínez (1993):

$$F_{2,c} = F^2(\sigma, \frac{V}{2}) + 2F(\sigma, \frac{V}{2}) \tilde{F}(\sigma, \frac{V}{2}) F(2\sigma, \frac{V}{2}) \tag{38}$$

where $F_{2,c}$ is the cumulative probability of collapse of a system of two parallel bars, each one with volume V/2, or a total volume V. In the case of one bar of volume V/2 we have:

$$F_{1,c} = F(\sigma, \frac{V}{2}) = 1 - \exp[-\frac{V}{2 V_0} \phi(\sigma)] \tag{39}$$

448

Figure 1 shows the curves $F_{1,c}$ and $F_{2,c}$ obtained when, for example, it is adopted $\phi(\sigma) = \sigma^4$ and $V_0 = V = 1$. Here it can be seen that, for the range $0 \le \sigma \le 0.5$, it is obtained $F_{1,c} > F_{2,c}$.

If there are two models, one in which the strength decreases when the volume increases and the other functioning in the inverse way, then a combination of both models can describe a intermediate situation such as the one given by $0 \le \lambda \le 1$.

If in the model that was proposed by Breysse, Breysse (1990), for the elementary springs indicated in that work, it is employed, as the cumulative probability function, the one of Weibull with the corresponding dependence of the volume, then the results obtained are similar to the ones of equations (38) and (39).

Hence, it is observed that only when $\lambda = 0$, corresponding to a Weibull statistics, this theory can be extended to a variable stress-field by means of an integral algorithm. All of resting models, i.e. those ones of crack nucleation empirical and of parallel systems have to be, in a variable stress-field, numerically treated.

In all of the developed theories, the hypothetic-deductive system has been used in such a way that all of this exposed to here has justification in the primary hypothesis. It is necessary to emphasize that this logical structure can justify the brittle and the beginning of yielding behaviour of solids, whichever the change in the strength when the volume varies.

Acknowledgements

The authors wish to thank the Fondo Nacional de Desarrollo Científico y Tecnológico (FONDECYT) for grants N° 1195-91 and 1931056.

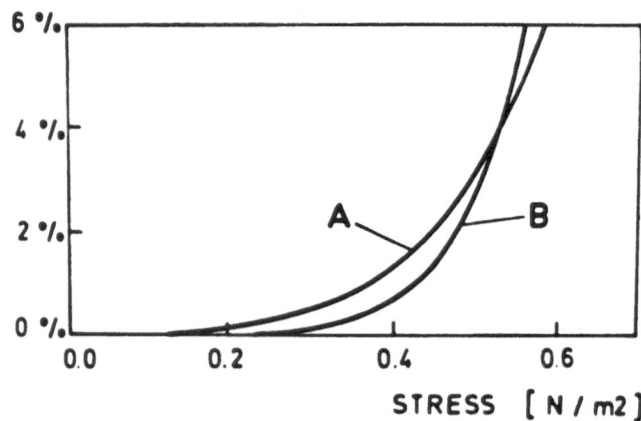

Fig.1. Cumulative probability of collapse of a system of one (A) and two (B) parallel bars.

4 References

Breysse, D. (1990) Un modèle probabiliste d'endommagement résultant d'une approche micro-macro. **Mat. and Struct.**, 23, 161-171.

Díaz, G. and Morales, M. (1988) Fracture statistics of torsion in glass cylinders. **J. Mater. Sci.**, 23, 2444-2448.

Jayatilaka, A. de S. and Trustrum, K. (1977) Statistical approach to brittle fracture. **J. Mater. Sci.**, 12, 1426-1430.

Jayatilaka, A. de S. and Trustrum, K. (1977) Application of a statistical method to brittle fracture in biaxial loading systems. **J. Mater. Sci.**, 12, 2043-2048.

Jayatilaka, A. de S. and Trustrum, K. (1978) Fracture of brittle materials in uniaxial compression. **J. Mater. Sci.**, 13, 455-457.

Kittl, P. and Camilo, G.M. (1981) Local probability of failure in statistical theory of brittle fracture. **Res Mech. Lett.**, 1, 115-118.

Kittl, P. and Günther, O. (1981) Volume size and fracture statistics of a compacted cement paste. **Res Mech. Lett.**, 1, 145-148.

Kittl, P., Castro, J.H.C., Knapp, W.J. and Camilo, G.M. (1982) The χ^2 proof for Weibull's distribution and the anisotropy of an electrical porcelain. **Latin Am. J. Metall. Mater.**, 2, 76-77.

Kittl, P. and Aldunate, R. (1983) Compression fracture statistics of compacted cement cylinders. **J. Mater. Sci.**, 18, 2947-2950.

Kittl, P. and Díaz, G. (1987) Earthquake risk and fracture statistics. **Phys. Earth Planet. Inter.**, 49, 222-224.

Kittl, P. and León, M. (1987) On fracture statistics of almost-brittle bodies. **Cerâmica**, 33, 55-62.

Kittl, P. and Díaz, G. (1988) Weibull's fracture statistics, or probabilistic strength of materials: state of the art. **Res Mech.**, 24, 99-207.

Kittl, P., León, M., Díaz, G. and Lillo, A. (1990) Probabilistic compressive strength of sound dry granite. **Rock Mech. Rock Engn.**, 23, 21-28.

Kittl, P., Díaz, G. and Martínez, V. (1993) On the parallel systems of brittle materials. Personal communication.

Kittl, P., Martínez, V., Díaz, G., Bölcich, J.C. and Fernández, L. (1993) Non-linear deformation probability of an ASTM-516 steel. **J. Mater. Sci. Lett.**, 12, 823-824

Oh, H. L. and Finnie, I. (1970) On the location of fracture in brittle solids. I. Due to static loading. **Int. J. Fract. Mech.**, 6, 287-300.

Weibull, W. (1939) A statistical theory of the strength of materials. **Ing. Vetenskaps Akad. Handl.**, 151 , 1-45.

A STOCHASTIC MODEL TO DESCRIBE PROBABILISTIC ASPECTS OF DAMAGE ACCUMULATION IN CONCRETE

H. MIHASHI
Dept. of Architecture, Tohoku University, Sendai 980, Japan

Abstract
This paper presents basic concepts of a stochastic model to describe probabilistic aspects of damage accumulation in concrete. This model is a kind of macromodel which takes into account the variability due to the heterogeneous structure and for that due to the thermodynamic randomness. Some examples of the application of this model are shown, which are degradation of concrete due to freezing and thawing action and deformation of concrete under a cyclic load.
Keywords: Concrete, Microstructure, Stochastic Model, Heterogeneity, Frost Damage, Fatigue.

1 Introduction

Usually material properties more or less deviate. When the deviation is negligible, deterministic treatments are feasible and the mean value is representative. Since concrete is a multiphase composite material, most of properties have probabilistic aspects. For analyzing the mechanical behavior of concrete, Wittmann (1983) suggested to subdivide the structure of concrete into three levels that are microlevel, mesolevel and macrolevel. Usually mechanical behavior of concrete is described on the macrolevel, which are used for engineering purpose.   Many kinds of mathematical models to describe various probabilistic aspects have been proposed.   To determine the probabilistic parameters, however, a number of specimens need to be tested and it is often difficult to find the physical meaning of each parameter.

Because the macrolevel probabilistic aspects are results of interrelation between microlevel and mesolevel variabilities, it is essential to establish a mathematical model to describe the variability on each level and to link these three different levels.

A stochastic model was developed by the author in the middle of 1970s for fracture of concrete (Mihashi and Izumi, 1977) in which the microlevel variability was described

451

D. Breysse (ed.), Probabilities and Materials, 451–460.
© 1994 Kluwer Academic Publishers.

Table 1. Sources of probabilistic aspects in mechanical
behavior of concrete

Microlevel	size, shape and orientation of pores and hydration products; thermodynamic effect due to temperature, moisture and rate of loading.
Mesolevel	size, shape and orientation of aggregates, voids and pre-existing cracks.
Macrolevel	loading condition.

according to the theory of rate process, the mesolevel
variability was represented by a probability density
function of local stress amplifier coefficient caused by the
heterogeneity. Both of these variabilities were linked
together to describe the probabilistic aspects of the
macrolevel strength on the basis of Markovian process model.
    Differences of this model from usual Weibull models are
following: 1) parameters have some physical meanings but
not simple mathematical ones; 2) considering the failure
process, more sophisticated models were developed to
describe the different probabilistic aspects between tensile
or flexural failure and compressive failure (Mihashi, 1983).
    This model is applicable to various damage problems such
as fatigue and frost damage of concrete, too. Since damage
properties are usually scattered in a wide range,
probabilistic models are essential to analyze the damage
mechanisms.
    In this paper, basic concepts of a stochastic model to
describe probabilistic aspects of damage accumulation in
concrete are presented. As examples of the application of
this model, degradation of concrete due to freezing and
thawing action and deformation of concrete under a cyclic
load are discussed.

2 Basic Concepts of Stochastic Model

Basic concepts of the stochastic model are composed of the
following characteristics: (1) 'probabilistic properties of
characteristic local stress' is introduced to take account
of randomness due to the heterogeneity; (2) 'stochastic
process theory' is introduced to represent thermodynamic
randomness in the rate of microstructural change.
    Strength of a system with a crack can be described as a
function of the crack length, toughness of the material and
the stress condition.   Let $E\gamma$ be the toughness and s be a
stress amplifier coefficient in the local area around the
crack.   For the purpose of simplicity, only the expected
largest crack length $2\bar{a}$ is considered.   Even if a specimen

has constant values of $E\gamma$ and $2\bar{a}$, s is randomly distributed over the specimen because of the heterogeneity and the orientation of the crack. Therefore the equivalent crack length becomes a random variable which includes all kinds of mechanical factors affecting the crack initiation and is given by eq.(1).

$$c = s\sqrt{\bar{a}/E\gamma} \tag{1}$$

The probability density function of the equivalent crack length: $f_c(c)$ is described as follows:

$$f_c(c) = \sqrt{E\gamma/\bar{a}}\ f_s(s) \tag{2}$$

where $f_s(s)$ is the probability density function of the stress amplifier coefficient in the local area around the crack.

Let's consider an element composed of n units each of which contains a crack of an equivalent length (Fig.1) and suppose that a certain accumulation of microstructural change at the crack tip causes cracking which contributes to the progress of damage process. On the basis of the rate process theory (Yokobori, 1974), the mean value of the probability of a microstructural change between a given time t and t+dt may be represented by eq.(3) (Mihashi and Wittmann, 1980).

$$\bar{\mu}(t)\cdot dt = \bar{Z}A(E\gamma)^{-\beta/2}\ \bar{a}^{\beta/2+1}\ \sigma(t)^{\beta}\int_0^{\infty} s^{\beta}f_s(s)\,ds\cdot dt \tag{3}$$

For the element containing n cracks, the rate of cracking is given by the following equation:

$$\bar{p}_{01}(t) = n\bar{\mu}(t) = L\sigma(t)^{\beta} \tag{4}$$

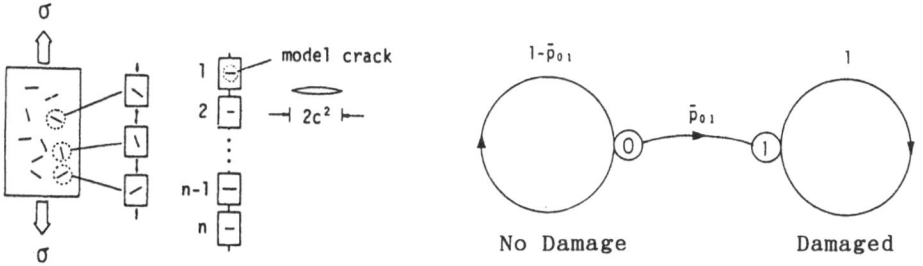

Element Model of Units

Fig.1. Model of element. Fig.2. Transition line graph.

where

$$L = n\bar{Z}A(E\gamma)^{-\beta/2}\bar{a}^{\beta/2+1}R \quad \text{and} \quad R = \int_0^\infty s^\beta f_s(s)ds \qquad (5)$$

The possible situation of the element is "0" which has no damage and "1" which is damaged (Fig.2). The survival probability of the element is represented by the following equation.

$$P_0(t) = \exp\{-\int_0^t \bar{p}_{01}(t)dt\} \qquad (6)$$

where $P_0(t)$ means the probability that no damage occurs before the time t in the element.

Specimens are usually composed of a large number of elements and the survival probability or soundness of the specimen depends on the system to link the elements.

3 Damage Accumulation in Concrete due to Freezing and Thawing Action

3.1 Theoretical model

Since damage accumulation in concrete due to a freezing and thawing action is considered to be caused by microcracking during the freezing process, the stochastic model can be applicable. In this case, capillary pores filled with water are the main source for this cracking. They are very small in comparison with the size of the specimen and microcracking occurs independently each other by freezing the contained water. Hence the system can be described with a parallel model. On the assumption that the specimen is composed of m elements which link each other in parallel, the survival probability of an element is supposed to represent the soundness of the whole specimen which is usually measured as the remained ratio of the dynamic modulus of elasticity.

Let $\bar{p}_{01}(N)$ be the mean value of the transition probability from the situation 0 to the situation 1 at the Nth cycle of freezing-and-thawing. The survival probability $P_0(N)$ of an element, which means the probability that the microcrack doesn't occur in the element before a given cycle N, can be given as follows:

$$P_0(N) = \exp\{-\int_0^N \bar{p}_{01}(N) \, dN\} \qquad (7)$$

Previous experimental studies show that the rate of microcracking due to freezing and thawing action is influenced by the freezing temperature, water content, material structure and the number of cycles. Therefore the transition probability $\bar{p}_{01}(N)$ which can be regarded as the rate of damage progress, may be described as a function of these variables as follows:

$$\bar{p}_{01}(N) = \alpha(T) \cdot \beta(H) \cdot \gamma(L) \cdot \delta(N) \tag{8}$$

where $\alpha(T)$ is a function of the freezing (lowest) temperature, $\beta(H)$ is a function of water content which is unity for a specimen in water and zero for an oven-dried specimen, $\gamma(L)$ is a function of the heterogeneity of the material and $\delta(N)$ is a function of the number of cycles. When $\bar{p}_{01}(N)$ is independent of N, a linear relation between $\ln P_0(N)$ and N is obtained as follows:

$$\ln P_0(N) = - \alpha(T) \cdot \beta(H) \cdot \gamma(L)N + const. \tag{9}$$

Hence probabilistic properties of the frost damage influenced by the freezing temperature, water content and the structure can be quantitatively described by this model.

3.2 Influence of freezing temperature

Thirty mortar prisms of 40x40x160 mm were tested in two different freezing temperature conditions. Ordinary Portland cement was used. Water-cement ratio was 0.60, sand-cement ratio was 3.20 and the flow was 212. Specimens were cured for 28 days in water of 20°C before testing. During the test, specimens were frozen in air of -18°C or -12°C and melt in water of +5°C. One cycle took about one and a half hours.

Fig.3 shows the influence of the freezing temperature on the damage progress which is described with the reduction of relative dynamic modulus of elasticity. Mean values are drawn with a solid line in the figure. In case of the freezing temperature of -12°C, three specimens were more

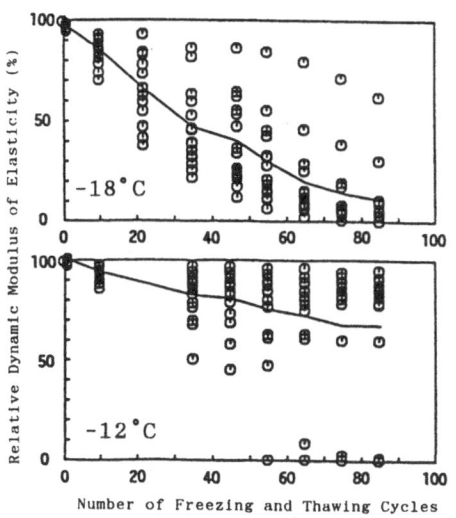

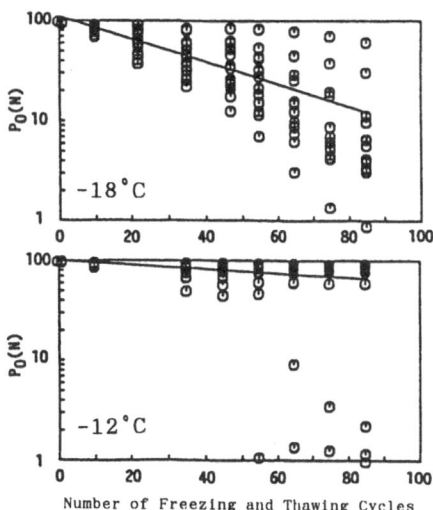

Fig.3
Freezing and thawing
test results.

Fig.4
Number of cycles vs. $P_0(N)$
under different temperatures.

Table 2.   Coefficient: $\alpha(T)\bar{\gamma}$

Freezing Temp.	Mean Value	Standard Deviat.	Peak Value
-12°C	0.0050	0.0085	0.0008
-18°C	0.0142	0.0053	0.0176

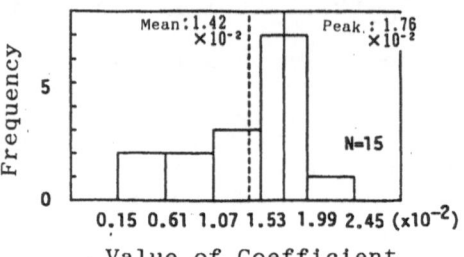

Value of Coefficient

Fig.5. Histogram of $\alpha(T)\bar{\gamma}$
in case of -18°C.

rapidly damaged after about 50 cycles.   Serious scaling was observed on the two square surfaces.   For such specimens, another model need to be applied.

Analyzed results of those experimental data with eq.(9) by means of the least square method are shown in Fig.4. Values of the coefficient: $\alpha(T)\bar{\gamma}$ for each freezing temperature were obtained as shown in Table 2, where $\beta(H)$ = 1.0 because the specimen was cured in water; $\bar{\gamma}$ = $\gamma(L)$ (that is constant).   The histogram of the coefficient is shown in Fig.5.   Now it has become possible to recognize quantitatively that   lower freezing temperature causes much faster damage progress than higher one.

3.3 Influence of heterogeneity

Concrete and mortar prisms were tested to study the influence of heterogeneity on the frost damage properties. Ordinary Portland cement was used.   Water-cement ratio (w/c=0.65) and sand-cement ratio (s/c=2.065) were kept constant for both test series of concrete and mortar. Specimen size was 100x100x400 mm and three specimens were tested for each series.   Specimens were cured in water of 20 C for 28 days before testing.   The freezing temperature was -18°C and the thawing temperature was +5°C.   It took about 4 and a half hours for one cycle.

Fig.6 shows the difference of frost damage progress between concrete and mortar, where each point is the mean value of three data.   The tested results were analyzed by eq.(9) and Fig. 7 was   obtained.   In this case, the value of $\bar{\alpha}\gamma(L)$ was -0.0127 for mortar and -0.053 for concrete, where $\bar{\alpha}$ is constant.   It means concrete was damaged about four times faster than mortar.

4 Deformation under Cyclic Loads

Predicting damage properties under cyclic loads has become more important for the serviceability limit state design. Although many experimental studies on fatigue of concrete under cyclic loads have been carried out, large scatters are

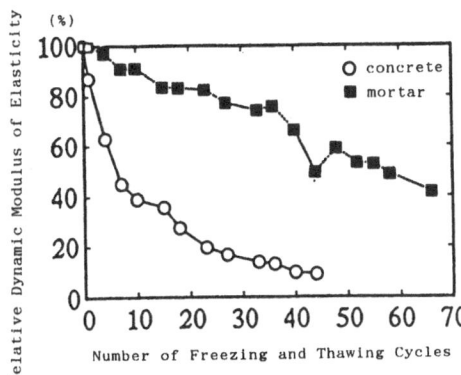

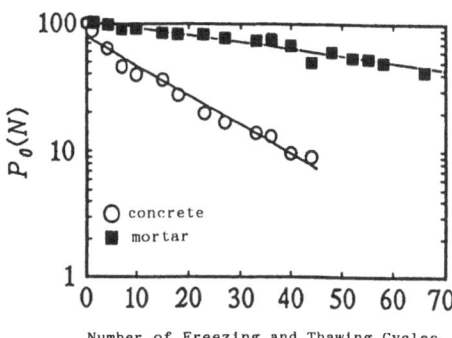

Fig.6
Freezing and thawing test results
of concrete and mortar.

Fig.7
Number of cycles vs $P_0(N)$
of concrete and mortar.

observed in the published results on fatigue properties and
any theoretical models have not been proposed to analyze the
probabilistic aspects. Since the fatigue of concrete is
supposed to be a progressive damage process, the present
model may be applicable (Mihashi, 1987).

If the progressive damage process is dominated by a crack
extension, a strain increment should be observed when a
crack with the length of 2c (Fig.8) occurs.

$$\epsilon_c = \frac{2\bar{v}}{\bar{a}} = \frac{2\sqrt{2}(1-\nu)}{\bar{a}G}\sigma c \tag{10}$$

where $\bar{a}$ is a distance between marked points. The strain
rate under a cyclic load may be given by eq.(11).

$$\dot{\epsilon} = \frac{d\epsilon_c}{dN} = \frac{2\sqrt{2}(1-\nu)}{\bar{a}G}\sigma\frac{dc}{dN} \tag{11}$$

The following relation is accepted for various kinds of
materials (Yokobori, 1974).

$$\frac{dc}{dN} \propto (\Delta K)^\alpha \tag{12}$$

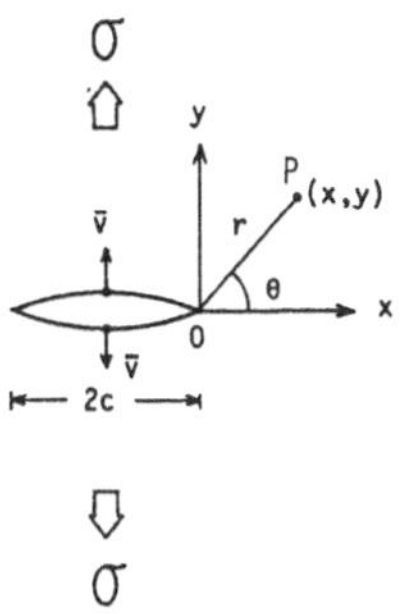

Fig.8
Geometry of single crack.

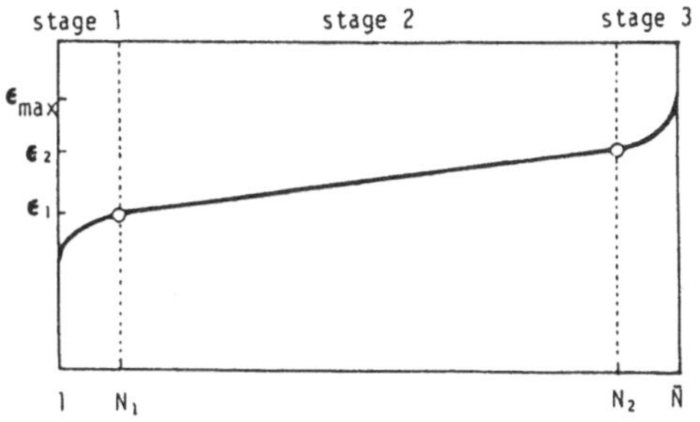

Fig.9
Total peak strain vs. number of
of loading cycles.

Table 3. Strain rate phenomena and mechanical condition

Stage	Observed Phenomena	Mech. Condition
1	$\dfrac{d\dot\epsilon}{dN} < 0$	$\dfrac{dc}{dN} < 0$ (unfeasible)
2	$\dfrac{d\dot\epsilon}{dN} = 0$	$\dfrac{dc}{dN} = 0$
3	$\dfrac{d\dot\epsilon}{dN} > 0$	$\dfrac{dc}{dN} > 0$

where α is a constant ($2 < \alpha < 7$), ΔK is equal to $\Delta\sigma\sqrt{\pi c}$ and $\Delta\sigma$ is the amplitude of the applied stress. Eqs. (11) and (12) suggest that the strain rate may be proportional to $c^{\alpha/2}$ and eq. (13) is obtained.

$$\frac{d\dot\epsilon}{dN} \propto \frac{\alpha}{2} c^{(\alpha-2)/2} \frac{dc}{dN} \qquad (13)$$

According to previous Sparks et al. (1973) and Tokumitsu et al. (1979), the relation between the maximum strain in each cycle and the number of cycles are subdivided into three stages as shown in Fig.9. Comparing eq. (13) with the diagram shown in Fig.9, the condition of dc/dN shown in Table 3 must be satisfied in each stage. The crack extension model, however, cannot satisfy the conditions of stages 1 and 2 .

Instead of the crack extension model, the following mechanism of the fatigue process is hypothesized. The process may be composed of three stages: (1) crack initiation around larger aggregates; (2) damage accumulation in the matrix and interfaces; (3) unstable crack extension to lead fracture.

The strain rate on the stage 1 may be proportional to the probability of crack initiation and the magnitude of the strain may be controlled by the number of mesocracks initiated in the weakest region such as interfaces. The following equation is obtained:

$$\dot\epsilon_1 = \bar\epsilon_0 L_1 \sigma^\beta \exp(-A_1\sigma^\beta N) \qquad (14)$$

$$\epsilon_1 = \frac{\bar\epsilon_0 L_1}{A_1} \{1 - \exp(-A_1\sigma^\beta N)\} \qquad (15)$$

where $\bar\epsilon_0$ is given by eq. (10) with $c = c_0$, L_1 is a function of the heterogeneity and β is a material constant affected by the atmosphere.

After releasing the strain energy in the weakest region by cracking, the damage accumulation process is the dominant mechanism to increase the strain on the stage 2 and the strain rate may be given by eq. (16).

$$\dot\epsilon_2 = k_0 \bar\epsilon \mu_s \qquad (16)$$

where k_0 is a material constant, $\bar{\epsilon}$ is the mean value of the strain increase per one cycle and μ_s is the probability to cause the strain increase per each cycle. Provided that the damage is due to microstructural changes which are controlled by the rate process dependent on stress, eq.(17) is obtained.

$$\mu_s \propto L_2 \sigma^\beta \tag{17}$$

According to eqs. (16) and (17), the following equations are obtained.

$$\dot{\epsilon}_2 = k_0 \bar{\epsilon} L_2 \sigma^\beta \tag{18}$$

$$\epsilon_2 = k_0 \bar{\epsilon} L_2 \sigma^\beta N \tag{19}$$

After the stage 2 when the matrix and interfaces are already damaged with the accumulation of microstructural changes, the fracture toughness of the matrix may be decreased. Since the stability of crack propagation is proportional to the remained toughness of the system, the probability for the system to reach the critical condition may be inversely proportional to the survival probability on the stage 2. If the unstable crack length is described as a function of the κth power of N, the following equations are obtained from eq.(10).

$$\epsilon_3 = k_1 \sigma N^\kappa \exp(A_2 \sigma^\beta N) \tag{20}$$

$$\dot{\epsilon}_3 = k_1 \sigma (A_2 \sigma^\beta N^\kappa + \kappa N^{\kappa-1}) \exp(A_2 \sigma^\beta N) \tag{21}$$

The strain rate and the whole strain at a certain cycle are given as the summation of those on three stages.

$$\dot{\epsilon} = \dot{\epsilon}_1 + \dot{\epsilon}_2 + \dot{\epsilon}_3 \tag{22}$$

$$\epsilon = \epsilon_{el} + \epsilon_1 + \epsilon_2 + \epsilon_3 \tag{23}$$

where ϵ_{el} means the elastic strain under the maximum stress.
Fig. 10 shows some simulated curves according to eqs. (22) and (23) for different atmosphere and stress level η, where β was calculated from published experimental data. These theoretical predictions are in good agreement with the experimental results of Cornelissen (1981) (Mihashi, 1987).

5 Conclusion

A stochastic model was presented to describe probabilistic aspects of damage accumulation in concrete. Since this model is based on relevant physical models, it may be useful to analyze the damage mechanisms. For more complex damage phenomena, comprehensive models such as 'changeable model' (Mihashi, 1983) need to be developed.

460

References

Cornelissen, H.A.W. and Timmer, G. (1981) Fatigue of plain concrete in uniaxial tension and in alternating tension-compression experiments and results. **Steven Report**, 5-81-7, Delft University of Technology, The Netherlands.

Mihashi, H. and Izumi, M. (1977) A stochastic theory for concrete fracture. **Cement and Concrete Research**, 7, 411-421.

Mihashi, H. and Wittmann, F.H. (1980) Stochastic approach to study the influence of rate of loading on strength of concrete. **Heron**, 25 (3), 1-54.

Mihashi, H. (1983) A Stochastic Theory for Fracture of Concrete. in **Fracture Mechanics of Concrete** (ed. F.H. Wittmann), Elsevier, 301-339.

Mihashi, H. (1987) Stochastic approach to study fatigue of concrete. **Engin. Fracture Mechanics**, 28 (5/6), 785-793.

Sparks, P.R. and Menzies, J.B. (1973) The effect of rate of loading upon the static and fatigue strength of plain concrete in compression. **Mag. Concr. Res.**, 25, 78-83.

Tokumitsu, Y. and Matsushita, H. (1979) Fatigue strength of plain concrete under repeated loading. **Concrete J.**, 17, 13-22 (in Japanese).

Wittmann, F.H. (1983) Structure of Concrete with Aspect to Crack Formation. in **Fracture Mechanics of Concrete** (ed. F.H. Wittmann), Elsevier, 43-74.

Yokobori, T. (1974) **Interdisciplinary Approach to Fracture and Strength of Solids**. Iwanami (in Japanese).

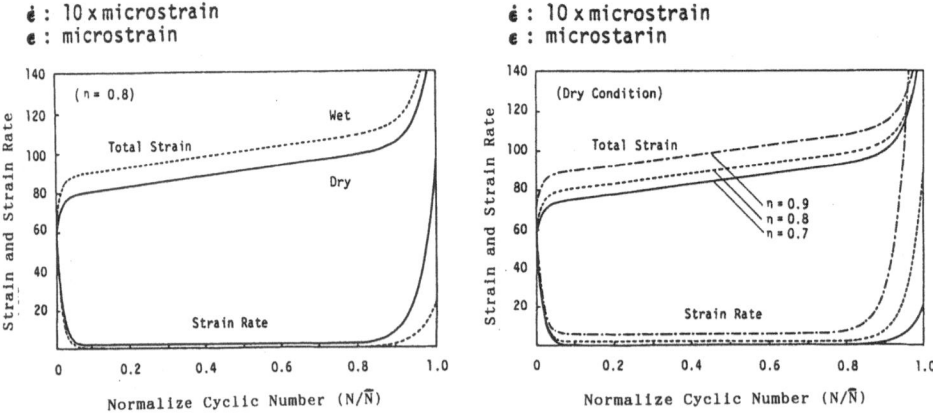

(a). Influence of humidity.   (b). Influence of peak stress.
Fig.10. Simulated curves of fatigue process (Mihashi, 1987).

STOCHASTIC FIELD THEORY IN MATERIALS ENGINEERING

M. SHINOZUKA
Princeton University, Princeton, U.S.A.

Abstract

This paper consists of three chapters. The first one deals with the problem of simulating a stochastic process or field using the spectral representation method. The stochastic process or field is simulated according to its prescribed power spectrum. The simulation is performed very efficiently using the Fast Fourier Transform technique. In the second chapter, a probabilistic model for the spatial variation of the strength of materials used for laminated orthotropic composites is introduced. The principal idea lies in the interpretation that the material strength can be idealized as a two-dimensional stochastic field. Under such an assumption, the strength of laminated orthotropic composite plates in uniaxial tension is found to follow a Weibull distribution function. Another conclusion is the demonstration of the statistical size effect. Finally, in the third chapter, a methodology is introduced to perform both the response variability and reliability analysis of plates subjected to plane stress or plane strain. A stochastic finite element approach based on the concept of weighted integrals is utilized. The variability response function of the system is established in closed form. Then, the response variability is calculated by a first-order Taylor expansion, while the safety index is computed using an advanced first-order second-moment approach.

Keywords : Stochastic process and field, simulation, composites, stochastic finite element method, variability response function.

1. Simulation of Stochastic Processes and Fields Using the Spectral Representation Method

1.1 Introduction

The spectral representation method was developed by Shinozuka and his associates (Shinozuka and Jan 1972, Shinozuka 1974, Shinozuka and Deodatis 1988 and 1991) to simulate stochastic processes, fields and waves according to their target power spectral density functions. The method can be applied to generate sample functions of one-dimensional stochastic processes, multi-dimensional stochastic fields, uni-variate or multivariate stochastic processes, stationary or non-stationary stochastic processes and homogeneous or non-homogeneous stochastic fields. One of the major advantages of the method is the fact that it can be used in conjunction

461

D. Breysse (ed.), Probabilities and Materials, 461–489.
© 1994 Kluwer Academic Publishers.

with the Fast Fourier Transform technique for superior computational efficiency.

1.2 Simulation of $1D - 1V$ Stationary Stochastic Processes

Let $f_0(t)$ be a $1D - 1V$ stationary stochastic process with mean zero and auto-correlation function $R_{f_0 f_0}(\tau)$. Then

$$E[f_0(t)] = 0 \tag{1}$$

$$E[f_0(t + \tau)f_0(t)] = R_{f_0 f_0}(\tau) \tag{2}$$

It is well known that $R_{f_0 f_0}(\tau)$ and the power spectral density function $S_{f_0 f_0}(\omega)$ of the process $f_0(t)$ are related through the Weiner-Khintchine transform pair:

$$S_{f_0 f_0}(\omega) = \frac{1}{2\pi} \int_{-\infty}^{\infty} R_{f_0 f_0}(\tau)e^{-i\omega\tau}d\tau \tag{3}$$

$$R_{f_0 f_0}(\tau) = \int_{-\infty}^{\infty} S_{f_0 f_0}(\omega)e^{i\omega\tau}d\omega \tag{4}$$

It follows immediately from Eq. 2 that $R_{f_0 f_0}(\tau)$ is an even function of τ, and consequently the power spectral density $S_{f_0 f_0}(\omega)$ is also an even function of ω in accordance with Eq. 3. Also, it can be shown that $S_{f_0 f_0}(\omega) \geq 0$.

Under these conditions, the stochastic process $f_0(t)$ can be simulated (Shinozuka and Deodatis 1991) by the following series, as $N \to \infty$.

$$f(t) = \sqrt{2} \sum_{j=1}^{N} \sqrt{2S_{f_0 f_0}(\omega_j)\Delta\omega} \cos(\omega_j t + \phi_j) \tag{5}$$

where

$$\omega_j = j\Delta\omega \qquad\qquad j = 1, 2, \cdots, N \tag{6}$$

An upper bound of the frequency

$$\omega_u = N\Delta\omega \tag{7}$$

is implicit in Eq. 5 where ω_u represents an upper cut-off frequency beyond which $S_{f_0 f_0}(\omega)$ may be assumed to be zero for either mathematical or physical reasons. In Eq. 5, ϕ_j are independent random phase angles uniformly distributed over the range $(0, 2\pi)$. Note that the simulated process is asymptotically Gaussian as N becomes large due to the central limit theorem.

Finally, it can be shown that the expected value and auto-correlation function of the simulated process $f(t)$ are identical to the corresponding targets, $E[f_0(t)] = 0$ and $R_{f_0 f_0}(\tau)$, respectively.

The simulated process given by Eq. 5 is ergodic, at least to the second moment, regardless of the size of N (Shinozuka and Deodatis 1991). The computer cost of digitally generating sample functions of process $f(t)$ can be dramatically reduced by applying the FFT (Fast Fourier Transform) technique to Eq. 5 (Shinozuka and Deodatis 1991).

1.3 Simulation of $nD - 1V$ Homogeneous Stochastic Fields

The simulation of $1D - 1V$ stationary stochastic processes using spectral representation, which was presented in the previous section, can be extended in a straightforward fashion to the simulation of $nD-1V$ (n-dimensional and uni-variate) homogeneous stochastic fields in the following way.

Consider an $nD - 1V$ homogeneous stochastic field $f_0(x_1, x_2, \cdots, x_n) = f_0(\mathbf{x})$ with mean zero:

$$E[f_0(\mathbf{x})] = 0 \qquad (8)$$

The autocorrelation function of $f_0(\mathbf{x})$ is defined by

$$R_{f_0 f_0}(\boldsymbol{\xi}) = E[f_0(\mathbf{x}_r) f_0(\mathbf{x}_s)] \qquad (9)$$

where $\mathbf{x}_r$ and $\mathbf{x}_s$ are position vectors in an n-dimensional space and $\boldsymbol{\xi}$ is the separation vector. For a homogeneous field, $R_{f_0 f_0}(\boldsymbol{\xi})$ is symmetric with respect to the separation vector $\boldsymbol{\xi}$ and therefore:

$$R_{f_0 f_0}(\boldsymbol{\xi}) = R_{f_0 f_0}(-\boldsymbol{\xi}) \qquad (10)$$

For some $nD - 1V$ homogeneous fields, the following equation is valid:

$$R_{f_0 f_0}(\boldsymbol{\xi}) = R_{f_0 f_0}(I_{\pm}\boldsymbol{\xi}) \qquad (11)$$

where $I_{\pm}$ is any $n \times n$ diagonal matrix whose diagonal components are either 1 or -1. Hence, Eq. 10 is a special case of Eq. 11 in which the diagonal members of

$I_\pm$ are all equal to -1. When Eq. 11 is valid, the stochastic field is referred to as a "quadrant field" (Vanmarcke 1984). Assuming that an n-fold Fourier transform of $R_{f_0 f_0}(\xi)$ exists, the spectral density function of $f_0(\mathbf{x})$ is defined as:

$$S_{f_0 f_0}(\kappa) = \frac{1}{(2\pi)^n} \int_{-\infty}^{\infty} R_{f_0 f_0}(\xi) e^{-i\mathbf{k}\cdot\xi} d\xi \qquad (12)$$

and its inverse transform is given by:

$$R_{f_0 f_0}(\xi) = \int_{-\infty}^{\infty} S_{f_0 f_0}(\kappa) e^{i\kappa\cdot\xi} d\kappa \qquad (13)$$

The preceding two equations represent the n-dimensional version of the Wiener-Khintchine transform pair, where $\kappa = [\kappa_1 \kappa_2 \cdots \kappa_n]^T$ is the wave number vector and $\kappa \cdot \xi$ is the inner product of κ and ξ, and, for simplicity:

$$\int_{-\infty}^{\infty} (\,)d\xi = \int_{-\infty}^{\infty} \overset{n-fold}{\cdots\cdots\cdots} \int_{-\infty}^{\infty} (\,)d\xi_1 d\xi_2 \cdots d\xi_n \qquad (14)$$

$$\int_{-\infty}^{\infty} (\,)d\kappa = \int_{-\infty}^{\infty} \overset{n-fold}{\cdots\cdots\cdots} \int_{-\infty}^{\infty} (\,)d\kappa_1 d\kappa_2 \cdots d\kappa_n \qquad (15)$$

It can be easily shown that:

$$S_{f_0 f_0}(\kappa) = S_{f_0 f_0}(-\kappa) \qquad (16)$$

and that the spectral density function is real. In addition, the following equation is obtained under the condition of Eq. 11:

$$S_{f_0 f_0}(\kappa) = S_{f_0 f_0}(I_\pm \kappa) \qquad (17)$$

This equation indicates that the value of $S_{f_0 f_0}(\kappa)$ is identical at a corresponding point in each quadrant ($\kappa = I_\pm \kappa$), hence the name "quadrant field."

Finally, it can be shown that the autocorrelation function $R_{f_0 f_0}(\xi)$ is non-negative definite and has a non-negative n-dimensional Fourier transform, i.e.:

$$S_{f_0 f_0}(\kappa) \geq 0 \qquad (18)$$

Based on these properties of $S_{f_0 f_0}(\boldsymbol{\kappa})$, the n-dimensional homogeneous stochastic field $f_0(\mathbf{x})$ can be simulated by a stochastic field $f(\mathbf{x})$ in the following fashion: consider an $nD - 1V$ homogeneous field $f_0(\mathbf{x})$ with mean zero and spectral density function $S_{f_0 f_0}(\boldsymbol{\kappa})$ which is of insignificant magnitude outside the region defined by:

$$-\kappa_u \le \kappa \le \kappa_u \tag{19}$$

where $\kappa_u = [\kappa_{1u} \kappa_{2u} \cdots \kappa_{nu}]^T$ with $\kappa_{iu} > 0 \; (i = 1, 2, \cdots, n)$. Denote the interval vector by:

$$(\Delta \kappa_1 \Delta \kappa_2 \cdots \Delta \kappa_n) = (\frac{\kappa_{1u}}{N_1} \frac{\kappa_{2u}}{N_2} \cdots \frac{\kappa_{nu}}{N_n}) \tag{20}$$

and then construct the simulated field $f(\mathbf{x})$ by the following series, as $N_1, N_2, \cdots, N_n \to \infty$ simultaneously:

$$f(\mathbf{x}) = \sqrt{2} \sum_{k_1=1}^{N_1} \sum_{k_2=1}^{N_2} \cdots \sum_{k_n=1}^{N_n} \sum_{I_1=1, I_i=\pm 1, i=2,3,\cdots,n}$$
$$\{2 S_{f_0 f_0}(I_1 \kappa_{1k_1}, I_2 \kappa_{2k_2} \cdots, I_n \kappa_{nk_n}) \Delta \kappa_1 \Delta \kappa_2 \cdots \Delta \kappa_n\}^{\frac{1}{2}}$$
$$\times \cos(I_1 \kappa_{1k_1} x_1 + I_2 \kappa_{2k_2} x_2 + \cdots + I_n \kappa_{nk_n} x_n + \phi_{k_1 k_2 \cdots k_n}^{I_1 I_2 \cdots I_n}) \tag{21}$$

where $\phi_{k_1 k_2 \cdots k_n}^{I_1 I_2 \cdots I_n} = $ independent random phase angles uniformly distributed between 0 and 2π, and

$$\kappa_{ik_i} = k_i \Delta \kappa_i, \qquad k_i = 1, 2, \cdots, N_i, \qquad i = 1, 2, \cdots, n. \tag{22}$$

The simulated field $f(\mathbf{x})$ is asymptotically Gaussian as $N_1, N_2, \cdots, N_n \to \infty$ simultaneously, again due to the central limit theorem. Note that a set of $I_1, I_2, \cdots, I_n$ indicates one of the 2^n quadrants of the wave number $\boldsymbol{\kappa}$ space. Because $S(\boldsymbol{\kappa}) = S(-\boldsymbol{\kappa})$, we need to cover only 2^{n-1} quadrants, half of the total 2^n, for simulation purposes. Thus I_1 is always chosen to be unity ($I_1 = 1$). This also implies that (i) there are 2^{2n-1} sets of $N_1 N_2 \cdots N_n$ random phase angles in the expression for $f(\mathbf{x})$ given by Eq. 21 and (ii) twice the spectral density function always appears in the same equation.

If the stochastic field is quadrant, then $S_{f_0 f_0}(I_1 \kappa_{1k_1}, I_2 \kappa_{2k_2}, \cdots, I_n \kappa_{nk_n})$ in Eq. 21 can be replaced by $S_{f_0 f_0}(\kappa_{1k_1}, \kappa_{2k_2}, \cdots, \kappa_{nk_n})$.

Referring to $2D - 1V$ homogeneous stochastic fields, Eq. 21 can be written explicitly as

$$f(\mathbf{x}) = \sqrt{2} \sum_{k_1=1}^{N_1} \sum_{k_2=1}^{N_2} \{[2S_{f_0 f_0}(\kappa_{1k_1}, \kappa_{2k_2})\Delta\kappa_1\Delta\kappa_2]^{\frac{1}{2}} \cos(\kappa_{1k_1} x_1 + \kappa_{2k_2} x_2 + \phi_{k_1 k_2}^{(1)})$$

$$+ [2S_{f_0 f_0}(\kappa_{1k_1}, -\kappa_{2k_2})\Delta\kappa_1\Delta\kappa_2]^{\frac{1}{2}} \cos(\kappa_{1k_1} x_1 - \kappa_{2k_2} x_2 + \phi_{k_1 k_2}^{(2)})\} \quad (23)$$

Furthermore, if the stochastic field is quadrant,

$$f(\mathbf{x}) = \sqrt{2} \sum_{k_1=1}^{N_1} \sum_{k_2=1}^{N_2} [2S_{f_0 f_0}(\kappa_{1k_1}, \kappa_{2k_2})\Delta\kappa_1\Delta\kappa_2]^{\frac{1}{2}}$$

$$\times \{\cos(\kappa_{1k_1} x_1 + \kappa_{2k_2} x_2 + \phi_{k_1 k_2}^{(1)})$$

$$+ \cos(\kappa_{1k_1} x_1 - \kappa_{2k_2} x_2 + \phi_{k_1 k_2}^{(2)})\} \quad (24)$$

where $\phi_{k_1 k_2}^{(1)} = \phi_{k_1 k_2}^{1,1}$ and $\phi_{k_1 k_2}^{(2)} = \phi_{k_1 k_2}^{1,-1}$ if the notation introduced in Eq. 21 is to be used.

It can be again shown that the expected value and autocorrelation function of the simulated field are the same as the target values, i.e., $E[f(\mathbf{x})] = E[f_0(\mathbf{x})] = 0$ and $R_{ff}(\boldsymbol{\xi}) = R_{f_0 f_0}(\boldsymbol{\xi})$. Also, the $nD - 1V$ simulated field is ergodic, at least to the second moment, and the computational cost for the digital generation of sample functions of the stochastic field $f(\mathbf{x})$ can be dramatically reduced by applying the FFT technique to the appropriate trigonometric series expression.

1.4 Simulation of $nD - mV$ Homogeneous Stochastic Fields

The simulation of $nD-mV$ (n-dimensional and m-variate) homogeneous stochastic fields, unlike the case of $nD - 1V$ fields, cannot be achieved by straightforward generalization of the $1D - 1V$ case, but can be done in the way shown below.

Consider a set of m homogeneous Gaussian n-dimensional stochastic fields $f_j^0(\mathbf{x})$ $(j = 1, 2, \cdots, m)$ with mean zero:

$$E[f_j^0(\mathbf{x})] = 0 \quad (25)$$

and with cross spectral density matrix $\mathbf{S}^0(\boldsymbol{\kappa})$ defined by:

$$\mathbf{S}^0(\boldsymbol{\kappa}) = \begin{bmatrix} S_{11}^0(\boldsymbol{\kappa}) & S_{12}^0(\boldsymbol{\kappa}) & \cdots & S_{1m}^0(\boldsymbol{\kappa}) \\ S_{21}^0(\boldsymbol{\kappa}) & S_{22}^0(\boldsymbol{\kappa}) & \cdots & S_{2m}^0(\boldsymbol{\kappa}) \\ \cdots & \cdots & \cdots & \cdots \\ S_{m1}^0(\boldsymbol{\kappa}) & S_{m2}^0(\boldsymbol{\kappa}) & \cdots & S_{mm}^0(\boldsymbol{\kappa}) \end{bmatrix} \quad (26)$$

where $S_{jk}^0(\kappa)$ is the n-dimensional Wiener-Khintchine transform of the so-called cross-correlation function $R_{jk}^0(\xi)$ $(j \neq k)$ or auto-correlation function $R_{jk}^0(\xi)$ $(j = k)$.

Due to the assumption of homogeneity:

$$R_{jk}^0(\xi) = R_{kj}^0(-\xi) , \tag{27}$$

the following expression can be obtained:

$$S_{jk}^0(\kappa) = \bar{S}_{kj}^0(\kappa) \tag{28}$$

where the super bar indicates the complex conjugate. The matrix $S^0(\kappa)$ is therefore Hermitian. It can be shown that $S^0(\kappa)$ is also non-negative definite.

Suppose now we can find a matrix $H(\kappa)$ which possesses an n-dimensional Fourier transform and satisfies the equation:

$$S^0(\kappa) = H(\kappa)\bar{H}(\kappa)^T \tag{29}$$

where $S^0(\kappa)$ is the specified target cross-spectral density matrix and T indicates the transpose.

To find the matrix $H(\kappa)$ in an efficient way, we assume that $H(\kappa)$ is a lower triangular matrix:

$$H(\kappa) = \begin{bmatrix} H_{11}(\kappa) & 0 & \cdots & 0 \\ H_{21}(\kappa) & H_{22}(\kappa) & \cdots & 0 \\ \cdots & \cdots & \cdots\cdots \\ H_{m1}(\kappa) & H_{m2}(\kappa) & \cdots & H_{mm}(\kappa) \end{bmatrix} \tag{30}$$

By substituting the above into Eq. 29, solutions are obtained as:

$$H_{kk}(\kappa) = \left[\frac{D_k(\kappa)}{D_{k-1}(\kappa)} \right]^{\frac{1}{2}} \qquad k = 1, 2, \cdots, m \tag{31}$$

where $D_k(\kappa)$ is the k-th principal minor of $S^0(\kappa)$ with D_0 being defined as unity, and

$$H_{jk}(\kappa) = H_{kk}(\kappa) \frac{S^0 \begin{pmatrix} 1, & 2, & \cdots, & k-1, & j \\ 1, & 2, & \cdots, & k-1, & k \end{pmatrix}}{D_k(\kappa)} \qquad \begin{array}{l} k = 1, 2, \cdots, m \\ j = k+1, \cdots, m \end{array} \qquad (32)$$

where

$$S^0 \begin{pmatrix} 1, & 2, & \cdots, & k-1, & j \\ 1, & 2, & \cdots, & k-1, & k \end{pmatrix} = \begin{vmatrix} S^0_{11} & S^0_{12} & \cdots & S^0_{1,k-1} & S^0_{1k} \\ S^0_{21} & S^0_{22} & \cdots & S^0_{2,k-1} & S^0_{2k} \\ \cdots & \cdots & \cdots & \cdots\cdots \\ S^0_{k-1,1} & S^0_{k-1,2} & \cdots & S^0_{k-1,k-1} & S^0_{k-1,k} \\ S^0_{j1} & S^0_{j2} & \cdots & S^0_{j,k-1} & S^0_{jk} \end{vmatrix} \qquad (33)$$

is the determinant of a submatrix obtained by deleting all the elements except the $(1, 2, \cdots, k-1, j) - th$ row and $(1, 2, \cdots, k-1, k) - th$ column of $S^0(\kappa)$. It is noted that the above decomposition is valid only when the matrix $S^0(\kappa)$ is Hermitian and positive definite as can be seen from Eq. 31.

Once $H(\kappa)$ is computed using Eqs. 31 and 32, instead of passing a white noise vector through filters, the field $f_j(x)$ can be simulated in a more efficient way by the following series, as $N_1, N_2, \cdots, N_n \to \infty$ simultaneously and under the assumption that the stochastic field possesses quadrant symmetry:

$$f_j(\mathbf{x}) = 2 \sum_{m=1}^{j} \sum_{l_1=1}^{N_1} \sum_{l_2=1}^{N_2} \cdots \sum_{l_n=1}^{N_n} \sum_{I_1=1, I_i=\pm 1, i=2,3,\cdots,n} |H_{jm}(\kappa_{1l_1}, \kappa_{2l_2}, \cdots, \kappa_{nl_n})|$$

$$\times \sqrt{\Delta\kappa_1 \Delta\kappa_2 \cdots \Delta\kappa_n} \cos[I_1\kappa_{1l_1} x_1 + I_2\kappa_{2l_2} x_2 + \cdots + I_n\kappa_{nl_n} x_n +$$

$$\theta_{jm}(\kappa_{1l_1}, \kappa_{2l_2}, \cdots, \kappa_{nl_n}) + \phi_{ml_1l_2\cdots l_n}^{I_1 I_2\cdots I_n}] \qquad (34)$$

where

$$(\Delta\kappa_1 \ \Delta\kappa_2 \cdots \Delta\kappa_n) = (\frac{\kappa_{1u}}{N_1} \frac{\kappa_{2u}}{N_2} \cdots \frac{\kappa_{nu}}{N_n}) \qquad (35)$$

$$\kappa_{il_i} = l_i \Delta\kappa_i \ ; \qquad l = 1, 2, \cdots, N_i \ ; \qquad i = 1, 2, \cdots, n \qquad (36)$$

$$\theta_{jm}(\kappa_{1l_1}, \kappa_{2l_2}, \cdots, \kappa_{nl_n}) = \tan^{-1}\left[\frac{ImH_{jm}(\kappa_{1l_1}, \kappa_{2l_2}, \cdots, \kappa_{nl_n})}{ReH_{jm}(\kappa_{1l_1}, \kappa_{2l_2}, \cdots, \kappa_{nl_n})}\right] \qquad (37)$$

$\phi_{m l_1 l_2 \cdots l_n}^{I_1 I_2 \cdots I_n}$ = independent random phase angles uniformly distributed between 0 and 2π. The simulated field $f_j(\mathbf{x})$ is asymptotically Gaussian as $N_1, N_2, \cdots, N_n \to \infty$ simultaneously, again due to the central limit theorem.

More significantly, we can show that a vector $[f_1(\mathbf{x}) f_2(\mathbf{x}), \cdots f_m(\mathbf{x})]^T$ of stochastic fields $f_j(\mathbf{x})$ $(j = 1, 2, \cdots, m)$ thus simulated is asymptotically Gaussian with the aid of the multivariate central limit theorem.

Finally, it can be shown once more that the expected value and autocorrelation function of the simulated field (using Eq. 34) are the same as the target values, i.e., $E[f_j(\mathbf{x})] = E[f_j^0(\mathbf{x})] = 0$ and $R_{jk}(\boldsymbol{\xi}) = R_{jk}^0(\boldsymbol{\xi})$; $j, k = 1, 2, \cdots, m$. Here again, the computational effort for the digital generation of sample functions of the stochastic field $f_j(\mathbf{x})$ can be substantially reduced by applying the FFT technique to Eq. 34.

2. Spatial Strength Variation of Laminated Orthotropic Composites

2.1 Introduction

The engineering materials used for the production of laminated orthotropic composites are known to have considerable spatial variations in their resisting strength and other properties. In the present study, particular emphasis is placed on the probabilistic modeling of the spatial variation of the material strength and on the resulting statistical size effect. The principal idea lies in the interpretation that the material strength can be idealized as a multi-dimensional stochastic field. Under such an assumption, the type of statistical distribution followed by the strength of laminated orthotropic composite plates in uniaxial tension can be estimated, with the aid of Monte Carlo simulation techniques. Two different failure criteria and three failure mechanisms are used in order to study their effect on the probabilistic characteristics of the strength of composite plates.

2.2 Estimation of Strength Distribution and Demonstration of Statistical Size Effect Using Monte Carlo Simulation

Consider a laminated orthotropic composite plate consisting of m laminae. Each lamina is divided into an equal number of n elements. Therefore the whole plate is divided into $n \cdot m$ elements. Figure 1 shows a special case where $m = 2$ and $n = 4 \cdot 6 = 24$. The procedure for the Monte Carlo simulation is the following:

Step I: Assign values for the material strengths in the three principal directions (longitudinal, transverse and shear, respectively) for each one of the $n \cdot m$ elements. This can be achieved by considering that for each lamina, the material strength can be represented by a two-dimensional stochastic field that can be digitally generated by a method described in the next section.

Step II: Considering now uniaxial stress of the laminated orthotropic composite plate (more specifically uniaxial tension), we can start from a very small stress-level σ and check whether one or more of the $n \cdot m$ elements fail under the given stress σ. The Tsai-Hill and Tsai-Wu failure criteria are used to perform this check (both criteria are described in detail in a subsequent section). It should be noted that this failure check is performed $n \cdot m$ times, once for each element.

Step III: If no failure is detected at the current stress level, the stress level is increased by $\Delta\sigma$ and the previous step is executed. The value of the externally applied uniaxial tension at which some element(s) fail(s) for the first time is considered as the strength of the whole plate in uniaxial tension. Three different failure mechanisms (number of elements that have to fail so that the whole plate fails) are used; all three are described in detail in a later section.

The above three-step procedure is repeated l times for the sake of the Monte Carlo simulation. Thus l values of the strength of the whole plate are obtained and their statistical distribution can be estimated using goodness of fit.

Finally, the whole procedure mentioned in this section is repeated for a plate having a larger area than the original one, in order to demonstrate the statistical size effect. At this point it should be noted that no stress redistribution is considered after the failure of some element(s). This assumption is made for the sake of simplicity of the numerical calculations involved in the Monte Carlo simulation. To examine the validity of such an assumption would be interesting future work.

2.3 Digital Generation of Two-Dimensional Random Fields

The idealization of each material strength as a two-dimensional stochastic field, encountered in Step 1 of the previous section, is done as follows:

$$\text{Axial strength in tension:} \quad X_t(x,y) = \bar{X}_t[1 + f_t(x,y)] \tag{38}$$
$$\text{Axial strength in compression:} \quad X_c(x,y) = \bar{X}_c[1 + f_c(x,y)] \tag{39}$$
$$\text{Transverse strength in tension:} \quad Y_t(x,y) = \bar{Y}_t[1 + g_t(x,y)] \tag{40}$$
$$\text{Transverse strength in compression:} \quad Y_c(x,y) = \bar{Y}_c[1 + g_c(x,y)] \tag{41}$$
$$\text{Shear strength:} \quad S(x,y) = \bar{S}[1 + h(x,y)] \tag{42}$$

where $\bar{X}_t$, $\bar{X}_c$, $\bar{Y}_t$, $\bar{Y}_c$ and $\bar{S}$ are mean values of the respective resisting strengths and $f_t(x,y)$, $f_c(x,y)$, $g_t(x,y)$, $g_c(x,y)$ and $h(x,y)$ are univariate, two-dimensional homogeneous stochastic fields. All five stochastic fields are assumed to be described by the same power spectrum:

$$S(\kappa_x, \kappa_y) = \sigma_0^2 \frac{d^2}{4\pi} exp[-\frac{d^2}{4}(\kappa_x^2 + \kappa_y^2)] \tag{43}$$

The following values are used for the various constants:

$$\bar{X}_t = 150,000 \text{ psi}, \qquad \bar{S} = 6,000 \text{ psi}, \qquad \bar{X}_c = 150,000 \text{ psi}, \qquad \sigma_0 = 0.15,$$

$$\bar{Y}_t = 4,000 \text{ psi}, \qquad d = 1 \text{ in.} \quad \text{and} \quad \bar{Y}_c = 20,000 \text{ psi}.$$

The above values of resisting strengths correspond to glass/epoxy. The digital generation of sample functions $f(x,y)$ of a two-dimensional stochastic field having power spectrum $S(\kappa_x, \kappa_y)$ is performed using Eq. 24 of the first section of this paper:

$$f(x,y) = \sqrt{2} \sum_{i=1}^{M} \sum_{j=1}^{M} [2S(\kappa_{1i}, \kappa_{2j}) \Delta \kappa_1 \Delta \kappa_2]^{1/2}$$

$$\times \{\cos(\kappa_{1i}x + \kappa_{2j}y + \phi_{ij}^{(1)}) + \cos(\kappa_{1i}x - \kappa_{2j}y + \phi_{ij}^{(2)})\} \tag{44}$$

where

$$\Delta \kappa_1 = \frac{\kappa_{1u}}{M} \quad \Delta \kappa_2 = \frac{\kappa_{2u}}{M} \tag{45}$$

$$\kappa_{1i} = i \Delta \kappa_1 \quad \kappa_{2j} = j \Delta \kappa_2 \tag{46}$$

and $\phi_{ij}^{(1)}$, $\phi_{ij}^{(2)}$ = random phase angles randomly and uniformly distributed between 0 and 2π. The following values are used for the various constants:

$$\kappa_{1u} = \kappa_{2u} = 4.92 \text{ per inch} \quad \text{and} \quad M = 60$$

2.4 Failure Criteria and Failure Mechanisms Used

The Tsai-Hill and Tsai-Wu failure criteria used in Step 2 of the Monte Carlo simulation are described now.

Tsai-Hill failure criterion (Jones 1975 and Soni 1983):

$$\frac{\cos^4 \theta}{X^2} + (\frac{1}{S^2} - \frac{1}{X^2})\cos^2 \theta \sin^2 \theta + \frac{\sin^4 \theta}{Y^2} = \frac{1}{\sigma^2} \tag{47}$$

where X, Y, and S are the strengths in the three principal directions (longitudinal, transverse and shear, respectively), assuming that the tensile and compressive strengths coincide, σ is the externally applied axial stress and θ is the angle between the principal direction X and the direction of application of the uniaxial stress σ.

Tsai-Wu failure criterion (Jones 1975 and Soni 1983):

$$\frac{\sigma^2 \cos^4 \theta}{X_t X_c} - \frac{\sigma^2 \sin^2 \theta \cos^2 \theta}{\sqrt{X_t X_c Y_t Y_c}} + \frac{\sigma^2 \sin^4 \theta}{Y_t Y_c} + \frac{\sigma^2 \sin^2 \theta \cos^2 \theta}{S^2} +$$

$$(\frac{1}{X_t} - \frac{1}{X_c})\sigma \cos^2 \theta + (\frac{1}{Y_t} - \frac{1}{Y_c})\sigma \sin^2 \theta = 1 \tag{48}$$

where X_t, X_c, Y_t, Y_c and S have all been defined in the previous section and σ and θ have been defined in the previous failure criterion. The three failure mechanisms used in Step 2 of the Monte Carlo simulation are described below.

Failure mechanism by one element:

It is assumed that failure of just one element of the plate causes failure of the whole plate.

Failure mechanism by two elements:

It is assumed that failure of two elements in a row causes failure of the whole plate.

Failure mechanism by three elements:

It is assumed that failure of three elements in a row causes failure of the whole plate.

2.5 Numerical Examples

In this section, a laminated orthotropic composite plate consisting of two laminae will be examined under the state of uniaxial tension. Each lamina is divided into $4 \cdot 6 = 24$ elements, exactly as shown in Fig. 1. Each one of the 48 elements of the plate is assumed to be 1 in. by 1 in. in area. Therefore, the area of the whole plate is $4 \cdot 6 = 24$ in². The fiber orientation is as follows:

$$\theta = +30° \quad \text{for the upper lamina}$$

$$\theta = -30° \quad \text{for the lower lamina}$$

The above mentioned plate will be analyzed by Monte Carlo simulation with $l = 50$ samples, for the following six combinations of failure criteria and failure mechanisms:

Case 1:   Tsai-Hill failure criterion and failure by 1 element
Case 2:   Tsai-Hill failure criterion and failure by 2 elements
Case 3:   Tsai-Hill failure criterion and failure by 3 elements
Case 4:   Tsai-Wu failure criterion and failure by 1 element
Case 5:   Tsai-Wu failure criterion and failure by 2 elements
Case 6:   Tsai-Wu failure criterion and failure by 3 elements

Each one of the above six cases produced 50 values for the resisting strength of the whole plate, from the Monte Carlo simulation. In order to study now the statistical size effect, the area of the original plate was quadrupled, however always keeping two laminae. Each one of them is now divided into $8 \cdot 12 = 96$ elements of 1 in. by 1 in. area each. Therefore, the area of the whole plate is now $8 \cdot 12 = 96$ in² (four times the original area). The fiber orientation remains the same. This new larger plate will be analyzed again by Monte Carlo simulation with $l = 50$ samples, for the following six combinations of failure criteria and failure mechanisms:

Case 7:   Tsai-Hill failure criterion and failure by 1 element
Case 8:   Tsai-Hill failure criterion and failure by 2 elements
Case 9:   Tsai-Hill failure criterion and failure by 3 elements
Case 10:   Tsai-Wu failure criterion and failure by 1 element
Case 11:   Tsai-Wu failure criterion and failure by 2 elements
Case 12:   Tsai-Wu failure criterion and failure by 3 elements

Each one of the above six cases produced again 50 values for the resisting strength of the whole plate, from the Monte Carlo simulation.

2.6 Statistical Analysis of the Results

The histograms constructed from the results of cases 1-12 are plotted in Figs. 2-7. There cannot be a single conclusion for all twelve cases concerning the skewness of the plotted results, since some of them are skewed to the right and some to the left. However, the statistical size effect is clearly seen in all six figures, as the mean value of the distribution moves to the left (mean value diminishes), when the area of the plate is quadrupled. In order to find the type of statistical distribution followed by the resisting strengths of the twelve cases examined, the results were plotted on four different probability papers. Specifically, on normal, log-normal, asymptotic of the first kind and Weibull probability papers. It was concluded that for all twelve cases the Weibull distribution fits best to the results. In Fig. 8, cases 2 and 8 are plotted in the Weibull probability paper. The statistical size effect is also clearly seen in this figure, as the results corresponding to the quadrupled area are to the left of those corresponding to the original area.

2.7 Comparison of Results With Experiments

A series of experimental results were provided by the U.S. Army Materials Technology Laboratory (Papirno 1985) for the sake of comparison with the results obtained from the Monte Carlo simulation. Unfortunately, the specimens used in the experiments were 8-ply, S2-Glass fiber reinforced composites with different dimensions from the ones used in the Monte Carlo simulation. Therefore, only general trends can be compared.

There were 49 experimental data, taken from a uniaxial tension test. The histogram of those 49 values is found in Fig. 9 and those values are plotted in the Weibull probability paper in Fig. 10. This histogram has many similarities to those of cases 11 and 12 as can be easily seen in Figs. 6 and 7.

2.8 Conclusions

The main objective of this study was to introduce a probabilistic model for the spatial variation of strength of the materials used for laminated orthotropic composites. The principal idea lay in the interpretation that the material strength could be idealized as a multidimensional stochastic field. In this way, it was concluded that the statistical distribution followed by the strength of laminated orthotropic composite plates in uniaxial tension is the Weibull distribution. Another conclusion was the statistical size effect which is a well-known phenomenon for materials of this kind.

3. Response Variability and Reliability of Plates Using the Weighted Integral Method

3.1 Introduction

The consideration of material and geometric uncertainties (or system stochasticity) has become increasingly important over the past several years. The limited number of analytic solutions for such problems indicates that a numerical approach based on some version of the stochastic finite element method (SFEM) is the only available tool. Recently, a SFEM based on the concept of weighted integrals has been introduced to calculate response variability and reliability of stochastic trusses and frames (Deodatis & Shinozuka, 1989; Deodatis, 1991; Deodatis & Shinozuka, 1991). Following this method, the continuous stochastic field describing the randomness of the system is accurately reduced to a number of random variables called weighted integrals. The method was extended to $2D$ plane stress/plane strain problems (using the constant stress/constant strain triangular element) in (Deodatis et al., 1991; Wall, 1991).

In this study, the weighted integral method (WIM) is further extended to plane stress/plane strain problems analyzed by rectangular finite elements.

3.2 Method of Analysis

For linear systems subjected to static loading, the equations of equilibrium can be written in matrix form as:

$$\mathbf{KU} = \mathbf{F} \tag{49}$$

where $\mathbf{K}$ = global stiffness matrix, $\mathbf{U}$ = global nodal displacement vector, and $\mathbf{F}$ = global load vector. In the following, a stochastic system under deterministic loading will be examined. The elastic modulus is considered to vary over the area of a certain finite element (e) as: $E^{(e)}(x,y) = E_0^{(e)}[1 + f^{(e)}(x,y)]$, where $E_0^{(e)}$ = mean value of elastic modulus and $f^{(e)}(x,y)$ = two-dimensional, zero-mean, homogeneous stochastic field. In orfer to avoid negative values of the elastic modulus, bounds are imposed on the stochastic field (see: Deodatis et al., 1991). Based on the principle of stationary potential energy, the stochastic element stiffness matrix for plane stress/plane strain problems using a four-node rectangular element is computed as:

$$\mathbf{K}^{(e)} = \mathbf{K}_0^{(e)} + X_1^{(e)}\Delta\mathbf{K}_1^{(e)} + X_2^{(e)}\Delta\mathbf{K}_2^{(e)} + \cdots + X_6^{(e)}\Delta\mathbf{K}_6^{(e)} \tag{50}$$

where all matrices on the right-hand-side are deterministic and the weighted integrals $X_k^{(e)}$ are random variables defined as:

$$X_k^{(e)} = \int_{A^{(e)}} x^i y^j f^{(e)}(x,y) dA^{(e)} \tag{51}$$

where $A^{(e)}$ denotes the area of the finite element. In Equation 51, subscript k is related to the powers i and j as: $(k \rightarrow (i,j)) : 1 \rightarrow (0,0), 2 \rightarrow (1,0), 3 \rightarrow (0,1), 4 \rightarrow (1,1), 5 \rightarrow (2,0), 6 \rightarrow (0,2)$.

To analyze the response variability, the displacement vector is approximated by a first-order Taylor expansion around the mean values of the weighted integrals. The resulting variance vector of the response displacements can be written in the following form:

$$Var(\mathbf{U}) = \int_{-\infty}^{\infty} \int_{-\infty}^{\infty} S_{ff}(\kappa_x, \kappa_y) V(\kappa_x, \kappa_y) d\kappa_x d\kappa_y \tag{52}$$

where S_{ff} is the spectral density function of the stochastic field $f(x, y)$ and V is the variability response function (VRF) (a concept first introduced in (Deodatis & Shinozuka, 1989)) given by:

$$
\begin{aligned}
V(\kappa_x, \kappa_y) = \sum_{e=1}^{N} \sum_{f=1}^{N} \sum_{k=1}^{6} \sum_{n=1}^{6} & diag(\mathbf{K}_0^{-1} \Delta \mathbf{K}_k^e \mathbf{U}_0) \mathbf{K}_0^{-1} \Delta \mathbf{K}_n^f \mathbf{U}_0 \\
& \times \{\cos(\Delta x_{fe} \kappa_x + \Delta y_{fe} \kappa_y)[Q_{ek} Q_{fn} + W_{ek} W_{fn}] \\
& - \sin(\Delta x_{fe} \kappa_x + \Delta y_{fe} \kappa_y)[Q_{fn} W_{ek} + Q_{ek} W_{fn}]\}
\end{aligned}
\tag{53}
$$

The subscript 0 denotes quantities calculated at the mean values of the weighted integrals, Δx_{fe} and Δy_{fe} are a measure of the relative position of elements (e) and (f) and Q and W involve trigonometric functions. The VRF is an indispensable tool in calculating upper bounds of hte response variability of stochastic systems.

In order to check the error introduced by the Taylor expansion approximation, numerical simulations are carried out based on the covariance matrix of the weighted integrals. Finally, the safety index is evaluated using the advanced first-order second-moment approach. The iterative algorithm proposed in (Madsen et al., 1986) is used. Herein, the limit state function for a nodal displacement is written in the following form:

$$
g_{u_j} = u_{jR} - u_j(\mathbf{X}) = 0
\tag{54}
$$

where u_{jR} = maximum allowable value of u_j, and $\mathbf{X}$ = vector containing weighted integrals.

3.3 Numerical Example

A square plate under uniform vertical loading is considered (see Fig. 11). The spectral density characterizing the stochastic field $f(x, y)$ is chosen to be

$$
S_{ff}(\kappa_x, \kappa_y) = \frac{\sigma_{ff}^2 b^2}{[\pi^2 (1 + b^2 \kappa_x^2)(1 + b^2 \kappa_y)]}
\tag{55}
$$

where σ_{ff} = standard deviation of random field (here: $\sigma_{ff} = 0.1$), and b = constant related to the correlation distance of $f(x, y)$. Fig. 12 displays the variability response function (VRF) of the vertical (v) and horizontal (h) displacements of the upper right node (see Fig. 11) as a function of wave numbers κ_x and κ_y. Fig. 13 shows the coefficient of variation (COV) of v and h as a function of correlation length parameter b. Results are based on the weighted integral method (wi) and the simulation procedure (si) mentioned above. The safety index is then calculated for the same two displacement components (Table 1). The maximaum allowable values u_{jR} are chosen as the mean value plus a multiple of the standard deviation of the respective displacement. Calculations are carried out for two values of b, i.e., $b = 1$ and $b = 5$.

3.4 Conclusions

The weighted integral method is applied to a $2D$ stochastic system (i.e., plane stress/plane strain) using a four-node rectangular element. For the present problem the error introduced by the Taylor expansion in the response variability is found to be very small. The reliability of the structure is evaluated in terms of a safety index based on the advanced first-orfer second-moment approach.

Table 1: Safety index for nodal displacements

u_{jR}	$\beta_{v(b=1)}$	$\beta_{v(b=5)}$	$\beta_{h(b=1)}$	$\beta_{h(b=5)}$
$m + 1\sigma$	1.94	1.33	3.36	1.83
$m + 2\sigma$	3.68	2.47	6.12	3.32
$m + 3\sigma$	5.22	3.47	8.43	4.57
$m + 4\sigma$	6.61	4.34	10.37	5.62
$m + 5\sigma$	7.86	5.11	12.03	6.51

4. References

Deodatis, G. (1991). "Weighted integral method. I: Stochastic stiffness matrix." *Journal of Engineering Mechanics*, Vol. 117(8), pp. 1851-1864.

Deodatis, G. and Shinozuka, M. (1989). "Bounds on response variability of stochastic systems." *Journal of Engineering Mechanics*, Vol. 115(11), pp. 2543-2563.

Deodatis, G. and Shinozuka, M. (1991). "Weighted integral method. II: Response variability and reliability." *Journal of Engineering Mechanics.*, Vol. 117(8), pp. 1865-1877.

Deodatis, G., Wall, W. A. and Shinozuka, M. (1991). "Analysis of two dimensional stochastic systems by the weighted integral method." *Computational Stochastic Mechanics* (eds: P. D. Spanos and C. A. Brebbia), Computational Mechanics Publ., Southampton-Boston, co-published by Elsevier Applied Science, London-New York, pp. 395-406.

Jones, R. M. (1975). *Mechanics of Composite Materials*, Washington, D.C., Scripta Book Company.

Madsen, H. O., Krenk, S. and Lind, N. C. (1986). *Methods of structural safety.* Prentice-Hall, Englewood Cliffs, N. J.

Papirno, R. (1985). "Algebraic Approximations of Stress/Strain Curves for Kevlar Reinforced Composites," *Journal of Testing and Evaluation*, Vol. 13, No. 2, pp. 115-122.

Shinozuka, M. and Jan, C-M (1972). "Digital Simulation of Random Processes and its Applications," *Journal of Sound and Vibration.* Vol. 25(1), pp. 111-128.

Shinozuka, M. (1974). "Digital Simulation of Random Processes in Engineering Mechanics with the Aid of FFT Technique," *Stochastic Problems in Mechanics.* (Eds

477

S. T. Ariaratnam and H. H. E. Leipholz), University of Waterloo Press, Waterloo, pp. 277-286.

Shinozuka, M. and Deodatis, G. (1988). "Stochastic Process Models for Earthquake Ground Motion," *Journal of Probabilistic Engineering Mechanics*. Vol. 3(3), pp. 114-123.

Shinozuka, M. and Deodatis, G. (1991). "Simulation of Stochastic Processes by Spectral Representation," *Applied Mechanics Reviews*, ASME, VOL. 44, No. 4, pp. 191-204.

Soni, S. R. (1983). "A New Look at Commonly Used Failure Theories in Composite Laminates," *Journal of the American Institute of Aeronautics and Astronautics*, pp. 171-179.

Vanmarcke, E. (1984). *Random Fields*, MIT Press, Cambridge.

Wall, W. A. (1991). "Entwicklung eines stochastischen finite Element Verfahrens fuer dynamische Analyse," Masters Thesis, University of Innsbruck, Innsbruck, Austria (in German).

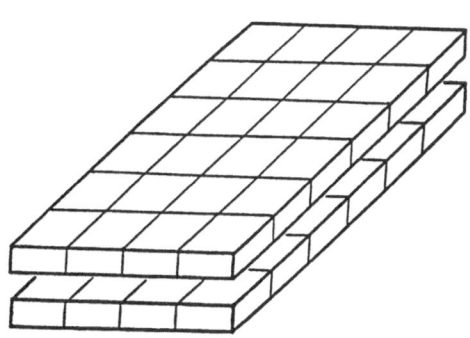

Fig. 1 Laminated Orthotropic Composite Plate Consisting of Two Laminae.

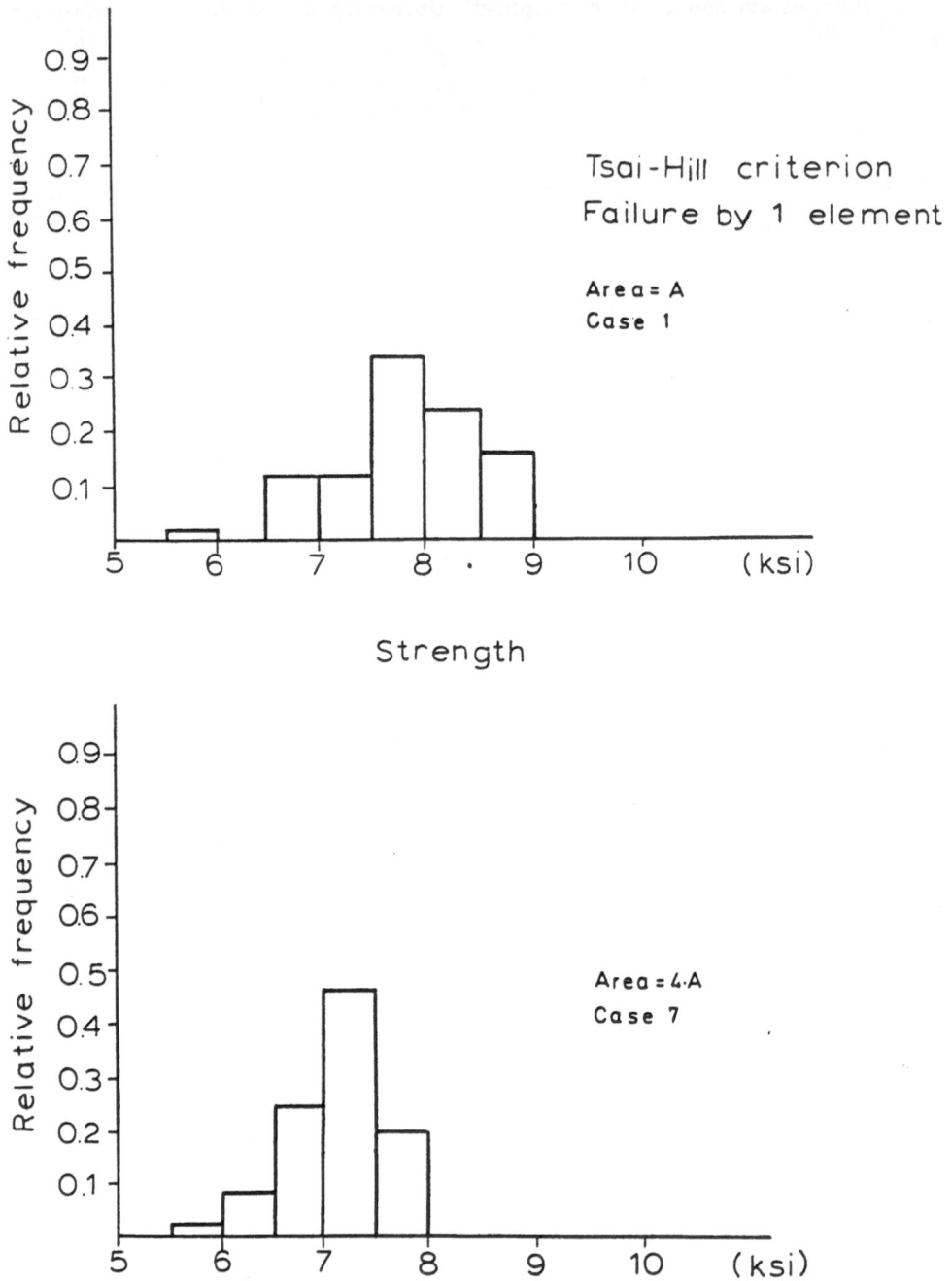

Fig. 2 Histograms for Cases 1 and 7.

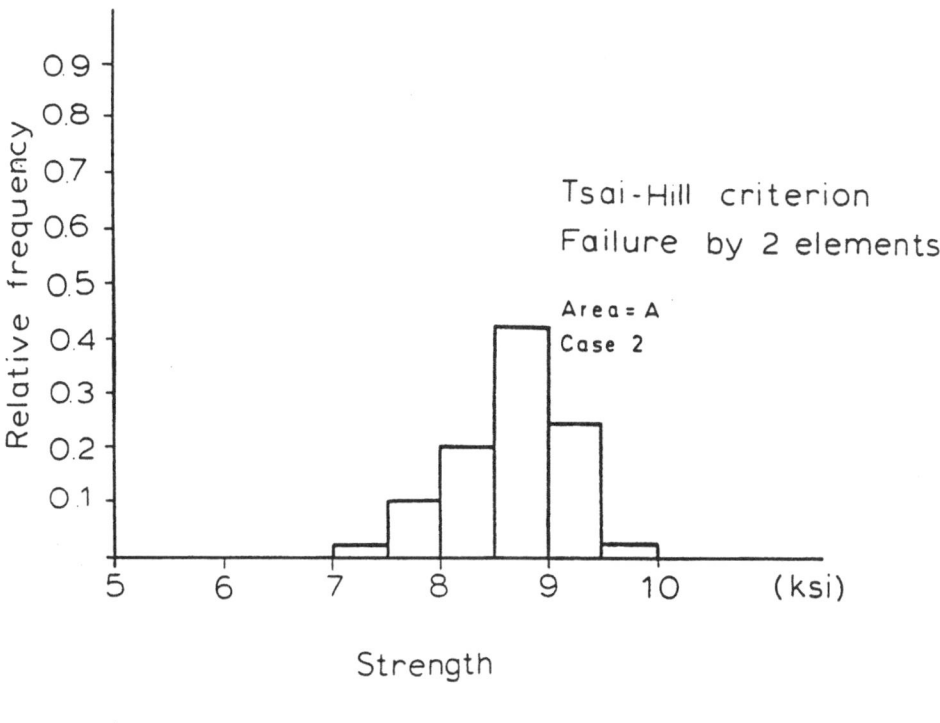

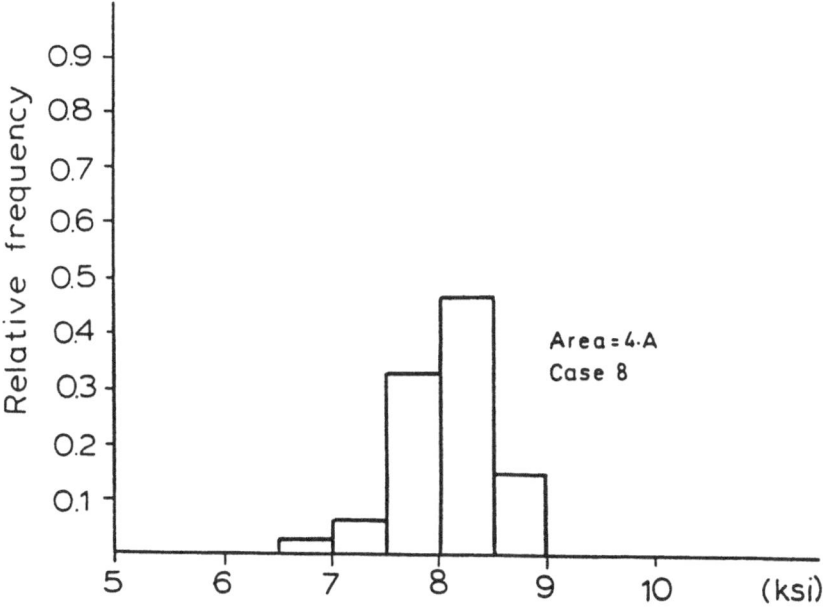

Fig. 3   Histograms for Cases 2 and 8.

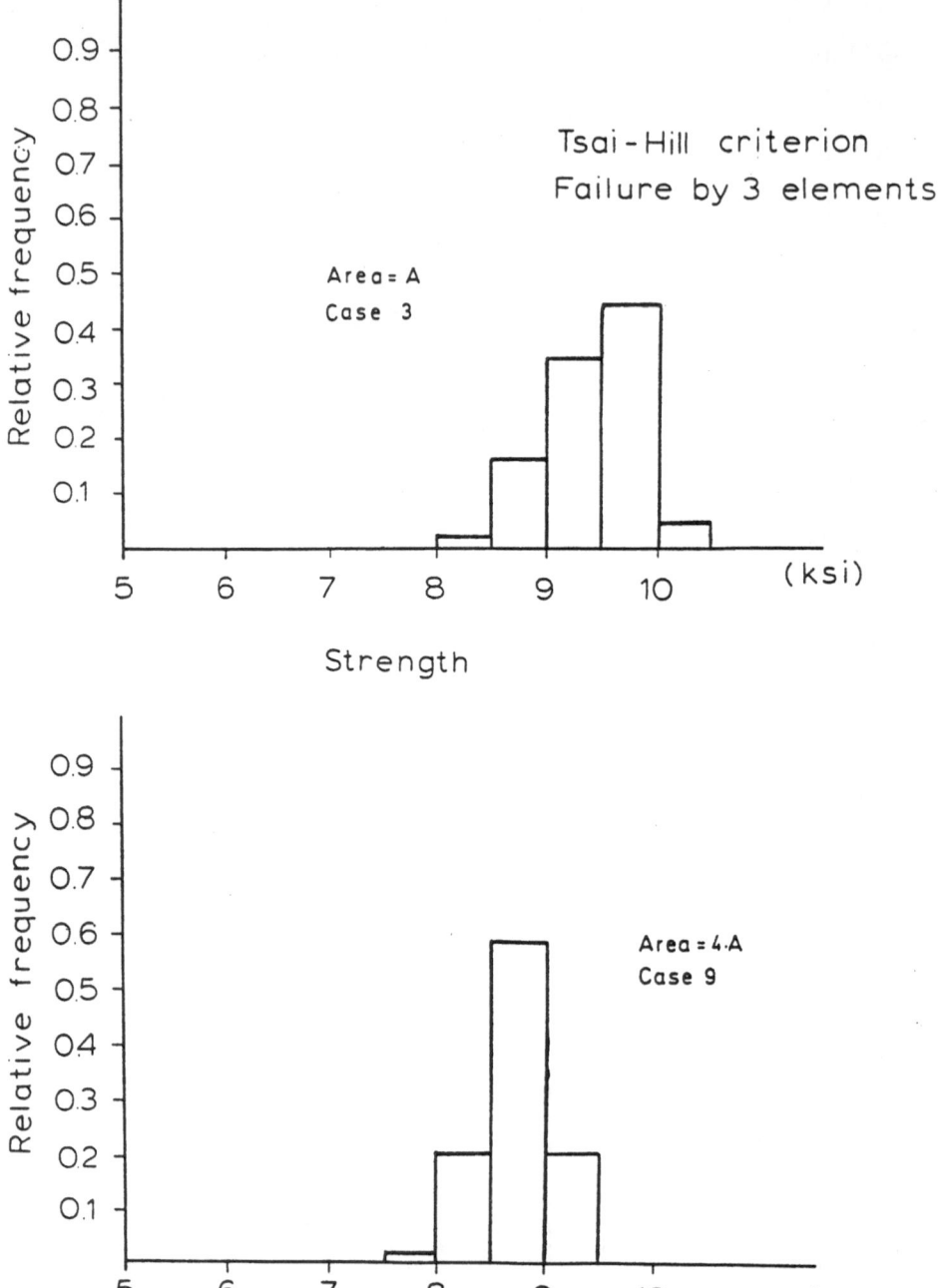

Fig. 4 Histograms for Cases 3 and 9.

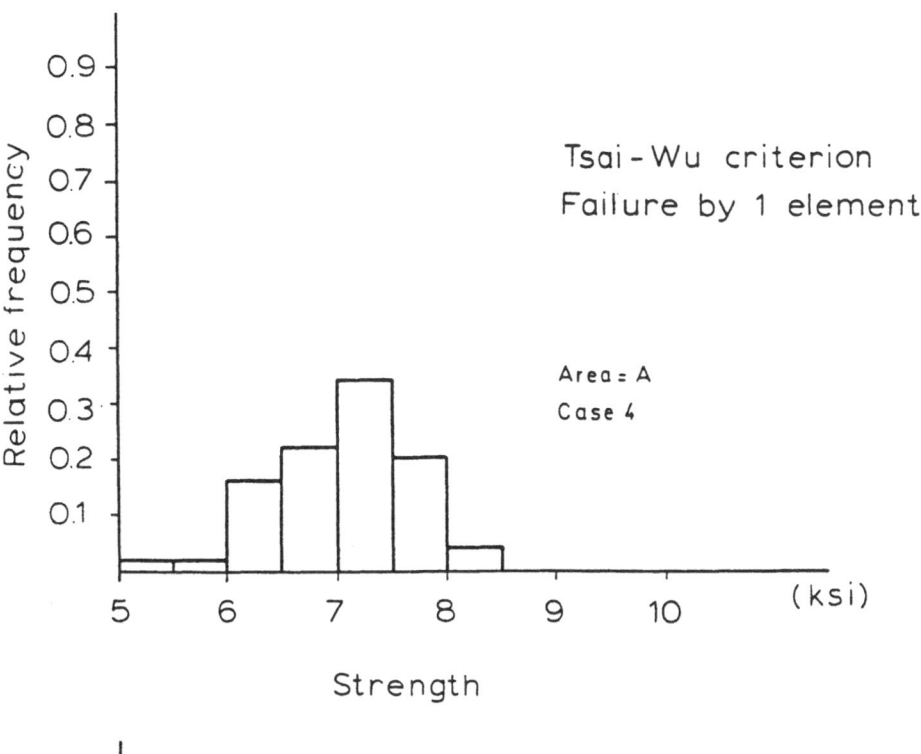

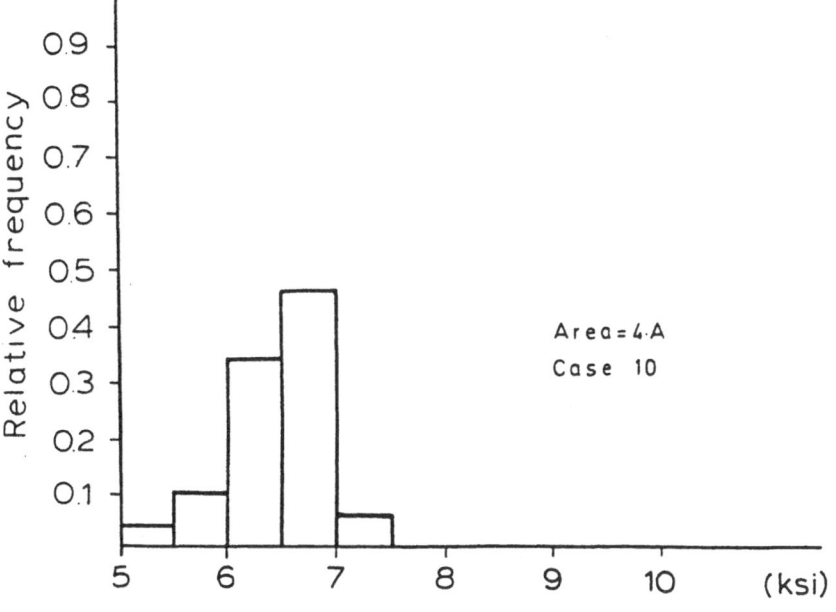

Fig. 5   Histograms for Cases 4 and 10.

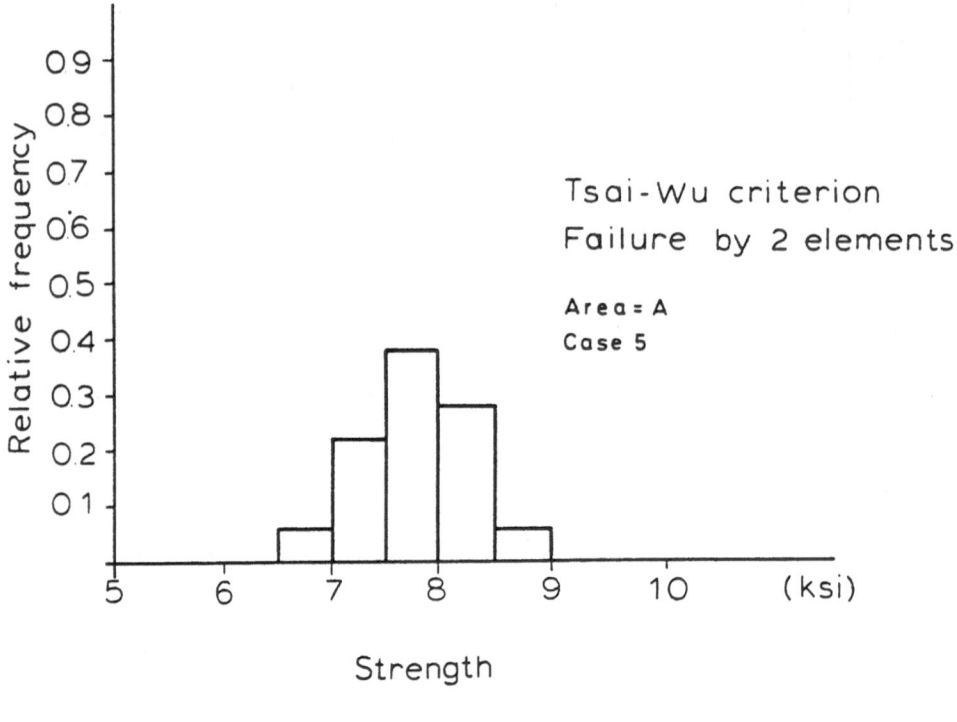

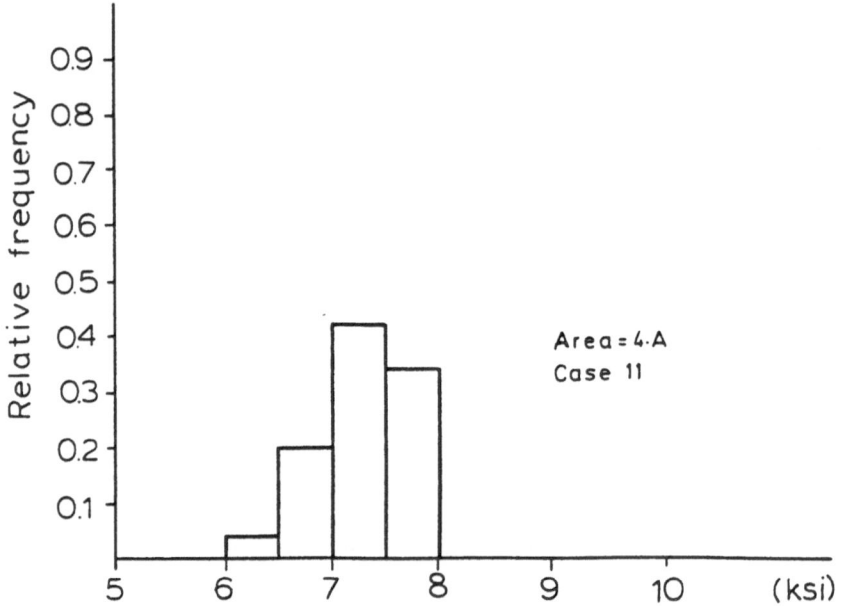

Fig. 6 Histograms for Cases 5 and 11.

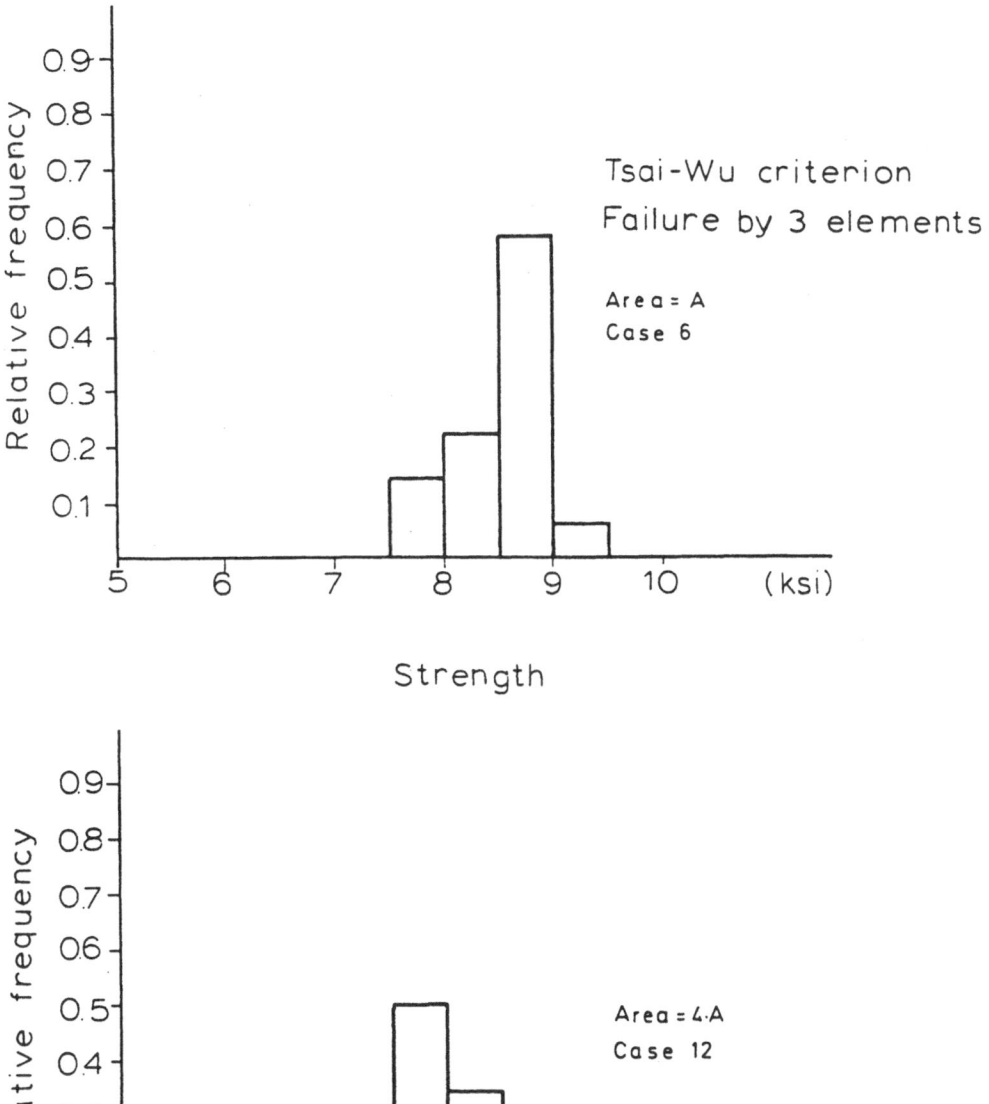

Fig. 7 Histograms for Cases 6 and 12.

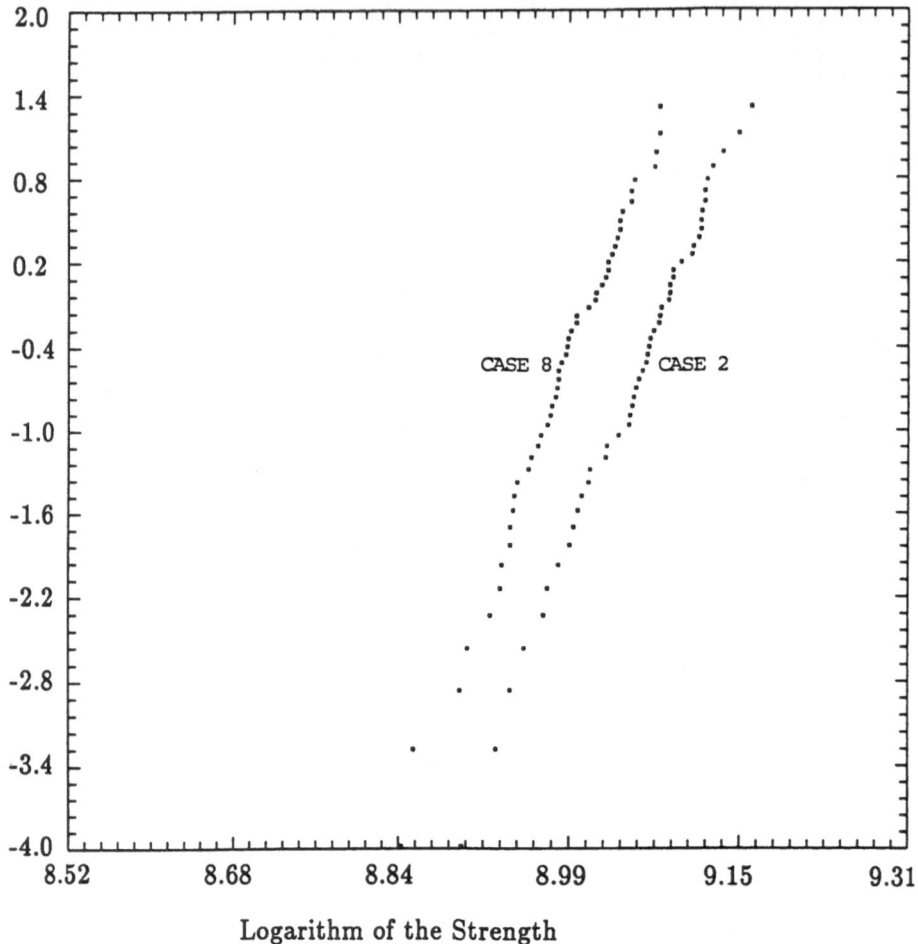

Fig. 8 Weibull Distribution Probability Paper for Cases 2 and 8.

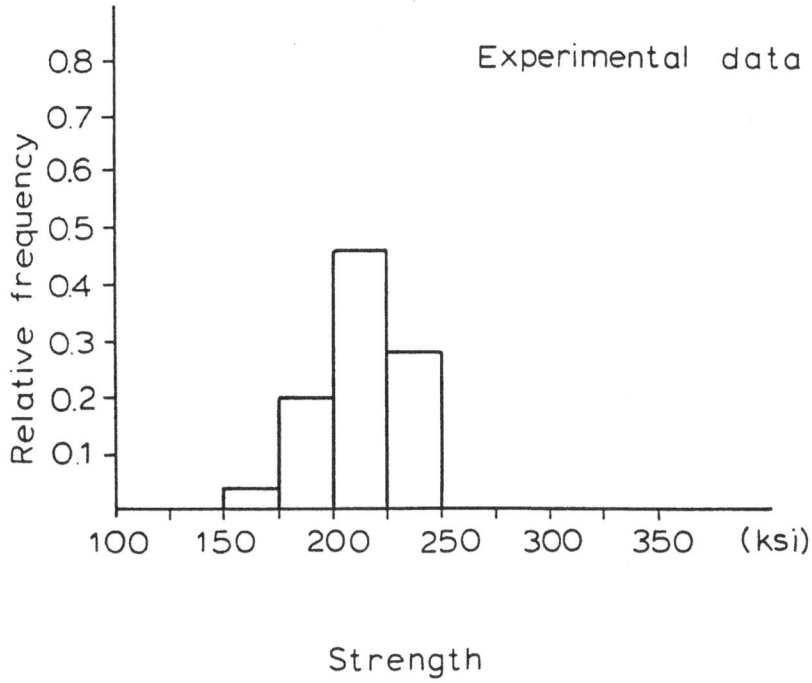

Fig. 9 Histogram of Experimental Data.

486

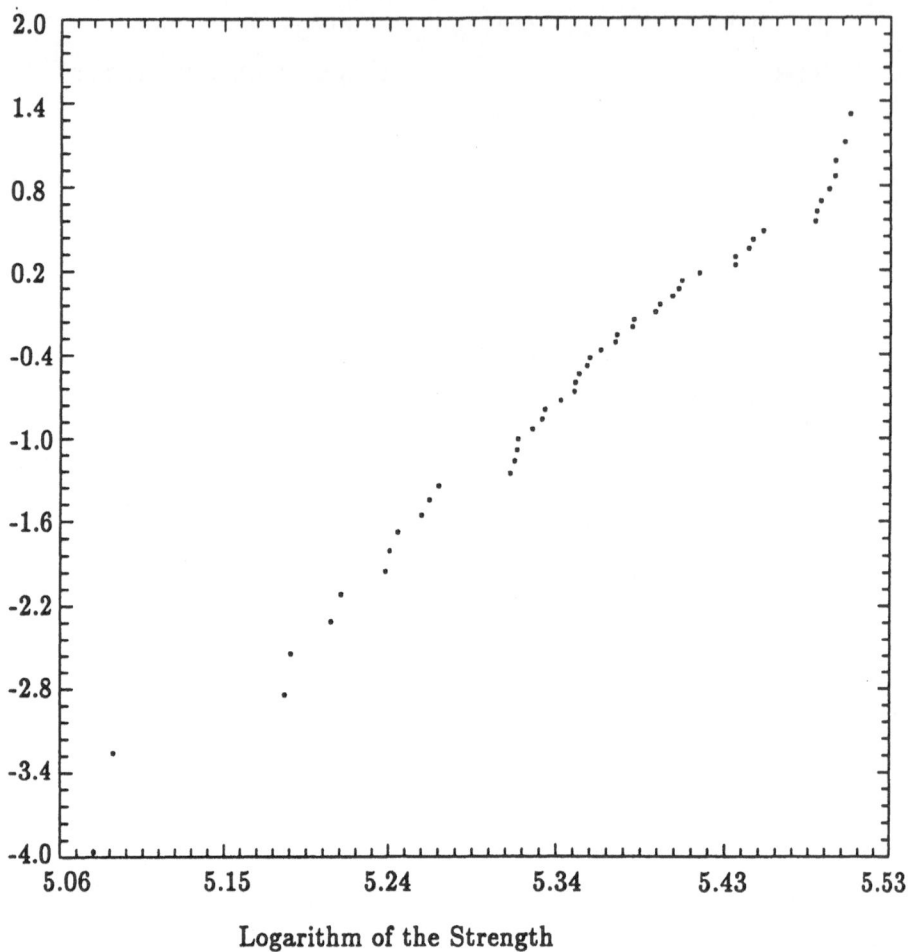

Fig. 10 Weibull Distribution Probability Paper for the Experimental Data.

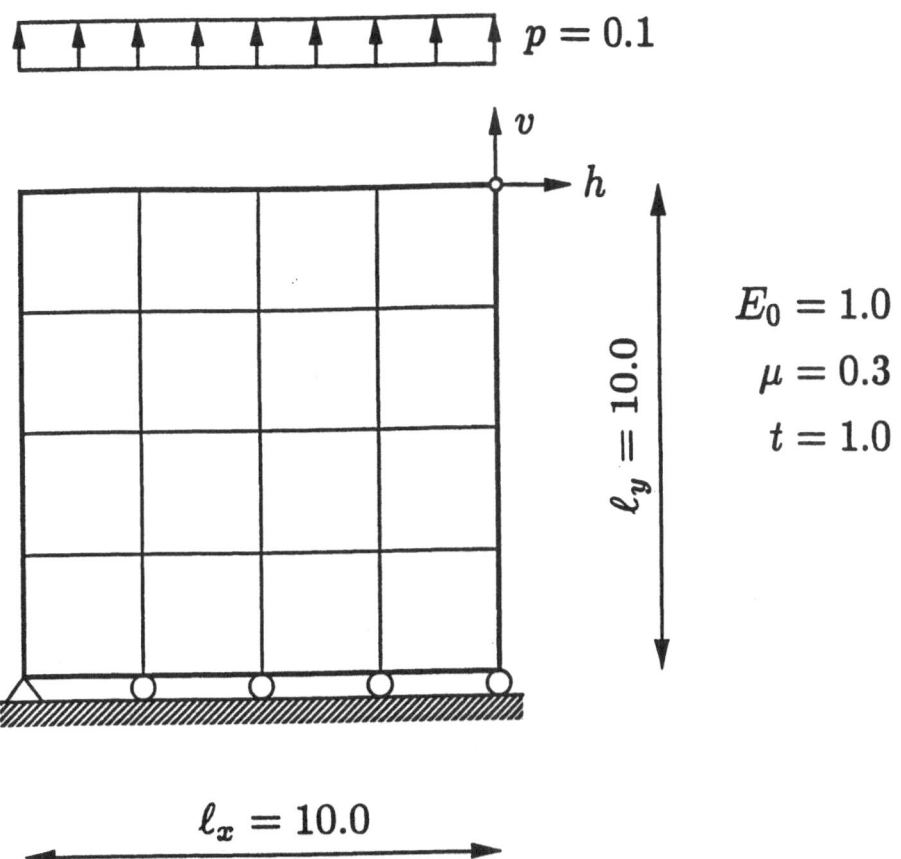

Fig. 11 Square Plate Under Uniform Vertical Loading.

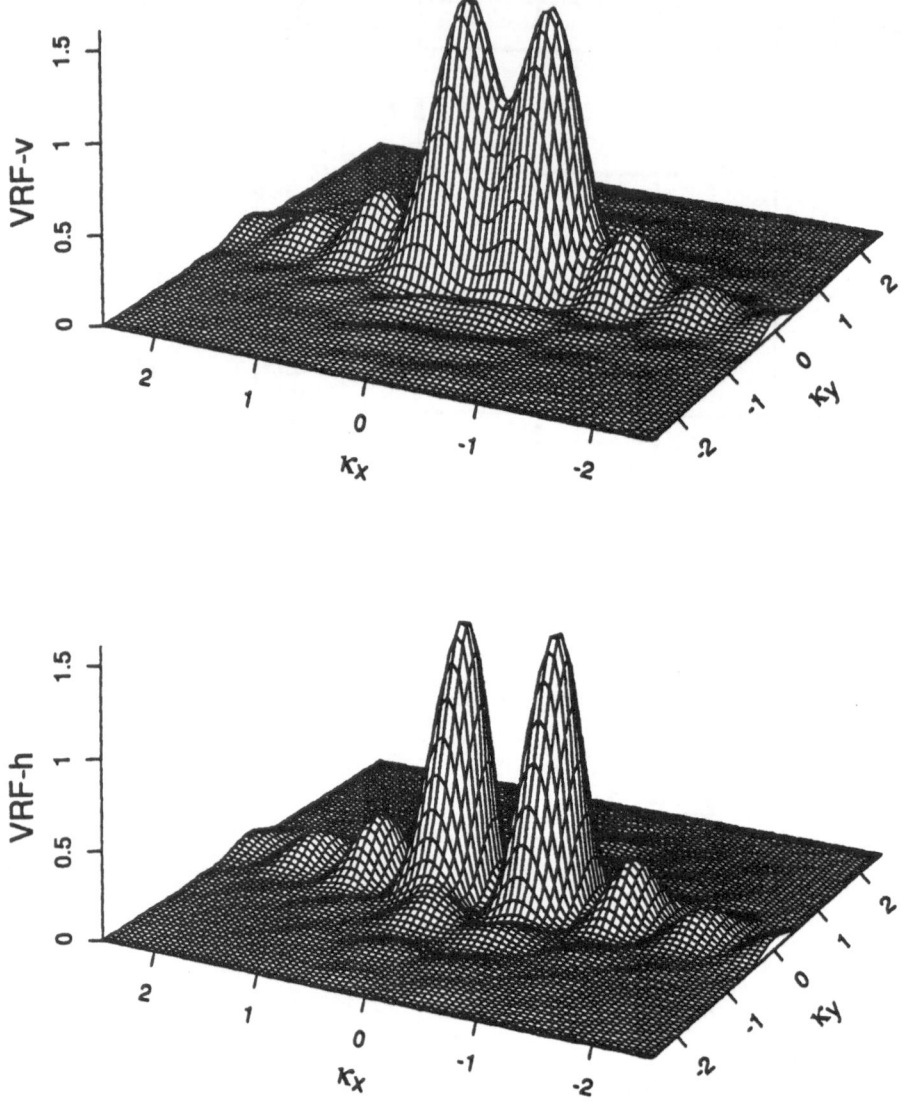

Fig. 12 Variability Response Function of Vertical and Horizontal Displacements.

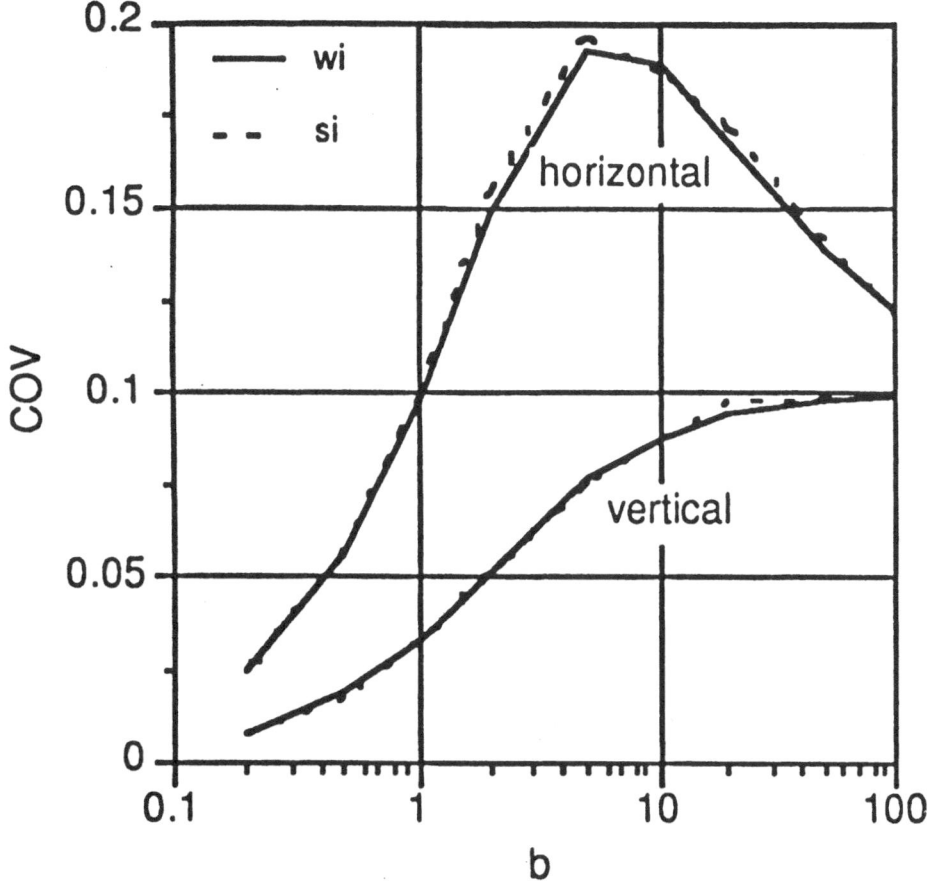

Fig. 13 Coefficient of Variation of Vertical and Horizontal Displacements as a Function of Correlation Length Parameter b.

STOCHASTIC APPROACHES FOR DAMAGE EVOLUTION
IN QUASI-BRITTLE MATERIALS

Jan Carmeliet
Catholic University of Leuven, Department of Civil Engineering, Belgium

René de Borst
Delft University of Technology, Faculty of Civil Engineering, The Netherlands/
Eindhoven University of Technology, Faculty of Mechanical Engineering, The ·
Netherlands

ABSTRACT

The damage process in quasi-brittle materials is a complex phenomenon in which heterogeneity plays an important role. This heterogeneity may imply that the exact failure mode can be highly dependent upon the precise spatial distribution of initial imperfections. To model this inhomogeneity stochastic material properties must be assumed in numerical simulations. However, the use of a stochastic approach does not resolve the issue of the change of character of the governing differential equations during progressive damage. As a consequence, the initial value problem becomes ill-posed at the onset of strain localisation and the finite element calculations suffer from a severe mesh dependence. A simulation technique that describes the true failure process must incorporate both a regularisation of the standard continuum during progressive damage and a stochastic description of the random continuum.

This statement will be substantiated in this contribution. To do so, we will present finite element analyses of direct tension tests with a standard local damage model and with a nonlocal damage model. The randomness in the damage process will be introduced by considering the initial damage threshold of the continuum damage model as a random field, characterized by a relevant distribution and autocorrelation function. The response statistics calculated by the Monte Carlo technique will be presented for two different levels of discretisation. The nonlocal and random field formulations rely both on the introduction of length parameter: the internal length of the nonlocal continuum and the correlation distance of the random field. The effect of the relative variation of the correlation distance and the internal length will also be investigated.

D. Breysse (ed.), Probabilities and Materials, 491–503.
© 1994 Kluwer Academic Publishers.

1. INTRODUCTION

Failure in quasi brittle disordered materials involves a progressive strain concentration in highly localised damage zones. This phenomenon is referred to as strain localisation. Strain localisation is initially driven by the growth of preexisting micro-defects in the disordered continuum and the development of several non-critical damage zones. In a second stage one critical damage zones develops, such as cracks in concrete, shear bands in soils and rock faults.

In a deterministic continuum approach the microstructural changes during the damage process can be modeled by means of a strain-softening material behaviour, i.e., after a certain limit a descending relation between stresses and strains (figure 1). An inherent property of a constitutive law with softening branch, when applied in a classical continuum framework, is that the damage tends to localize. However, a deficiency of this approach is that the field equations cease to be elliptic at the start of localisation. As a consequence, the mathematical model no longer is a proper description of the real damage behaviour (de Borst et al. 1993). With respect to this loss of well-posedness of the problem, finite element analysis exhibits a severe mesh dependence. To overcome this deficiency two different approaches may be applied. Firstly, we have a discontinuum approach where the crack discontinuity is modeled by interface elements showing a softening force-displacement behaviour (Rots 1988). The second is to enrich to standard continuum model. These so-called regularisation methods or non-standard models include the nonlocal damage model (Pijaudier-Cabot and Bazant 1987), the Cosserat continuum (de Borst and Mühlhaus 1991b), rate dependent and gradient models (de Borst et al 1993).

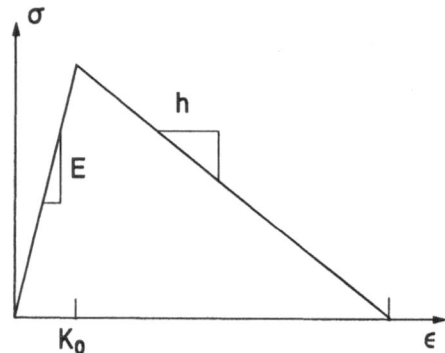

Figure 1. Linear strain-softening diagram

An important property of quasi brittle materials is the inherent heterogeneity and the presence of initial damage. When these uncertainties have a limited influence on the global response of the structure, the deterministic non-standard models are generally sufficient. These models need to define a internal length scale l which measures the distance over which a strong micro-structural interaction will persists in the continuum. Although the internal length reflects a physical phenomenon of stochastic microstructural interplay, the parameter is generally considered as a deterministic property. By the introduction of a internal length the non-standard models are capable of describing the deterministic size effects observed in many experiments.

When the damage evolution, the structural behaviour and reliability of mechanical structures are highly directed by the internal disorder of the material, it does not remain possible to avoid the modelling of the microstructural uncertainties and their interactions. A possible approach is based upon the mapping of the micro-structural geometry and topology on a numerical discrete support like the finite element method and then to perform numerical simulation. However, these models suffers from objectivity regarding the discrete representation by the finite elements and will exhibit spurious results (Breysse et al. 1993). Moreover, these models lead to very heavy computations and are yet not applicable to real engineering structures.

When the main priority is to analyze the damage behaviour of real engineering structures, a more appropriate approach of accounting for material disorder is by considering the macro-scale material properties to be randomly distributed in space (Carmeliet and Hens 1993). A suitable framework to describe the continuous spatial distribution of the random variables and their interdependence is based on the random field theory (e.g. Van Marcke et al. 1986). This approach introduces a second length parameter, the correlation distance θ, which describes the correlative characteristic of the random continuum. A promising feature of these models is the possibility of obtaining full statistical information regarding the structural reliability, e.g. the calculation of very small probabilities of failure. Adequate finite element reliability models are recently proposed including first order and second order reliability methods (Der Kiureghian and Ke 1988, Der Kiureghian and Liu 1989). However, it will be shown that the use of a stochastic approach does not resolve the above mentioned issue of the loss of well-posedness of the governing differential equations during the damage process. It will be shown that a numerical technique that describes the true failure process properly within a continuum framework must incorporate both a regularisation of the standard continuum and a stochastic description of the random continuum.

The paper starts by reviewing the governing equations of the deterministic nonlocal continuum damage model. It will be shown that the nonlocal theory provides a proper regularization technique to capture strain localisation.
In the next section the stochastic approach based on the random field theory is laid out. The initial damage in the continuum is modeled as a univariate homogeneous autocorrelated two-dimensional random field. Finally the role of the length parameters introduced, the internal length l and the correlation distance θ, is investigated.

2. DETERMINISTIC APPROACH

A possible way to remedy the ill-posedness of the boundary value problem at the onset of localisation is the nonlocal damage concept. In the nonlocal theory the constitutive equations are formulated such that they allow for dependence on the variables of the whole body (Edelen 1976). In the nonlocal damage model as developed by Pijaudier-Cabot and Bazant (1987), the nonlocal concept is applied to the internal damage variable D, while the other variables remain local. In this section, we will show that application of the nonlocal concept to damage can regularize the damage localisation problem.

The constitutive law of the continuum with isotropic damage may be written in the form :

$$\sigma = (1 - D) \, C \, \epsilon \tag{1}$$

where C is the initial elasticity tensor of the virgin material, σ and ϵ the stress and strain tensor, D the internal damage variable. From an initial equilibrium state given by the stress σ_0, the corresponding strain ϵ_0 and value of damage D_0, the rate constitutive relation can be derived as :

$$\dot{\sigma} = (1 - D_0) \, C \, \dot{\epsilon} - \dot{D} \, C \, \epsilon_0 \tag{2}$$

with the conditions for the growth of damage $\dot{D}$:

$$\text{if } f(\bar{\epsilon}) = 0 \text{ and } \dot{f}(\bar{\epsilon}) = 0 \quad \text{then } \dot{D}(x) = \frac{\partial F(\bar{\epsilon})}{\partial \bar{\epsilon}} \dot{\bar{\epsilon}} \tag{3}$$

$$\text{if } f(\bar{\epsilon}) < 0 \text{ or if } f(\bar{\epsilon}) < 0 \text{ and } \dot{f}(\bar{\epsilon}) < 0 \quad \text{then } \dot{D}(x) = 0$$

with F the evolution law for damage and f is the damage loading function:

$$f(\bar{\epsilon}) = \bar{\epsilon} - K(\bar{\epsilon}) \tag{4}$$

K is the damage threshold. In the initial state the damage threshold K is taken equal to the initial damage threshold K_0, afterwards K is assumed to be the maximum value of $\bar{\epsilon}$ ever reached at the considered point $x = [x, y, z]^T$ of the solid during loading history.

The damage loading function f and damage evolution F are specified through the nonlocal strain measure $\bar{\epsilon}$. The definition of $\bar{\epsilon}$ is based on the hypothesis of the attenuating neighbourhood. It is known that in quasi-brittle materials the effect of long range strains on the local damage evolution will attenuate fast with the distance. Thus we take :

$$\bar{\epsilon}(x) = \frac{1}{V_r(x)} \int_V \tilde{\epsilon}(x + \tau) \, \alpha(\tau) \, dV \tag{5}$$

with $\tau = [\tau_x, \tau_y, \tau_z]^T$ the separation vector between two points x and $x+\tau$, and α a attenuating weight function. We note that the weight function must be subjected to continuity (smoothness) conditions, in order to avoid unacceptable fluctuations of the damage variable. Motivated by this, a squared exponential weight function is introduced, which remains continuous in all points :

$$\alpha(\tau) = e^{-\left[\frac{|\tau|^2}{2l^2}\right]}$$
(6)

with l the so-called internal length. In (Pijaudier-Cabot and Benallal 1993), it is shown that the squared exponential weight function (in contrast to a uniform weight function) admits a non-trivial solution to the bifurcation problem on the onset of localisation. V_r in eq. 5 is the representative volume:

$$V_r(x) = \int_V \alpha(\tau)\, dv$$
(7)

The local equivalent strain $\tilde{e}$ in eq. 5 is defined as the accumulated tensile strain in the material (Mazars 1986):

$$\tilde{e} = \sqrt{\sum_{i=1}^{3} \left(<e_i>_+\right)^2}$$
(8)

$$<\epsilon_i>_+ = 0 \quad \text{if } \epsilon_i > 0$$

$$<\epsilon_i>_+ = \epsilon_i \quad \text{if } \epsilon_i \leq 0$$

where ϵ_i are the principal strains.

To demonstrate the essential deficiency of the standard continuum model to capture localisation and the proper regularisation by the nonlocal damage model, we solve the two-dimensional plane stress problem of a beam subjected to three point bending (for data see Carmeliet and Hens 1993). The effect of mesh refinement on the response is studied using three meshes of 12, 60 and 120 layered four noded plate elements with 30 layers. In figure 2 the global load-displacement curves calculated by the standard ($l=0$) and the nonlocal damage model ($l\neq0$) are compared. The nonlocal results differ markedly from the local results.

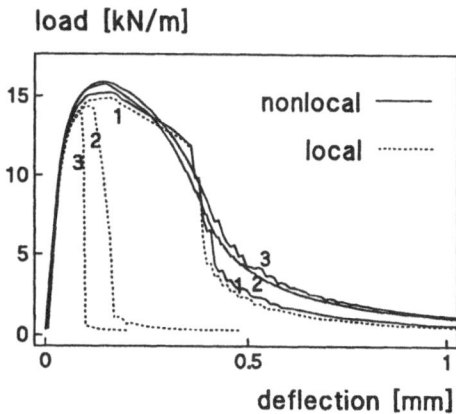

Figure 2. Load-deflection diagram for three point bending test. Comparison on local and nonlocal results for three meshes : 12(1), 60(2) and 120(3) layered plate elements.

The local solution exhibits a spurious mesh-dependence and an improper convergence: a decreasing maximum load and energy dissipation upon mesh refinement. At the contrary, in the nonlocal model we observe a proper convergence to a unique solution upon mesh refinement.

The difference between local and nonlocal solutions is much more significant, when we look at the damage distribution in the beam at maximum load for the mesh of 120 elements (figure 3). In the local solution the damage remains almost confined to the layered element in the middle of the beam. Upon mesh refinement the width of the damage zone tends to zero resulting in a failure without dissipation of energy. From a physical point of view, this is unrealistic. In the nonlocal solution, the damage propagates over a large number of elements and is independent on mesh refinement, which confirms findings of Pijaudier-Cabot and Bazant (1987). The same conclusion can be drawn from other non-standard continuum models (Pamin 1993). These models bear the same essential feature: the introduction of an internal length avoids spurious and discontinuous strain localisation.

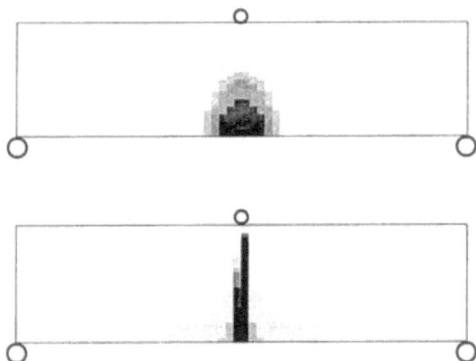

Figure 3. Damage distribution at maximum load for the nonlocal and local solution (120 elements)

3. STOCHASTIC APPROACH

As already mentioned in the introduction, heterogeneity may play an important role in the localisation process in quasi-brittle materials. Therefore, stochastic approaches seem to be natural. They give useful information in the form of higher order moments or full distributions of the intrinsic properties regarding failure: failure strength and dissipation of energy (e.g. Carmeliet and Hens 1992a). They also offer a framework to analyze failure probability or reliability of mechanical structures. Another use is the study of statistical size, volume and shape effects not captured by the gradient or nonlocal approach. However, a fundamental question to answer regarding localisation phenomena is *'Does a statistical description of the standard continuum resolve the ill-posedness of the continuum model at localisation ?'*. This question becomes more imperative if we consider that the description of a heterogenous continuum by correlated random variables introduces analogously as the non-standard approach an internal length in the form of the correlation distance θ. This distance represents the length beyond which the spatial correlation ceases to be important and may significantly influence the interaction between the defects during the damage process.

The example of a tensile specimen with random initial damage is well suited to study this fundamental issue. We assume that the initial damage threshold is randomly distributed over the solid and can be represented by a non-Gaussian correlated random field. For the non-Gaussian field a three-parameter Weibull distribution function is assumed:

$$f_{K_0}(K_0) = \lambda \mu (K_0 - K_0^{\min})^{\mu-1} \exp\left[- \lambda \left(K_0 - K_0^{\min}\right)^{\mu}\right] \qquad (9)$$

with λ, μ the Weibull-parameters and $K_0^{\min}$ the lower bound of the initial damage threshold. The experimental identified material data are taken from (Carmeliet 1992b) : $\lambda = 6.56 \ 10^5$, $\mu = 1.6$, $K_0^{\min} = 0.66 \ 10^{-4}$. Furthermore, we assume the field homogeneous and isotropic, which implies that the autocorrelation function $\rho(\tau)$ can be expressed in terms of the separation vector $\tau = [\tau_x, \tau_y]^T$ between the points x and x+τ. The autocorrelation function is assumed to be similar to the weight function of the nonlocal damage model (squared exponentional function):

$$\rho(\tau) = e^{-\left(\frac{|\tau|^2}{2d^2}\right)} \qquad (10)$$

with d a parameter representing the decay of correlation. The correlation distance for the squared exponentional function is defined as $\theta = \sqrt{2\pi}\,d$.

For finite element analysis involving random field properties, it is necessary to discretise the continuous random field into random vector representations. This discretisation involves the division of the structure into several stochastic elements and the representation of the stochastic field within the elements by random variables. A critical review of existing discretisation methods is presented by Li and Kiureghian (1992). In this paper, we will use the midpoint method (Der Kiureghian and Ke 1988). To accurately describe the random field, the size of the stochastic element is taken quarter of the correlation distance.

A sample tensile specimen with initial damage is generated following a cutting procedure. Firstly a sample of 200x200 mm^2 with random initial damage is digitally generated according to the method of (Yamazaki and Shinozuka 1988) (figure 4a). Then out of this continuum a tensile specimen with dimensions of 100x25 mm^2 is cut at a random position (figure 4b). This cutting procedure omits the problem of defining statistical boundary condition or considering boundary layer effects. In the reference simulations the correlation parameter has been assigned the value d=5 mm or a correlation distance θ=13 mm.

For the finite element discretisation two different meshes are used: 8x32 and 16x64 elements. During the discretisation process, the size of the stochastic element is not changed. This means that a stochastic element is respectively a block of 1 or 4 finite elements with identical probabilistic properties. Furthermore, we assume a constant linear softening diagram with a softening modulus h=-0.1.E and the elastic modulus E=20000 MPa. The assumption of a deterministic softening modulus is relaxed in (Carmeliet and Hens 1993), where the initial damage threshold and softening modulus are considered as correlated random variables.

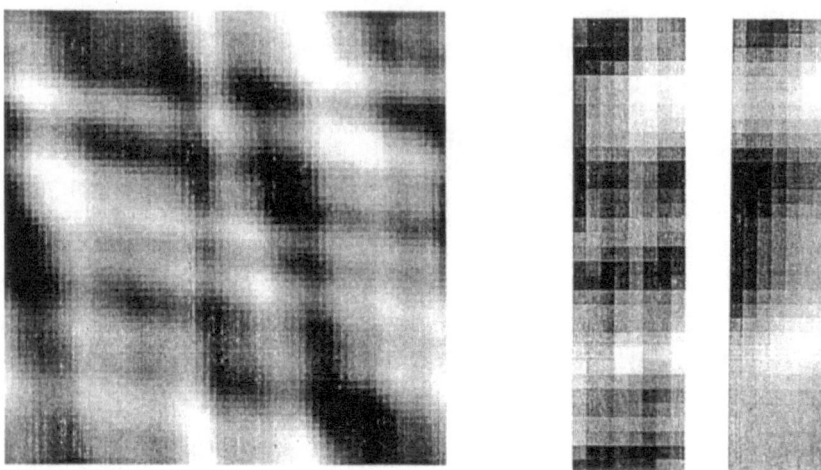

Figure 4. Sample of 200x200 mm^2 with generated initial damage. The correlation parameter d is 5 mm. Cut-out tensile specimens of 100x25 mm^2 for respectively d=5 and d=10 mm. Increasingly dark regions express a increasing initial damage threshold (stronger regions).

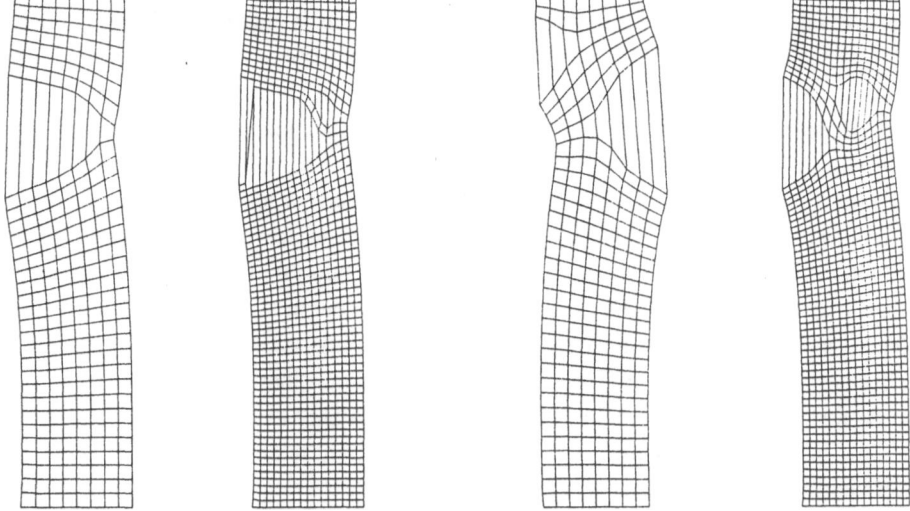

Figure 5. Deformation patterns of two tensile specimens at maximum load for the local continuum.

Figure 5a shows the deformation pattern at maximum load for the standard damage model. We observe a spurious localisation pattern with the localisation concentrated in a single band of elements which generally follows the mesh lines and occasionally jumps from one row to the next. The finite element solution tries to capture a line crack and the results are clearly mesh dependent.   These observation were already found using discrete deterministic models (Rots 1988) or standard rate independent models (Sluys 1992). Moreover, we observe that upon mesh refinement complete different deformation patterns can be attained (figure 5b). These examples point out that the introduction of the correlation distance θ does not resolve the improper behaviour of the standard continuum and that regularization techniques like the nonlocal damage concept have to be used. For the nonlocal calculations, we choose the internal length $l=5$ mm equal to the correlation parameter d. Figures 6a and 6b show the deformation patterns respectively at maximum load and at complete failure. Before the maximum load is reached, the damage process is characterized by diffuse microcracking and by the development of two non-critical damage zones. Comparing figure 4b and 6a, we remark that the location of the damage zones completely reflect the zones of the lowest initial damage threshold, which are due to the correlation function periodically located over the specimen. At complete failure the damage now does not proceed along the element lines and is no longer confined to one row of elements. The damage is spread out over a damage zone independent of the degree of mesh refinement.

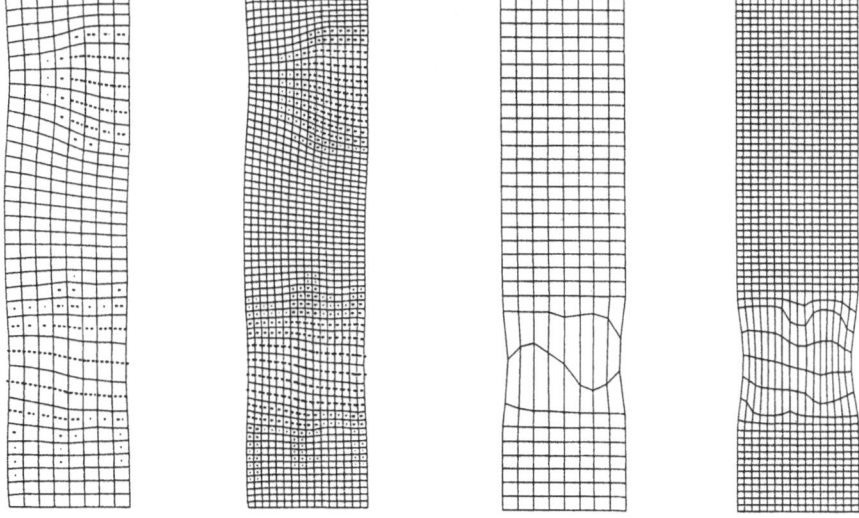

Figure 6. Deformation and damage patterns of tensile specimen a maximum load and at complete failure for the nonlocal continuum (d=5mm, l=5mm).

The width of the damage zone w is almost 17 mm. This much lower then calculated by the deterministic nonlocal damage model, which can analytically determined as $w=2\pi l$ (de Borst and Mühlhaus 1991a) or for the considered data w=31 mm. The reason for this is that the width of the failure zone in the correlated probabilistic continuum will be significantly influenced by the width of the periodical weak initial damage zones. Figure 7, which gives the overall stress-displacement response of the tensile specimen, shows that the results of the nonlocal model are insensitive with respect to the discretisation. The simulations with 256 and 1024 elements yield the same result.

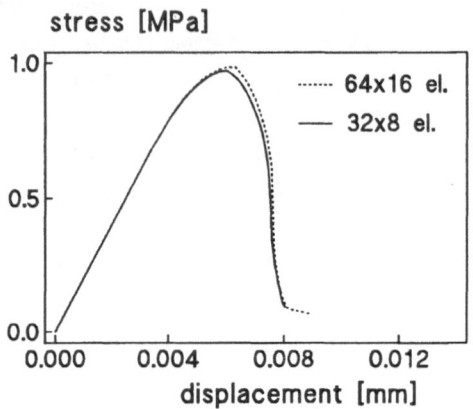

Figure 7. Stress-displacement diagram for a tensile specimen (d=5mm, l=5mm).

So far, we have shown that a probabilistic continuum description of damage localisation must account for the loss of the well-posedness of the governing differential equations by including in the standard model a regularisation technique. A major problem now lies in physically combining the two internal lengths introduced: the internal length l of the nonlocal continuum and the correlation distance θ of the random field. Both lengths result from the transition of a micro-level to macro continuum level and can be viewed as measures for the distance of which a strong micro-structural interaction will persist. If an internal length l much larger then the correlation distance θ is chosen, the correlative random structure may alter or disappear due to the averaging effects of the nonlocal model.

The effect of the relative variation of the correlation distance θ and the internal length l on the localisation process is investigated. Figure 8 compares the global stress-displacement curves for various values of the internal length of the nonlocal model (l=5 and l=2.5 mm) and of the correlation parameter (d=5 and d=10 mm). The influence of the correlation distance θ is limited compared with the internal length l. A smaller internal length results in lower value of the tensile strength and more brittle post-failure behaviour. This seems logical when we compare the deformation patterns at complete failure for two values of l (figure 9). A smaller internal length involves localisation in a smaller damage zone, resulting in a decrease of the maximum load and of the energy dissipation during damage.

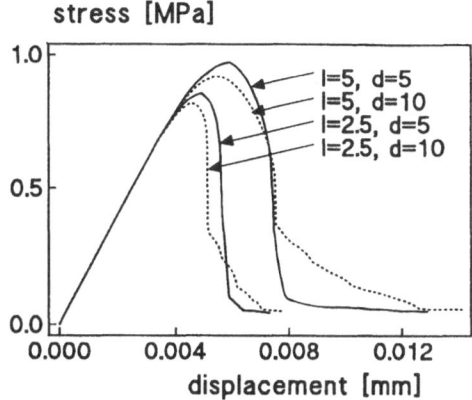

Figure 8. Stress-displacement diagram for various values of the correlation parameter d and the internal length *l*.

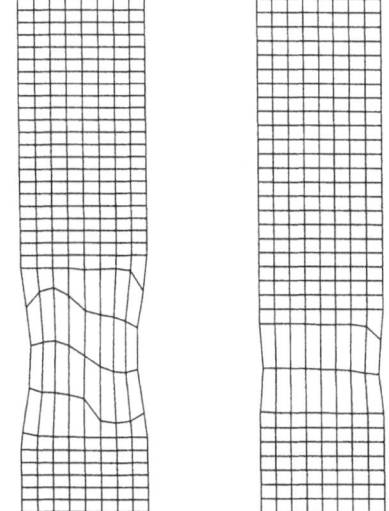

Figure 9. Deformation pattern at complete failure for the nonlocal continuum (*l*=5 mm and 2.5 mm).

Figure 10 compares the cumulative distribution of the maximum failure load for various values of the internal length of the nonlocal model (*l*=5 and *l*=2.5 mm) and of the correlation parameter (d=5 and d=10 mm). The cumulative distributions were calculated from the responses of 100 samples using the Monte Carlo technique. We observe that for a higher correlation distance an increase of the variance of the maximum load is observed. Intuitively, this observation seems reasonable. A high correlation distance involves a lower frequency of periodical weak damage zones. Therefore, the chance of finding a similar weak zone in a specimen with constant size will decrease. This results in an increasing variance of the maximum load.

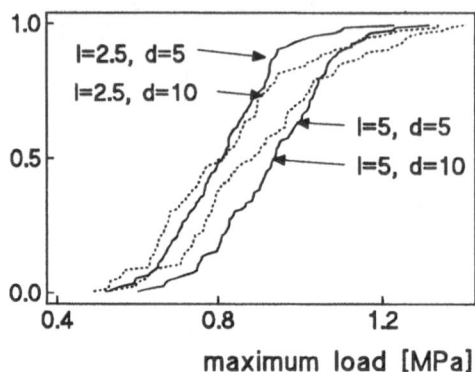

Figure 10. Cumulative distribution of maximum load for various values of the internal length l and the correlation parameter d.

4. Conclusion

In this contribution it was shown that one has to be very cautious on the computational results of probabilistic damage models, which rely on a stochastic formulation of the classical continuum framework. Various examples pointed out that a probabilistic description of the damage threshold does not solve the difficulties associated to strain-softening in a classical standard continuum. The initial value problem becomes ill-posed at the onset of localisation and as a consequence spurious localisation patterns and a severe mesh dependence were found. In order to formulate a probabilistic continuum model that describes the true localisation process properly in quasi-brittle materials, one must incorporate both a regularisation of the standard continuum and a stochastic description of the disordered continuum. In both formulations a length parameter is introduced: the internal length of the nonlocal continuum and the correlation distance of the random field. Both length parameters rely on the same physical background: microstructural phenomena situated at a level below the continuum level. The main interest seems to be a good modelling of the relation between the internal length and correlation distance. Another open question with respect with these parameters is the proper experimental determination.

5. REFERENCES

Breysse, D., Fokwa, D. and Drahy, F. (1993) "Spatial variability in concrete: nature, structure and consequences." *Applied Mechanics Review*, accepted.

Borst, R. de, Mühlhaus, H.-B (1991[a]). "Continuum Models for Discontinuous Media." *Conf. on Fracture Processes in Concrete, Rock and Ceramics*, J.G.M. van Mier et al., eds.,

Chapman & Hall, London, 601-618.

Borst, R. de and Mühlhaus, H.-B. (1991[b]) "Simulation of strain localisation: A reappraisal of the Cosserat continuum." *Eng. Comput.*, 8, 317-332.

Borst, R. de, Sluys, L.J., Mühlhaus H.-B., Pamin, J. (1993) "Fundamental issues in finite element analyses of localisation of deformation". *Eng. Comput.*, 10, 99-121.

Carmeliet, J., Hens, H. (1992[a]). "Fracture of a Fabric Reinforced Mortar Based on a Stochastic Approach to Initial Damage." *Conf. on Localized Damage II: Fatigue and Fracture Mechanics*, M.H. Aliabadi et al., eds., Computational Mechanics Publications, Southampton, pp. 283-298.

Carmeliet, J. ([1]992[b]). "*Durability of fiber-reinforced renderings for outside insulation: a probabilistic approach based on the nonlocal continuum damage mechanics.*" PhD thesis, Catholic University of Leuven, Belgium (in Dutch).

Carmeliet, J., Hens, H (1993) "A Probabilistic nonlocal continuum damage model for strain-softening materials with random field properties." *J. Engrg. Mech., ASCE*, accepted

Der Kiureghian, A., and Ke, J-B. (1988). "The Stochastic Finite Element Method in Structural Reliability." *Probabilistic Engineering Mechanics*, 3, 83-91.

Der Kiureghian, A., and Liu, P.-L (1989). "*First and Second-Order Finite Element Reliability Methods.*" Computational Mechanics of Probabilistic and Reliability Analysis, W.K. Liu and T. Belytschko,eds., Elmepress Int., Lausanne, Swiss., 9-46.

Edelen, G.B. (1976) "*Nonlocal Field Theories*", Continuum Physics, Volume IV-Polar and Nonlocal Field Theories, Part II, A.C. Eringen, ed., Academic Press, NY., 76-204.

Li, C. C., and Der Kiureghian, A. D. (1992) "An Optimal Discretization of Random Fields." *Report No. UCB/SEMM-92/04*, Department of Civil Engineering, University of California at Berkeley.

Mazars, J., Pijaudier-Cabot, G. (1989). "Continuum Damage Theory - Application to Concrete." *J. Engrg. Mech., ASCE*, 115(2), 345-365.

Pamin, J. (1993). "*Mesh Sensitivity in gradient dependent softening plasticity: II. Application of Rankine and Drucker-Prager yield criteria*". Report 25-2-93-2-14, Delft University of technology, Delft.

Pijaudier-Cabot, G., Bažant, Z.P (1987) "Nonlocal damage theory". *J. Engrg. Mech., ASCE*, 113(10), 1512-1533.

Pijaudier-Cabot, G., Benallal, A. (1993). "Strain Localisation and Bifurcation in a Nonlocal Continuum." *Int. J. Solids Structures*, 30, 1761-1775

Sluys, L.J. (1992). "*Wave propagation, Localisation and Dispersion in Softening Solids.*" Dissertation, Delft University of Technology, Delft.

Rots, J.G. (1988). "*Computational Modelling of Concrete Fracture.*" Dissertation, Delft University of Technology, Delft.

Van Marcke, E. (1983). *Random Fields: Analysis and Synthesis.* The MIT Press, Cambridge, Mass.

Yamazaki, F., Shinozuka, M. (1988). "Digital Generation of Non-Gaussian Stochastic Fields.", *J. Engrg. Mech., ASCE*, 114(7), 1183-1197.

HOW TO MANAGE THE SPATIAL VARIABILITY OF NATURAL SOILS

A. BOLLE
Université de Liège, Liège, Belgium

Abstract
In a first section this paper discusses the spatial variability of the
properties in natural soils and rocks. Simple statistics are proposed
to reduce large quantities of measurements to a small number of
numerical parameters, without losing the information about the spatial
distribution of the analyzed property. The autocorrelation curve
corresponding to an equivalent homogeneous medium appears to be
totally inconsistent with observed actual curves for CPT and pressuremeter tests.
The numerical methods taking into account the spatial variability
require to divide the domain into a large number of elements, and to
attribute random (and correlated) properties to each of these
elements. An original method is presented in the second part, allowing
the use of large numbers of random variables, with a limited computing
effort.
In the third section, the proposed method is applied to geotechnical
problems. The conclusions stress on the various ways the results are
influenced by the spatial variability of the material properties,
depending on the type of problem.
Keywords: Spatial Variability, Probabilistic Approach, Statistics,
Soil Properties.

1 Introduction

The usual way to manage the variations of the soil and rock properties
consists in the choice of **design values** for these properties. That
choice is sometimes based on statistics (average values), but it is
more often the result of the geotechnical engineer's experience, with
rather conservative assumptions. The design procedures (for foundations, retaining structures, ...) use these **average values** to produce
the well known **safety factors**. The uncertainties attached to these
data, as well as to the design model, are seldom considered. As a
matter of fact, these uncertainties are included into a customary
safety factor, the meaning of which is: a limit value, within a given
design procedure, leading to an acceptable rate of failure.
A better way is the probabilistic approach, considering all the
possible sources of uncertainty (input data, model, ...) and expressing the results as random variables. The probability to observe a

505

D. Breysse (ed.), Probabilities and Materials, 505–516.

failure, or a loss of serviceability, can then be estimated from the probability density functions of these random variables. Leaving out the uncertainties related to the model itself, the main difficulties in applying that procedure appear to be (1) the modelling of the uncertainties of the soil properties and (2) the probabilistic computations involving a large number of random variables as input values.

The aim of this paper is neither to make a complete review of the existing methods covering these topics, nor to develop a theoretical argumentation. It is rather to propose practical procedures applicable to the most actual problems, and to show on some examples how to avoid the implicit errors included in some **too simple** procedures.

2 The spatial variability of soil properties

2.1 The origins of the soil variability
The observed variability of the measured soil characteristics can be imputed to several origins:

The measuring errors, systematic and/or purely random;
The scale effects, resulting e.a. from the unavoidable averaging procedure involved in the measurement of any physical property;
The **real** variations of the observed property.

Using accurate techniques, such **real** variations can be observed even in highly homogeneous layers. They are obviously related to the geological processes the material have undergo and they can therefore exhibit some regular patterns corresponding to seasonal (daily, annual, secular, ...) events.

Theoretically, the scale of these variations is not bounded. A soil observed at the scale of the individual grains exhibits a very large variability (solid, void, solid, void, ...). At smaller scales, we will observe the crystalline structure, the molecules, the atoms, ... However there is a practical lower bound, close to the size of the **averaging volume** involved in the measuring method.

The upper limit must coincide with the volume of the **homogeneous** layer, to avoid to mix different kinds of materials.

A good model for the spatial variability will cover the full range starting from the **measurement averaging volume** up to the layer size.

2.2 The autocorrelation model
Soils exhibit specific structural characteristics as well as locally irregular almost random behavior. The local fluctuations consist of small-scale anomalies attributed to the randomness associated with geological details, microstructures, nugget effects or sampling errors (Christakos, 1987). This randomness necessitates the use of probability concepts rather than analytical expressions.

The hypothesis usually employed is the **second order homogeneity** of the data : constant mean and correlation between points depending only of the separation vector. The weaker hypothesis of **intrinsity** (or incremental homogeneity) means that the variance is finite and is the same at any point. Under these assumptions, the spatial variability of experimental data can be represented by various functions of the distance between the points : semi-variogram, covariogram or auto-correlation plot.

For the most part of the practical cases, the following simple transformations can be applied to the original experimental data :

(a) eliminate the regional trends by fitting an **average regional function** and subtracting it from the original data,
(b) normalize the remaining variations around that **average regional function** (constant variance and distribution close to the normal curve).

The (a) and (b) transformations can be repeated and combined in several ways to obtain the desired result: a **stationary** random variable, with a constant (zero) local average and a constant local variance.

The next step is to estimate the autocorrelation as a function of the distance between two points. For soil mechanics purposes that distance is usually measured vertically along a bore hole or along any vertical profile obtained by an other technique. In some cases, significant variations are observed along particular directions and the autocorrelation can be evaluated along these axes.

2.3 One-dimensional example

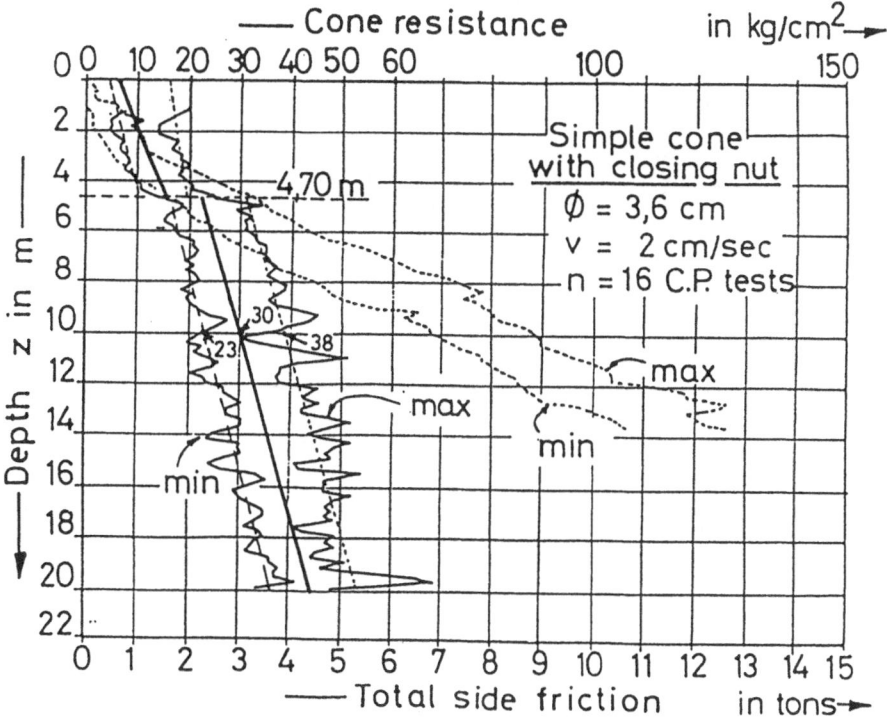

Fig.1. CPT tests - Envelopes of the measured values.

The first example is based on the results of 22 CPT tests (Cone Penetration Test) carried out in the northern part of Belgium, in a homogeneous clay layer (De Beer et al. 1977). The figure 1 shows the envelopes of all the measured values for the cone resistance and the total side friction.

Within the clay layer between 5 and 20 meter depth, the average cone resistance varies almost linearly and a strait line can represent the **regional trend**. The latter is easily obtained using a least square procedure. The remaining variations around these average lines can be considered as stationary. Two examples of autocorrelation plots are presented in figure 2 (Maertens 1990).

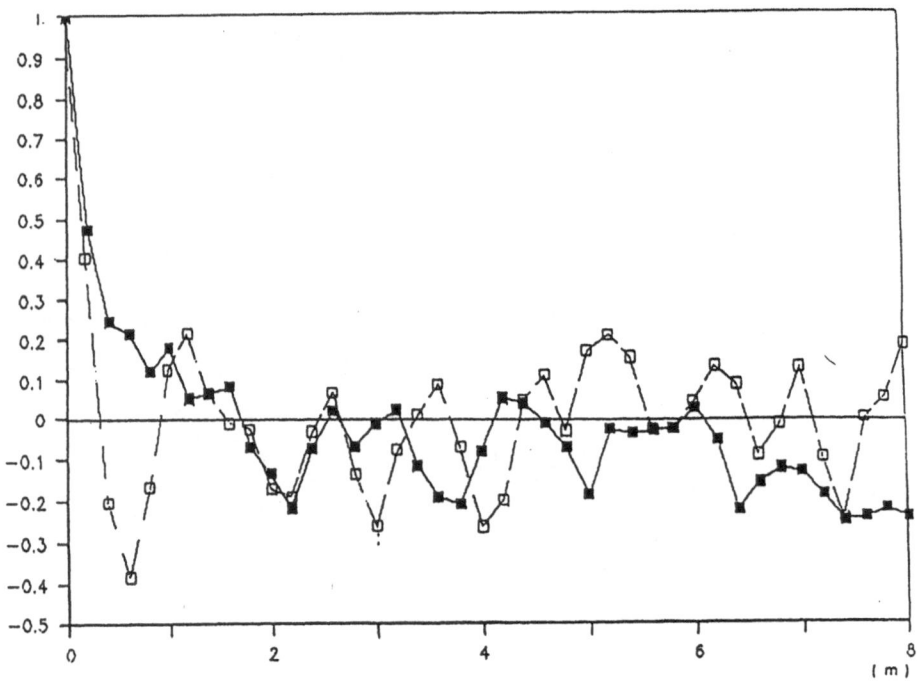

Fig.2. CPT - Autocorrelation vs vertical distance

The influence distance, defined as the first intersection with the axis, varies widely from one test to the other, within the range 0.3 - 1.6 meter. For larger vertical distances, the autocorrelation function oscillates around the axis, with a **wave length** close to 1.5m.

All the autocorrelation plots were averaged and the resulting diagram is shown in figure 3. The shapes are smoother and we can observe, after a rapid decay up to 0.4m, almost regular oscillations.

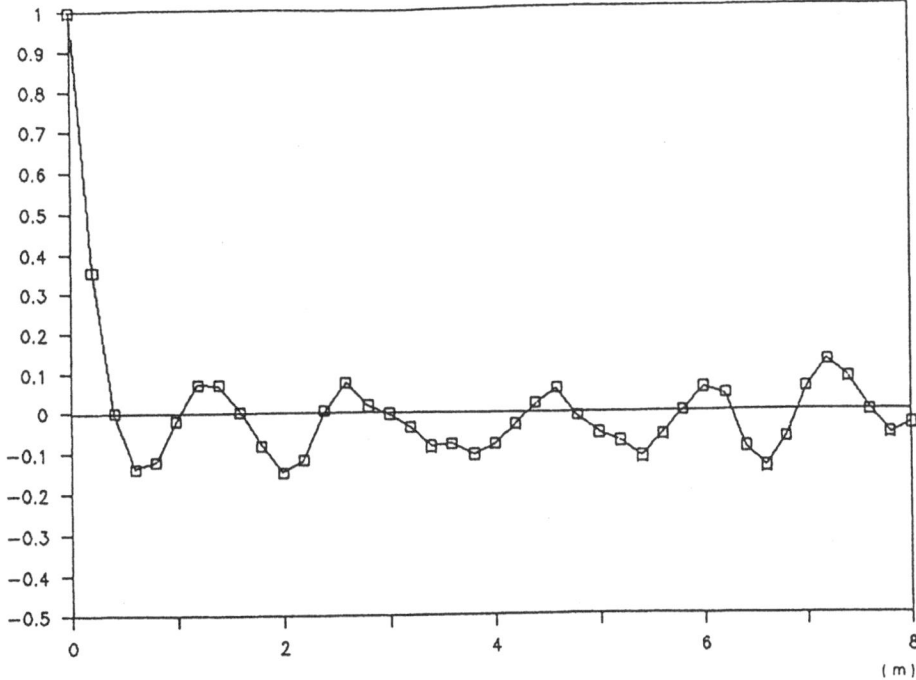

Fig.3. CPT - Averaged autocorrelation

2.4 Three-dimensional example
The second example deals with a very heterogeneous weathered rock mass
the deformability of which was measured by a large number of pressuremeter tests in vertical and inclined bore holes (Bolle et al. 1989).

Not any geological structure was clearly apparent and the whole
rock mass was therefore considered as a **roughly homogeneous** volume of
isotropic material.

Due to the large variations in the nature and the weathering of the
rocks, the distribution of the pressuremeter modulus appears strongly
dissymmetric, and a first transform (4th order root) is used to obtain
a more symmetric distribution. The regional trend appears to be three-
dimensional, with significant variations along each axis.

A stationary random variable, with zero mean and constant variance,
is then obtained by subtracting that regional trend, and the isotropic
autocorrelation diagram is presented on figure 4.

The influence distance can be estimated between 2.5m, assuming of a
strait line decay, and 9m when using the first intersection with the
axis. Oscillations with several superimposed wave lengths are also
observed.

510

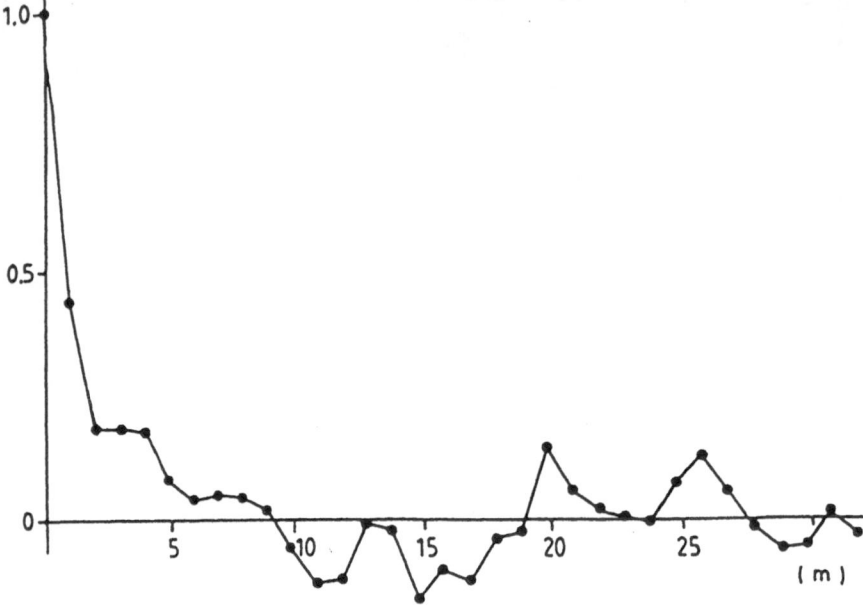

Fig.4. Isotropic autocorrelation (pressuremeter modulus)

3 Tools

3.1 Modelling the random properties of a soil mass

The examples above illustrate the usually observed autocorrelation for
the properties of a natural soil (or rock) mass: a more or less rapid
decay from unity to zero and following oscillations around zero. The
absolute values of these oscillations remain generally below the
confidence limit and they are not significant.

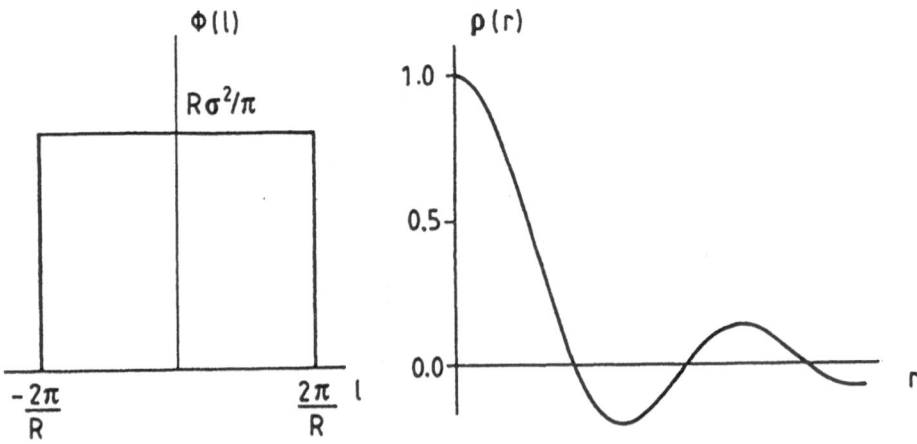

Fig.5. **Low pass noise** autocorrelation model

Using the Maximum Entropy Formalism, Baker & Zeitoun (1987) suggested the use of a **low pass noise** power spectrum to model the autocorrelation function, giving an oscillating shape as in figure 6, when the
given information consists of estimates of the average value, standard
deviation and autocorrelation distance. Reasonable estimates of these
three numbers are the best we can hope for under usual circumstances.

Similar oscillating shapes can alternatively be obtained by modelling an internal structure formed by regular patterns of two (or more)
sub-layers with different average characteristics. The larger observed
wave length corresponds to the thickness of one complete pattern and
the negative values relate to the relative thicknesses of the sub-
layers (Bolle 1990).

An isotropic model must use the three basic pieces of information:
average value (including a regional trend if necessary), variance and
one parameter characterizing the autocorrelation function. A three-
dimensional model will include trend and autocorrelation parameters
for each axis.

Simplified models, similar to the **three layers model** proposed by
Asaoka and Matsuo (1983), are only able to reproduce an actual autocorrelation curve, but can not be applied for any other purpose.

The only acceptable method is to divide the whole volume into N
small elements (small in comparison with the size of the problem) and
to assign to each element values of the property X defined by the N
random variables:

$$\mathbb{X} = \{x_1, x_2, \ldots, x_i, \ldots, x_N\} \tag{1}$$

The probability distribution function of each element of $\mathbb{X}$ is build
according to the estimates of the average (regional trend and transformations included, if any) and variance.

The spatial variability is expressed by the N by N autocorrelation
matrix $\mathbb{R}(X)$ relating each element to each other. If the distance
between two elements is larger than the influence distance, the
properties of these two elements are assumed to be independent and the
corresponding value in $\mathbb{R}$ is set to zero. For smaller distances a
correlation value ranging from 0 to 1 is computed according to a
chosen model (linear, exponential, ...).

3.2 Reducing the size of the problem

The N correlated random variables $\mathbb{X}$ can easily be converted into a new
set of N independent random variables using the following procedure:

(a) Transform the $\mathbb{X}$ variables into $\mathbb{T}$ random variables, all with
zero mean and unit variance:

$$\mathbb{X} = \mathbb{A}.\mathbb{T} + \mathbb{B} \tag{2}$$

The matrix $\mathbb{A}$ is the diagonal matrix of the standard-deviations of
$\mathbb{X}$, while $\mathbb{B}$ is the vector of the average values of $\mathbb{X}$, and the correlation matrix remains unchanged by that one-by-one transformation :

$$\mathbb{R}(T) \equiv \mathbb{R}(X) \tag{3}$$

(b) Diagonalize the correlation matrix. The resulting diagonal matrix $\mathbb{L}$ is formed by the N eigenvalues of $\mathbb{R}(T)$:

$$\mathbb{L} = \text{Diag}\{\lambda_1, \lambda_2, \ldots, \lambda_i, \ldots, \lambda_N\} \qquad (4)$$

The matrix $\mathbb{L}$ can be interpreted as the covariance matrix of a new set of N random variables $\mathbb{U}$, which are **independent** because all the off-diagonal elements are equal to zero. These $\mathbb{U}$ variables have zero mean values and their variances are equal to the eigenvalues of $\mathbb{R}(T)$. That transformation of $\mathbb{T}$ into $\mathbb{U}$ consists in a rotation around the origin in the N-dimensional space, in such a way that the $\mathbb{U}$ axes coincide with the **principal axes**. The linear coefficients used for that rotation are the corresponding eigenvectors of $\mathbb{L}$, arranged in the matrix $\mathbb{D}$:

$$\mathbb{T} = \mathbb{D}.\mathbb{U} \qquad (5)$$

Customary numerical methods provide the eigenvalues (variances) in a decreasing order, and the resulting $\mathbb{U}$ random variables are then automatically sorted by decreasing variances. The sum of these N variances is always equal to N, which is the sum of the N diagonal elements of $\mathbb{R}(T)$, each equal to unity. The variances of the first random variables of $\mathbb{U}$ will therefore be larger than unity, while these of the last elements will be smaller than unity. In most practical cases, for large values of N, the variance vanishes from $N^* < N$. The last elements of $\mathbb{U}$, with indices larger than N^*, are no more to be considered as random variables, but rather as deterministic values without any significant uncertainty:

$$\mathbb{U} = \{\mathbb{U}^* \vdots \mathbb{U}^0\} = \{u_1, u_2, \ldots, u_{N*} \vdots 0, 0, \ldots, 0\} \qquad (6)$$

A set of N correlated random variables $\mathbb{X}$ appears then as linear combinations of a smaller set $\mathbb{U}^*$ of N^* independent (uncorrelated) random variables:

$$\mathbb{X} = \mathbb{A}.\mathbb{D}.\mathbb{U} + \mathbb{B} = \mathbb{A}.\mathbb{D}.\{u_1, u_2, \ldots, u_{N*} \vdots 0, 0, \ldots, 0\} + \mathbb{B} \qquad (7)$$

3.3 Probabilistic computations with a large number of random variables

The probabilistic design procedure of structures involving soil (deep or shallow foundation, retaining structure, ..) can use a customary deterministic method, but with random variables as input data. These input data are e.a. the applied loads, the characteristics of the materials and the soil properties.

Due to the necessity to divide the soil mass in a large number of small elements to be able to model the spatial variability, as stated above, the number of input random variables can reach rather high values.

The customary computing methods (Monte-Carlo simulation, Taylor's series development, ..) are generally not suited to deal with such problems and the point estimates methods as proposed by Rosenblueth (1975) are very effective, but they are limited to small sets of variables (Bolle 1986).

The so-called **independent disturbances method** (Bolle 1988) was developed with a view to minimizing the computing effort while optimizing the use of the given information. The output quantities, which are obviously random variables, are estimated by their first statistical moments (mean value, variance, skewness, ...). These moments are computed as the sum of:

(1) a **central value**, with the assumption that all the input variables are deterministic;
(2) N **disturbing moments**, defined as the effects, on the statistical moments, of the randomness of each of the N independent random variables.

The basic assumption states that the effects of the randomness of each independent random variable are also independent from each other, and can then be simply added. The disturbing moments are computed separately for each random variable, assuming all the other variables deterministic. The computing technique uses an optimized three-point estimate. For N input random variables, normally (Gauss) distributed, the number of simulations is 2N+1, which is close to the number of available pieces of information, i.e. N average values and N variances.

For a function $Y(X)$ of the input random variables X, the **central value** is simply:

$$Y_c = Y(\overline{X}) = Y(A.D.\{0,0,..,0\} + B) \tag{8}$$

The **disturbed values** of the function Y are separately computed for each independent basic variable of rank (i), all the other elements of U being then considered as deterministic:

$$Y_i^+ = Y(A.D.\{0,\ldots,+u_i,\ldots,0\} + B) \tag{9}$$

$$Y_i^- = Y(A.D.\{0,\ldots,-u_i,\ldots,0\} + B) \tag{10}$$

using the point-estimates given by:

$$u_i = (3\,\lambda_i)^{1/2} \tag{11}$$

The **relative disturbed values** are introduced:

$$d_i^+ = Y_i^+ - Y_c \qquad \text{and} \qquad d_i^- = Y_i^- - Y_c \tag{12}$$

and the average value of Y can then be estimated by:

$$\overline{Y} \simeq Y_c + \frac{1}{6} \sum_{i=1}^{N} (d_i^+ + d_i^-) \tag{13}$$

Similar expressions can be developed for the second and the third statistical moments of the function Y:

$$\sigma^2(Y) \simeq \frac{1}{36} \sum_{i=1}^{N} \left(5d_i^{+^2} - 2d_i^+ . d_i^- + 5d_i^{-^2} \right) \tag{14}$$

$$\mu_3(Y) \simeq \frac{1}{54} \sum_{i=1}^{N} \left(5d_i^{+^3} - 3d_i^+ . d_i^- . (d_i^+ + d_i^-) + 5d_i^{-^3} \right) \tag{15}$$

Expressions are also available for estimating the fourth statistical moment of $Y(X)$ and the covariance $Cov(Y,Z)$ of two functions $Y(X)$ and $Z(X)$ of the input random variables X (Bolle 1990).

For most applications the sums may be limited to the first N* terms, because the influence of the last (N-N*) disturbances becomes negligible.

4 Examples

4.1 Foundation settlement
The foundation of a large arch bridge is established on a very weathered and heterogeneous rock mass, and the settlement could be of critical effect on the behavior of the structure (Bolle et al. 1989). The mechanical properties of the rock mass are presented in (2.4) above. The observed coefficient of variation for the pressuremeter modulus is rather large, about 70%.

According to the proposal above, the rock mass was divided in small elements with random mechanical properties, and the **independent disturbances method** was used to estimate the settlements for different loading cases.

The estimated average settlements appear close to the values computed by a deterministic approach, but surprisingly their coefficients of variation remain around 5%, much lower than that of the material properties.

4.2 Slope stability
Stojanovic (1991) reviews the existing methods and develops a probabilistic analysis of earth slopes. The random properties of the soil mass are modeled as autocorrelated random variables. The safety factor is evaluated according to the well known Bishop's method, the circular slip surface itself being considered as random.

Unfortunately, in order to reduce the number of random variables, the autocorrelated random properties of the soil mass are reduced to the three layers model proposed by Asaoka and Matsuo (1983). As already stated (3.1) such a model is only able to reproduce a few conditions of covariance, but can not be applied for any other purpose. Neither the variance, nor the autocorrelation curve, follows the expected pattern when varying the influence distance.

The safety factor Fs of the slope is strongly influenced by that choice. The figure 6 compares the standard-deviations of Fs (a) given by Stojanovic, (b) established by the author according to the proposal of Asaoka et al., and obtained by the proposed method with two different autocorrelation functions: (c) linear and (d) exponential.

In order to simplify the problem and to highlight a side-effect, the slope surface is considered unique and identical in all cases.

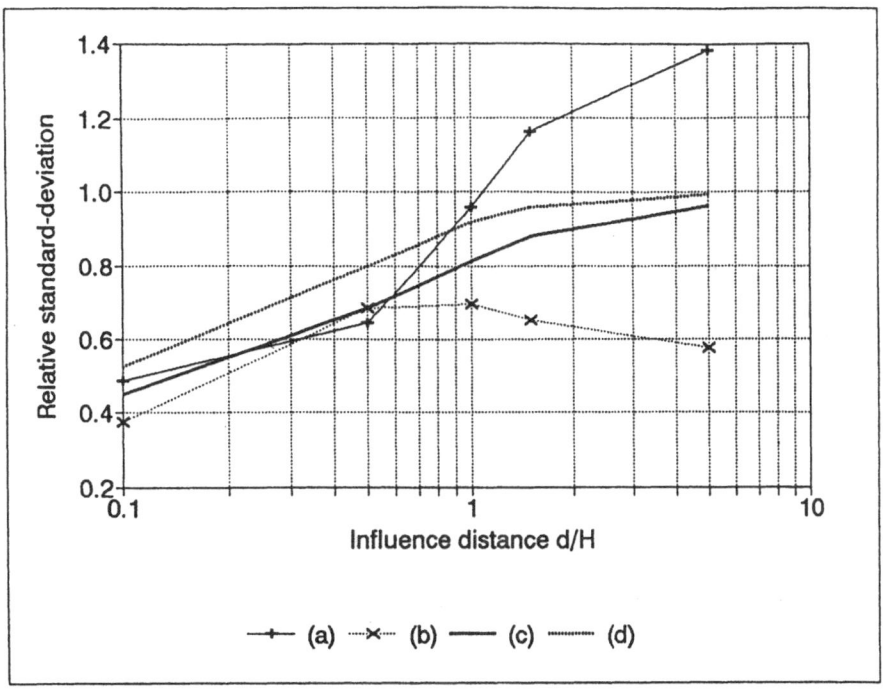

Fig.6. Standard-deviation of the safety factor of a slope

On the left side of the diagram, all the results are in good agreement: the soil tends to be a purely random material and the slope
surface **integrates** its properties, leading to a vanishing variability
of Fs.

When the influence (autocorrelation) distance increases, the
results diverge. As cohesion is the only random variable, in a
perfectly autocorrelated material, the variability (coefficient of
variation) of Fs must tend to the variability of that variable, as the
curves (c) and (d) do.

The differences between the results (a) and (b) are due to a side-
effect of that oversimplified model: the geometry of the critical
slope surface is strongly attracted towards the most variable layer,
increasing artificially the variability of Fs beyond that of the
cohesion.

5 Conclusions

The spatial variability of the properties of the natural soils has
always to be considered in all problems. As a matter of fact, a layer
of soil formed by an homogeneous random material exhibits an autocorrelation function everywhere equal to unity, totally inconsistent will
all observed actual curves. The only way to represent that spatial
variability is to divide the soil mass in small elements and to

express the correlation relating each element to each other.

In some problems, implied **integration** processes will reduce the variability and deterministic methods are well suited. A typical example is the settlement of a large shallow foundation.

In other kinds of problems, the geometry of the phenomenon can be influenced by the structural properties of the soil mass which are included in the spatial variability characteristics. A good example is the localization of a critical slope surface. Deterministic methods are no more applicable for these problems, in which the variability of the input variables could be amplified.

Tools are proposed for a probabilistic approach: (1) simple models for the spatial variability, based on the autocorrelation principle, and (2) numerical methods able to cope with large numbers of random variables. Furthermore, in most pratical cases, these numerous random variables can be reduce to a small number of basic variables.

6 References

Asaoka, A. and Matsuo, M. (1983) A simplified procedure for probability-based ϕ = 0 stability analysis. **Soils and Foundations**, Vol.23/1, pp. 8-18.

Baker, R. and Zeitoun, D.G. (1987) Soil variability and the maximum entropy principle. **I.C.A.S.P. 5**, Vancouver, pp. 642-649.

Bolle, A. (1986) Etude probabiliste de la stabilité d'une pente rocheuse. **Seminar on Reliability-based Design in Civil Eng.**, E.P.F. Lausanne, June 1986.

Bolle, A. (1988) Approche probabiliste en mécanique des sols avec prise en compte de la variabilité spatiale. **Thèse No 743**, E.P.F. Lausanne.

Bolle, A. Bonnechère, F. and Crémer, J.M. (1989) Probabilistic approach of the settlement of an arch bridge. **XII ICSMFE,** Rio de Janeiro.

Bolle, A. (1990) Méthodes statistiques en infrastructures. **Lecture notes at the University of Liège** (unpublished).

Christakos, G. (1987) A stochastic approach in modelling and estimating geotechnical data. **International Journal for Numerical Methods in Geomechanics**, Vol.11, pp. 79-102.

De Beer, E. Lousberg, E. Wallays, M. Carpentier, R. De Jaeger, J. and Paquay, J. (1977) Bearing capacity of displacement piles in stiff fissured clays. **Verslag over navorsingen n°39 van het I.W.O.N.L.**, Brussels, March 1977.

Maertens, J. (1990) Aanwending van de statistiek voor de verwerking van grondmechanische karakteristieken. **Geoproba 90** (eds Belgian Group of the ISSMFE), Brussels.

Stojanovic, M. (1991) Probabilistic analysis of earth slopes. **Thèse No 965**, E.P.F. Lausanne.

Rosenblueth, E. (1975) Point estimates for probability moments. **Proc. Nat. Acad. Sc. USA**, Vol.72, No 10.

STATISTICAL ANALYSIS OF THE MECHANICAL PROPERTIES OF MASONRY WALLS BY COMPUTATIONAL MODELS

A.BARATTA
G.VOIELLO
G.ZUCCARO
Dept. of "Scienza delle Costruzioni", Naples, Italy

Abstract
The statistical parameters of the single/joint probability
laws of the mechanical properties of masonry elements are
investigated by making use of theoretical models of the
constitutive elements of masonry. In particular, standard
tests on masonry blocks are simulated. The model is
elaborated by extrapolation of numerical point results. The
correlation between mortar and brick (or stone) properties
is investigated, for a number of material and mortar
properties. The influence of inaccuracy in predicting the
relative magnitude of the mechanical properties of the
components is evaluated. Some comparison with experimental
results in the literature is attempted.
Keywords: Statistic, Mechanical Properties, Masonry Walls

1 Introduction

In many cases, when checking the state of existing
buildings, the mechanical properties of masonry structural
members are of interest in order to make an effective
interpretation of the disease and/or in order to check the
stress/strain levels through analytical models.

Unfortunately, most times it is not possible to extract
significant specimens from the masonry body, either because
of the size needed for such specimens to be tested in
laboratories, or because it may considered impolitic to
damage an historic building. It is more practical, most
times, to take small pieces (i.e. core samples) able to
yield the properties of the components but not of the
whole. With this view, it may be interesting to develop a
procedure able to produce mechanical properties of the whole
starting from knowledge of the components' behaviour.

In this paper, the problem is approached by simulating
tests on masonry panels of consistent size, made by
components with known parameters. The paper is limited to
the analysis of the "equivalent homogeneous" elastic
parameters of masonry panels, but the basic philosophy is

517

D. Breysse (ed.), Probabilities and Materials. 517–528.

518

that the procedure can be applied for any other parameters
(e.g. the compressive strength, the shear strength, the
fracture point, and so on), and for any other kind of
masonry tissue.

The simulated test results are probabilistically
elaborated in way to evaluate the influence of possible
misunderstanding in the evaluation of the mortar properties,
that is the most difficult to be drawn out.

2 The simulated tests

The masonry panel examined is made by regular tuff stone
blocks of the size 36x18x24 cm. connected according to a
skintled rhythm by mortar joints of variable thickness. The
size of the panel is assumed to be 5 stones on the vertical
direction times 3 stones on the horizontal direction

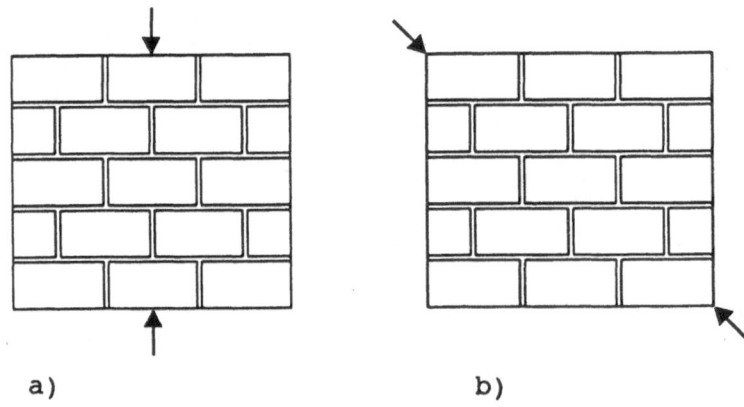

        a)   b)

Fig. 1: The sample masonry panel :a) The normal compression
test; b) The diagonal shear test

The panel is assumed to be submitted to a normal
compression test, for the evaluation of the E_p Young's
modulus and to a diagonal test for the evaluation of the G_p
shear modulus. The suffix "p" means that the above
parameters are referred to the behaviour of the panel as a
whole, considered equivalent to an homogeneous panel of the
same size.

The elastic properties of the tuff are denoted by E_t, the
Young modulus and v_t the Poisson coefficient. E_m and v_m
denote the same parameters for the mortar. It has been
verified that different values of v do not influence
significantly the behaviour of the panel as regards the

global elastic parameters. The two test schemes have been
modeled by a finite element model assembling mortar elements
with tuff elements. All elements are of the "linear stress
field" type (Fig.2)

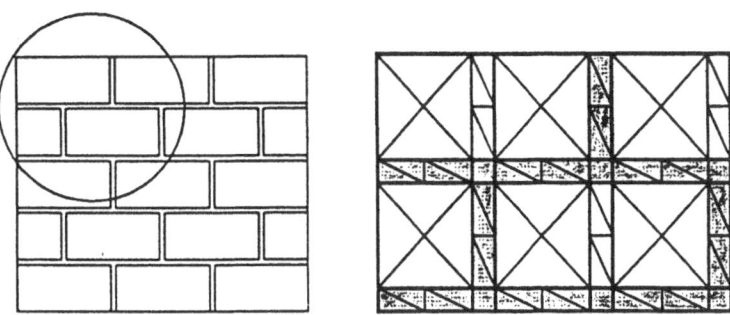

Fig.2: The FEM mesh

The analysis is carried on with reference to adimensional
parameters α and β given by:

$$\alpha = E_m/E_t \quad ; \quad \beta = E_p/E_t \tag{1}$$

for the normal test and

$$\alpha = G_m/G_t \quad ; \quad \beta = G_p/G_t \tag{2}$$

for the diagonal test.

The results of the numerical tests, correlating the
values of α and β are extrapolated for different thickness
of the mortar joints, by an expression of the type

$$\alpha(\beta;s) = \alpha_0(s) \exp [\alpha_1(s)\beta] \tag{3}$$

where $\alpha_0(s)$ and $\alpha_1(s)$ are interpolated in turn by polynomial
relations

$$\alpha_0(s) = \exp(a_0+a_1s+a_2s^2+a_3s^3)$$
$$\tag{4}$$
$$\alpha_1(s) = \exp(b_0+b_1s+b_2s^2+b_3s^3)$$

From the above correlation $\alpha(\beta;s)$ it is possible to get
the value of β (the global normal or shear modulus) after a

nominal estimate α_n of α is entered. In order to take into account the uncertainty in the evaluation of α, it is assumed that it corresponds to the realization of a random variable with expected value and variance given by

$$\bar{\alpha} = \alpha_n$$

$$\sigma_\alpha^2 = V_\alpha^2 \cdot \bar{\alpha}^2$$

(5)

where V_a is the coefficient of variation of α_n, i.e. a measure of the accuracy that the operator attributes to the estimate.

This accuracy can be translated to the expected accuracy in β, by making use of the well known first-order approximations

$$\beta_n(s) = \beta(\alpha_n) + \frac{1}{2} \beta'(\alpha_n) \sigma_\alpha^2$$

(6)

$$\sigma_\beta^2(s) = [\beta'^2(\alpha_n) + \beta(\alpha_n) \beta''(\alpha_n)] \sigma_\alpha^2 + \frac{3}{4} \beta''^2(\alpha_n) \sigma_\alpha^4$$

(7)

where $\quad \beta'(\alpha) = \dfrac{d\beta(\alpha)}{d\alpha} \quad$ and $\quad \beta''(\alpha) = \dfrac{d^2\beta(\alpha)}{d\alpha^2}$

whence one can evaluate the influence of V_α on the uncertainty in the evaluation of β that is measured by V_β.

3 The normal compression test

In this section the results for the rule to evaluate E_p from the normal compression test are reported.
First in Fig. 3 the results of the FEM models for different thickness s ranging from 0.5cm to 2cm. are plotted

by small squares connected by a solid line, superposed with the interpolation expression (3).

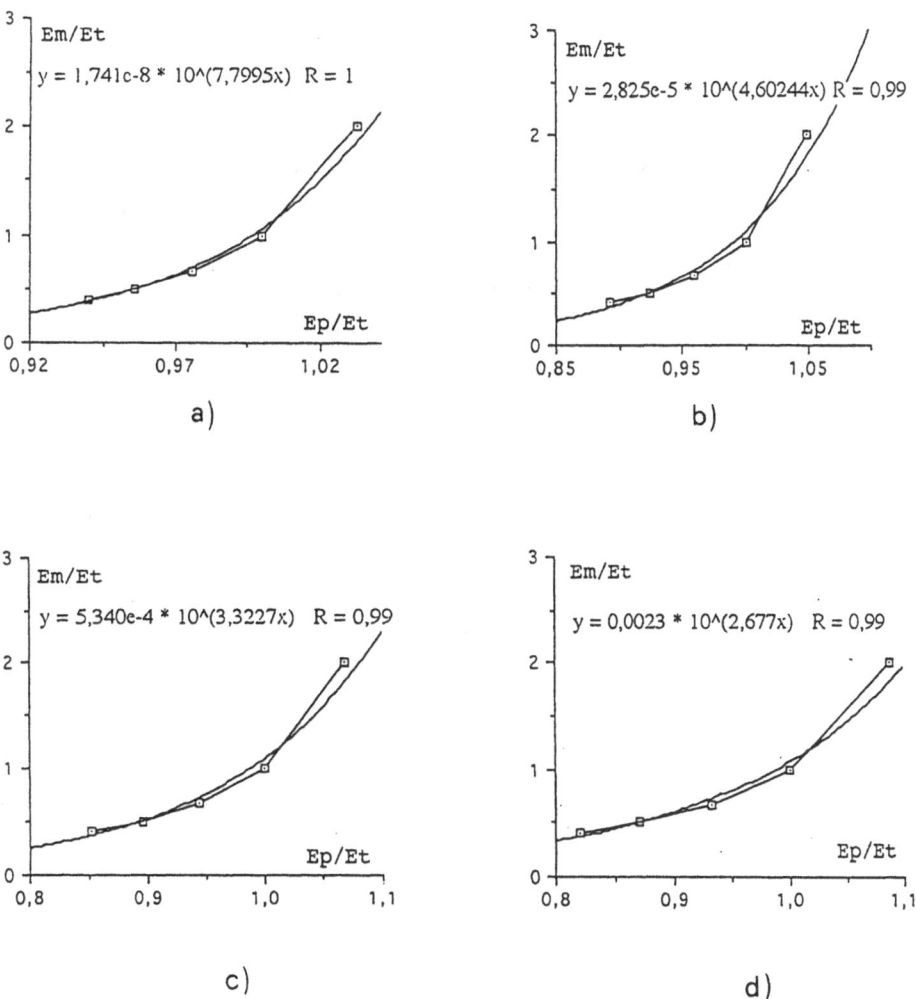

Fig.3:Results of the FEM calculus and interpolation formulas
    a) s=0.5cm; b) s=1.0 cm; c) s=1.5cm;   d) s=2.0 cm

In Fig. 4 the values of the interpolation coefficients for $\alpha_0(s)$ and $\alpha_1(s)$ are plotted, together with the interpolation coefficients from eqs.(4).

522

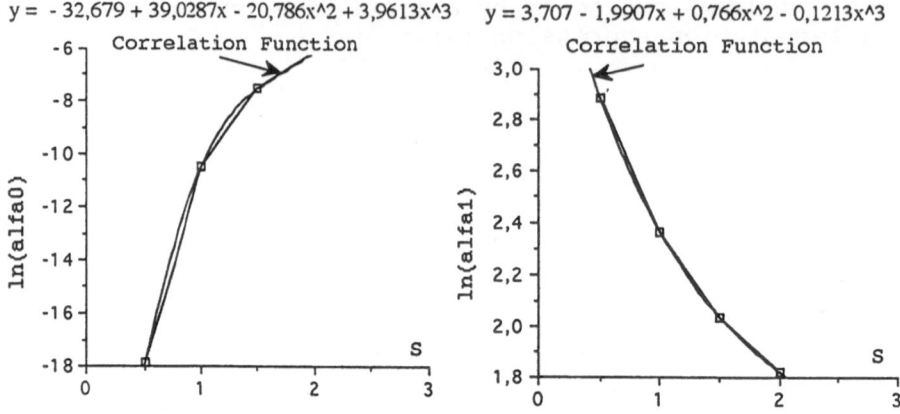

$y = -32,679 + 39,0287x - 20,786x^2 + 3,9613x^3$ $\qquad$ $y = 3,707 - 1,9907x + 0,766x^2 - 0,1213x^3$

Fig.4: Coefficients of eq.(4) and interpolation lines

In Fig. 5 the values of the expected value of β versus the nominal estimate of α are plotted for a number of values of V_α and for different values of s.

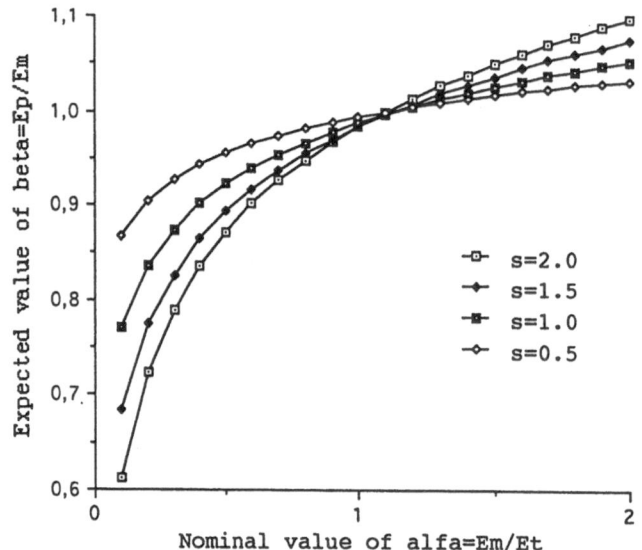

Fig.5: Expected β versus α_n

It can be noted that the uncertainty in α_n does not affect, practically, the nominal value of β.

In Fig. 6, finally, the uncertainty in the evaluation of β, i.e. the calculated V_β, is plotted versus the nominal value of α, for different values of V_α and of the mortar thickness s.

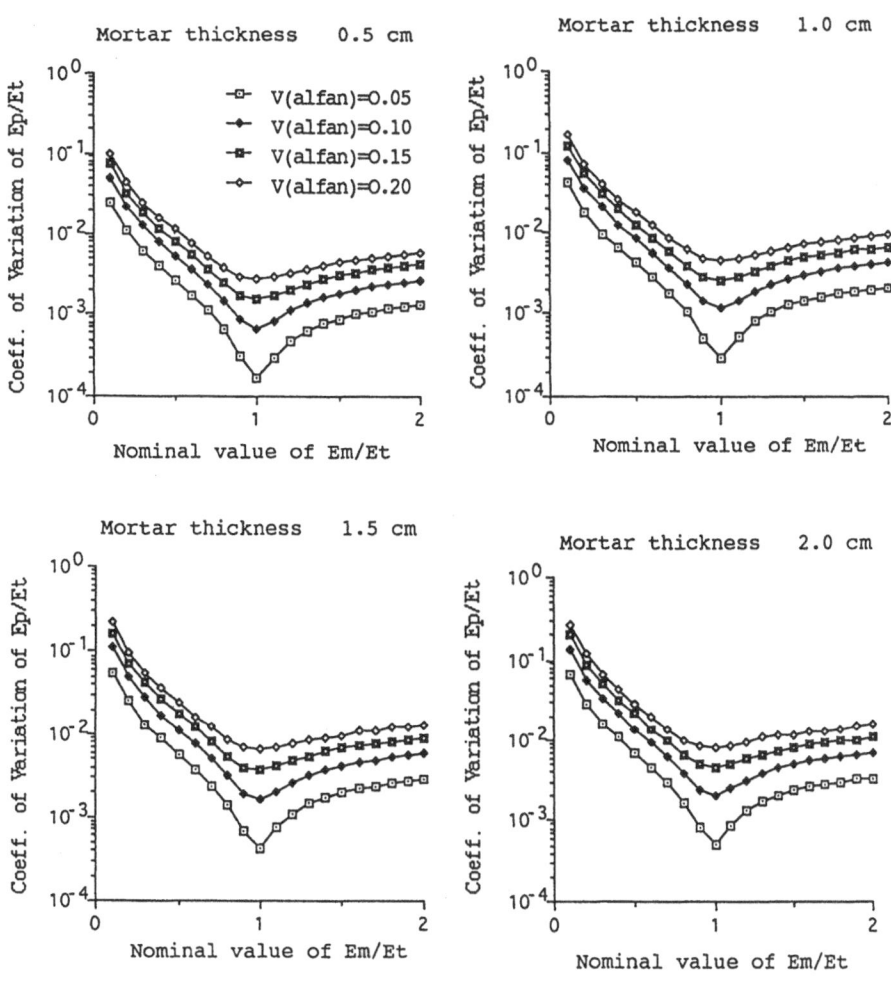

Fig.6: V_β versus α_n for V_α in the range 0.05 - 0.20

From the above results, it is possible to understand that the uncertainty in β is always smaller than the uncertainty on α, and that, anyway,this uncertainty increases with increasing the mortar thickness, but it is always negligible

when the modulus of the mortar is not very different from
the one of the tuff stones.

4 The diagonal shear test

In this section the results for the rule to evaluate G_p
from the diagonal compression test are reported.
As in the normal test case, Fig. 7 shows the results of
the FEM models for different mortar thickness. and the
interpolation of the data obtained by using expression (3).

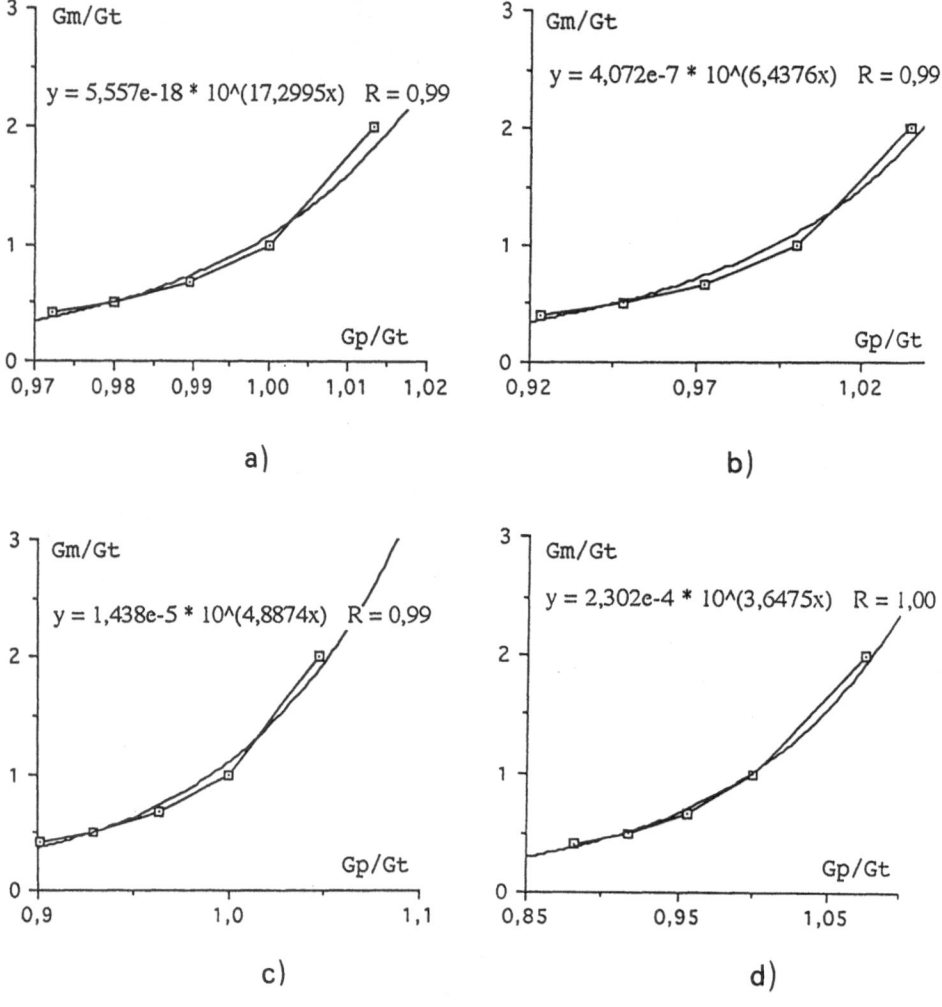

Fig.7:Results of the FEM calculus and interpolation formulas
a) s=0.5cm; b) s=1.0 cm; c) s=1.5cm; d) s=2.0 cm

In Fig. 8 the values of the interpolation coefficients for
$\alpha_0(s)$ and $\alpha_1(s)$ are plotted, together with the interpolation
coefficients from eqs.(4).

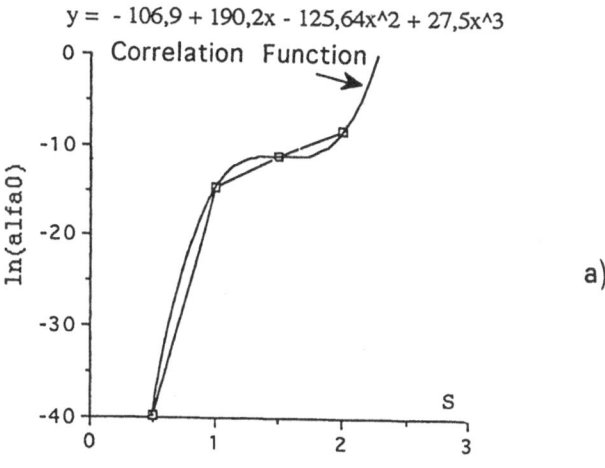

$y = -106,9 + 190,2x - 125,64x^2 + 27,5x^3$

a)

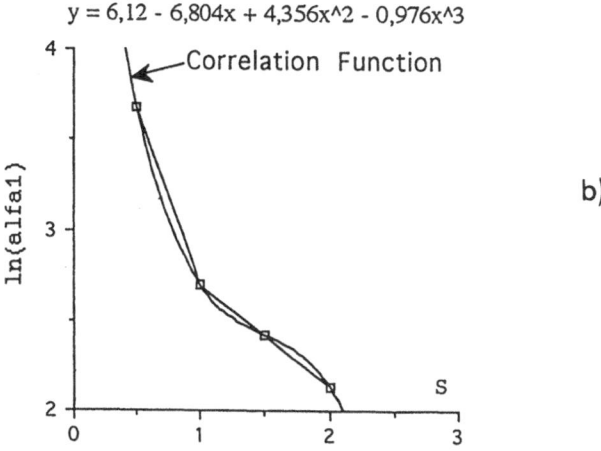

$y = 6,12 - 6,804x + 4,356x^2 - 0,976x^3$

b)

Fig.8: Coefficients of eq.(4) and interpolation lines
a) α_0 vs. s; b) α_1 vs. s

In Fig. 9 the values of the expected value of β versus the
nominal estimate of α are plotted for a number of values of
V_α and for different values of s.

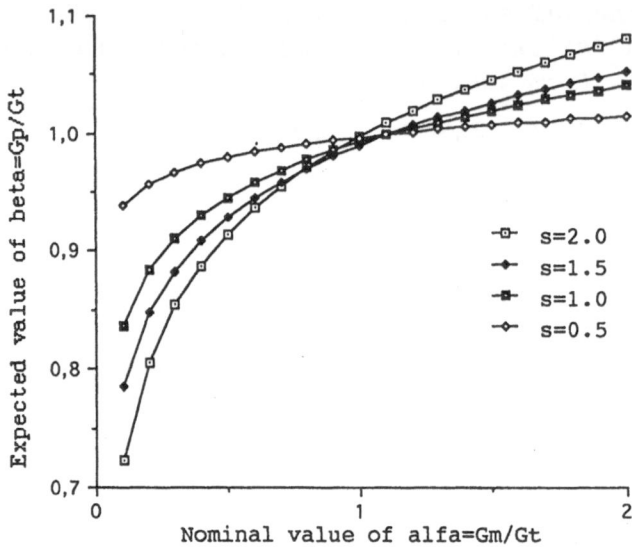

Fig.9: Expected β versus α_n

It can be noted, as in the normal test case, that the uncertainty in α_n does not affect, practically, the nominal value of β.

In Fig. 10, finally, the uncertainty in the evaluation of β, i.e. the calculated V_β, is plotted versus the nominal value of α, for different values of V_α and of the mortar thickness s.

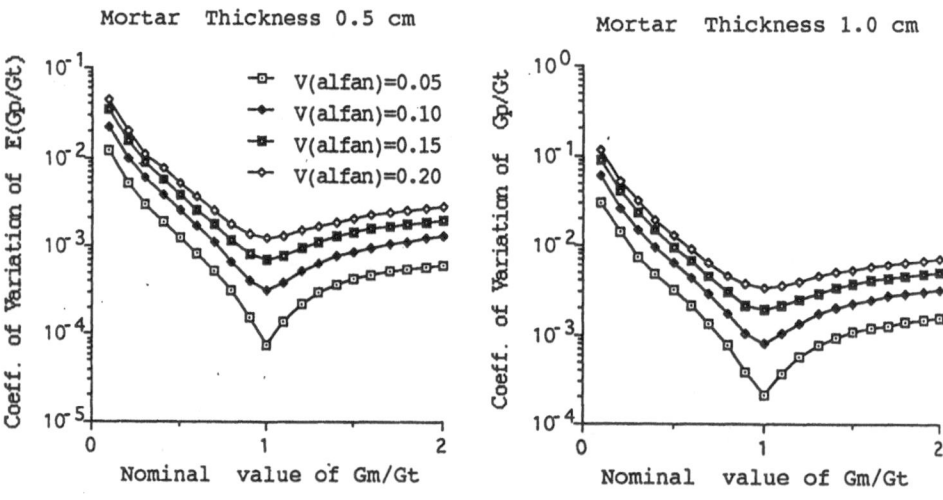

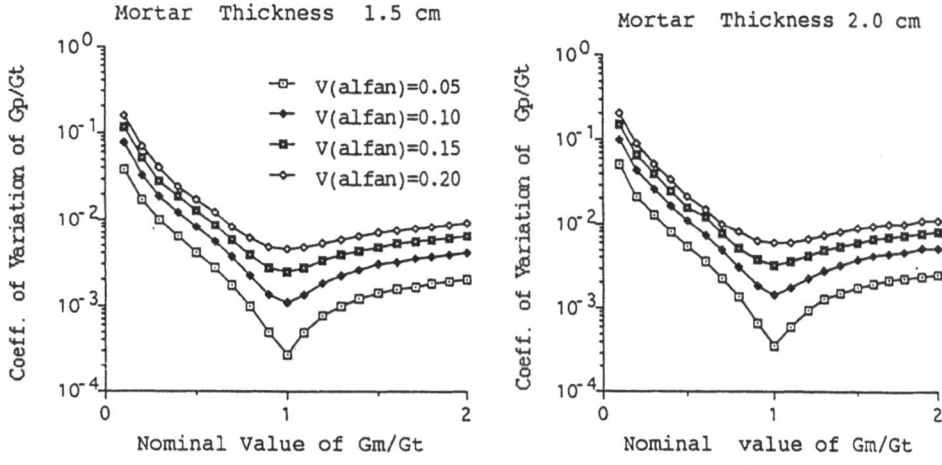

Fig.10: V_β versus α_n for V_α in the range 0.05 - 0.20

5 Comparison with previous correlation formulas

In order to check the results obtained through the
statistical approach, it has been conducted a deterministic
computation by using consolidate theories to determine the
elastic properties of the masonry panel once the mechanical
characteristic of the mortar and of the stone are known.
In particular the "equivalent models theory" has been
applied. The Young's modulus of the panel Ep is determined
as function of Em and of Et. The panel is shaped by a set of
elementary unit of single stone with the relevant mortar
(see Fig.11). Linear behaviour of material is assumed and no
lateral strenght due to different lateral deformation of the
materials is considered.

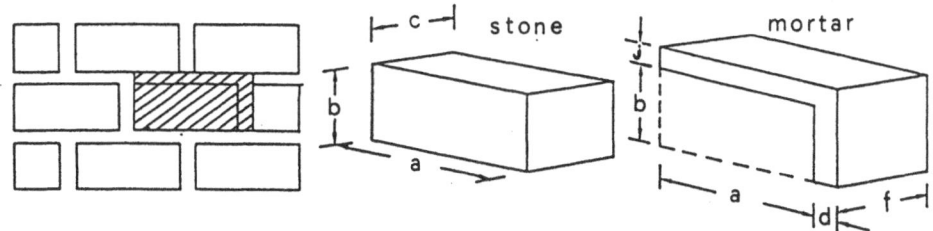

Fig.11: The elementary unit
[Sacchi Landriani G.,Riccioni, R. (1982) Comportamento
Statico e Sismico delle Strutture Murarie, CLUP]

528

The expression considered is:

$$E_p = \frac{(b+j) \cdot (a+d) \cdot f \cdot E_m (a \cdot c \cdot E_t + d \cdot f \cdot E_m)}{A_{eq}[(a+d) \cdot f \cdot b \cdot E_m + a \cdot c \cdot j \cdot E_t + d \cdot f \cdot j \cdot E_m]} \tag{8}$$

where A_{eq} represent the equivalent cross section area

The comparison between the values of E_p obtained by applying eq.(8) and the values calculated through the statistical approach has showed a satisfactory agreement.

6 Conclusions

The procedure outlined in the paper is intended to apply elementary probabilistic methods to evaluate the reliability of numerical estimates of composed materials properties by FEM numerical models. The procedure has been applied to the evaluation of the elastic stiffness properties of a masonry material, composed by stones and mortar. It is expected that it works in a similar way for other problems, as quoted in the Introduction.
The results obtained in the paper prove that a good estimate can be expected by the procedure, even if some parameter can be subject to a very significant uncertainty, so giving the possibility to trust on a reliable estimate of the global behaviour of the wall.

7 References

Augusti G., Baratta A., Casciati, F., (1984) **Probabilistic Methods in Structural Engineering**, Chapman Hall ed., London
Jessop,E.L.,Shrive,N.G. (1978) **Elastic and Creep Properties of Masonry**, Proc. of North American Masonry Conf., University of
Colorado, Boulder, Colorado
Hilsdorf,H.K. (1969) **Investigation into the failure mechanism of brick masonry loaded in axial compression.** Houston, Texas: Gulf Publishing co; Designing, engineering and contruction with masonry products.
Sacchi Landriani G.,Riccioni, R. (1982) **Comportamento Statico e Sismico delle Strutture Murarie**, CLUP ed. Milan.

Acknowledgment
This research was supported by a grant from the Italian Ministry of University and Research (MURST 40%).

A REPORT ON ROUND TABLE 3: APPLICATIONS

L. FARAVELLI
Department of Structural Mechanics, University of Pavia, Pavia, Italy

G. ROUSSELIER
EDF-DER, Moret sur Loing, France

Abstract
Fields of ongoing and potential application of a probabilistic modeling of material
are reviewed and discussed. Emphasis is put on the technical problem, the failure
of deterministic approaches and the different nature of the uncertainties to be han-
dled. A basic inconvenience is found on the long-term investigative stage required
by most of the material behaviour of interest. The use of judgement evaluations
during the design followed by monitoring and updating during construction and
service of the structural system is suggested.
Keywords : Applications of Uncertainty Modelling, Damage Accumulation, Environment, Nuclear Waste, Management, Simulation.

1 INTRODUCTION

The round table was co-chaired by the authors. In order to drive and to stimulate
the discussion, three further panelists were invited to illustrate their field of interest and to summarize their expertise on engineering applications of probabilistic
methods:

- Professor H.H. Einstein, of the Department of Civil Engineering at MIT,
  Boston, USA;

- Dr. A. Millard of the French Commission of Nuclear Energy, CEA/ DMT/
  LAMS, Gif sur Yvette, France;

- Dr. H. Riesch-Oppermann of the Nuclear Research Center in Karlsruhe, Germany.

L. Faravelli introduces the panelists and, on behalf of the organizing committee,
thanks the industry delegates for the supports that made possible the round table

D. Breysse (ed.), Probabilities and Materials, 529–534.
© 1994 *Kluwer Academic Publishers.*

within the NATO workshop. The technical introduction is given by G. Rousselier; his talk is followed by the ones of L. Faravelli and of the three panelists in the order listed above.

2 PROPOSED KEY-POINTS

The field of activity of G. Rousselier is related to those aspects of the Mechanics of Materials and of the Structural Mechanics which interact with the Nuclear Industry. In this field there are several problems where a deterministic approach fails and the adoption of probabilistic methods becomes a need. Three typical examples, among others, are introduced:

A variety of components, such as pipes, pressure vessels and so on, undergoes the crack nucleation/propagation sequence typical of Fracture Mechanics: one of the main points to be accounted for in their design is the large variability of the material properties which influence the physical phenomenon.

The management of nuclear waste is today approached by its confinement into concrete containers and geomaterials (granite, clay, salt): the time-evolution of the material is of primary importance in order to avoid radionucleid contamination.

When speaking of safety factors to be applied to the loads, a cost-benefit optimization policy requires one also provides their explanation in term of safety; only the understanding of the scatter of the material properties can help toward this goal.

Two main problems are envisaged within the practical application of a probabilistic approach:

1. the statistical analyses on material data would require large samples, especially when extrapolation to low-probability tails is pursued; by contrast, a limited amount of laboratory tests can be developed and/or made available;

2. when numerical simulations are adopted, instead of the laboratory tests or toward a uncertainty propagation, high computational costs are met: simplification would be welcome.

L. Faravelli shows two applications quite different between them. They teach, however, the same lesson. The engineer often meets situations in which a better understanding of the material behaviour is required. However, there is no time for waiting for the results of the laboratory tests and/or further theoretical developments: a decision must be taken immediately. The uncertainty on the material evolution in time is better approached by monitoring and, in case, by an IMR (inspection, maintenance and repair) policy.

For a wide-diameter floating tunnel, for instance, an internal steel pipe is coupled with reinforced concrete in order to provide stiffness; an external steel pipe is then added (to form a sandwich section) for contrasting the water aggressiveness. The external steel could undergo corrosion and a further removable layer, isolated by air, (the coffer-dam) is eventually the solution. It is designed to act against impact, but the possibility of substituting it provides a by-pass of the large scatter of the corrosion behaviour by means of a simple maintenance policy. Similarly, the anchorage cables are conveniently made in kevlar, but nobody knows the viscoelastic behaviour of this material when very large strands are adopted. Since there is not time for long-duration creep tests, the solution still relies on a maintenance approach.

The second problem concerns a gas turbine: heat transmission, high temperature and pressure fields are dominated by a stochastic finite elements thermo-elastic analysis but, at the end, ... the creep safety check does not account on a reliable material model. Again a maintenance policy offers perhaps the right solution.

In summary, a better characterization of the materials should be recommended, but, when it is lacking, one can only rely on the expertise arising from the maintenance process.

H.H. Einstein illustrates his experience in soil mechanics applications. Along the planning-design-construction cycle there are several goals which may receive help from the adoption of probabilistic methods, namely: safety, serviceability, economics, aesthetic and environment. The relevant decision process requires to be iterated several times both during the construction and the life of the engineering system.

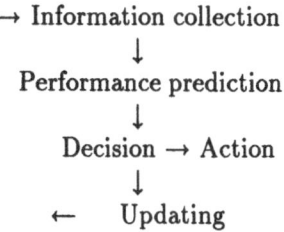

$\rightarrow$ Information collection
$\downarrow$
Performance prediction
$\downarrow$
Decision $\rightarrow$ Action
$\downarrow$
$\leftarrow$ Updating

A. Millard, then, discusses some problems related with the storage of high level nuclear wastes. At present the solution proposed for the long term storage consists in underground repositories. In France the following rocks are considered as possible candidates: salt, clay, granite. Each of these media has advantages and drawbacks and leads to different issues with regards to the various problems posed

532

(construction phase, operation phase, safety evaluation ...). One of the main problems is to evaluate the probabilities of transfers of radionucleids from the repository to the biosphere for different scenarios. Many items are affected with uncertainty: properties of the geological medium, failure modes of the canisters, major events like climate variations and so on. Two issues must be addressed in the investigation of the rock properties:

1. the rock properties exhibit spatial variations which can be important, as for example the permeabilities. When these informations are available, a stochastic approach is usually adopted to model this variability;

2. most of the times the data are not sufficient to characterize the domain under investigation. In this case, one may proceed to sensitivity analyses to appraise the uncertainty of data. Some new methods, like fuzzy logic or solution of problems with incomplete data, might be of help.

Note that 1. means randomness and 2. uncertainty: they are the two main aspects of any problem requiring the adoption of a probabilistic approach. According to the problem to solve, many different scales can be considered:

- the scale L_1 of the domain under investigation. In the case of a near field study it is of the order of ten meters; in the case of a far field study, it is of the order of some kilometers.

- the scale L_2 of the heterogeneity, which is the scale corresponding to measures showing the spatial variability of the material properties.

- the scale L_3 of homogeneization, which is the scale associated with the representative elementary volume used to find the properties of the continuum equivalent to the real medium.

In practice these scales usually verify the following inequalities:

$$L_1 >> L_2 >> L_3 \tag{1}$$

Probabilistic methods are required to handle the randomness exhibited by the material properties at the two scales L_2 and L_3. According to the various media, the problem to be solved can be of hydraulics, of mechanics or mixed.

H. Riesch-Oppermann tries to answer two questions related to his research on the material behaviour: why probabilities? and how to use probabilities? The first answer is just a list of problems where deterministic models fail or are inefficient:

- the description of the scatter in material properties which arises from phenomenological bases;

- the relation of microscale mechanisms with the macroscopic behaviour;

- the handling of microstructural aspects (e.g. the anisotropy) and geometrical uncertainties towards a stochastic finite element description;

- the prediction of the lifetime (reliability, durability) of ceramics components.

As a consequence two different probabilistic approaches are envisaged: i) the use of appropriate tools to describe the material behaviour, i.e. micromechanical models, geomechanical models; ii) the use of probabilistic lifetime models, i.e. creep damage modelling, thermal fatigue damage. The mechanics of fibre reinforced materials is the object of ongoing research and the use of Voronoi mosaics in this field is currently investigated.

3 COMMENTS AND QUESTIONS

The attendance discussion concentrates on two different ideas: the intrinsic value of the probabilistic approach and the difficulty of planning long term material researches providing results of actual use in engineering applications.

The first topic is considered by Professor Soulié of the Polytechnic of Montreal (Canada). The spatial variability in his field, soil mechanics, naturally leads to a probabilistic model, but the final decision is reached averaging with personal technical expertise. The panelist H. Einstein partially disagrees with him since the technical decision maker must interact with more general decision makers, as politicians or businesmen, and this people is not sensitive to concepts as safety factors, but understand a probabilistic language. Professor Bolle of the University of Liege (Belgium) remembers the long term Bayesian approach which characterized the old concept of safety factor and recalls the need for incorporating model uncertainty in any probabilistic approach. Safety factors concern usual problems of usual size making use of usual materials, but probabilistic methods are a need for new problems characterized by unusual situations. Dr. Mebarki of ENS, Cachan (France) make some comments on the model errors. The Eurocodes are especially sensitive to this aspect. In particular, the Eurocodes try to obtain a more and more homogeneous engineering approach in soil and structural mechanics.

Professor Casciati of the University of Pavia (Italy) emphasizes the attitude of the designer to pretend immediate answers to all question he meets during his design decision making process. In engineering applications, new material problems are often encountered but there is not time to wait for its scientific investigation. If a resercher decides to proceed on such a subject, however, the results of his work will surely have a scientific value, but, often, they are reached when the technological development have already launched new materials and/or operative situations. Professor Breysse of ENS, Cachan (France), agrees on this subject and gives an example related with the durability of the canisters introduced by the panelist A. Millard. He recalls the problem of model uncertainty. A durability analysis must model and account for external accidents, strain fields, mesocracking, opening/closing of cracks. The adoption of a probabilistic framework is an evident need, but too many problems are still very far from a solution. Dr. Jeulin of ENSMP, Fontainbleau (France) underlines how hierarchical models are used in introducing randomness and uncertainty. The problem is: where to stop? He concludes by the remark that many parameters are not well known in material research and the introduction of always new products is enlarging the gap between material scientists and the designers of engineering systems.

4 CONCLUDING REMARKS

A probabilistic approach to the material property description is becoming more and more usual; this is especially evident in civil and nuclear engineering applications. They are introduced everywhen deterministic models fail and/or engineer expertise is lacking.

In the last case an unsolved loop is often met: a material is found or adopted in nonstandard conditions; physical and laboratory investigations should be required but a decision must be reached without waiting for the analysis results.

There is a great need of database and/or expertise in the field of advanced technologies; they should be pursued by regarding not only the planning, design and construction, but also the service lifetime as aspects of competence of the designer.

INDEX